Protein Digestion-Derived Peptides

Chemistry, Bioactivity, and Health Effects

Protein Digestion-Derived Peptides

Chemistry, Bioactivity, and Health Effects

Edited by

Cristina Martínez-Villaluenga

Institute of Food Science, Technology and Nutrition, Spanish National Research Council, Madrid, Spain

Blanca Hernández-Ledesma

Department of Bioactivity and Food Analysis, Institute of Food Science Research (CIAL, CSIC-UAM, CEI UAM + CSIC), Madrid, Spain

ELSEVIER

ACADEMIC PRESS

An imprint of Elsevier

Academic Press is an imprint of Elsevier
125 London Wall, London EC2Y 5AS, United Kingdom
525 B Street, Suite 1650, San Diego, CA 92101, United States
50 Hampshire Street, 5th Floor, Cambridge, MA 02139, United States

ISBN: 978-0-443-19141-1

For Information on all Academic Press publications
visit our website at https://www.elsevier.com/books-and-journals

Publisher: Jonathan Simpson
Acquisitions Editor: Nicole Denis
Editorial Project Manager: Emerald Li
Production Project Manager: Kumar Anbazhagan
Cover Designer: Greg Harris

Typeset by MPS Limited, Chennai, India

Contents

List of contributors

María Cristina Añón
Laboratory of Food Proteins Research, Development and Innovation (LIDiPA), Center of Research and Development in Food Criotechnology, National Council of Scientific and Techical Research (CONICET), Scientific Research Commission of Buenos Aires Province (CICPBA), Faculty of Exact Sciences, National University of La Plata (UNLP), La Plata, Argentina

Antonella Baldi
Department of Veterinary Medicine and Animal Sciences (DIVAS), University of Milan, Lodi, Italy

Débora Parra Baptista
Department of Food Engineering and Technology, School of Food Engineering, University of Campinas, Campinas, São Paulo, Brazil

Begoña Bartolome
CSIC, Spanish National Research Council, CIAL-Institute of Food Science and Research, Madrid, Spain

Lucia Cassani
Department of Analytical Chemistry and Food Science, Universidade de Vigo, Nutrition and Bromatology Group, Instituto de Agroecoloxía e Alimentación (IAA) – CITEXVI, Vigo, España

Raffaella Colombo
University of Pavia, Department of Drug Sciences, Pavia, Italy

Milena Corredig
Department of Food Science, CiFood Center, Aarhus University, Skejby, Denmark

Javier Echave
Department of Analytical Chemistry and Food Science, Universidade de Vigo, Nutrition and Bromatology Group, Instituto de Agroecoloxía e Alimentación (IAA) – CITEXVI, Vigo, España

Elsa Marisa Ferreira Vieira
REQUIMTE/LAQV/ISEP, Polytechnic of Porto – School of Engineering, Porto, Portugal

María Fraga-Corral
Department of Analytical Chemistry and Food Science, Universidade de Vigo, Nutrition and Bromatology Group, Instituto de Agroecoloxía e Alimentación (IAA) – CITEXVI, Vigo, España

Ilaria Frosi
University of Pavia, Department of Drug Sciences, Pavia, Italy

María Concepción García
Departamento de Química Analítica, Química Física e Ingeniería Química, Universidad de Alcalá, Madrid, Spain; Instituto de Investigación Química “Andrés M. del Río”, Universidad de Alcalá, Madrid, Spain

Susan Fiorella Garcia Filleria
Laboratory of Food Proteins Research, Development and Innovation (LIDiPA), Center of Research and Development in Food Criotechnology, National Council of Scientific and Techical Research (CONICET), Scientific Research Commission of Buenos Aires Province (CICPBA), Faculty of Exact Sciences, National University of La Plata (UNLP), La Plata, Argentina

Mirna Lúcia Gigante
Department of Food Engineering and Technology, School of Food Engineering, University of Campinas, Campinas, São Paulo, Brazil

Carlotta Giromini
Department of Veterinary Medicine and Animal Sciences (DIVAS), University of Milan, Lodi, Italy; CRC, Innovation for Well-Being and Environment, Università degli Studi di Milano, Milano, Italy

Pilar Gómez-Cortés
Department of Bioactivity and Food Analysis, Institute of Food Science Research (CIAL, CSIC-UAM, CEI UAM+CSIC), Madrid, Spain

Ruixian Han
School of Food Science & Nutrition, University of Leeds, Leeds, United Kingdom

Alan Javier Hernandez-Alvarez
School of Food Science & Nutrition, University of Leeds, Leeds, United Kingdom

Ester Hernández-Corroto
Departamento de Química Analítica, Química Física e Ingeniería Química, Universidad de Alcalá, Madrid, Spain

Blanca Hernández-Ledesma
Department of Bioactivity and Food Analysis, Institute of Food Science Research (CIAL, CSIC-UAM, CEI UAM+CSIC), Madrid, Spain

Luis Jiménez Munoz
Department of Food Science, CiFood Center, Aarhus University, Skejby, Denmark

Davide Lanzoni
Department of Veterinary Medicine and Animal Sciences (DIVAS), University of Milan, Lodi, Italy

María Luisa Marina
Departamento de Química Analítica, Química Física e Ingeniería Química, Universidad de Alcalá, Madrid, Spain; Instituto de Investigación Química "Andrés M. del Río", Universidad de Alcalá, Madrid, Spain

Diana Martin
Sección Departamental de Ciencias de la Alimentación, Facultad de Ciencias, Universidad Autónoma de Madrid, Madrid, Spain; Departamento de Producción y Caracterización de Nuevos Alimentos, Instituto de Investigación en Ciencias de la Alimentación (CIAL) (CSIC–UAM), Madrid, Spain

Cristina Martínez-Villaluenga
Institute of Food Science, Technology and Nutrition, Spanish National Research Council, Madrid, Spain

Martin Mondor
Saint-Hyacinthe Research and Development Centre, Agriculture and Agri-Food Canada, Saint-Hyacinthe, QC, Canada; Institute of Nutrition and Functional Foods (INAF), Université Laval, Quebec, QC, Canada

Leticia Mora
Department of Food Science, Institute of Agrochemistry and Food Technology, Spanish National Research Council (CSIC), Paterna, Spain

M. Victoria Moreno-Arribas
CSIC, Spanish National Research Council, CIAL-Institute of Food Science and Research, Madrid, Spain

Agustina Estefania Nardo
Laboratory of Food Proteins Research, Development and Innovation (LIDiPA), Center of Research and Development in Food Criotechnology, National Council of Scientific and Techical Research (CONICET), Scientific Research Commission of Buenos Aires Province (CICPBA), Faculty of Exact Sciences, National University of La Plata (UNLP), La Plata, Argentina

Joaquín Navarro del Hierro
Sección Departamental de Farmacia Galénica y Tecnología Alimentaria, Facultad de Veterinaria, Universidad Complutense de Madrid, Madrid, Spain

Adele Papetti
University of Pavia, Department of Drug Sciences, Pavia, Italy

Samuel Paterson
Department of Bioactivity and Food Analysis, Institute of Food Science Research (CIAL, CSIC-UAM, CEI UAM+CSIC), Madrid, Spain

Elena Peñas
Institute of Food Science, Technology and Nutrition, Spanish National Research Council, Madrid, Spain

Miguel Angel Prieto
Department of Analytical Chemistry and Food Science, Universidade de Vigo, Nutrition and Bromatology Group, Instituto de Agroecoloxía e Alimentación (IAA) – CITEXVI, Vigo, España

Alejandra Viviana Quiroga
Laboratory of Food Proteins Research, Development and Innovation (LIDiPA), Center of Research and Development in Food Criotechnology, National Council of Scientific and Techical Research (CONICET), Scientific Research Commission of Buenos Aires Province (CICPBA), Faculty of Exact Sciences, National University of La Plata (UNLP), La Plata, Argentina

Raffaella Rebucci
Department of Veterinary Medicine and Animal Sciences (DIVAS), University of Milan, Lodi, Italy

Mariela Rodriguez
Laboratory of Food Proteins Research, Development and Innovation (LIDiPA), Center of Research and Development in Food Criotechnology, National Council of Scientific and Techical Research (CONICET), Scientific Research Commission of Buenos Aires Province (CICPBA), Faculty of Exact Sciences, National University of La Plata (UNLP), La Plata, Argentina

Ana Clara Sabbione
Laboratory of Food Proteins Research, Development and Innovation (LIDiPA), Center of Research and Development in Food Criotechnology, National Council of Scientific and Techical Research (CONICET), Scientific Research Commission of Buenos Aires Province (CICPBA), Faculty of Exact Sciences, National University of La Plata (UNLP), La Plata, Argentina

Kenji Sato
Division of Applied Biosciences, Graduate School of Agriculture, Kyoto University, Kyoto, Japan

Adriana Scilingo
Laboratory of Food Proteins Research, Development and Innovation (LIDiPA), Center of Research and Development in Food Criotechnology, National Council of Scientific and Techical Research (CONICET), Scientific Research Commission of Buenos Aires Province (CICPBA), Faculty of Exact Sciences, National University of La Plata (UNLP), La Plata, Argentina

Aurora Silva
Department of Analytical Chemistry and Food Science, Universidade de Vigo, Nutrition and Bromatology Group, Instituto de Agroecoloxía e Alimentación (IAA) – CITEXVI, Vigo, España; REQUIMTE/LAQV, Instituto Superior de Engenharia do Porto, Instituto Politécnico do Porto, Porto, Portugal

Amanda Silva do Carmo
Department of Food Engineering and Technology, School of Food Engineering, University of Campinas, Campinas, São Paulo, Brazil

Jesus Simal-Gandara
Department of Analytical Chemistry and Food Science, Universidade de Vigo, Nutrition and Bromatology Group, Instituto de Agroecoloxía e Alimentación (IAA) – CITEXVI, Vigo, España

Santiago Emmanuel Suárez
Laboratory of Food Proteins Research, Development and Innovation (LIDiPA), Center of Research and Development in Food Criotechnology, National Council of Scientific and Techical Research (CONICET), Scientific Research Commission of Buenos Aires Province (CICPBA), Faculty of Exact Sciences, National University of La Plata (UNLP), La Plata, Argentina

Valeria Tironi
Laboratory of Food Proteins Research, Development and Innovation (LIDiPA), Center of Research and Development in Food Criotechnology, National Council of Scientific and Techical Research (CONICET), Scientific Research Commission of Buenos Aires Province (CICPBA), Faculty of Exact Sciences, National University of La Plata (UNLP), La Plata, Argentina

Fidel Toldra
Department of Food Science, Institute of Agrochemistry and Food Technology, Spanish National Research Council (CSIC), Paterna, Spain

Jianbo Xiao
Department of Analytical Chemistry and Food Science, Universidade de Vigo, Nutrition and Bromatology Group, Instituto de Agroecoloxía e Alimentación (IAA) – CITEXVI, Vigo, España; International Joint Research Laboratory of Intelligent Agriculture and Agri-products Processing, Jiangsu University, Zhenjiang, P.R. China

Ahmed A. Zaky
Department of Food Technology, Food Industries and Nutrition Research Institute, National Research Centre, Cairo, Egypt

Bin Zhang
College of Food and Bioengineering, Henan University of Science and Technology, Luoyang, P.R. China

Preface

In an era where the link between diet and health is at the forefront of our collective consciousness, the intricate world of protein digestion-derived peptides takes center stage. *Protein Digestion-Derived Peptides: Chemistry, Bioactivity, and Health Effects* serves as a beacon, illuminating the latest international breakthroughs in research that are poised to transform our understanding of the profound impact gastrointestinal digestion has on our well-being.

Within the pages of this book, we embark on a journey that traverses the boundaries of fundamental and applied research. With a keen focus on the interplay between the human digestive system and the bioactive peptides concealed within our daily meals, we delve into the intricacies of this transformative relationship. As we explore this dynamic connection, we unravel the mysteries of how gastrointestinal digestion can release peptides from food sources with the potential to shape our health.

At the core of our exploration are the models meticulously designed to simulate the physiological conditions of the human organism during food digestion and absorption. These models, a triumph of scientific innovation, are the key to understanding how our bodies interact with the proteins we consume. We investigate how various factors, including the nature and properties of food matrices, the physicochemical and functional characteristics of proteins, and individual pathophysiological variables, influence the digestibility, bioaccessibility, and bioavailability of the peptides generated in this intricate process.

Beyond the scientific methodology, we delve into the practical implications of our findings. In meticulous detail, we dissect food proteins of diverse origins, from traditional to alternative sources, unveiling the troves of bioactive peptides they harbor. The book uncovers the biological activities of these peptides and their potential to bolster our health and well-being. By connecting the dots between digestion, bioactivity, and health, we pave the way for innovative strategies to prevent and treat chronic diseases that exact a heavy toll on our society.

This book represents a groundbreaking contribution to the field, offering a comprehensive exploration of the complex interplay between the diet proteins and our overall health. It is our hope that this book will not only inform but inspire further research and innovation in the quest to harness the potential of bioactive peptides to improve our lives and combat the health challenges of our time. With this book, we take a significant step toward a future where our dietary choices become a powerful tool in the fight against chronic diseases that plague our modern world.

The aim of this book is to serve as an indispensable resource for food professionals, researchers, academics, and scientists who are deeply engaged in the ever-evolving realms of nutrition, food science, and product development. Additionally, it offers valuable insights and information for students pursuing careers in related fields, ensuring that the next generation of experts is well-equipped with a thorough understanding of this critical subject matter.

We extend our heartfelt gratitude to all our contributors, whose unwavering commitment, expertise, and dedication have been the driving force behind the creation of this book. Their exceptional efforts have elevated this work to new heights, and we are profoundly thankful for their contributions.

Cristina Martínez-Villaluenga
Blanca Hernández-Ledesma

CHAPTER

Food protein digestion by in vitro static approaches 1

Raffaella Colombo, Ilaria Frosi and Adele Papetti
University of Pavia, Department of Drug Sciences, Pavia, Italy

1.1 Introduction

The study of bioaccessibility and bioavailability of nutrients and nonnutrients food molecules is an important topic to estimate and predict the potential in vivo effects of food compounds. Many in vitro and ex vivo models have been set-up and applied as preliminary approaches to simulate in vivo physiological digestion conditions, testing absorption, metabolism, and interactions of food bioactive components.

In vitro methods can be subdivided into noncell-based (biochemical assays) and cell-based (cellular assays with animal or human cells) which can often be used in a complementary multiapproach; in fact, cells represent the first simulation of a physiological environment with its potential dynamic aspects (shearing, mixing, etc.) and give data closer to in vivo behavior than biochemical methods, but they cannot support extreme conditions as occur in vivo at gastric and intestinal level (especially pH values), which can be well reproduced in noncell-based systems.

In ex vivo assays, the intestinal tissue or its parts are isolated from an organism (animal or human) and maintained in artificial conditions. Notwithstanding they represent a valid alternative to animal models for testing food compounds' bioactivity, they remain mostly used only for absorption experiments in screening studies (Verhoeckx et al., 2015).

Nowadays, in vitro static methods represent simple, rapid, and less expensive approaches even if compared to the dynamic ones they are not able to reproduce some key factors of the digestion process, such as motility, dynamic circulation (physical/shear forces and mixing), and changes in biochemical parameters (hydration, pH, and electrolyte composition at different gastrointestinal tracts). Only a few dynamic models have been available and set-up because of their complexity and high cost. The most known human gastric simulator (also namely the Riddet model), dynamic gastric model, and TNO gastrointestinal model (TIM) have been able to reproduce gastric and/or intestinal peristaltic movements, thus representing a very important key factor to mimic more accurately in vivo conditions (pH changes and gradual secretion). In addition, TIM was able to simulate the conditions of duodenum, jejunum, and ileum, which slightly differ for pH

Protein Digestion-Derived Peptides. DOI: https://doi.org/10.1016/B978-0-443-19141-1.00001-7

value (from 6.5 to 7.2), differently from static models, in which a fixed pH value (6.5 or 7.5) is used (Kong & Singh, 2010; Minekus et al., 1995, 1999; Wickham & Faulks, 2013; Wickham et al., 2012). Dynamic models have been particularly used to study complex food matrices, clarifying the relation between food structure and digestion/digestibility (Lucas-González et al., 2018).

However, in contrast to the more complex dynamic systems, the static methods can be set-up and validated to obtain approved international procedures suitable for comparing interlaboratories results (Hur et al., 2011; Lucas-González et al., 2018).

Static/dynamic in vitro models could be applied to all common foods, mainly vegetables, dairy, and bakery foods, highlighting how bioaccessibility, bioavailability, and bioactivity are affected by food matrix and release kinetic (Lucas-González et al., 2018; Wang et al., 2021).

In the evolution of static approaches, a fundamental step was represented by COST Action INFOGEST (2011–15), which involved more than 40 countries (Europe, United States, Canada, etc.) and almost 500 researchers with the aim to create an international network working to promote the harmonization of food digestive methods (http://www.cost-infogest.eu). COST Action INFOGEST protocol (2014) (more than 2870 citations in Scopus database on February 2023) and its improved advancement, namely INFOGEST 2.0 (2019) (more than 960 citations in Scopus database on February 2023) elaborated an international consensus for static digestion protocols, setting standardized experimental conditions (gastrointestinal fluid composition, enzymes or phospholipids addition, time of each digestion phase) to allow a harmonization of the numerous in vitro digestion models present in literature up now (Brodkorb et al., 2019; Colombo et al., 2021; Minekus et al., 2014; Van De Wiele et al., 2007). The main changes of the INFOGEST 2.0 protocol, which represented an evolution of the 2014 action, mainly regarded the enzyme modification of gastric phase, which is particularly important for protein digestion (Brodkorb et al., 2019; Colombo et al., 2021). These methods have been internationally accepted and currently represent the reference models, also useful to compare new proposed digestion methodologies (Colombo et al., 2019; Colombo et al., 2023; Lucas-González et al., 2018). In addition, they become very promising approaches to mimic the physiological digestion process and consequently to predict in vivo nutrient and nonnutrient bioavailability (Bohn et al., 2018; Egger et al., 2017).

In particular, INFOGEST protocols have shown their large versatility and over the years they have been widely applied to different food matrices, structures, nutrients, and components (carbohydrates, proteins/peptides, lipids, minerals, carotenoids, and polyphenols) with some implementations and adaptations also to various subjects (age and pathological or diet-related conditions), providing specific validated food digestion models (Calvo-Lerma et al., 2019; Colombo et al., 2021; Egger et al., 2016; Ménard et al., 2014; Ménard et al., 2018) (Fig. 1.1).

In the last years, other static methods have been proposed and the most significant innovation regarded the addition of a colon step to complete the in vitro simulation of digestion process and obtain more predictive results of the in vivo

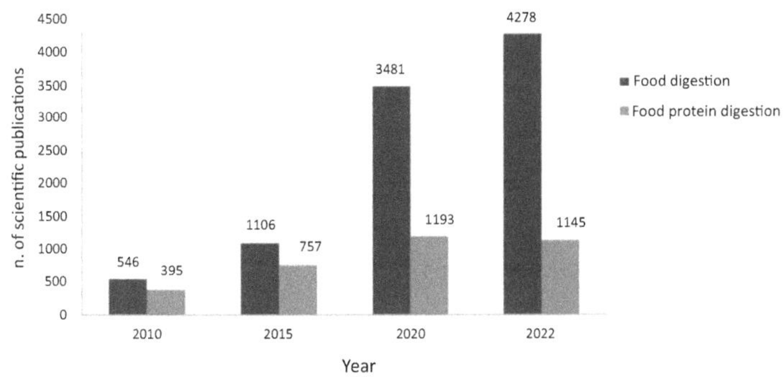

FIGURE 1.1 Food digestion/food protein digestion.

Overview on the trend of the scientific publications over years.

PubMed 10.02.2023 keywords: "Food digestion" and "Food protein digestion".

conditions. This step consists of the addition of a simulated colon fluid to the intestinal digested in the static protocols previously set-up (biochemical assays) or of the use intestinal cellular models, such as Caco-$_2$ cells (cell-based assays), often combined with biochemical approaches (Amigo & Hernández-Ledesma, 2020; Colombo et al., 2021; Hamzalıoğlu & Gökmen, 2016).

Finally, the relevance of in vitro digestion models is yet correlated to a great discussion about the use of animal models for studying food digestion in humans. In fact, nowadays in vivo experiments using animals are not only expensive and complex but also subject to ethical issues and not always so predictive of human physiological conditions (Meigs et al., 2018; Russell & Burch, 1960).

1.2 Food protein digestion

Food proteins are not only nutrients but also source of peptides that possess very important healthy properties. Therefore the study of food proteins is based both on the evaluation of their nutritional quality (amino acid composition and bioaccessibility) and on the technological and industrial processes which highly affect the digestion and therefore have a great impact on peptide health properties and bioactivity. The main challenges of bioactive peptides used as valuable ingredients of functional foods and/or nutraceuticals involve their production methods, their gastrointestinal stability, and intestinal absorption. Thus to know peptides' bioaccessibility and bioavailability can become mandatory to optimize the production of bioactive peptides (Chakrabarti et al., 2018; Mulet-Cabero et al., 2019). Protein digestion, both enzymatic and nonenzymatic, represents a fundamental tool in protein identification, characterization, and quantification, and requires many steps and long reaction time, with the continue search of new challenging opportunities over time. Proteases are the enzymes responsible for protein digestion. In particular, a single-step digestion requires trypsin, chymotrypsin, and peptidases, and allows a higher protein digestibility, while a two-step digestion uses pepsin and pancreatic enzymes (α-amylase, lipase, and colipase) giving rise to less digestible peptides. The possibility to have in vitro simple enzymatic models is mandatory to simulate and predict in vivo digestive process of complex food protein matrices and, consequently, protein metabolization and peptide bioactivity. Food matrix/structure and peptides bioaccessibility represent a critical point, which can affect protein interactions/reactions with other food nutrients/components and by consequence hydrolysis and type of peptides generated. For example, protein structure and amino acid bioaccessibility and bioavailability can be affected by polyphenols during the digestion process and, in turn, polyphenols stability, availability, and antioxidant activity can be correlated with protein interaction (Oliveira et al., 2018).

To quantify protein hydrolysis degree and kinetic, many analytical techniques, such as gel electrophoresis, spectrophotometric methods, high-performance liquid chromatography, and size exclusion chromatography, have been used in combination with INFOGEST protocol. In particular, the study of protein digestion also represents an important starting point for proteomics, useful for a successful protein identification, and this is strictly correlated to the development of rapid and sensitive automated digestion systems. The consequent analysis of undigested peptides by Mass Spectrometry (MS or MS/MS) approaches clarifies protein identification and modification and gives information on potential allergen epitopes or bioactive sequences (Santos-Hernández et al., 2020; Switzar et al., 2013).

Gastrointestinal digestion and protein hydrolysis represent a very important issue for protein bioactivity and turnover, as they can actively contribute to peptide degradation and/or formation of inactive sequences of amino acids, depending on proteolysis degree and protein stability. Notwithstanding some food proteins are effective in their intact form, the most interesting bioactive species are peptides derived from digestion, hydrolysis, or fermentation, and their level in human organism must be well-balanced to exert benefits, avoiding negative effects (Amigo & Hernández-Ledesma, 2020; Chakrabarti et al., 2018; Contreras et al., 2013).

It is well known the role of stomach in protein digestion. In fact, hydrolyzed peptides and amino acids stimulate the gastric cells, probably with a mechanism that involves G-protein-coupled receptors and dietary L-glutamate (Glu). Glu is the only amino acid regularly ingested in a free form and seems to positively influence the gastric response (secretion and kinetics) to protein digestion. A diet extremely rich in protein, mainly plant proteins, causes a sort of gastric adaptation, accelerating gastric emptying with a probable key role of endogenous cholecystokinin (CCK). CCK is a brain-gut peptide able to control food intake and gastric emptying (Zolotarev, 2014). Also at the intestinal level, a diet rich in complex proteins improves protein digestion, by directly acting on the small intestinal epithelium in which an enterocyte adaptation occurs (Hansen et al., 2015). Therefore protein digestion begins in the stomach, but all proteins are essentially absorbed as small peptides (dipeptides and tripeptides) or amino acids in the duodenum or proximal jejunum of the small intestine. The peptides and/or amino acids are absorbed by diffusion or active transport (sodium and adenosine triphosphate—ATP), depending on their side chains and reach the liver through the enterohepatic circulation. In fact, the oral enzymes do not significantly act on protein hydrolysis and digestibility; on the contrary, the amount and activity over time of gastric and intestinal enzymes represent the key factor. Therefore nutrient hydrolysis mainly takes place at the intestinal level, even if also gastric conditions could be very important for protein digestion. Today the role of the stomach as the primary site of protein digestion is well known and proteins can be classified in digestion-resistant or digestion-labile according to their digestibility at the gastric level (Callahan et al., 2020; Verhoeckx et al., 2015).

Notwithstanding the issue between protein stability and allergenicity remained controversial, the resistance to proteolysis could convert a protein into a potential allergenic factor. In fact, specific properties of protein, such as the molecular size (lower than 70 kDa) and the stability to pH and enzymatic proteolysis, are essential to exhibit allergenicity, allowing to a food protein to reach the specific immune inductive sites. In particular, the stability to gastric digestion seems to be very important and let to classify proteins in allergens and nonallergens (Bøgh & Madsen, 2016; Pali-Schöll et al., 2018; Untersmayr & Jensen-Jarolim, 2006; Verhoeckx et al., 2015).

Among the applied conditions in the in vitro digestion simulation, pH gastric value proved to be fundamental for pepsin activity. Therefore to study gastrointestinal optimal conditions, pH-stat titration represents an interesting approach in the absence of lipids, whose hydrolysis can contribute to change pH. Low pH ($pH \leq 4.5$) is the ideal value for pepsin and dynamic conditions seem to better increase enzyme kinetics, simulating better the physiological environment (Mat et al., 2016; Mat et al., 2018).

Protein digestibility (kinetics and amount) can be highly modified according to the digestion model applied and the main factors affecting it are the gastric pH, pepsin activity and the salt concentrations (Torcello-Gómez et al., 2020). Protein digestion of food complex matrices has been mainly studied applying semidynamic/dynamic models, which allowed to obtain the entire digestion process under similar physiological conditions (Lucas-González et al., 2018). In particular, in literature, the main dynamic approach to study protein digestion, set-up around the 90s, has been TIM model. It consists of a multicompartmental dynamic model, in which the bioaccessibility is determined by measuring the dialyzed protein nitrogen percentage (dialysis membranes) in relation to the meal protein nitrogen. TinyTIM, a simplification of TIM approach was promising also as a good alternative to animal studies. It has a single small intestinal environment and allowed a rapid determination of protein digestibility and (limiting) amino acids with very predictive results of two important parameters: protein digestibility-corrected amino acid score and digestible indispensable amino acid score (DIAAS). TIM platforms are considered advanced dynamic systems in which many parameters, such as gastrointestinal volume and digesta residence time, can be controlled, also simulating different human physiopathological conditions (Havenaar et al., 2013; Minekus et al., 1995; Van Der Lugt et al., 2020).

On the contrary, in vitro static digestion models can be better adapted to different food proteins and easily harmonized and standardized, and although they are limited to purified food components or simple food matrices, they had a wide range of applications over time, from food/nutritional to toxicological and pharmaceutical fields (Verhoeckx et al., 2015). In addition, notwithstanding the wide use of animal models (mainly rat and pig) to study protein digestibility, their real advantages in reproducing the human digestion process remain still controversial, especially when compared with the in vitro approaches (Deglaire & Moughan, 2012; Mansilla et al., 2020). In particular, in vitro static digestion methods have been widely used to investigate the protein digestibility, pointing out a great potential in the prediction of in vivo behavior (Butts et al., 2012).

Our focus will be on an overview of the in vitro static digestion approaches (mainly biochemical methods) used to investigate food protein digestion in the last 10 years (Fig. 1.2). The impact of INFOGEST protocol (2014) and its evolution INFOGEST 2.0 (2019) will be further explored and the most significant issues and challenges of static approaches over time will be reported and discussed. Other static systems in which a more complete digestion process has been proposed (addition of colon phase) will be considered, both when biochemical-based and cell-based assays. In particular, cell-based assays are very promising when used in combination with biochemical methods for their capability to investigate not only the whole digestion but also absorption and toxicity, being systems able to better simulate simil-physiological conditions.

1.3 Static models for food protein digestion investigations

1.3.1 Static approaches before the COST Action INFOGEST protocol: main issues

Before the COST Action INFOGEST network, many different gastrointestinal in vitro digestion approaches have been proposed in food/nutritional research,

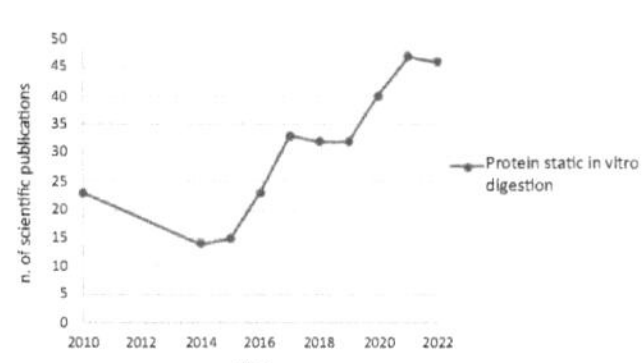

FIGURE 1.2 Protein static in vitro digestion.

Trend of the scientific publications over years.

PubMed 10.02.2023 keywords: "Protein static in vitro digestion".

each with remarkable differences in some principal critical points (Colombo et al., 2021; Fu et al., 2002; Ulleberg et al., 2011):

- number of digestion phases;
- simulated digestion fluids composition;
- digestion time in each phase;
- enzymes concentration;
- presence of additives;
- mechanical treatments.

In general, three main digestion phases, that is, oral, gastric, and small intestinal-duodenum, have been reproduced, even if the oral phase has been generally omitted for liquid food (Oomen et al., 2002); the importance of this phase has been later clarified by INFOGEST protocols (see paragraphs below). The colon stage has always been excluded over the years and notwithstanding its significance has been later discussed by INFOGEST 2.0, today it still represents a critical key point and a potential issue to be discussed in particular for protein digestion (Brodkorb et al., 2019; Colombo et al., 2021).

Simulated digestion (gastric and/or intestinal) fluids differed in pH value, composition, and volume/sample ratio (Verhoeckx et al., 2015). In particular, gastric phase pH is fundamental for protein digestion, as already stated, as low pH values cause a decrease in nutrient bioaccessibility. In vivo, gastric pH ranges from 1.1 to 4.0 in the different tracts, but all the static models were not able to reproduce this in vivo stomach behavior, which is characterized by a pH decrease following a gradual emptying (Oomen et al., 2002).

Fluid composition is correlated to the content and ionic force of mineral salts and human in vivo investigations indicated the fundamental constituents (potassium, sodium, calcium, ammonium, magnesium, chloride, carbonate/bicarbonate, dihydrogen phosphate) to be used to reproduce the simulated salivary fluid (SSF), the simulated gastric fluid (SGF), and the simulated intestinal fluid (SIF) (Lentner, 1981). In particular, pH and composition in sodium and potassium chloride of intestinal fluids were considered important key factors, which differ in specific human populations (from infants to adults) and therefore have to be suitably considered in reproducing different digestion models (Colombo et al., 2021).

Another important issue concerning the digestive simulated fluids was the need to have commonly recognized preparations, which can be stable over time during the digestion process. In fact, the definition of samples and simulated fluids storage time and temperature as well as the specific pH value and/or the potential addition of enzyme inhibitors, such as protease inhibitors in the presence of protein samples, represented the key factor to lead to standardized digestion protocols (Ulleberg et al., 2011). It is also very important to calculate the right dilution ratio of food sample in the final mixture of simulated fluids, in relation to the addition of each fluid constituent, that is, electrolytes, enzymes, additives, and also hydrochloric acid or sodium hydroxide used to reach the target pH of each digestion phase. All the constituents have to be prepared starting from stock solutions and a correct volume of water should be added to obtain the right final digestion mixtures. In addition, enzyme activities should be expressed in units per mL of final digestion mixture (Colombo et al., 2021).

For all the food components the time for each digestion phase (oral, gastric, and intestinal) had an important role not only to reproduce the physiological food residence time in each district but also to let the right contact time with the enzymes (Colombo et al., 2021). Each main digestion phase (gastric and intestinal) could last from 1 to 3 hours. As regards proteins, the time of the intestinal step could be critical. For this reason, some authors proposed an extension of the intestinal phase until 5 hours, especially for protein digestion to reproduce a complete process (as, e.g., for cheese and cured beef meat peptides), even if generally 2 hours were considered a sufficient time to simulate the intestinal behavior (Bordoni et al., 2011; Marcolini et al., 2015; Oomen et al., 2002).

Regarding enzymes, different sources were used, animal or also human-derived, showing consequently a different activity. In general, commercially available porcine, bovine, and rabbit enzymes have been used, while human-derived remained the less used (Colombo et al., 2021; Oomen et al., 2002). Among the animal-derived enzymes, the porcine one is the most used because of its low cost and high homology with human enzymes (Fujinaga et al., 1995; Ulleberg et al., 2011). Another important issue was represented by the enzyme concentration. For an ideal protein digestion, in the gastric phase, over years different pepsin amounts were used till the selection of 2000 U/mL of final digestion mixture as the best concentration, as also confirmed later by INFOGEST protocol (Minekus et al., 2014; Verhoeckx et al., 2015).

The presence of additives was generally due to the need to reproduce lipid-based models. They could be specific enzymes (i.e., gastric lipase), emulsifiers (mucin or bile salts), or phospholipids, and in protein food digestion they can be considered necessary only if the considered food matrix is also a good source of lipids.

The addition of mucin (proteins produced from minor salivary glands, with high viscosity and lubricant action) to the oral phase has been suggested only by few authors, and it could be important only for a solid food homogenization process and therefore its role is not generally fundamental (Oomen et al., 2003; Sarkar et al., 2009; Versantvoort et al., 2005).

Calcium is known to be an essential cofactor of digestive enzymes, by influencing their digestion rate, particularly in the presence of lipids, and over time its addition at least in the main digestive phases (gastric and intestinal) in in vitro approaches has been recommended for all nutrients (Hur et al., 2011; Oomen et al., 2003). In particular, calcium chloride has also been used for its gelling properties, in particular in the stabilization of trypsin and in the regulation of other ions (sodium and potassium) essential for intestinal motility and nutrient absorption (Papaleo et al., 2005).

Mechanical treatments include the agitation process, which is necessary to mix food and fluid mixtures, reproducing digestion peristalsis, even remotely. Different agitation approaches (sonication, mechanical stirring, centrifugation) and parameters (type of shaking, speed, and time) have been reported in literature highlighting the need of standardization. Different filtration methods have also been added by using different systems (filters and membranes); filtration is mainly suggested in the initial steps of the in vitro digestion (chyme solutions) or to remove the particles floating in solution, if necessary. Conversely, the well-known fixed parameter in all the in vitro digestion methods was the physiological temperature (37°C) (Oomen et al., 2002).

1.3.2 COST Action INFOGEST consensus method: main novelties

The COST Action INFOGEST protocol (2014) has been developed and evolved to meet the need of comparing results interlaboratories to create a harmonized static in vitro digestion method for food studies. In addition, the possibility to create standardized models for different age human groups or physiological states has been particularly useful in protein studies, as the rate of protein digestion and metabolization is strictly correlated to human conditions.

INFOGEST protocol consisted of three standardized phases (oral, gastric, and small intestinal) without a colonic step. This represents an important limitation of this protocol, which should be associated with other methods (see paragraph 1.4) able to reproduce the colonic fermentation, as microbial flora could have a key role in the bioaccessibility of proteins (Colombo et al., 2021).

In the study of food proteins, such as milk proteins, INFOGEST protocol has been often used in comparison to in vitro dynamic digestion systems and in vivo experiments (animal models), demonstrating a potential predictivity and a good approximation of what really could happen in vivo (Egger et al., 2019; Nehir El et al., 2015; Sanchón et al., 2018).

This method also highlighted that protein hydrolysis and, accordingly, digestibility are strongly dependent from food matrix composition and protein isolate solubility. In fact, for example, it was possible to confirm that whey proteins had higher digestibility than zein protein, as already stated in clinical trials (Sousa et al., 2020). Other INFOGEST protocol applications specific for proteins are discussed in section 1.4.

The three main digestion phases (oral, gastric, and small intestinal-duodenum) previously reported for digestion models applied before INFOGEST protocol were maintained, clarifying the need of a rapid oral phase (2 minutes) for solid foods and suggesting the addition of this stage also for liquid foods, if they contain carbohydrates (starch).

Simulated digestion fluids (SSF, SGF, and SIF) have been standardized for pH value, composition (electrolytes enzymes and concentration, acid/base added to adjust pH value in each phase), and volume. The final standardized protocol is reported in Table 1.1.

The recommended digestion time for each step (gastric, and intestinal) was 2 hours, also in the case of protein samples for which a longer intestinal digestion

Table 1.1 Main procedures and development of the two INFOGEST protocols for oral, gastric, and duodenal phases.

	INFOGEST	INFOGEST 2.0
Preliminary procedures		• Purchase of specific reagents and chemicals • Suggestions on specific equipment • pH test adjustment experiment (5 h) • Preparation of a number of digestion tubes for each time points • Enzyme activity and bile assays (4 days)
Preparation	Stock solutions (gastrointestinal fluids, enzymes, additives)	• Detailed digestion spreadsheets (2 h)
Oral phase (30 min)	Solid meal SSF (1:1) • 15.1 mL KCl (0.5 M) • 3.7 mL KH_2PO_4 (0.5 M) • 6.8 mL $NaHCO_3$ (1 M) • 0.5 mL $MgCl_2(H_2O)_6$ (0.15 M) • 0.06 mL $(NH_4)_2CO_3$ (0.5 M) 0.09 mL HCl (6 M) Salivary amilase, 75 U/mL $CaCl_2$, 1.5 mM pH 7.0, 2 min, 37°C	The same conditions, but: this phase must be introduced also for liquid meal • Detailed procedure • Prewarm of SSF, 37°C
Gastric phase (2–3 h)	Oral bolus SGF (1:1) • 6.9 mL KCl (0.5 M) • 0.9 mL KH_2PO_4 (0.5 M) • 12.5 mL $NaHCO_3$ (1 M) • 0.4 mL $MgCl_2(H2O)_6$ (0.15 M) • 0.5 mL $(NH_4)_2CO_3$ (0.5 M) • 11.8 mL NaCl (2 M) 1.3 mL HCl (6 M) Pepsin, 2000 U/mL $CaCl_2$, 0.15 mM pH 3.0, 2 h, 37°C	The same conditions, but: • Detailed procedure • Prewarm of SGF, 37°C + Specific preparation of pepsin (10 mM Tris, 150 mM NaCl, pH 6.5) + Gastric lipase, 60 U/mL

(Continued)

Table 1.1 Main procedures and development of the two INFOGEST protocols for oral, gastric, and duodenal phases. *Continued*

	INFOGEST	INFOGEST 2.0
Intestinal (duodenum) phase (2–3 h)	Gastric chyme SIF (1:1) • 6.8 mL KCl (0.5 M) • 0.8 mL KH_2PO_4 (0.5 M) • 42.5 mL $NaHCO_3$ (1 M) • 1.1 mL $MgCl_2(H_2O)_6$ (0.15 M) • 9.6 mL NaCl (2 M) 0.7 mL HCl (6 M) Pancreatin (based on trypsin activity 100 U/mL) (in alternative: + Trypsin, 100 U/mL) + Chymotrypsin, 25 U/mL + Pancreatic lipase, 2000 U/mL + Colipase, 2:1 molar ratio with lipase + Pancreatic amylase, 200 U/mL + $CaCl_2$, 0.6 mM + Bile salts, 10 mM pH 7.0, 2 h, 37°C	The same conditions, but: • Detailed procedure • Prewarm of SIF, 37°C

time has been proposed, as previously discussed. In fact, a complete digestion time usually takes 4 hours (Dressman et al., 1990).

Enzymes, such as salivary α-amylase, gastric pepsin, and intestinal pancreatin (enzyme extract), have been recommended of porcine origin. Porcine intestinal pancreatin can be also replaced by a mixture of individual enzymes: porcine trypsin, pancreatic amylase, pancreatic lipase and colipase, and bovine chymotrypsin.

Regarding additives, the addition of lipase in the gastric step of protein digestion can be omitted, as it is necessary for lipid sample digestion, as already mentioned. The addition of bile salts to the intestinal phase has been recommended, differently from the gastric phase because in vivo only in a limited number of subjects low bile acids concentration (0.2 mM) was found in the gastric fluid. Bile (porcine or bovine origin) can be in fresh or extract form. The presence of bile particularly influences the lipolysis rate. If phospholipids or surfactants are not present in the tested food, phospholipids should be optionally added to the gastric step also for protein-rich food.

To maintain fluid and enzyme stability, the just addition of proteases can be enough in a system containing only proteins.

The addition of calcium chloride (dihydrate form) to each digestion phase (also in SSF) was indicated by INFOGEST protocol. In particular, it should be directly added to each final mixture in each digestion step (containing the simulated digestion fluid and food) and not to digestive fluids stock solutions as precipitation problems can occur.

Final enzyme/additive concentrations are reported in Table 1.1.

The mixing during all three digestion phases was recommended, but it has not been standardized and the use of shakers or rotators has been indicated.

1.3.3 INFOGEST protocol development and improvements (INFOGEST 2.0 method): main challenges

One of the main critical steps of INFOGEST protocol was the determination of the standard units of enzyme activity (α-amylase, pepsin, trypsin, chymotrypsin, and pancreatic lipase): in fact, a great variability in results occurring interlaboratories could be due to different enzyme concentrations used. Enzyme activity assays have been already described by (Minekus et al., 2014) and better schematized in the INFOGEST 2.0 giving more details, mainly about specific reagents, chemicals, and solvents to be used to dissolve enzymes, buffer concentration, pH, the preparation of additives, mixing mode, temperature and time to be applied, and how to monitor the enzymes activity. In particular, for protein digestion it is necessary to standardize pepsin activity assays as it is a key-factor to reproduce a correct gastric digestion (Egger et al., 2016).

The main improvements of INFOGEST 2.0 method were focused on the standardization of some procedures necessary to carefully set-up every step of the protocol, starting from the preparation of all solutions (simulated digestive fluids, enzymes, and bile salts) for which precise spreadsheets have been provided to calculate the volumes needed on food amount basis. In this context, the initial preparation of food sample and SSF solution became mandatory to determine the food-to-enzyme ratio in the inner digestion process. Therefore the oral step, which previously was only suggested as optional, became fundamental, and an easier preparation process was standardized. In addition, the oral phase made the bolus consistency more similar to the in vivo behavior.

For protein digestion, a particular preparation of pepsin solution was also suggested to stabilize the enzyme activity, which was not reproducible: the use of Tris buffer and sodium chloride (pH 6.5) instead of sodium hydroxide and sodium chloride improved pepsin stability. Conversely, in the previous protocol, pepsin was directly prepared at pH 3, considering the final pH of gastric phase and the pH value at which the enzyme activity is maximum (pH 2–4) (Minekus et al., 2014). This highlighted the strong correlation between the enzyme stability/activity and the type of additives and pH selected to prepare the stock solutions.

During digestion from gastric to intestinal phase, the addition of 1 M sodium bicarbonate or 1 N sodium hydroxide stops the pepsin activity, because it causes a pH shift (from 3 to 7) similarly to what happens in the duodenum tract. For samples that are not stable to a pH variation, this step can be replaced by the addition of pepstatin A, a selective inhibitor of pepsin. On the contrary, at the end of digestion process, the addition of Pefabloc SC (4-(2-aminoethyl) benzenesulfonyl fluoride hydrochloride) was recommended to irreversibly inhibit trypsin and chymotrypsin.

Before starting with the digestion process, Brodkorb et al. (2019) suggested to perform a preliminary experiment, namely the pH-test adjustment experiment, to set the exact volume of acid or base (at known concentration) to reach the desired pH. These volumes were fundamental to predict the final volumes of each digestion phase. It is also important to underline that food particle size and consistence affect the rate to obtain a stable pH and therefore this test can also give an idea of the experimental timeline. Another important issue regarding the food stability and sensitivity to light or oxidation should also be preliminarily considered before starting with the set-up of all procedures.

The main novelties of INFOGEST 2.0 method in comparison to INFOGEST protocol are reported in Table 1.1.

In INFOGEST 2.0, a specific important issue for proteins/peptides bioaccessibility and bioavailability was discussed in relation to the role of brush-border membrane (BBM) enzymes (e.g., lactase–phlorizin–hydrolase) in trypsinogen activation. Over years, the addition of such enzymes has been suggested by some authors, the lack of commercial enzymes, and the need to isolate them from animal intestine (fresh intestine or intestinal extract) let the problem unsolved (Mamone et al., 2015; Picariello et al., 2010).

In addition, as the INFOGEST protocol has been set-up only for adult digestion, the possibility to extend this protocol to specific population groups (infant, adolescent, and elderly) and also to patients with digestive tract pathologies (e.g., obesity, cystic fibrosis, gastric bypass surgery, or exocrine pancreatic insufficiency) was also discussed, thus becoming a very important issue (Shani-Levi et al., 2017).

In the case of infants, different conditions of emptying half-time, pH changes (the main change in pH regards the gastric phase from pH 3 for an adult to pH 5.3 in infants), digestive fluids composition and enzyme/bile salts amounts should be carefully considered (Ménard et al., 2018).

1.4 Main applications of INFOGEST protocols to protein digestion

The main applications of INFOGEST protocols in the study of protein bioaccessibility and absorption are related to dairy products and meat or to specific allergens and plant-based (PB) proteins (Zhou et al., 2023).

1.4.1 Dairy products

Milk proteins have been widely studied with INFOGEST protocols obtaining results in a good agreement with those obtained in vivo (Brodkorb et al., 2019; Egger et al., 2019; Nehir El et al., 2015; Sanchón et al., 2018; Sousa et al., 2020). Bovine milk contains two main types of proteins: micellar caseins (β-casein and α_{s1}-casein) and whey proteins (manly β-lactoglobulin and α-lactalbumin). With a good correlation with in vivo experiments, caseins were unstable to the gastric digestion and rapidly degraded, while β-lactoglobulin and α-lactalbumin metabolization mainly occurred at intestinal level, being resistant to the action of pepsin because rich in proline or negatively charged peptides. This study also pointed out which milk-derived peptides are able to exert biological activities and which are immunogenic (Sanchón et al., 2018). In fact, thanks to the simulation of the gastrointestinal digestion milk peptides hydrolysis degree was correlated mainly with antioxidant and antidiabetic bioactivities. For example, digested kefir and goat milk exhibited higher antioxidant and antidiabetic activities than the undigested samples (Nehir El et al., 2015).

A very recent work by Atallah et al. compared some milk matrices differing for micellar caseins, whey proteins, calcium, and total nitrogen content and obtained by applying different industrial processes. For the first time, the need to optimize porcine pancreatin concentration according to the protein content of each matrix was highlighted (Atallah et al., 2020). An INFOGEST-based infant model set-up by Ménard et al. pointed out that gastric digestion of β-casein and β-lactoglobulin did not occur, mainly because of the higher infants pH in comparison to adults; conversely, the infant intestinal digestion seemed to be more rapid and efficient for these two proteins. Casein was differently metabolized, too. In particular, at the gastric level, it was hydrolyzed, but with a slow kinetics (Ménard et al., 2018).

An interesting application of INFOGEST was the study of particular formulations consisting of milk proteins, polysaccharides, and oil. The obtained results indicated that milk proteins can stabilize emulsions during in vitro gastrointestinal digestion and that the protein content has an important role in controlling gastric proteolysis and subsequent lipolysis. This is especially important in food processing when formulations for specific disease (obesity), are produced, because the lipid digestion results delayed (Liu et al., 2022).

1.4.2 Egg products

In literature, the applications of the INFOGEST protocol to egg products are few. An important issue regards the egg proteins (mainly ovoalbumin) digestion is the study of its allergenicity. In fact, eggs (mainly white eggs and only to a lesser extend yolks) are together with milk the most common allergenic foods, in particular for children. It is well known that egg allergenicity is linked to its protein resistance to gastroduodenal digestion, so protein digestion studies could be very important for understanding the egg products impact on health. Specific protein structural characteristics, such as quaternary structure, disulfide bridges, and binding, or proteolytic products generated during digestion can make a protein/peptide allergenic. In addition, egg protein digestion can be affected by food processing and cooking methods, as for example for egg white protein for which digestibility increase following heat treatments, probably due to the protein unfolding/denaturation and the higher accessibility to intermolecular peptide bonds by digestive enzymes (Rao et al., 2020; Santos-Hernández et al., 2020; Shin et al., 2013). INFOGEST protocol has been applied by Santos-Hernández et al. to investigate the egg white gastrointestinal-derived peptides and the relative amino acids profiles with the aim to identify not only allergen epitopes but also bioactive peptides able to induce hormones with antidiabetic effect (Santos-Hernández et al., 2020). Egg peptides possess many biological activities, mainly antioxidant, antiinflammatory, antihypertensive, and antidiabetic. To investigate this aspect, INFOGEST protocol has been used by Rao et al., to study the effect of pH and heat on egg protein solubility/aggregation and digestibility in relation to the released peptides and their bioactivities. Heat (60°C) was a key factor to improve the protein digestibility and the formation of antioxidant peptides generated during the digestion process (Rao et al., 2020).

1.4.3 Meat products

Few studies on in vitro digestion approaches applied to cooked or dry-cured meat did not use a complex digestive process, such as INFOGEST protocol, and let to the identification of peptides differing in bioactivities (ACE inhibitory and antioxidant activities) according to the digestibility protocol used (Escudero et al., 2012; Sangsawad et al., 2017; Zhu et al., 2016).

Differently, a recent interesting study used INFOGEST protocol in combination with a peptidomic approach (LC–MS/MS) for the identification of peptides obtained by the digestion of cooked pork, beef, chicken, and turkey proteins and for the characterization of their biological activities. It is known that cooking method and temperature could influence meat protein antioxidant properties. For example, cooking at high temperatures (180°C) increased meat antioxidant activity because of the protein degradation. In this work, different types of meat (all grilled at 140°C) were compared for their digestibility and by consequence for their peptide bioactivities. The intestinal step highlighted a different amount and also different degradation kinetic of the released amino groups; in particular, beef meat showed the highest values differently from the results obtained in previous studies using different digestion models. Radical scavenging activity and lipid peroxidation inhibitory effects, as well as ACE/dipeptidyl peptidase-IV (DPP-IV) inhibitory activities (antioxidant and antidiabetic effect) were also tested. The results indicated that pork meat proteins, notwithstanding their quite lower digestibility in comparison to beef meat proteins, were the richest in antioxidant and antidiabetic peptides with DPP-IV inhibitory activity. (Martini et al., 2019).

Conversely, chicken meat peptides exhibited the highest ACE-inhibitory activity. These results can be related to the different release of peptide sequences by the meat types, which are independent from the digestibility results. Another interesting application of INFOGEST highlighted how Bresaola meat protein bioaccessibility changed according to ripening, which increased the digested peptides bioactivity (antioxidant and antidiabetic effect), thus confirming that peptide profile is highly affected by meat processing procedures (ripening, but also thermal treatment or drying). The bioactivities could be correlated to the release of myofibrillar proteins and peptides by the muscle tissue during digestion. The antidiabetic effect was mainly due to a direct activity of the peptides on α-amylase and α-glucosidase, which are the digestive enzymes involved in the carbohydrate metabolization (Picone et al., 2019).

1.4.4 Allergenic proteins

One important issue is the already mentioned link between protein digestion stability and allergenicity. In fact, some peptides in foods as pasta, hazelnuts, and peanuts are known to be potentially allergenic, as they are resistant to digestion. They were studied with INFOGEST protocol and an important result was the relation between food consistency and proteolysis. In fact, a solid consistency resulted more prone to proteolysis than liquid food as evident from the high degree of proteolysis of whey proteins when subjected from gastrointestinal enzymes (Di Stasio et al., 2017; Korte et al., 2017; Prodić et al., 2018; Wang et al., 2017). A very recent study pointed out that the raw and roasted hazelnuts, which have different protein profiles, undergo a similar pepsin digestion, which slightly increases the overall immunoreactivity in both samples, confirming the role of gastric digestion on allergenicity. This study also highlighted how a food matrix rich in lipid can protect protein digestibility and therefore their allergenicity (Prodić et al., 2022).

INFOGEST platform was also used to study immunogenic gluten peptides, highlighting how the type of food processing affected the proteolysis and the content of gliadins and therefore the type and quantity of antigens of celiac disease. For example, sourdough fermentation is a powerful tool to improve nutritional and functional properties in flours and wheat-based foods like bread, and it is useful in the food processing for celiac-safe and healthy wheat products. This type of fermentation increases the extractability/solubility of gliadins, altering their digestion, and decreasing the formation of some immunogenic gluten peptides (Ogilvie et al., 2021).

1.4.5 Plant-based proteins

Actually, plants are well studied as potential source able to replace animal-based foods. The interest in plants increased, as a vegetarian diet seems to have a great positive impact on health, mainly reducing cardiovascular diseases and cancer. Key factors remain the differences in nutrients composition and gastrointestinal behavior (digestibility and bioavailability) between animal and PB foods (Sridhar et al., 2022;

Zhou et al., 2023). Proteins are the most important functional ingredient in animal/PB products and one important issue is the protein structure, which is fibrous in animal food, differently from vegetable food in which the proteins are globular. The different protein structure leads to differences in some aspects from appearance and texture to cookability, and nowadays, food industry works on the search for mechanical processing to convert globular plant proteins into fibrous materials (McClements & Grossmann, 2021). Plant-based proteins are mainly isolated from cereals (wheat) and legumes (lentil, soy, pea, and bean) and widely used to formulate meat analogs and protein-rich oilseeds (mainly sunflower). Digestion and absorption rates of animal/PB proteins are different: for example, soy proteins are digested faster than casein proteins. On the contrary, the digestibility of some PB proteins could be negatively affected by the antinutritional factors, which are more present in vegetable foods (Lynch et al., 2018).

Plant-derived peptides were less studied in relation to their bioactivities and few bioactive peptides from these sources have been reported. In fact, PB food components are well known for their antioxidant activity, and this does not concern the protein content, but compounds, such as catechins, flavanols, phenolic acids, and anthocyanins (Abeyrathne et al., 2022). Some plant-derived peptides seem to have a promising effect in controlling food intake and appetite. This action could be regulated by a direct effect on δ opioid receptor (green leaves-derived peptides), inhibiting high-fat diet intake or on μ-opioid receptor (mainly soy-derived peptides), suppressing the small intestine transit. Soy-derived peptides also act with another mechanism, that is common to wheat-derived peptides, stimulating the ghrelin secretion. Ghrelin is an endogenous hormonal peptide involved in appetite, increasing gastric secretion and motility (Kaneko et al., 2014; Kaneko, 2021).

Today there is a lack of knowledge also on gastrointestinal fate of PB meat analogs and a comparison with meat products is very difficult. Recently the INFOGEST protocol has been applied to clarify PB meat protein digestion; these products have been produced adding PB proteins (pea protein isolate, rice protein, or lentil flour) to a meat model (fortification) and the simulated digestion process highlighted that, for example, the addition of lentil to beef steaks can cause after digestion a high release of small peptides (<500 Da) rich in essential amino acids, such as isoleucine, lysine, phenylalanine, and valine (Baugreet et al., 2019). A comparison between PB ground beef (containing soy proteins) and ground beef indicated that PB protein hydrolysis occurs more rapidly than ground beef protein at the gastric level and this can be correlated both to the different protein structure and also to the formation of a viscous mixture between soy proteins and dietary fibers (often useful to give particular functional and nutritional properties to PB meat analogs) that are also responsible for a strong satiety response (Zhou et al., 2021). In fact, an increase in shear viscosity in the stomach affects the rate of digestion. Another example of INFOGEST application regarded a very recent study about the comparison of digestibility and DIAAS values of milk-, plant- (soy and wheat), and insect-based proteins, clarifying a comparable or higher protein content of plant sources, but a lower quality of these proteins in comparison to milk source. The aim was to highlight the potentiality to combine different protein sources to avoid the risk of a shortage of high-quality animal proteins, ensuring an optimal nutritional protein quality (Komatsu et al., 2023).

1.5 Other static models including the addition of a colon step: main potentialities

1.5.1 Biochemical assays

Hamzalıoğlu and colleagues have been the first researchers to adapt INFOGEST model (2014 version) by adding a colon phase containing a protease solution and Viscozyme L (pH 8.0) after the duodenal step. They applied this complete digestion system to study the formation of α-dicarbonyl compounds (glyoxal, methylglyoxal, 3-deoxyglucosone, and glucosone), which can interact with proteins, mainly with arginine and lysine residues, causing the formation of advanced glycated end products, responsible for many age-related disorders (such as diabetes and Alzheimer's disease) (Hamzalıoğlu & Gökmen, 2016). This study pointed out the importance of the colon step and the microbial fermentation for food components' bioavailability and bioactivation, which can increase thanks to active metabolites produced by colon enzymes. In addition, this model can be versatile with different applications in the study of microbiota.

The procedure set-up by Hamzalıoğlu et al. opened the possibility to perform a complete digestion in the study of the α-dicarbonyl compounds linked with proteins. In fact, the α-dicarbonyl compounds can have an important role in food macronutrient metabolism and gut microbiota, modifying proteins (Brighina et al., 2021; Cruz et al., 2022). In addition, the search for new agents with a scavenging effect against α-dicarbonyl compounds has resulted in interesting results for protein metabolites (glycine and serine), which act against methylglyoxal toxicity because of the formation of imidazole salts (Hu et al., 2022).

1.5.2 Cell-based assays

Since the 1990s Caco-$_2$ cells (human colon carcinoma cells) have been considered a conventional model to reproduce the intestinal barrier because and it is useful to evaluate the cell transport and uptake of different products (food components and drugs). Caco-$_2$ cell monolayer simulates the structure and functions of the human intestinal epithelium, whose internal surface is covered by a monolayer of epithelial cells. In fact, Caco-$_2$ cells exert a spontaneous enterocyte-like differentiation after culturing as a monolayer for about 20 days and a brush border hydrolytic ability (Jochems et al., 2018; Lennernäs, 2007).

This cell line has been widely used and is still a promising approach to reproduce the human intestine, despite its low permeability, which is due to a lower expression of brush border enzymes (proteinases and peptidases) in comparison with the in vivo intestinal barrier, thus causing a reduction in particular in the peptide metabolism (Amigo & Hernández-Ledesma, 2020). Recently a $Caco\text{-}_2$ cells clone (TC7) has proved particularly promising to predict intestinal absorption and transport, as it consists of a more homogeneous cellular population with developed intercellular junctions able to better predict intestinal absorption (García Fillería & Tironi, 2021).

Another human colon carcinoma cell line (HT-29) and its subpopulation of mucus-secreting cells (HT29-MTX) have been occasionally used in combination with $Caco\text{-}_2$ cells or as a good alternative to study peptides absorption. Differently from $Caco\text{-}_2$ cells, HT-29 cells exhibit relevant characteristics, such as the mucus layer formation, which represents the first barrier, thus interfering in particular with the peptide absorption (Amigo & Hernández-Ledesma, 2020; Contreras et al., 2012; Jochems et al., 2018; Wikman-Larhed & Artursson, 1995; Xu et al., 2017).

It is important to highlight that most studies published in literature after the INFOGEST consensus, that combine noncell-based and cell-based experiments, have been performed using indifferently digestion protocols ante INFOGEST or post-INFOGEST, thus complicating the comparison of results obtained interlaboratories (Sangsawad et al., 2018; Wang et al., 2016; Xing et al., 2018).

In this section, we only report some examples of the most used intestinal cell models, mainly $Caco\text{-}_2$ cells, coupled to the INFOGEST protocol as an alternative to the Hamzalıoğlu et al. procedure, which provided the addition of a noncell-based colon phase, as previously discussed (Hamzalıoğlu & Gökmen, 2016).

The combination of biochemical assays as INFOGEST and cell-based models allowed to evaluate the food-derived bioactive peptides (FBPs) absorption which is directly correlated to the potential peptide biological activity/toxicity. The knowledge of FBPs bioavailability is also fundamental to identify industrial strategies (such as the use of permeation enhancers or enzyme inhibitors) able to improve FBPs delivery (Peng et al., 2022).

For example, the antioxidant activity of food peptides, which is usually measured by biochemical assays, should also be evaluated considering their cellular absorption and mechanism to predict the real in vivo effects. Therefore the coupling with digestion experiments coupled with absorption assays can better clarify how the bioactivity is correlated to the digested samples. More recently, food-derived antioxidant peptides have gained much attention mainly because of their safety and above 800 peptide sequences with antioxidant activity were actually known. Considering that peptides become active upon their release following enzymatic or chemical hydrolysis or microbial fermentation, bioaccessibility studies using simulated digestion models are mandatory (Fan et al., 2022; Juárez-Chairez et al., 2022; Zhu et al., 2022).

Interesting studies performed on the antioxidant peptides isolated from soybean protein hydrolysate showed an absorption of the intact sequences of the released peptides across $Caco\text{-}_2$ cell monolayers with an increase in their antioxidant activity after the digestion process. Therefore this hydrolysate can be considered a promising source of functional food ingredients able to regulate the oxidation in food samples (Zhang et al., 2018).

The coculture of Caco-$_2$ and HT-29 has been used to better evaluate the bioavailability of peptides from whey proteins, exploiting the HT-29 mucus secretion that improves the Caco-$_2$ model. The results identified some new bioactive peptides, which have not been previously reported as antioxidants and they had also immunomodulatory effects (Corrochano et al., 2019).

Caco-$_2$ model has been widely used to evaluate the absorption of antihypertensive food peptides, which mainly act through ACE inhibition, and to clarify the mechanism of intestinal absorption because it represents the ideal model (Amigo & Hernández-Ledesma, 2020). Most studies have been carried out using only Caco-$_2$ cells without the previous application of a digestion protocol, only focusing on the uptake and transepithelial transport in particular of new formulations of FBPs to improve their bioavailability and to develop bioactive ingredients for the functional food market (Guo et al., 2019; He et al., 2018; Zhang et al., 2019)

1.5.3 Static batch fermentation

The INFOGEST protocol can also be implemented with anaerobic mini reactors with intestinal or fecal microbial cells of animal (mainly pig) or human origin, which represent promising high-throughput colonic fermentation systems. They are mostly suitable for rapid screening experiments, as the main problem of these systems consists in the generation of microbial metabolites when incubated for long time, which decreases the microbial activity. This set-up can be useful to obtain more accurate results (Colombo et al., 2021; Venema & Van Den Abbeele, 2013) and it has been mainly used to investigate fiber and polyphenols bioavailability, but also promising perspectives for dietary proteins have emerged (Pérez-Burillo et al., 2021). In fact, gut microbiota is involved in the digestion of dietary protein and protein fermentation can cause the formation of toxic metabolites, mainly branched-chain fatty acids, amines, ammonia, hydrogen sulfide, and phenolic and indolic compounds; therefore these models can help to clarify the protein metabolism fate (Venema & Van Den Abbeele, 2013; Zhao et al., 2019).

Another interesting application regards the evaluation of the probiotics effect on protease and peptidase activity. In fact, the intestinal microflora has an important role in affecting proteolysis, thus increasing the absorption of small peptides and amino acids (by acting on the absorption and transport across the epithelium) and also decreasing protein fermentation and toxic metabolites production (Wang & Ji, 2018).

These approaches have been also used to test the effect of probiotics on gut microbiota composition and metabolism in the presence of a bacterial infection, an antibiotic treatment, or an antimicrobial agent based on proteins, such as casein hydrolysates (Venardou et al., 2021; Zihler et al., 2011).

1.6 Conclusions and future perspectives

Nowadays, INFOGEST protocols eventually integrated with a colonic step or coupled with cell-based systems represent the key static approaches to study protein and generally food digestion. An important issue to be considered is that notwithstanding the existence of these accepted International protocols, the variability of the gastrointestinal digestion models proposed in literature is still very extensive and this makes the comparison of results interlaboratories very hard and difficult. An interesting future perspective could be a new advanced static model able to offer standardized guidelines and experimental conditions with a more general consensus.

In addition, actually, the research is focused on the set-up and validation of dynamic platforms, which represent a real evolution compared to static assays for the possibility of studying directly in real-time a simil-physiological digestion (bioaccessibility, absorbance, metabolism, and bioactivity) of food components, in particular proteins. They are still very complex to set-up, but these more rapid systems can be used as complementary tools of static methods, as the need of a multimethod approach seems to be mandatory to have results in vitro, closer to those obtained in vivo. In particular, fluidic systems, also namely bioreactors, which are based on microfluidic or millifluidic principles, represent very promising approaches able to reproduce also multicompartmental cellular systems, in which similar physiological conditions (flow, permeability) and connections (cross-talk between organs) can be reproduced (Colombo et al., 2019; Giusti et al., 2014; Iori et al., 2012; Oleaga et al., 2016; Vinci et al., 2012; Colombo et al., 2023). The higher costs of dynamic models in comparison to the static ones can be compensated as they are less time consuming and can already reduce in vivo experiments, particularly during preliminary investigations (screening tests), or even represent a promising future alternative to animal models with a wide range of applications from food analysis to drug discovery.

However, the ideal solution remains to perform multiapproach experiments, in which advanced and standardized static and dynamic techniques can be complementary used to obtain more predictive results of the in vivo digestion and metabolism.

References

Abeyrathne, D. D. N. S., Nam, K., Huang, X., & Ahn, D. U. (2022). Plant- and animal-based antioxidants' structure, efficacy, mechanisms, and applications: A review. *Antioxidants*, *11*, 1025. Available from https://doi.org/10.3390/antiox11051025.

Amigo, L., & Hernández-Ledesma, B. (2020). Current evidence on the bioavailability of food bioactive peptides. *Molecules (Basel, Switzerland)*, *25*(19), 4479. Available from https://doi.org/10.3390/molecules25194479.

Atallah, N., Deracinois, B., Boulier, A., Baniel, A., Jouan-Rimbaud Bouveresse, D., Ravallec, R., Flahaut, C., & Cudennec, B. (2020). In vitro assessment of the impact of industrial processes on the gastrointestinal digestion of milk protein matrices using the InfoGest protocol. *Foods*, *9*(11), 1580. Available from https://doi.org/10.3390/foods9111580.

Baugreet, S., Gomez, C., Auty, M.A.E, Kerry, J.P., Hamill, R.M., & Brodkorb, A. (2019). In vitro digestion of proteinenriched restructured beef steaks with pea protein isolate, rice protein and lentil flour following sous vide processing. *Innov. Food Sci. Emerg. Technol. 54*, 152–161. Available from https://doi.org/10.1016/j.ifset.2019.04.005.

Bohn, T., Carriere, F., Day, L., Deglaire, A., Egger, L., Freitas, D., ... Dupont, D. (2018). Correlation between in vitro and in vivo data on food digestion. What can we predict with static in vitro digestion models?, *Crit. Rev. Food Sci. Nutr., 59*, 2239–2261. Available from https://doi.org/10.1080/10408398.2017.1315362.

Bordoni, A., Picone, G., Babini, E., Vignali, M., Danesi, F., Valli, V., Di Nunzio, M., Laghi, L., & Capozzi, F. (2011). NMR comparison of in vitro digestion of Parmigiano Reggiano cheese aged 15 and 30 months. *Magnetic Resonance in Chemistry, 49*(1), S61–S70. Available from https://doi.org/10.1002/mrc.2847.

Brighina, S., Poveda Turrado, C., Restuccia, C., Walton, G., Fallico, B., Oruna-Concha, M.G., & Arena, E. (2021). Detrimental effect on the gut microbiota of 1,2-dicarbonyl compounds after in vitro gastro-intestinal and fermentative digestion, *Food Chemistry, 341*, 128237.

Brodkorb, A., Egger, L., Alminger, M., Alvito, P., Assunção, R., Ballance, S.,...Recio, I. (2019). Infogest static in vitro simulation of gastrointestinal food digestion, *Nat. Protoc., 14*, 991–1014.

Butts, C. A., Monro, J. A., & Moughan, P. J. (2012). In vitro determination of dietary protein and amino acid digestibility for humans. *British Journal of Nutrition, 108*(2), S282–S287. Available from https://doi.org/10.1017/S0007114512002310.

Bøgh, K.L., & Madsen, C.B. (2016). Food Allergens: Is There a Correlation between Stability to Digestion and Allergenicity? *Crit Rev Food Sci Nutr, 56*, 1545–1567.

Callahan, A., Leonard, H., & Powell, T. (2020). Nutrition: Science and everyday application. *Open Oregon Educational Resources*.

Calvo-Lerma, J., Fornés-Ferrer, V., Heredia, A., & Andrés, A. (2019). In vitro digestion models to assess lipolysis: The impact of the simulated conditions of gastric and intestinal pH, bile salts and digestive fluids. *Food Research International, 125*108511. Available from https://doi.org/10.1016/j.foodres.2019.108511.

Chakrabarti, S., Guha, S., & Majumder, Kv (2018). Food-derived bioactive peptides in human health: Challenges and opportunities. *Nutrients, 10*(11), 1738. Available from https://doi.org/10.3390/nu10111738.

Colombo, R., Ferron, L., Frosi I., & Papetti, A. (2021). Advances in static in vitro digestion models after the COST action Infogest consensus protocol. *Food Funct., 12*, 7619–7636.

Colombo, R., Ferron, L., Frosi, I., & Papetti, A. (2021). Advances in static: In vitro digestion models after the COST action Infogest consensus protocol. *Food and Function, 12* (17), 7619–7636. Available from https://doi.org/10.1039/d1fo01089a, http://pubs.rsc.org/en/journals/journal/fo.

Colombo, R., Paolillo, M., Frosi, I., Ferron, L., & Papetti, A. (2023). Effect of the simulated digestion process on the chlorogenic acid trapping activity against methylglyoxal. *Food & Function, 14*(1), 541–549. Available from https://doi.org/10.1039/d2fo02778j.

Colombo, R., Paolillo M., & Papetti, A. (2019). A new millifluidic based gastrointestinal platform to evaluate the effect of simulated dietary methylglyoxal intakes, *Food Funct., 10*, 4330–4338.

Contreras, M. M., Sancho, A. I., Recio, I., & Mills, C. (2012). Absorption of casein antihypertensive peptides through an in vitro model of intestinal epithelium. *Food Digestion, 3*, 16–24. Available from https://doi.org/10.1007/s13228-012-0020-2.

Contreras, M. M., Sanchez, D., Sevilla, M. A., Recio, I., & Amigo, L. (2013). Resistance of casein-derived bioactive peptides to simulated gastrointestinal digestion. *The International Dairy Journal, 32*, 71–78. Available from https://doi.org/10.1016/j.idairyj.2013.05.008.

Corrochano, A.R., Ferraretto, A., Arranz, E., Stuknyte, M., Bottani, M., O'Connor, P.M., ... Giblin, L. (2019). Bovine whey peptides transit the intestinal barrier to reduce oxidative stress in muscle cells, *Food Chem., 288*, 306–314.

Cruz, N., Flores, M., Urquiaga, I., & Ávila, F. (2022). Modulation of 1,2-dicarbonyl compounds in postprandial responses mediated by food bioactive components and mediterranean diet. *Antioxidants, 11*(8), 1513. Available from https://doi.org/10.3390/antiox11081513.

Deglaire, A., & Moughan, P. J. (2012). Animal models for determining amino acid digestibility in humans—A review. *British Journal of Nutrition, 108*(2), S273–S281. Available from https://doi.org/10.1017/S0007114512002346.

Di Stasio, L., Picariello, P., Mongiello, M., Nocerino, R., Berni Canani, R., Bavaro, S., Mamone, G. (2017). Peanut digestome: identification of digestion resistant IgE binding peptides. *Food Chem. Toxicol., 107*, 88–98.

Dressman, J. B., Berardi, R. R., Dermentzoglou, L. C., Russell, T. L., Schmaltz, S. P., Barnett, J. L., & Jarvenpaa, K. M. (1990). Upper gastrointestinal (GI) pH in young, healthy men and women. *Pharmaceutical Research, 7*(7), 756–761. Available from https://doi.org/10.1023/A:1015827908309.

Egger, L., Ménard, O., Delgado-Andrade, C., Alvito, P., Assunção, R., Balance, S., ... Portmann, R. (2016). The harmonized InfoGest in vitro digestion method: From knowledge to action, *Food Res. Int., 88*, 217–225.

Egger, L., Schlegel, P., Baumann, C., Stoffers, H., Guggisberg, D., Brügger, C., ... Portmann, R. (2017). Physiological comparability of the harmonized InfoGest in vitro digestion method to in vivo pig digestion, *Food Res. Int., 102*, 567–574.

Egger, L., Ménard, O., Baumann, C., Duerr, D., Schlegel, P., Stroll, P., ... Portmann, R. (2019). Digestion of milk proteins: comparing static and dynamic in vitro digestion systems with in vivo data, *Food Res. Int., 118*, 32–39.

Escudero, E., Toldrá, F., Sentandreu, M. A., Nishimura, H., & Arihara, K. (2012). Antihypertensive activity of peptides identified in the in vitro gastrointestinal digest of pork meat. *Meat Science, 91*, 382–384. Available from https://doi.org/10.1016/j.meatsci.2012.02.007.

Fan, H., Liu, H., Zhang, Y., Zhang, S., Liu, T., & Wang, D. (2022). Review on plant-derived bioactive peptides: biological activities, mechanism of action and utilizations in food development, *Journal of Future Foods, 2*, 143–159.

Fu, T. J., Abbott, U. R., & Hatzos, C. (2002). Digestibility of food allergens and nonallergenic proteins in simulated gastric fluid and simulated intestinal fluid—A comparative study. *Journal of Agricultural and Food Chemistry, 50*(24), 7154–7160. Available from https://doi.org/10.1021/jf020599h.

Fujinaga, M., Chernaia, M. M., Mosimann, S. C., James, M. N. G., & Tarasova, N. I. (1995). Crystal structure of human pepsin and its complex with pepstatin. *Protein Science, 4*(5), 960–972. Available from https://doi.org/10.1002/pro.5560040516.

García Fillería, S., & Tironi, V. (2021). Intracellular antioxidant activity and intestinal absorption of amaranth peptides released using simulated gastrointestinal digestion with Caco-2 TC7 cells. *Food Bioscience, 41*, 101086. Available from https://doi.org/10.1016/j.fbio.2021.101086.

Giusti, S., Sbrana, T., La Marca, M., Di Patria, V., Martinucci, V., Tirella, A., ... Ahluwalia, A. (2014). A novel dual-flow bioreactor simulates incrased fluorescein permeability in epithelial tissue barriers, *Biotechnol. J., 9*, 1175–1184.

Guo, Y., Jiang, X., Xiong, B., Zhang, T., Zeng, X., Wu, Z., ... Pan, D. (2019). Production and transepithelial transportation of angiotensin-I-converting enzyme (ACE)-inhibitory peptides from whey protein hydrolyzed by immobilized Lactobacillus helveticus proteinase, *J. Dairy Sci., 102*, 961–975.

Hamzalıoğlu, A., & Gökmen, V. (2016). Investigations on the reactions of α-dicarbonyl compounds with amino acids and proteins during in vitro digestion of biscuits. *Food & Function*, *7*(6), 2544–2550. Available from https://doi.org/10.1039/c5fo01555c.

Hansen, S.A., Ashley, A., & Chung, B.M. (2015). Complex dietary protein improves growth through a complex mechanism of intestinal peptide absorption and protein digestion, *JPEN J Parenter Enteral Nutr, 39*, 95–103.

Havenaar, R., Anneveld, B., Hanff, L. M., De Wildt, S. N., De Koning, B. A. E., Mooij, M. G., Lelieveld, J. P. A., & Minekus, M. (2013). In vitro gastrointestinal model (TIM) with predictive power, even for infants and children? *International Journal of Pharmaceutics*, *457* (1), 327–332. Available from https://doi.org/10.1016/j.ijpharm.2013.07.053.

He, Y.T., Li, T.T., Chen, J.X., She, X.X., Ren, D.F., Lu, J. Transport of ACE inhibitory peptides Ile-Gln-Pro and Val-Glu-Pro derived from Spirulina platensis across Caco-2 monolayers. *J. Food Sci., 83*, 2586–2592.

Hu, J., Jiang, K., Huang, C., Zheng, J., Zhou, H., Ou, J., & Ou, S. (2022). Glycine and serine markedly eliminate methylglyoxal in the presence of formaldehyde via the formation of imidazole salts. *Food Chemistry*, *369*130952. Available from https://doi.org/10.1016/j.foodchem.2021.130952.

Hur, S. J., Lim, B. O., Decker, E. A., & McClements, D. J. (2011). In vitro human digestion models for food applications. *Food Chemistry*, *125*(1), 1–12. Available from https://doi.org/10.1016/j.foodchem.2010.08.036.

Iori, E., Vinci, B., Murphy, E., Marescotti, M. C., Avogaro, A., Ahluwalia, A., & Federici, M. (2012). Glucose and fatty acid metabolism in a 3 tissue in-vitro model challenged with normo- and hyperglycaemia. *PLoS One*, *7*(4)e34704. Available from https://doi.org/10.1371/journal.pone.0034704.

Jochems, P.G.M., Garssen, J., van Keulen, A.M., Masereeuw, R., & Jeurink, P.V. (2018). Evaluating human intestinal cell lines for studying dietary protein absorption, *Nutrients, 10*, 322.

Juárez-Chairez, M.F., Cid-Gallegos, M.S., Meza-Márquez, O.G., & Jiménez-Martínez, C. (2022). Biological functions of peptides from legumes in gastrointestinal health. A review legume peptides with gastrointestinal protection. *J Food Biochem, 46*, e14308.

Kaneko, K. (2021). Appetite regulation by plant-derived bioactive peptides for promoting health. *Peptides*, *144*170608. Available from https://doi.org/10.1016/j.peptides.2021.170608.

Kaneko, K., Mizushige, T., Miyazaki, Y., Lazarus, M., Urade, Y., Yoshikawa, M., Kanamoto, R., & Ohinata, K. (2014). δ-Opioid receptor activation stimulates normal diet intake but conversely suppresses high-fat diet intake in mice. *American Journal of Physiology. Regulatory, Integrative and Comparative Physiology*, *306*, R265–R272. Available from https://doi.org/10.1152/ajpregu.00405.2013.

Komatsu, Y., Tsuda, M., Wada, Y., Shibasaki, T., Nakamura, H., & Miyaji, K. (2023). Nutritional evaluation of milk-, plant-, and insect-based protein materials by protein digestibility using the InfoGest digestion method. *Journal of Agricultural and Food Chemistry*, *71*, 2503–2513. Available from https://doi.org/10.1021/acs.jafc.2c07273.

Kong, F., & Singh, R. P. (2010). A human gastric simulator (HGS) to study food digestion in human stomach. *Journal of Food Science*, *75*(9), E627–E635. Available from https://doi.org/10.1111/j.1750-3841.2010.01856.x.

Korte, R., Bräcker, J., & Brockmeyer, J. (2017). Gastrointestinal digestion of hazelnut allergens on molecular level: Elucidation of degradation kinetics and resistant immunoactive peptides using mass spectrometry. *Molecular Nutrition & Food Research*, *61* (10)1700130. Available from https://doi.org/10.1002/mnfr.201700130.

Lennernäs, H. (2007). Intestinal permeability and its relevance for absorption and elimination. *Xenobiotica; the Fate of Foreign Compounds in Biological Systems*, *37*(10-11), 1015–1051. Available from https://doi.org/10.1080/00498250701704819.

Lentner, C (1981). *Geigy Scientific Tables, Volume 1: Units of Measurement, Body Fluids, Composition of the Body, Nutrition*. Basle: CIBA-GEIGY.

Liu, W., Lad, M., & Foster, T. (2022). In vitro digestion of designed emulsions based on milk protein and guar gum systems. *Food & Function*, *13*(11), 6022–6035. Available from https://doi.org/10.1039/d2fo00592a.

Lucas-González, R., Viuda-Martos, M., Pèrez-Alvarez, J.A., & Fernàndez-Lòpez, J. (2018). In vitro digestion models suitable for foods: opportunities for new fields of application and challenges, *Food Res. Int.*, *107*, 423–436.

Lynch, H., Johnston, C., & Wharton, C. (2018). Plant-based diets: Considerations for environmental impact, protein quality, and exercise performance. *Nutrients*, *10*, 1841. Available from https://doi.org/10.3390/nu10121841.

Mamone, G., Nitride, C., Picariello, G., Addeo, F., Ferranti, P., & Mackie, A. (2015). Tracking the fate of pasta (T. Durum Semolina) immunogenic proteins by in vitro simulated digestion. *Journal of Agricultural and Food Chemistry*, *63*(10), 2660–2667. Available from https://doi.org/10.1021/jf505461x.

Mansilla, W.D., Marinangeli, C.P.F., Cargo-Froom, C., Franczyk, A., House, J.D., Elango, R. ... Shoveller, A.K. (2020). Comparison of methodologies used to define the protein quality of human foods and support regulatory claims, *Appl. Physiol. Nutr. Metab.*, *45*, 917-926.

Marcolini, E., Babini, E., Bordoni, A.M., Di Nunzio, M., Laghi, L., Maczó, G. ... Capozzi, F. (2015). Bioaccessibility of the bioactive peptide carnosine during in vitro digestion of cured beef meat, J. Agric. Food Chem., 63, 4973–4978.

Martini, S., Conte, A., & Tagliazucchi, D. (2019). Comparative peptidomic profile and bioactivities of cooked beef, pork, chicken and turkey meat after in vitro gastrointestinal digestion. *Journal of Proteomics*, *208*103500. Available from https://doi.org/10.1016/j.jprot.2019.103500.

Mat, D.J.L., Le Feunteun, S., Michon, C., & Souchon, I. (2016). In vitro digestion of foods using pH-stat and the InfoGest protocol: Impact of matrix structure on digestion kinetics of macronutrients, proteins and lipids. *Food Research International*, *88*, 226–233.

Mat, D.J.L., Cattenoz, T., Souchon, I., Michon, C. & Le Feunteun, S. (2018). Monitoring protein hydrolysis by pepsin using pH-stat: in vitro gastric digestions in static and dynamic pH conditions. *Food Chem.*, *239*, 268–275.

McClements, D.J., & Grossmann, L. (2021). The science of plant-based foods: constructing next-generation meat, fish, milk, and egg analogs. *Compr. Rev. Food Sci. Food Saf.*, *20*, 4049–4100.

Meigs, L., Smirnova, L., Rovida, C., Leist, M., & Hartung, T. (2018). Animal testing and its alternatives - the most important omics is economics. *ALTEX Alternativen zu Tierexperimenten*, *35*(3), 275–305. Available from https://doi.org/10.14573/altex.1807041.

Minckus, M., Marteau, P., Havenaar, R., & Huis in't Veld, J. H. J. (1995). A multicompartmental dynamic computer-controlled model simulating the stomach and small intestine.

Alternatives to Laboratory Animals, *23*(2), 197–209. Available from https://doi.org/10.1177/026119299502300205.

Minekus, M., Smeets-Peeters, M., Bernalier, A., Marol-Bonnin, S., Havenaar, R., Marteau, P., … Huis in't Veld, J.H., (1999). A computer-controlled system to simulate conditions of the large intestine with peristaltic mixing, water absorption and absorption of fermentation products, *Appl. Microbiol. Biotechnol.*, *53*, 108–114.

Minekus, M., Alminger, M., Alvito, P., Ballance, S., Bohn, T., Bourlieu, C., Carrière, F., Boutrou, R., Corredig, M., Dupont, D., Dufour, C., Egger, L., Golding, M., Karakaya, S., Kirkhus, B., Le Feunteun, S., Lesmes, U., Macierzanka, A., Mackie, A., … Brodkorb, A. (2014). A standardised static in vitro digestion method suitable for food – An international consensus. *Food and Function*, *5*(6), 1113–1124. Available from https://doi.org/10.1039/c3fo60702j.

Mulet-Cabero, A. I., Mackie, A. R., Wilde, P. J., Fenelon, M. A., & Brodkorb, A. (2019). Structural mechanism and kinetics of in vitro gastric digestion are affected by process-induced changes in bovine milk. *Food Hydrocolloids*, *86*, 172–183. Available from https://doi.org/10.1016/j.foodhyd.2018.03.035.

Ménard, O., Cattenoz, T., Guillemin, H., Souchon, I., Deglaire, A., Dupont, D., & Picque, D. (2014). Validation of a new in vitro dynamic system to simulate infant digestion. *Food Chemistry*, *145*, 1039–1045. Available from https://doi.org/10.1016/j.foodchem.2013.09.036.

Ménard, O., Bourlieu, C., De Oliveira, S.C., Dellarosa, N., Laghi, L., Carrière, F., … Deglaire, A. (2018). A first step towards a consensus static in vitro model for simulating full-term infant digestion, *Food Chem.*, *240*, 338–345.

Nehir El, S., Karakaya, S., Simsek, S., Dupont, D., Menfaatli, E., & Eker, A.T. (2015). In vitro digestibility of goat milk and kefir with a new standardised static digestion method (InfoGest cost action) and bioactivities of the resultant peptides, *Food Funct.*, *6*, 2322–2330.

Ogilvie, O. J., Gerrard, J. A., Roberts, S., Sutton, K. H., Larsen, N., & Domigan, L. J. (2021). A case study of the response of immunogenic gluten peptides to sourdough proteolysis. *Nutrients*, *13*, 1906. Available from https://doi.org/10.3390/nu13061906.

Oleaga, C., Bernabini, C., Smith, A.S.T., Srinivasan, B., Jackson, M., McLamb, W. … Hickman, J.J. (2016). Multi-Organ toxicity demonstration in a functional human in vitro system composed of four organs, *Sci. Rep.*, *6*, 20030.

Oliveira, A., Amaro, A.L., & Pintado, M. (2018). Impact of food matrix components on nutritional and functional properties of fruit-based products, *Current Opinion in Food Science*, *22*, 153–159.

Oomen, A. G., Hack, A., Minekus, M., Zeijdner, E., Cornelis, C., Schoeters, G., Verstraete, W., Van De Wiele, T., Wragg, J., Rompelberg, C. J. M., Sips, A. J. A. M., & Van Wijnen, J. H. (2002). Comparison of five in vitro digestion models to study the bioaccessibility of soil contaminants. *Environmental Science and Technology*, *36*(15), 3326–3334. Available from https://doi.org/10.1021/es010204v.

Oomen, A. G., Rompelberg, C. J. M., Bruil, M. A., Dobbe, C. J. G., Pereboom, D. P. K. H., & Sips, A. J. A. M. (2003). Development of an in vitro digestion model for estimating the bioaccessibility of soil contaminants. *Archives of Environmental Contamination and Toxicology*, *44*(3), 281–287. Available from https://doi.org/10.1007/s00244-002-1278-0.

Pali-Schöll, I., Untersmayr, E., Klems, M., & Jensen-Jarolim, E. (2018). The effect of digestion and digestibility on allergenicity of food. *Nutrients*, *10*(9), 1129. Available from https://doi.org/10.3390/nu10091129.

Papaleo, E., Fantucci, P., & De Gioia, L. (2005). Effects of calcium binding on structure and autolysis regulation in trypsins. A molecular dynamics investigation. *Journal of Chemical Theory and Computation*, *1*(6), 1286–1297. Available from https://doi.org/10.1021/ct050092o.

Peng, S., Song, H., Chen, Y., Li, S., & Guan, X. (2022) Oral Delivery of Food-derived Bioactive Peptides: Challenges and Strategies, *Food Reviews International, 39*, 8, 5297–5325.

Picariello, G., Ferranti, P., Fierro, O., Mamone, G., Caira, S., Di Luccia, A., Monica, S., & Addeo, F. (2010). Peptides surviving the simulated gastrointestinal digestion of milk proteins: Biological and toxicological implications. *Journal of Chromatography B: Analytical Technologies in the Biomedical and Life Sciences*, *878*(3-4), 295–308. Available from https://doi.org/10.1016/j.jchromb.2009.11.033.

Picone, G., De Noni, I., Ferranti, P., Nicolai, M.A., Alamprese, C., Trimigno, A., … Capozzi, F. (2019). Monitoring molecular composition and digestibility of ripened bresaola through a combined foodomics approach, *Food Research International, 115*, 360–368.

Prodić, I., Smiljanić, K., Nagl, C., Ballmer-Weber, B., Hoffmann-Sommergruber, K., & Veličković, T. Ć. (2022). InfoGest digestion assay of raw and roasted hazelnuts and its impact on allergens and their IgE binding activity. *Foods*, *11*(18), 2914. Available from https://doi.org/10.3390/foods11182914.

Pérez-Burillo, S., Molino, S., Navajas-Porras, B., Valverde-Moya, A.J., Hinojosa-Nogueira, D., López-Maldonado, A., & Rufián-Henares, J.A. (2021). An in vitro batch fermentation protocol for studying the contribution of food to gut microbiota composition and functionality. *Nat Protoc, 16*, 3186–3320.

Prodić, I., Stanic-Vucinic, D., Apostolovic, D., Mihailovic, J., Radibratovic, M., Radosavljevic, J., Burazer, L., Milcic, M., Smiljanic, K., van Hage, M., & Cirkovic Velickovic, T. (2018). Influence of peanut matrix on stability of allergens in gastric-simulated digesta: 2S albumins are main contributors to the IgE reactivity of short digestion-resistant peptides. *Clinical & Experimental Allergy*, *48*(6), 731–740. Available from https://doi.org/10.1111/cea.13113.

Rao, P.S., Nolasco, E., Handa, A., Naldrett, M.J., Alvarez, S., & Majumder, K. (2020). Effect of pH and heat treatment on the antioxidant activity of egg white protein-derived peptides after simulated in-vitro gastrointestinal digestion. *Antioxidants, 9*, 1114.

Russell, W. M. S., & Burch, R. L. (1960). The principles of humane experimental technique. *The Medical Journal of Australia*, *1*. Available from https://doi.org/10.5694/j.1326-5377.1960.tb73127.x.

Sanchón, J., Fernàndez-Tomé, S., Miralles, B., Hernández-Ledesma, B., Tomé, D., Gaudichon, C., & Recio, I. (2018). Protein degradation and peptide release from milk proteins in human jejunum. Comparison with in vitro gastrointestinal simulation, *Food Chem., 239*, 486–494.

Sangsawad, P., Roytrakul, S., & Yongsawatdigul, J. (2017). Angiotensin converting enzyme (ACE) inhibitory peptides derived from the simulated in vitro gastrointestinal digestion of cooked chicken breast. *Journal of Functional Foods*, *29*, 77–83. Available from https://doi.org/10.1016/j.jff.2016.12.005.

Sangsawad, P., Roytrakul, S., Choowongkomon, K., Kitts, D.D., Chen, X.-M., Meng, G., ... Yongsawatdigula, J. (2018). Transepithelial transport across Caco-2 cell monolayers of angiotensin converting enzyme (ACE) inhibitory peptides derived from simulated in vitro gastrointestinal digestion of cooked chicken muscles. *Food Chem. 251*, 77–85.

Santos-Hernández, M., Miralles, B., Amigo, L., & Recio, I. (2020). Peptidomic data of egg white gastrointestinal digests prepared using the Infogest Harmonized Protocol. *Data in Brief*, *31*105932. Available from https://doi.org/10.1016/j.dib.2020.105932.

Sarkar, A., Goh, K. K. T., & Singh, H. (2009). Colloidal stability and interactions of milk-protein-stabilized emulsions in an artificial saliva. *Food Hydrocolloids*, *23*(5), 1270–1278. Available from https://doi.org/10.1016/j.foodhyd.2008.09.008.

Shani-Levi, C., Alvito, P., Andrés, A., Assunção, R., Barberá, R., Blanquet-Diot, S., ... Lesmes, U. (2017). Extending in vitro digestion models to specific human populations: Perspectives, practical tools and bio-relevant information. *Trends Food Sci. Technol. 60*, 52–63.

Shin, M., Han, Y., & Ahn, K. (2013). The influence of the time and temperature of heat treatment on the allergenicity of egg white proteins. *Allergy, Asthma Immunol. Res.*, *5*, 96–101. Available from https://doi.org/10.4168/aair.2013.5.2.96.

Sousa, R., Portmann, R., Dubois, S., Recio, I., & Egger, L. (2020). Protein digestion of different protein sources using the InfoGest static digestion model. *Food Research International*, *130*108996. Available from https://doi.org/10.1016/j.foodres.2020.108996.

Sridhar, K., Bouhallab, S., Croguennec, T., Renard, D., & Lechevalier, V. (2022). Recent trends in design of healthier plant-based alternatives: Nutritional profile, gastrointestinal digestion, and consumer perception. *Critical Reviews in Food Science and Nutrition*, 1–16. Available from https://doi.org/10.1080/10408398.2022.2081666.

Switzar, L., Giera, M., & Niessen, W.M. (2013). Protein Digestion: An Overview of the Available Techniques and Recent Developments, *J Proteome Res., 12*, 1067–1077.

Torcello-Gómez A., Dupont, D., Jardin, J., Briad-Biorn, V., Deglaire, A., Risse, K., ... Mackie, A. (2020). The pattern of peptides released from dairy and egg proteins is highly dependent on the simulated digestion scenario, *Food Funct, 11*, 5240–5256.

Ulleberg, E. K., Comi, I., Holm, H., Herud, E. B., Jacobsen, M., & Vegarud, G. E. (2011). Human gastrointestinal juices intended for use in in vitro digestion models. *Food Digestion*, *2*(1-3), 52–61. Available from https://doi.org/10.1007/s13228-011-0015-4.

Untersmayr, E., & Jensen-Jarolim, E. (2006). The effect of gastric digestion on food allergy. *Current Opinion in Allergy and Clinical Immunology*, *6*(3), 214–219. Available from https://doi.org/10.1097/01.all.0000225163.06016.93.

Van De Wiele, T. R., Oomen, A. G., Wragg, J., Cave, M., Minekus, M., Hack, A., Cornelis, C., Rompelberg, C. J. M., De Zwart, L. L., Klinck, B., Van Wijnen, J., Verstraete, W., & Sips, A. J. A. M. (2007). Comparison of five in vitro digestion models to in vivo experimental results: Lead bioaccessibility in the human gastrointestinal tract. *Journal of Environmental Science and Health - Part A Toxic/Hazardous Substances and Environmental Engineering*, *42*(9), 1203–1211. Available from https://doi.org/10.1080/10934520701434919.

Van der Lugt, T., Venema, K., van Leeuwen, S., & Vrolijk, M.F. (2020). A. Opperhuizen and A. Basta, Gastrointestinal digestion of dietary advanced glycation endproducts using an in vitro model of the gastrointestinal tract (TIM-1), *Food Funct., 11*, 6297–6307.

Venardou, B., O'Doherty, J.V., McDonnell, M.J., Mukhopadhya, A., Kiely, C., Ryan, M.T, & Sweeney, T. (2021). Evaluation of the in vitro effects of the increasing inclusion levels of yeast β-glucan, a casein hydrolysate and its 5 kDa retentate on selected bacterial populations and strains commonly found in the gastrointestinal tract of pigs, *Food Funct., 12*, 2189–2200.

Venema, K., & Abbeele, P.V.D. (2013). Experimental models of the gut microbiome, *Best Pract. Res. Clin. Gastroenerol., 27*, 115–126.

Verhoeckx, K., Cotter, P., Expósito., Kleiveland, C., Lea, T., Mackie, A., & Wichers. (2015). *The Impact of food bio-actives on gut health: In vitro and ex vivo models.* Springer, 978–3.

Versantvoort, C. H. M., Oomen, A. G., Van De Kamp, E., Rompelberg, C. J. M., & Sips, A. J. A. M. (2005). Applicability of an in vitro digestion model in assessing the bioaccessibility of mycotoxins from food. *Food and Chemical Toxicology, 43*(1), 31–40. Available from https://doi.org/10.1016/j.fct.2004.08.007.

Vinci, B., Murphy, E., Iori, E., Meduri, F., Fattori, S., Marescotti, M. C., Castagna, M., Avogaro, A., & Ahluwalia, A. (2012). An in vitro model of glucose and lipid metabolism in a multicompartmental bioreactor. *Biotechnology Journal, 7*(1), 117–126. Available from https://doi.org/10.1002/biot.201100177.

Wang, B., Xie, N., & Li, B. (2016). Charge properties of peptides derived from casein affect their bioavailability and cytoprotection against H2O2-induced oxidative stress, *J. Dairy Sci., 99*, 2468–2479.

Wang, B., Timilsena, Y. P., Blanch, E., & Adhikari, B. (2017). Mild thermal treatment and in-vitro digestion of three forms of bovine lactoferrin: Effects on functional properties. *International Dairy Journal, 64*, 22–30. Available from https://doi.org/10.1016/j.idairyj.2016.09.001, http://www.elsevier.com/inca/publications/store/4/0/5/8/6/0/index.htt.

Wang, J., & Ji, H. (2018). Influence of probiotics on dietary protein digestion and utilization in the gastrointestinal tract. *Current Protein & Peptide Science, 20*(2), 125–131. Available from https://doi.org/10.2174/1389203719666180517100339.

Wang, R., Mohammadi, M., Mahboubi, A., & Taherzadeh, M.J. (2021) In-vitro digestion models: a critical review for human and fish and a protocol for in-vitro digestion in fish, *Bioengineered, 12*, 3040–3064

Wickham, M., & Faulks, R. (2013). *Dynamic gastric model.* United States Patent. 8.

Wickham, M.J.S., Faulks, R.M, Mann, J., & Mandalari, G. (2012). The design, operation, and application of a dynamic gastric model, *Dissolution Technol., 19*, 15–22.

Wikman-Larhed, A., & Artursson, P. (1995). Co-cultures of human intestinal goblet (HT29-H) and absorptive (Caco-2) cells for studies of drug and peptide absorption. *European Journal of Pharmaceutical Sciences, 3*(3), 171–183. Available from https://doi.org/10.1016/0928-0987(95)00007-Z.

Xing, L., Liu, R., Tang, C., Pereira, J., Zhou, G., & Zhang, W. (2018). The antioxidant activity and transcellular pathway of Asp-Leu-Glu-Glu in a Caco-2 cell monolayer. *International Journal of Food Science & Technology, 53*(10), 2405–2414. Available from https://doi.org/10.1111/ijfs.13771.

Xu, Q., Fan, H., Yu, W., Hong, H., & Wu, J. (2017). Transport study of egg-derived antihypertensive peptides (LKP and IQW) using Caco-2 and HT29 coculture monolayers, *J. Agric. Food Chem., 65*, 7406–7414.

Zhang, Q., Tong, X., Qi, B., Wang, Z., Li, Y., Sui, X., & Jiang, L. (2018). Changes in antioxidant activity of Alcalase-hydrolyzed soybean hydrolysate under simulated gastrointestinal digestion and transepithelial transport, *J. Funct. Foods, 42*, 298–305.

Zhang, T., Su, M., Jiang, X., Xue, Y., Zhang, J., Zeng, X., ... Pan, D. (2019). Transepithelial transport route and liposome encapsulation of milk-derived ACE-inhibitory peptide Arg Leu-Ser-Phe-Asn-Pro. *J. Agric. Food Chem. 67*, 5544–5551.

Zhao, J., Zhang, X., Liu, H., Brown, M.A., & Qiao, S. (2019). Dietary Protein and Gut Microbiota Composition and Function, *Curr Protein Pept Sci., 20*, 145–154.

Zhou, H., Hu, Y., Tan, Y., Zhang, Z., & McClements, D. J. (2021). Digestibility and gastrointestinal fate of meat versus plant-based meat analogs: An in vitro comparison. *Food Chemistry, 364*130439. Available from https://doi.org/10.1016/j.foodchem.2021.130439.

Zhou, H., Tan, Y., & McClements, D. J. (2023). Applications of the InfoGest in vitro digestion model to foods: A REVIEW. *Annual Review of Food Science and Technology, 14* (1). Available from https://doi.org/10.1146/annurev-food-060721-012235.

Zhu, C. Z., Zhang, W. G., Zhou, G. H., & Xu, X. L. (2016). Identification of antioxidant peptides of Jinhua ham generated in the products and through the simulated gastrointestinal digestion system. *Journal of the Science of Food and Agriculture, 96*, 99–108. Available from https://doi.org/10.1002/jsfa.7065.

Zhu, Y., Lao, F., Pan, X., & Wu, J. (2022). Food protein-derived antioxidant peptides: Molecular mechanism, stability and bioavailability. *Biomolecules, 12*(11), 1622. Available from https://doi.org/10.3390/biom12111622.

Zihler, A., Gagnon, M., Chassard, C., & Lacroix, C. (2011). Protective effect of probiotics on Salmonella infectivity assessed with combined in vitro gut fermentation-cellular models. *BMC Microbiology, 11*. Available from https://doi.org/10.1186/1471-2180-11-264.

Zolotarev, V.A. (2014). Dietary free amino acids and the gastric phase of digestion, *Curr Pharm Des., 20*, 2731–2737.

CHAPTER

Modeling elderly gastrointestinal digestion for the evaluation of the release of bioactive peptides

2

Débora Parra Baptista, Amanda Silva do Carmo and Mirna Lúcia Gigante

Department of Food Engineering and Technology, School of Food Engineering, University of Campinas, Campinas, São Paulo, Brazil

2.1 Introduction

The elderly population is growing at an unprecedented rate, both in absolute numbers and proportionally compared to other age groups. The share of the world population aged 65 years or over corresponds to approximately 9.3%, and projections indicate that this group is expected to reach 16% of the population by 2050 (UN, 2020). With the increase in life expectancy of the global population, governmental agencies face the challenge of ensuring healthy aging of the population (Fernandes et al., 2021).

Aging is a natural and multidimensional process that includes physiological, pathological, social, and psychological changes (Amarya et al., 2015). Nutrition plays a crucial role in health promotion during aging, and the consumption of protein-rich foods is one of the main nutritional recommendations for older adults (Volkert et al., 2022). The consumption of adequate levels of protein is fundamental in the prevention of sarcopenia, an age-related progressive loss of muscle mass (Boirie & Guillet, 2018).

In addition to the nutritional contribution related to protein consumption, dietary proteins are considered sources of bioactive peptides. Bioactive peptides are fragments of proteins that can modulate physiological processes in different metabolic routes in the human body. These peptides may present various biological activities, such as antihypertensive, antimicrobial, anticoagulant, anticancer, anti-inflammatory, and antidiabetic activities (Rivero-Pino, 2023). These peptides are present in the structure of the parent proteins in their inactive latent form and can be released during food processing or by enzymatic action during the gastrointestinal digestion of proteins in our body (Korhonen & Pihlanto, 2006).

Biologically active peptides have the potential to contribute to the maintenance of health and quality of life during the aging process. Among the bioactive peptides, those with antihypertensive potential are the most widely studied and

Protein Digestion-Derived Peptides. DOI: https://doi.org/10.1016/B978-0-443-19141-1.00002-9

are of particular interest to the health of the elderly since the prevalence of hypertension increases considerably with aging (Gobbetti et al., 2002; Oliveros et al., 2020). In addition, bioactive peptides can contribute to bone health and can also act on brain functions with a preventive effect on age-related cognitive decline (Katayama & Nakamura, 2019; Shang et al., 2018). For these compounds to have a beneficial impact on the human body, they must reach their target organs (Amigo & Hernández-Ledesma, 2020). The absorption of nutrients and bioactive compounds in the human body depends on the conditions imposed by the gastrointestinal tract when food is digested, and in vitro methods mimicking gastrointestinal digestion are tools that enable the elucidation of the process of the food matrix disintegration and release of compounds during the digestive process, as well as the determination of the bioaccessibility of food bioactive compounds (Picariello et al., 2013; Shani-Levi et al., 2017).

The aging process is usually accompanied by several changes in digestive conditions, including increased gastric and intestinal transit times and reduced secretion of digestive fluids, enzymes, and bile salts (Shani-Levi et al., 2017). These changes may impact the proteolysis during gastrointestinal digestion (Melchior et al., 2023; Zhang et al., 2023) and, consequently, alter the bioactive peptides profile at the end of the digestion (Wang et al., 2022b). Thus studies that aim to evaluate the bioaccessibility of bioactive peptides with a focus on health promotion for older adults should consider the age-related changes in the digestive process.

This chapter will discuss the main age-related changes in the digestive process, the digestion models already proposed to mimic the elderly gastrointestinal tract, bioactive peptides with relevance for the health promotion of this specific population, and the in vitro studies that evaluated the bioaccessibility of bioactive peptides considering the elderly digestive conditions.

2.2 Age-related changes in the gastrointestinal tract

Gastrointestinal digestion is a complex process in which food undergoes mechanical transformations, which reduce the size of food matrix particles, and enzymatic transformations, which hydrolyze macromolecules leading to the release of smaller constituents that can be further absorbed. The disintegration processes of food occur mainly in the mouth and stomach, while the enzymatic and absorption processes occur mainly in the intestinal phase of digestion (Guerra et al., 2012).

The aging process leads to several changes in the human body, including alterations in the gastrointestinal tract. These alterations include changes in the mechanical breakdown of food, food transit, chemical digestion, and intestinal wall functionality. Thus age-related changes in the gastrointestinal tract can ultimately affect nutrient absorption (Rémond et al., 2015). However, it is important to emphasize that the aging process does not follow a pattern, and the age-related changes in the gastrointestinal tract are particular to each individual and might be related to specific health conditions. Thus this chapter will discuss changes in the gastrointestinal tract of the elderly in general, focusing on the main changes associated with the aging process.

2.2.1 Oral phase

The process of food digestion begins in the mouth with various operations. Oral food processing includes the first bite, chewing, the formation of a food bolus, and swallowing (Chen, 2009). Although this phase of digestion is relatively short compared to the other phases of gastrointestinal digestion, chewing influences the overall digestive process (Guerra et al., 2012). In this phase of digestion, the food is mixed with saliva, which has a pH close to neutrality and contains amylase, which can hydrolyze starch, and mucins, glycoproteins that are important for the food bolus formation (Boland, 2016).

Physiological changes associated with aging can alter oral structures and, consequently, the mechanical processing in the oral phase of digestion. Aging may be accompanied by bone loss, teeth loss, gum tissue loss, a higher incidence of caries, and changes in oral muscles, which may contribute to difficulty in chewing food properly (Chernoff, 2015). A reduction in salivary flow and changes in salivary composition, mainly salts and amylase concentrations, was also observed in elderly individuals, which can impair the formation of the food bolus (Chernoff, 2015; Ichikawa et al., 2011; Nagler & Hershkovich, 2005).

In general, two scenarios are possible when considering the oral phase of digestion in the elderly. In individuals with good oral health, without teeth loss or reduced salivary flow, the chewing process is adapted with more cycles and a longer duration of the chewing sequence to compensate for the loss of force in the bite and muscle tissue and achieve the formation of a food bolus suitable for swallowing. However, this adaptation is not satisfactory in individuals with compromised oral health (e.g., with teeth loss or wearing full dentures). Thus in this scenario, an impact on the bolus formation is observed (Mishellany-Dutour et al., 2008; Rémond et al., 2015), which can affect the subsequent phases of digestion and the bioaccessibility of nutrients (Ribes et al., 2023).

2.2.2 Gastric phase

After swallowing, food bolus passes through the esophagus and is conducted to the stomach, a digestive compartment composed of three regions: fundus, body, and antrum. The upper part of the stomach acts as a reservoir in which the food bolus is mixed with the gastric fluid, containing the enzymes pepsin and gastric lipase, and hydrochloric acid, responsible for lowering the pH from the range of 6–5 to 1.5. In the antrum, peristaltic waves are responsible for fragmenting the food structure into tiny particles that are conducted to the next phase of digestion through the pylorus in a process called gastric emptying (Boland, 2016; Guerra et al., 2012).

Although there is no evidence that aging significantly affects gastric functions, age-associated conditions, and diseases can affect the functioning of the stomach (Chernoff, 2015). That is the case of atrophic gastritis and *Helicobacter pylori* infections, which are more prevalent in the elderly and are associated with a decline in acid secretion in the stomach (Rémond et al., 2015). However, reduced pepsin secretion was associated with aging, independent of atrophic gastritis and *H. pylori* infection (Feldman et al., 1996).

Gastric emptying plays a key role in the kinetics of nutrient absorption and, consequently, in the use of nutrients in physiological functions (Rémond et al., 2015). Data available in the literature reveal controversial results about the effect of aging on gastric emptying (Salles, 2007). Madsen and Graff (2004) observed that age did not affect gastric emptying time when comparing healthy adults and elderly individuals. However, Moore et al. (1983) observed that, although the rate of gastric emptying of solid foods did not present a significant difference between the elderly and adults, the elderly presented slower gastric emptying of liquids when compared to adults. On the other hand, Clarkston et al. (1997) reported slower gastric emptying of solid foods and liquids in the elderly compared to adults, while Serra-Prat et al. (2013), when evaluating the effects of aging and frailty, a geriatric syndrome of functional decline of several systems, on gastrointestinal motility, found that age alone did not affect gastric emptying time. However, frail elderly persons had longer gastric emptying compared to nonfrail elderly (Serra-Prat et al., 2013).

2.2.3 Small intestinal phase

Following the digestive process, gastric chyme is conducted to the small intestine, which has two main functions: breakdown of macromolecules and absorption of water and nutrients (Guerra et al., 2012). The small intestine comprises three parts: duodenum, jejunum, and ileum. In the duodenum, the acid chyme obtained in the stomach is neutralized through bicarbonate secretion, and pancreatin, a mixture of enzymes that act in the digestion of proteins, lipids, and carbohydrates, is also secreted in this phase of digestion. Additionally, bile acts in the emulsification of lipids, facilitating enzymatic action. The jejunum and ileum act as reactors for enzymatic action. These regions of the small intestine have large surface areas to favor the absorption of compounds (Boland, 2016).

Both age and health conditions can affect the functioning of the small intestine (Chernoff, 2015). Regarding pancreatic secretion at this phase of digestion, studies have shown a reduction in the flow rate of this secretion and the concentration of bicarbonate and enzymes (chymotrypsin and lipase) in pancreatic juice in the elderly when compared to adults (Laugier et al., 1991; Vellas et al., 1988). Aging can also affect liver functions, reducing bile salt secretion (Schmucker, 2005).

The effect of aging on small bowel motility is not much explored in the literature (Rémond et al., 2015). Available studies did not find a significant effect of aging on transit time in the small intestine when comparing the digestion of healthy adults and elderly individuals (Madsen, 1992; Madsen & Graff, 2004).

2.2.4 Large intestinal/colonic phase

Compounds not absorbed in the small intestine are transported to the large intestine through the ileocaecal junction. This phase of digestion has as its main functions the absorption of water, electrolytes, and bile salts, fermentation of fibers and other residual not previously digested components, formation, storage, and elimination of feces (Boland, 2016; Chernoff, 2015; Guerra et al., 2012).

The gastrointestinal microbiota is established after birth and undergoes an evolution process with increased richness and diversity, reaching its greatest complexity in adulthood, a phase in which relative stability is observed. With aging, an important change in the microbiota occurs, and this process depends on several factors, including changes in lifestyle, diet, infectious and inflammatory processes, use of medication, and the environment in which the older adult lives (Mello et al., 2016; Ottman et al., 2012). Mueller et al. (2006) evaluated the fecal microbiota of adult and elderly populations in different European countries and observed age-related changes in the microbiota. However, these age-associated changes differed between populations in different countries, each with a characteristic colonization pattern, revealing a possible relationship between the evolution of the microbiota and the different eating habits of each region.

Claesson et al. (2012) evaluated the fecal microbiota of the elderly and found that housing conditions, which affect eating habits, play an important role in the fecal microbiota. Individuals who lived in long-term residential care had a lower diversity of the microbiota and a higher prevalence of the phylum Bacterioidetes compared to the higher prevalence of the phylum Firmicutes in older adults living in the community. The loss of community-associated microbiota was correlated with the increased frailty of the elderly (Claesson et al., 2012).

Considering the diversity of factors that can influence the alteration of the microbiota with aging, high interindividual variability is observed in the microbiota of the elderly (Claesson et al., 2011); what is the challenge for the establishment of a pattern of the microbiota of the elderly. The ratio between Bacteroidetes and Firmicutes is considered a parameter that reveals the status of the intestinal microbiota since these are the most abundant phyla in the gut. However, in the case of the elderly, the effect of aging on the Firmicutes/Bacteroidetes ratio is controversial (Biagi et al., 2012; Biagi et al., 2013; Rémond et al., 2015). Overall, despite the interindividual variability, aging results in a substantial reduction in the biodiversity of the intestinal microbiota (Biagi et al, 2012).

In addition to the effects on the microbiota, aging can be accompanied by adverse health conditions in the large intestine, such as cancer, diverticulosis, and constipation. Constipation, a condition associated with reduced intestinal motility,

Table 2.1 Summary of the age-related physiological changes in the gastrointestinal tract that can affect food digestion.

Digestion phase	Age-associated factors that may affect digestion
Oral phase (mouth) Responsible for the first bite, chewing, and formation of the food bolus and swallowing	Bone loss Teeth loss Gum atrophy Higher incidence of caries Alteration of the oral musculature Reduced salivary flow Change in α-amylase concentration
Gastric phase (stomach) Contact with gastric fluid composed of pepsin, gastric lipase, and hydrochloric acid	A decline in hydrochloric acid secretion in the stomach associated with atrophic gastritis and *Helicobacter pylori* infection Decreased secretion of pepsin Change in gastric emptying
Small intestinal phase (duodenum, jejunum, and ileum) Breakdown of macromolecules and absorption of water and nutrients	Decreased pancreatic secretion and concentration of bicarbonate and enzymes (chymotrypsin and lipase) in pancreatic juice Decreased secretion of bile salts
Large intestinal/colonic phase Absorption of water and electrolytes, fermentation of fibers and other components not digested by the colonic microbiota, reabsorption of bile salts, formation, storage, and elimination of feces	Modification of the microbiota Reduced intestinal motility

is a common complaint in the elderly and may be associated with several conditions, such as neurological diseases, drug reactions, systemic disease, fluid and food intake, and physical inactivity (Chernoff, 2015; Salles, 2007).

Table 2.1 summarizes the main age-related physiological alterations in the gastrointestinal tract that can ultimately affect the digestive process.

2.3 In vitro simulation of elderly gastrointestinal digestion

The complexity of gastrointestinal physiology is already a challenge for developing in vitro digestion models that mimic the sequence of modifications that food undergoes during its passage through the gastrointestinal tract. The available models are based on physiological conditions described in the literature to simulate digestive conditions in vitro, allowing pertinent evaluations in each study (Shani-Levi et al., 2017). In the specific case of the elderly, considering the individual character of the changes that the gastrointestinal tract can undergo due to aging and diseases associated with this process, the initial challenge of the digestion simulation is the definition of the conditions to be simulated (e.g., elderly with alterations in the oral phase of digestion, elderly with achlorhydria). Several static and dynamic models have already been used for in vitro simulation of the digestion of different food matrices using conditions that simulate the gastrointestinal digestion of the elderly (Table 2.2). Most of the studies available in the literature that evaluate elderly digestion in vitro are based on the protocol of young adult digestion simulation proposed by INFOGEST, an international network of scientific cooperation (Brodkorb et al., 2019; Minekus et al., 2014), with modifications to simulate age-related changes in digestive conditions (Aalaei et al., 2021; Hernández-Olivas, Muñoz-Pina, Andrés, et al., 2020; Hernández-Olivas, Muñoz-Pina, Sánchez-García, et al., 2020; Hernández-Olivas et al., 2021, Hernández-Olivas et al., 2022; Melchior et al., 2023; Plante, McCarthy, & O'Halloran, 2021; Plante, McCarthy, Lacey, et al., 2021; Ribes et al., 2023).

Although most studies that simulate elderly gastrointestinal digestion use adapted versions of the same standardized protocol, there is a great disparity between the conditions used concerning gastric (2–3 h) and intestinal (2–4 h) digestion times, pH values, type and activity of enzyme and type and concentration of bile salts. The pH of the gastric phase varied from 3 to 6 in static models and from 1.6 to 6.2 at the beginning of dynamic gastric digestions (Table 2.2). Little variation is observed in the pH of the intestinal phase (pH 6.5–7), regardless of the model used. Variations in the enzymes' composition and activities were also observed. Most studies used only pepsin in the gastric phase (750–2000 U/mL), while some studies associated gastric lipase with pepsin in this phase of digestion (pepsin 750 U/mL and gastric lipase 255 U/mL or pepsin 1200 U/mL and gastric lipase 12 U/mL). In the intestinal phase, most studies used pancreatin (46–100 U/mL), but the use of trypsin (46 U/mL) together with α-chymotrypsin (23 U/mL), or both associated with pancreatic lipase (1400 U/mL) and α-amylase (100 U/mL) was also observed. The addition of bile salts occurred mainly through the addition of commercial bile (5–50 mM), but a combination of sodium glycodeoxycholate (2.67 mM) and taurocholic acid sodium salt hydrate (2.67 mM) was also used in some studies.

Studies that employ an in vitro digestion model to compare the digestive conditions of the elderly and adults use altered oral, gastric, and/or intestinal conditions in comparison to the digestive conditions of adults established in each study to simulate older adults' digestion (Table 2.3). Changes in digestive conditions used in the oral phase include variation in amylase activity and KCl concentration in saliva, alteration of chewing cycles, and agitation rate. In the gastric and intestinal phases, they include changes in pH, variation in enzyme activity and bile salts concentration, variation in the rate of agitation or frequency of peristalsis, and changes in the duration of each phase.

Table 2.2 Examples of in vitro digestive conditions simulating elderly digestion reported in the literature.

Food matrices	Digestion model	Oral phase	Gastric phase			Intestinal phase				References
			Duration	pH	Enzymes	Duration	pH	Enzymes	Bile	
Milk, yogurt, fresh and aged cheeses	Three static models were applied to simulated altered digestive conditions associated to aging: E1 (oral altered conditions), E2 (oral and gastric altered conditions) and E3 (oral, gastric and intestinal altered conditions)	Not performed for milk and yogurt. The oral phase of cheeses' digestion was performed *in vivo* by a volunteer with normal dentition. 10 chewing cycles were performed to simulate the elderly conditions (E1, E2 and E3)	2 h	According to each tested condition. E1: pH 3 E2: pH 6 E3: pH 6	According to each tested condition. E1: Pepsin 2000 U/mL E2: Pepsin 1500 U/mL E3: Pepsin 1500 U/mL	According to each tested condition. E1: 2 h E2: 2 h E3: 4 h	7	According to each tested condition. E1: Pancreatin 100 U/mL E2: Pancreatin 100 U/mL E3: Pancreatin 50 U/mL	According to each tested condition. E1: Bovine bile 10 mM E2: Bovine bile 10 mM E3: Bovine bile 5 mM	Hernández-Olivas, Muñoz-Pina, Sánchez-García, et al. (2020)
Pasteurized and UHT milk	Static model simulating elderly gastric digestion	In vitro, according to INFOGEST without the addition of amylase, during 2 min at 100 rpm	1h[a]	3	Pepsin 1400 U/mL	Not performed	Not performed	Not performed	Not performed	Aalaei, Khakimov, Gobba (2021)
Untreated raw milk in comparison with milk samples after application of high-pressure processing	Static model simulating elderly gastric digestion	In vitro, according to INFOGEST without the addition of amylase, during 2 min at 100 rpm	30 min[a]	3	Pepsin 1400 U/mL	Not performed	Not performed	Not performed	Not performed	Aalaei, Khakimov, Gobba (2021)

UHT skim milk	Dynamic in vitro human stomach (DIVHS) system simulating elderly gastric digestion	Milk was mixed with simulated salivary fluid (SSF) for 2 min at 100 rpm	2 h	Fasting pH 3, gastric pH monitored during study	Pepsin 3000 U/mL SGF[b]	2 h	7	Pancreatin 100 U/mL SIF[c]	–	Zhang et al. (2023)
Cheese-Fruit Combinations	Static model simulating altered elderly digestion conditions	In vitro, according to INFOGEST	2 h	3	Pepsin 750 U/mL	2 h	6.5	Pancreatin 100 U/mL	10 mM[d]	Plante, McCarthy, Lacey, et al. (2021)
Cheeses with different texture and fat contents	Static model simulating altered elderly digestion conditions	In vitro, according to INFOGEST	2 h	3	Pepsin 750 U/mL	2 h	6.5	Pancreatin 100 U/mL	Porcine bile 10 mM	Plante, McCarthy, & O'Halloran (2021)
Enteral liquid formula, lactoferrin, β-lactoglobulin, α-lactalbumin, and whey protein isolate	Dynamic system composed by two reactors connected to mimic the upper gastrointestinal system	Enteral formula: static oral digestion consisting of the mixture of simulated salivary fluid containing α-amylase (100 U/mL) with the enteral formula (1:1 v/v) followed by vortexing for 3 s. Oral phase not performed for lactoferrin, β-lactoglobulin, α-lactalbumin, and whey protein isolate digestion.	3 h	Initial pH of 6.2, gastric pH gradient controlled by software	Pepsin 750 U/mL Gastric Lipase 255 U/mL	3 h	6.5	Trypsin 46 U/mL α-Chymotrypsin 23 U/mL Pancreatic lipase 1400 U/mL α-amylase (100 U/mL)	Sodium glycodeoxycholate 2.67 mM Taurocholic acid sodium salt hydrate 2.67 mM	Levi et al. (2017)

(Continued)

Table 2.2 Examples of in vitro digestive conditions simulating elderly digestion reported in the literature. *Continued*

Food matrices	Digestion model	Oral phase	Gastric phase			Intestinal phase				References
			Duration	pH	Enzymes	Duration	pH	Enzymes	Bile	
Lactoferrin, β-lactoglobulin, and α-lactalbumin solutions	Dynamic system composed by two reactors connected to mimic the upper gastrointestinal system	Not performed	3 h	pH gradient from 6.2 to 2.0	Pepsin 750 U/mL	3 h	6.5	Trypsin 46 U/mL α-Chymotrypsin 23 U/mL	Sodium glycodeoxycholate 2.67 mM Taurocholic acid sodium salt hydrate 2.67 mM	Levi & Lesmes (2014)
Whey isolate, and wheat, pea, and rice concentrates	Static model simulating altered elderly digestion conditions	In vitro, according to INFOGEST 2.0, with higher α-amylase activity (150 U/mL) and KCl concentration (30.1 mmol/L)	3 h	4.5	Pepsin 1500 U/mL	3 h	6.6	Pancreatin 46 U/mL	Porcine bile 59.4 mg/mL	Melchior et al. (2023)
Cooked legumes (chickpea, lentils, soya bean and white bean) and alternative grains (oats, spelt and quinoa)	Three static models were applied to simulated altered digestive conditions associated to aging: E1 (oral altered conditions), E2 (oral and gastric altered conditions) and E3 (oral, gastric and intestinal altered conditions)	Not performed for milk and yogurt. The oral phase of cheeses' digestion was performed *in vivo* by a volunteer with normal dentition. 10 chewing cycles were performed to simulate the elderly conditions (E1, E2 and E3) except for soya beans that were submitted to 15 chewing cycles	2 h	According to each tested condition. E1: pH 3 E2: pH 6 E3: pH 6	According to each tested condition. E1: Pepsin 2000 U/mL E2: Pepsin 1500 U/mL E3: Pepsin 1500 U/mL	According to each tested condition. E1: 2 h E2: 2 h E3: 4 h	7	According to each tested condition. E1: Pancreatin 100 U/mL E2: Pancreatin 100 U/mL E3: Pancreatin 50 U/mL	According to each tested condition. E1: Bovine bile 10 mM E2: Bovine bile 10 mM E3: Bovine bile 5 mM	Hernández-Olivas et al. (2021)

Bread (*baguette tradition*)	Oral digestion using masticatory apparatus AM^2 followed by static model of gastric and intestinal digestion	Performed using the masticatory apparatus considering normal and deficient mastication. Normal mastication (NM) and deficient mastication differed in terms of force (DfM), saliva (DsM), and the combination of both altered conditions (DfsM)	2 h	3.54	Pepsin 1200 U/mL Gastric lipase 12 U/mL	2 h	7	Pancreatin 66.3 U/mL	Porcine bile 6.7 mM	Ribes et al. (2023)
Cooked fishes (salmon, sardine, sea bass and hake)	Three static models were applied to simulated altered digestive conditions associated to aging: E1 (oral altered conditions), E2 (oral and gastric altered conditions) and E3 (oral, gastric and intestinal altered conditions).	The oral phase of fishes' digestion was performed *in vivo* by a volunteer with normal dentition. 10 chewing cycles were performed to simulate the elderly conditions (E1, E2 and E3)	2 h	According to each tested condition. E1: pH 3 E2: pH 6 E3: pH 6	According to each tested condition. E1: Pepsin 2000 U/mL E2: Pepsin 1500 U/mL E3: Pepsin 1500 U/mL	According to each tested condition. E1: 2 h E2: 2 h E3: 4 h	7	According to each tested condition. E1: Pancreatin 100 U/mL E2: Pancreatin 100 U/mL E3: Pancreatin 50 U/mL	According to each tested condition. E1: Bovine bile 10 mM E2: Bovine bile 10 mM E3: Bovine bile 5 mM	Hernández-Olivas, Muñoz-Pina, Andrés, et al. (2020)

(*Continued*)

Table 2.2 Examples of in vitro digestive conditions simulating elderly digestion reported in the literature. *Continued*

Food matrices	Digestion model	Oral phase	Gastric phase			Intestinal phase				References
			Duration	pH	Enzymes	Duration	pH	Enzymes	Bile	
Poultry meatballs	Masticatory apparatus AM^2	The masticatory apparatus was set to simulate deficient mastication associated to aging: deterioration of the masticatory sequence for force, motility of oral elements, and lack of saliva addition	Not performed	Not performed	Not performed	Not performed	Not performed	Not performed	Not performed	Peyron et al. (2019)
Cooked chicken breast, turkey breast, pork loin and beef entrecote	Three static models were applied to simulated altered digestive conditions associated to aging: E1 (oral altered conditions), E2 (oral and gastric altered conditions) and E3 (oral, gastric and intestinal altered conditions)	The oral phase of meats' digestion was performed *in vivo* by a volunteer with normal dentition. 15 chewing cycles were performed to simulate the elderly conditions (E1, E2 and E3)	2 h	According to each tested condition. E1: pH 3 E2: pH 6 E3: pH 6	According to each tested condition. E1: Pepsin 2000 U/mL E2: Pepsin 1500 U/mL E3: Pepsin 1500 U/mL	According to each tested condition. E1: 2 h E2: 2 h E3: 4 h	7	According to each tested condition. E1: Pancreatin 100 U/mL E2: Pancreatin 100 U/mL E3: Pancreatin 50 U/mL	According to each tested condition. E1: Bovine bile 10 mM E2: Bovine bile 10 mM E3: Bovine bile 5 mM	Hernández-Olivas et al. (2022)

Chicken, beef, pork and soy proteins. Obs.: A protein powder (>90%) was obtained from the proteins sources studied	Static model simulating conditions of elderly population with achlorhydria	Not performed	2 h	6	Pepsin 1500 U/mL	4 h	7	Pancreatin 50 U/mL	Bovine bile 50 mM	Wang et al. (2022a)
Chicken, beef, pork and soy proteins. Obs.: A protein powder (>90%) was obtained from the proteins sources studied	Static model simulating conditions of elderly population with achlorhydria	Not performed	2 h	6	Pepsin 1500 U/mL	4 h	7	Pancreatin 50 U/mL	Bovine bile 50 mM	Wang et al. (2022b)
Traditional Tibetan food (*tsampa*)	Dynamic in vitro human stomach system (DHS-IV). simulating elderly gastric digestion	*Tsampa* was mixed with simulated salivary fluid (SSF) containing α-amylase (225 U/mL SSF) for 30 s	3 h	Fasting pH 1.6, gastric pH monitored during study	Pepsin 3000 U/mL SGF[b]	Not performed	Not performed	Not performed	Not performed	Li et al. (2022)

[a]*Gastric phase was monitored during this period.*
[b]*SGF, simulated gastric fluid.*
[c]*SIF, simulated intestinal fluid.*
[d]*Bile origin not informed.*

Table 2.3 Summary of reported adaptations of in vitro gastrointestinal digestion models for simulating elderly digestion compared to adults' digestive conditions.

Digestion phase	Modified condition compared to adult simulated digestive conditions	References
Oral	Higher amylase activity	Levi et al. (2017), Li et al., (2022), and Melchior et al. (2023)
	Higher KCl concentration in saliva	Levi et al. (2017) and Melchior et al. (2023)
	Reduction of chewing cycles	Hernández-Olivas, Muñoz-Pina, Andrés, et al. (2020); Hernández-Olivas, Muñoz-Pina, Sánchez-García, et al. (2020) and Hernández-Olivas et al. (2021, 2022)
	Reduced mixing rate	Aalaei, Khakimov, Gobba (2021), Aalaei, Khakimov, Gobba (2021), and Melchior et al. (2023)
Gastric	Higher pH	Hernández-Olivas, Muñoz-Pina, Andrés, et al. (2020), Hernández-Olivas, Muñoz-Pina, Sánchez-García, et al. (2020), Hernández-Olivas et al. (2021, 2022), Levi and Lesmes, 2014, Levi et al. (2017), Melchior et al. (2023), Wang et al. (2022a,b), and Zhang et al. (2023)
	Reduced enzyme(s) activities	Aalaei, Khakimov, Gobba (2021); Aalaei, Khakimov, Gobba (2021), Hernández-Olivas, Muñoz-Pina, Andrés, et al. (2020), Hernández-Olivas, Muñoz-Pina, Sánchez-García, et al. (2020), Hernández-Olivas et al. (2021, 2022), Levi and Lesmes (2014), Levi et al. (2017), Li et al. (2022), Melchior et al. (2023), Wang et al. (2022a,b), and Zhang et al. (2023)
	Longer duration	Levi and Lesmes (2014), Levi et al. (2017), and Melchior et al. (2023)
	Reduced mixing rate or peristaltic frequency	Aalaei, Khakimov, Gobba (2021), Levi et al. (2017), Khakimov, Gobba (2021), Levi and Lesmes (2014), Li et al. (2022), Melchior et al. (2023), and Zhang et al. (2023)
	Reduced secretion rate of HCl	Li et al. (2022) and Zhang et al. (2023)

(Continued)

Table 2.3 Summary of reported adaptations of in vitro gastrointestinal digestion models for simulating elderly digestion compared to adults' digestive conditions. *Continued*

Digestion phase	Modified condition compared to adult simulated digestive conditions	References
Small intestinal	Higher pH	Levi & Lesmes (2014); Levi et al. (2017)
	Lower pH	Melchior et al. (2023)
	Reduced enzyme(s) activities	Hernández-Olivas, Muñoz-Pina, Andrés, et al. (2020), Hernández-Olivas, Muñoz-Pina, Sánchez-García, et al. (2020), Hernández-Olivas et al. (2021, 2022), Levi and Lesmes (2014), Levi et al. (2017), Melchior et al. (2023), Wang et al. (2022a,b), and Zhang et al. (2023)
	Reduced bile concentration	Hernández-Olivas, Muñoz-Pina, Andrés, et al. (2020), Hernández-Olivas, Muñoz-Pina, Sánchez-García, et al. (2020), Hernández-Olivas et al. (2021, 2022), Levi and Lesmes (2014), Levi et al. (2017), Melchior et al. (2023), and Wang et al. (2022a,b)
	Longer duration	Hernández-Olivas, Muñoz-Pina, Andrés, et al. (2020), Hernández-Olivas, Muñoz-Pina, Sánchez-García, et al. (2020), Hernández-Olivas et al. (2021, 2022), Levi and Lesmes (2014), Levi et al. (2017), Melchior et al. (2023), and Wang et al. (2022a,b)
	Reduced mixing rate	Melchior et al. (2023)
	Increased Ca^{2+} concentration	Zhang et al. (2023)

The various adaptations of in vitro models of gastrointestinal digestion for simulating the digestion of the elderly make it challenging to compare studies from different research groups and highlight the need to define standardized conditions. To meet this challenge, the INFOGEST network recently proposed a static in vitro digestion model adapted to the general older adult population (Menard et al., 2023). This protocol was established in international consensus and all the parameters were based on solid evidence available in the literature. The alterations in the older adult protocol, when compared to the young adult protocol, include a 50% increase in the gastric digestion period, totaling 3 h; reduction in enzymes' activities (pepsin: 1200 U/mL; gastric lipase: 36 U/mL; pancreatin: 80 U/mL), reduced bile concentration (6.7 mM) and altered calcium concentration in the simulated intestinal fluid (1 mM).

2.4 Bioactive peptides and aging

With the population aging, maintaining the health and quality of life of older adults has become a global challenge. The elderly population is more vulnerable to malnutrition and hypofunction of various organs. Additionally, a higher incidence of cardiovascular disease, osteoporosis, osteoarthritis, sarcopenia, and cognitive decline has been associated with aging (Fernandes et al., 2021; Shang et al., 2018). In this context, there is a growing demand for the development of functional foods and food-derived bioactive compounds that can act in the prevention and treatment of diseases and the health promotion of older adults.

Dietary proteins play an essential role in the health of the elderly. In addition to providing energy and amino acids for body protein synthesis, they are sources of bioactive peptides (Shang et al., 2018). Both the essential amino acid composition and the digestibility are factors affecting the nutritional quality of dietary proteins. High-quality proteins contain all essential amino acids above the FAO/WHO/UN reference levels and good digestibility, comparable to or better than the digestibility of egg white and milk proteins (Damodaran, 2017). Although, in general, animal proteins have better nutritional quality than plant proteins, both are potential sources of bioactive peptides. Bioactive peptides are specific protein fragments, usually containing 2–20 amino acids, which may positively impact human health. These peptides have a variety of targets in the human body and can act on the immune, cardiovascular, digestive, and nervous systems (Korhonen & Pihlanto, 2006; Ye et al., 2022). Proteins, in general, can have potentially bioactive peptides encoded in their native structure. These peptides are not active when present in the structure of the parent protein but can be released through hydrolysis, both in food processing and during gastrointestinal digestion. Once released, if these peptides reach the intestinal lumen, they can be absorbed and transported to target sites, acting on several metabolic pathways (Li et al., 2019; Rivero-Pino, 2023). The biological functions of bioactive peptides are related to several factors, such as molecular size and weight, the specific sequence of amino acids, amino acids present in the C-terminal and N-terminal regions, and hydrophobic and hydrophilic properties of the amino acid chain (Sarmadi & Ismail, 2010; Ye et al., 2022).

Although still not much explored for this purpose, bioactive peptides have good potential to contribute to the health of the elderly since many biological activities attributed to these compounds may play a role in the prevention or treatment of several diseases with greater prevalence in the elderly population or that present greater risk for the elderly. Among the bioactive activities of the peptides already reported with potential benefit for the health of the elderly, we can highlight antihypertensive, anticoagulant, antiinflammatory, antidiabetic, anticancer activities, food intake regulatory activity, as well as antioxidant, hypocholesterolemic, bone formation regulating activity, the preventive effect on age-related cognitive and memory decline and neurodegenerative diseases, and activities on the prevention of metabolic syndrome (Arroume et al., 2016; Ayyash et al., 2018; Fan et al., 2022; Hernández-Ledesma et al., 2011; Katayama et al., 2021; Li et al., 2019; Ye et al., 2022).

Among the proteins of animal origin, milk proteins are considered the main sources of bioactive peptides (Korhonen, 2009; Park, 2009). Fermented dairy products have been extensively studied regarding the presence of bioactive peptides and bioactive potential associated with the presence of these peptides (Galli et al., 2019; Pepe et al., 2016; Saito et al., 2000; Sánchez-Rivera et al., 2014). In these products, milk proteins are hydrolyzed by lactic acid bacteria during the fermentation process. In the case of ripened cheeses, caseins are hydrolyzed by the action of natural milk enzymes, residual coagulant, and microbial enzymes leading to the formation of peptides (Hernández-Ledesma et al., 2011). Among peptides derived from dairy proteins, those with antihypertensive activity associated with the inhibition of the angiotensin-converting enzyme (ACE) are the most explored in the literature (Gobbetti et al., 2002). ACE is part of the renin–angiotensin system, a pathway that regulates blood pressure. It acts as a catalyst for the conversion of angiotensin I to angiotensin II, a vasoconstrictor, and catalyzes the hydrolysis of bradykinin, a vasodilator peptide. Both reactions lead to the contraction of blood vessels and the consequent increase in blood pressure. Thus, inhibition of ACE activity can potentially regulate blood pressure (García et al., 2013; Udenigwe & Mohan, 2014). Peptides with mineral-binding activity may contribute to calcium absorption and potentially contribute to bone health (Shang et al., 2018), preventing osteoporosis. Osteoporosis, a bone disease characterized by decreased bone content and quality, has a high incidence in the elderly population. Loss of calcium and collagen are factors associated with the development of this disease (Huang et al., 2022). A study with an animal model conducted by Tsuchita et al. demonstrated that phosphopeptides derived from caseins might favor intestinal calcium absorption (Tsuchita et al., 2001). Similarly, phosphopeptides derived from egg yolk phosvitine also showed in vitro calcium binding activity (Jiang & Mine, 2001). A recent study evaluated osteogenic peptides derived from tilapia collagen and demonstrated the potential of these peptides to prevent osteoporosis (Huang et al., 2022).

Recent changes in the behavior of consumers that seek to reduce the consumption of animal-derived foods and increase the consumption of plant-derived foods justify the growing interest in plant-derived peptides. Although less studied than peptides of animal origin, peptides derived from plant proteins can present several bioactive activities, such as antihypertensive, hypocholesterolemic, immunomodulatory, antimicrobial, anticancer, antioxidant, and antidiabetic activities (Fan et al., 2022; García et al., 2013; Görgüç et al., 2020). Several plant proteins are described as sources of bioactive peptides (legumes, cereals, algae, mushrooms, garlic, and others) (García et al., 2013). In addition, plant-derived peptides have attracted particular attention from the scientific community because they can potentially improve memory and prevent cognitive decline associated with aging (Katayama et al., 2021; Katayama & Nakamura, 2019). However, more evidence about bioavailability and permeability through the blood-brain barrier is needed to prove the effectiveness of these peptides as neurological treatment options (Katayama et al., 2021).

2.5 In vitro studies of bioactive peptides bioaccessibility using elderly gastrointestinal conditions

For a food with bioactive properties to exert physiological functions, the bioactive compounds present in the food matrix or produced during its digestion must be absorbed and reach their sites of action at levels necessary to exert the intended beneficial effects (Rein et al., 2013). The bioactivity of a compound depends on its bioaccessibility, which corresponds to the fraction of a given compound that is released from the food matrix during the digestive process and becomes accessible for absorption in the small intestine, to be biotransformed by the gut microbiota or even to exert local functions through interaction with digestive receptors and cells (Fernández-Tomé & Hernández-Ledesma, 2020; Guerra et al., 2012; Rodrigues et al., 2022). To perform systemic functions, bioactive compounds need to be absorbed in the intestine to become bioavailable for use in physiological functions (Fernández-García et al., 2009).

Thus, knowledge of bioactive compounds' presence and/or concentration in a food matrix is insufficient to predict its health benefits. Assessing the bioaccessibility and bioavailability of nutrients and bioactive compounds is essential for understanding the relationship between food and nutrition (Rodrigues et al., 2022). Several factors can affect the bioaccessibility and, consequently, the bioavailability of bioactive compounds, including the composition of the food matrix, its physicochemical properties, such as pH, temperature, and texture, as well as synergistic/antagonistic relationships between different components (Rein et al., 2013).

Although in vivo models are considered the gold standard for the evaluation of bioaccessibility, absorption, bioavailability, metabolism, and excretion of compounds of interest, in vitro studies are widely used tools because they are less labor intensive, faster and have fewer ethical concerns compared to in vivo methods (Shani-Levi et al., 2017).

In vitro assays have already been widely used in studies to monitor proteolysis during gastrointestinal digestion and the bioaccessibility of peptides in various food matrices (Baptista et al., 2020; Cattaneo et al., 2017; Jin et al., 2016; Liu & Pischetsrieder, 2017; Sánchez-Rivera et al., 2014). However, most studies employed protocols that simulate the digestive conditions of healthy adults. Considering the potential of bioactive peptides as promoters of healthy aging, the study of the bioaccessibility and bioavailability of these compounds in different matrices and in digestive conditions that simulate elderly gastrointestinal digestion is essential to demonstrate the potential benefits of these compounds to older adults. Recently, in vitro digestion models under conditions that simulate the gastrointestinal digestion of the elderly and adults were used to compare the proteolysis that occurs during the digestion of different food matrices, such as UHT and pasteurized milk (Aalaei, Khakimov, Gobba, & Ahrne, 2021), high-pressure treated milk (Aalaei, Khakimov & Gobba, 2021), dairy products

(Hernández-Olivas, Muñoz-Pina, Sánchez-García, et al., 2020), meat (Hernández-Olivas et al., 2022; Wang et al., 2022b), pea protein-fortified sponge cake (Assad-Bustillos et al., 2020), fish (Hernández-Olivas, Muñoz-Pina, Andrés, et al., 2020), legumes (Hernández-Olivas et al., 2021), and cereal grains (Hernández-Olivas et al., 2021).

The effect of aging on the protein hydrolysis that occurs during digestion has been explored by a series of studies that evaluated the proteolysis of different food matrices during their passage through the gastrointestinal tract considering different digestive changes associated with aging (E1: altered oral conditions; E2: altered oral and gastric conditions; and E3: altered oral, gastric, and intestinal conditions) compared to control conditions that simulate the digestion of healthy adults (Hernández-Olivas, Muñoz-Pina, Andrés, et al., 2020; Hernández-Olivas, Muñoz-Pina, Sánchez-García, et al., 2020; Hernández-Olivas et al., 2021, 2022). The authors observed lower proteolysis in the simulation of digestion of different species of fish (salmon, sardine, sea bass, and hake) and dairy products when age-related changes in digestive conditions were used (E1, E2, and E3), compared to digestive conditions of adults (Hernández-Olivas, Muñoz-Pina, Andrés, et al., 2020; Hernández-Olivas, Muñoz-Pina, Sánchez-García, et al., 2020). However, when evaluating the digestion of legumes (chickpeas, lentils, soybeans, and white beans) and cereals/pseudocereals (oats, spelt, and quinoa), the extent of proteolysis was only affected when suboptimal intestinal conditions (E3) were employed (Hernández-Olivas et al., 2021). These results emphasize the importance of evaluating the in vitro digestion of different food matrices.

To evaluate the effect of aging on the bioaccessibility of bioactive peptides, a recent study evaluated the digestion of meat proteins (chicken, beef, and pork) and soy protein using two static in vitro models of gastrointestinal digestion simulation under conditions that mimic the gastrointestinal tract of healthy adults and older adults with achlorhydria. The authors observed that the digestion model (adult or elderly) affected the digestibility and peptide profile formed during digestion, with a greater number of potentially bioactive peptides detected in the digests obtained under digestive conditions of adults compared to the peptide profile detected in the digestion under elderly gastrointestinal conditions. Thus, the research revealed that the changes associated with the aging process affect not only the digestion of proteins but also the formation of bioactive peptides. This difference in protein digestion under the studied conditions is possibly associated with the lower enzymatic activities used in the elderly digestion conditions, the suboptimal pH for the action of pepsin used in the gastric digestion of older adults with achlorhydria (pH 6), and the effect of the low pH used in the digestion of adults (pH 3) in protein structures that may favor the exposure of hydrolysis sites (Wang et al., 2022b).

The digestibility of different meats (chicken, turkey, pork, and beef) was also evaluated by Hernández-Olivas et al. (2022) in a study that compared different digestive alterations associated with aging (E1: altered oral conditions; E2: altered oral and gastric conditions; and E3: altered oral, gastric, and intestinal conditions) in comparison to control conditions simulating the digestion of healthy adults. Although age-related changes in the oral and gastric conditions (E1 and E2) did not affect the proteolysis of meat, the authors observed that age-related changes in the oral, gastric, and intestinal conditions (E3) resulted in a reduction in proteolysis. Additionally, the change in gastric and intestinal conditions associated with aging reduced the ACE-inhibitory activity of the digests obtained after the intestinal phase. Thus, in agreement with the study conducted by Wang et al. (2022b), the results obtained by Hernández-Olivas et al. (2022) suggested lower bioactivity in the digesta obtained under conditions that simulate the digestion of the elderly, possibly due to the lower formation of bioactive peptides in these digestive conditions.

Interestingly, in a study using a dynamic in vitro model to simulate gastroduodenal digestion in adults and the elderly, Levi et al. (2017) observed a greater generation of bioactive peptides from β-lactoglobulin and α-lactoalbumin in the digestive conditions of adults. However, greater formation of bioactive peptides from lactoferrin was observed in the digestive conditions of the elderly.

Briefly, the studies available in the literature reveal possible differences in the proteolysis that occurs during the in vitro simulation of gastrointestinal digestion under conditions that simulate the digestion of adults and the elderly. It is noteworthy that there is no modification pattern, and this change depends on the food matrix and protein under study, since both greater and lower formation of bioactive peptides have been reported in digestive conditions of the elderly, which reinforces the importance of studies to evaluate different protein sources and food matrices.

2.6 Conclusion

Aging is a complex multidimensional process commonly accompanied by several changes in the human body that can change physiological functions, such as the digestive process. Age-related changes in the gastrointestinal tract can affect protein hydrolysis during digestion and, consequently, the bioaccessibility and bioavailability of bioactive compounds.

Several studies focusing on the evaluation of proteolysis during elderly digestion were recently published. The variability in the results obtained emphasizes both the importance of evaluating peptides' bioaccessibility in different food matrices and the need to standardize the in vitro digestion conditions for the elderly population since the variability of conditions used in different studies does not allow the comparison between results. Considering this challenge, the INFOGEST network recently published a static in vitro digestion model adapted to the general older adult population. This protocol was elaborated in an international consensus and will allow the generation of comparable data concerning elderly in vitro digestion studies.

The available in vitro studies that evaluated elderly gastrointestinal digestion used both static and dynamic models. The association between in vitro models of gastrointestinal digestion and cell culture studies using Caco-2 cells is a promising approach to simulate the absorption of bioactive compounds in the intestinal epithelium. However, as far as we know, this methodological approach has not yet been used in studies that simulate the digestive conditions of the elderly and could be considered a future perspective in the study of biopeptides bioavailability.

References

Aalaei, K., Khakimov, B., Gobba, C., & Ahrne, L. (2021). Digestion patterns of proteins in pasteurized and ultra-high temperature milk using in vitro gastric models of adult and elderly. *Journal of Food Engineering*, *292*, 110305. Available from https://doi.org/10.1016/j.jfoodeng.2020.110305.

Aalaei, K., Khakimov, B., Gobba, C., & Ahrné, L. (2021). Gastric digestion of milk proteins in adult and elderly: Effect of high pressure processing. *Foods*, *10*, 786. Available from https://doi.org/10.3390/foods10040786.

Amarya, S., Singh, K., & Sabharwal, M. (2015). Changes during aging and their association with malnutrition. *Journal of Clinical Gerontology and Geriatrics*, *6*(3), 78–84. Available from https://doi.org/10.1016/j.jcgg.2015.05.003.

Amigo, L., & Hernández-Ledesma, B. (2020). Current evidence on the bioavailability of food bioactive peptides. *Molecules (Basel, Switzerland)*, *25*, 4479. Available from https://doi.org/10.3390/molecules25194479.

Arroume, N., Froidevaux, R., Kapel, R., Cudennec, B., Ravallec, R., Flahaut, C., Bazinet, L., Jacques, P., & Dhulster, P. (2016). Food peptides: Purification, identification and role in the metabolism. *Current Opinion in Food Science*, *7*, 101–107. Available from https://doi.org/10.1016/j.cofs.2016.02.005.

Assad-Bustillos, M., Palier, J., Rabesona, H., Choiset, Y., Della Valle, G., & Feron, G. (2020). Role of the bolus degree of structure on the protein digestibility during in vitro digestion of a pea protein-fortified sponge cake chewed by elderly. *Journal of Texture Studies*, *51*, 134–143. Available from https://doi.org/10.1111/jtxs.12486, https://onlinelibrary.wiley.com/doi/10.1111/jtxs.12486.

Ayyash, M., Al-Nuaimi, A. K., Al-Mahadin, S., & Liu, S.-Q. (2018). In vitro investigation of anticancer and ACE-inhibiting activity, α-amylase and α-glucosidase inhibition, and antioxidant activity of camel milk fermented with camel milk probiotic: A comparative study with fermented bovine milk. *Food Chemistry*, *239*, 588–597. Available from https://doi.org/10.1016/j.foodchem.2017.06.149.

Baptista, D. P., Salgaço, M. K., Sivieri, K., & Gigante, M. L. (2020). Use of static and dynamic in vitro models to simulate Prato cheese gastrointestinal digestion: effect of Lactobacillus helveticus LH-B02 addition on peptides bioaccessibility. *LWT – Food Science and Tecnology*, *134*, 110229. Available from https://doi.org/10.1016/j.lwt.2020.110229.

Biagi, E., Candela, M., Fairweather-Tait, S., Franceschi, C., & Brigidi, P. (2012). Ageing of the human metaorganism: the microbial counterpart. *Age*, *34*, 247–267. Available from https://doi.org/10.1007/s11357-011-9217-5.

Biagi, E., Candela, M., Turroni, S., Garagnani, P., Franceschi, C., & Brigidi, P. (2013). Ageing and gut microbes: Perspectives for health maintenance and longevity. *Pharmacological Research*, *69*, 11–20. Available from https://doi.org/10.1016/j.phrs.2012.10.005.

Boirie, Y., & Guillet, C. (2018). Fast digestive proteins and sarcopenia of aging. *Current Opinion in Clinical Nutrition and Metabolic Care*, *21*(1), 37–41. Available from https://doi.org/10.1097/MCO.0000000000000427, https://journals.lww.com/co-clinicalnutrition/Abstract/2018/01000/Fast_digestive_proteins_and_sarcopenia_of_aging.9.aspx.

Boland, M. (2016). Human digestion – A processing perspective. *Journal of the Science of Food and Agriculture*, *96*(7), 2275–2283. Available from https://doi.org/10.1002/jsfa.7601, https://onlinelibrary.wiley.com/doi/10.1002/jsfa.7601.

Brodkorb, A., Egger, L., Alminger, M., Alvito, P., Assunção, R., Ballance, S., Bohn, T., Bourlieu-Lacanal, C., Boutrou, R., Carrière, F., Clemente, A., Corredig, M., Dupont, D., Dufour, C., Edwards, C., Golding, M., Karakaya, S., Kirkhus, B., Le Feunteun, S., ... Recio, I. (2019). INFOGEST static in vitro simulation of gastrointestinal food digestion. *Nature Protocols*, *14*, 991–1014. Available from https://www.nature.com/articles/s41596-018-0119-1.

Cattaneo, S., Stuknytė, M., Ferraretto, A., & De Noni, I. (2017). Impact of the in vitro gastrointestinal digestion protocol on casein phosphopeptide profile of Grana Padano cheese digestates. *LWT*, *77*, 356–361. Available from https://doi.org/10.1016/j.lwt.2016.11.069.

Chen, J. (2009). Food oral processing—A review. *Food Hydrocolloids*, *23*, 1–25. Available from https://doi.org/10.1016/j.foodhyd.2007.11.013.

Chernoff, R. (2015). Effects of aging on the digestive system. In Timothy L. Kauffman, Ron Scott, John O. Barr, Michael L. Moran. (Ed). A Comprehensive Guide to Geriatric Rehabilitation (Third Edition), (3, pp. 45–49) Churchill Livingstone, Elsevier Ltd. Available from https://doi.org/10.1016/B978-0-7020-4588-2.00008-5.

Claesson, M. J., Cusack, S., O'Sullivan, O., Greene-Diniz, R., Weerd, H., Flannery, E., Marchesi, J. R., Falush, D., Dinan, T., Fitzgerald, G., Stanton, C., van Sinderen, D., O'Connor, M., Harnedy, N., O'Connor, K., Henry, C., O'Mahony, D., Fitzgerald, A. P., Shanahan, F., ... O'Toole, P. W. (2011). Composition, variability, and temporal stability of the intestinal microbiota of the elderly. *Proceedings of the National Academy of Sciences of the United States of America*, *108*, 4586–4591. Available from https://doi.org/10.1073/pnas.1000097107.

Claesson, M. J., Jeffery, I. B., Conde, S., Power, S. E., O'Connor, E. M., Cusack, S., Harris, H. M. B., Coakley, M., Lakshminarayanan, B., O'Sullivan, O., Fitzgerald, G. F., Deane, J., O'Connor, M., Harnedy, N., O'Connor, K., O'Mahony, D., van Sinderen, D., Wallace, M., Brennan, L., ... O'Toole, P. W. (2012). Gut microbiota composition correlates with diet and health in the elderly. *Nature*, *488*, 178–183. Available from https://doi.org/10.1038/nature11319.

Clarkston, W. K., Pantano, M., Morley, J. E., Horowitz, M., Littlefield, J. M., & Burton, F. R. (1997). Evidence for the anorexia of aging: gastrointestinal transit and hunger in healthy elderly vs. young adults. *American Journal of Physiology*, *272*(1), R243–R248. Available from https://doi.org/10.1152/ajpregu.1997.272.1.R243.

Damodaran, S., & Parkin, K. L. (2017). Fennema's Food Chemistry. 5th Edition. p. 1123, Boca Raton, CRC Press. Available from https://doi.org/10.1201/9781315372914.

Fan, H., Liu, H., Zhang, Y., Zhang, S., Liu, T., & Wang, D. (2022). Review on plant-derived bioactive peptides: Biological activities, mechanism of action and utilizations in food development. *Journal of Future Foods*, *2*, 143–159. Available from https://doi.org/10.1016/j.jfutfo.2022.03.003.

Feldman, M., Cryer, B., McArthur, K. E., Huet, B. A., & Lee, E. (1996). Effects of aging and gastritis on gastric acid and pepsin secretion in humans: A prospective study. *Gastroenterology*, *110*, 1043–1052. Available from https://doi.org/10.1053/gast.1996.v110.pm8612992.

Fernandes, J.-M., Fernandes Araújo, J., Miguel Vieira, J., Cristina Pinheiro, A., & Augusto Vicente, A. (2021). Tackling older adults' malnutrition through the development of tailored food products. *Trends in Food Science & Technology*, *115*, 55–73. Available from https://doi.org/10.1016/j.tifs.2021.06.028.

Fernández-García, E., Carvajal-Lérida, I., & Pérez-Gálvez, A. (2009). In vitro bioaccessibility assessment as a prediction tool of nutritional efficiency. *Nutrition Research*, *29*, 751–760. Available from https://doi.org/10.1016/j.nutres.2009.09.016.

Fernández-Tomé, S., & Hernández-Ledesma, B. (2020). Gastrointestinal digestion of food proteins under the effects of released bioactive peptides on digestive health. *Molecular Nutrition & Food Research*, *64*, 1–12. Available from https://doi.org/10.1002/mnfr.202000401.

Galli, B. D., Parra Baptista, D., Giacometti Cavalheiro, F., Negrão, F., Nogueira Eberlin, M., & Lúcia Gigante, M. (2019). Peptide profile of Camembert-type cheese: Effect of heat treatment and adjunct culture Lactobacillus rhamnosus GG. *Food Research International*, *123*, 393–402. Available from https://doi.org/10.1016/j.foodres.2019.05.009.

García, M. C., Puchalska, P., Esteve, C., & Marina, M. L. (2013). Vegetable foods: A cheap source of proteins and peptides with antihypertensive, antioxidant, and other less occurrence bioactivities. *Talanta*, *106*, 328–349. Available from https://doi.org/10.1016/j.talanta.2012.12.041.

Gobbetti, M., Stepaniak, L., De Angelis, M., Corsetti, A., & Di Cagno, R. (2002). Latent bioactive peptides in milk proteins: proteolytic activation and significance in dairy processing. *Critical Reviews in Food Science and Nutrition*, *42*(3), 223–239. Available from https://doi.org/10.1080/10408690290825538.

Görgüç, A., Gençdağ, E., & Yilmaz, F. M. (2020). Bioactive peptides derived from plant origin by- products: biological activities and techno-functional utilizations in foods developments – A review. *Food Research International*, *136*, 109504. Available from https://doi.org/10.1016/j.foodres.2020.109504.

Guerra, A., Etienne-Mermin, L., Livrelli, V., Denis, S., Blanquet-Diot, S., & Alric, M. (2012). Relevance and challenges in modeling human gastric and small intestine digestion. *Trends in Biotechnology*, *30*(11), 591–600. Available from https://doi.org/10.1016/j.tibtech.2012.08.001.

Hernández-Ledesma, B., Contreras, M. D., & Recio, I. (2011). Antihypertensive peptides: production, bioavailability and incorporation into foods. *Advances in Colloid and Interface Science*, *165*(1), 23–35. Available from https://doi.org/10.1016/j.cis.2010.11.001.

Hernández-Olivas, E., Muñoz-Pina, S., Andrés, A., & Heredia, A. (2020). Impact of elderly gastrointestinal alterations on in vitro digestion of salmon, sardine, sea bass and hake: Proteolysis, lipolysis and bioaccessibility of calcium and vitamins. *Food Chemistry*, *326*, 127024. Available from https://doi.org/10.1016/j.foodchem.2020.127024.

Hernández-Olivas, E., Muñoz-Pina, S., Andrés, A., & Heredia, A. (2021). Age-related gastrointestinal alterations of legumes and cereal grains digestibility. *Food Bioscience*, *41*, 101027. Available from https://doi.org/10.1016/j.fbio.2021.101027.

Hernández-Olivas, E., Muñoz-Pina, S., García-Hernández, J., Andrés, A., & Heredia, A. (2022). Impact of common gastrointestinal disorders in elderly on in vitro meat protein digestibility and related properties. *Food Bioscience*, *46*, 101560. Available from https://doi.org/10.1016/j.fbio.2022.101560.

Hernández-Olivas, E., Muñoz-Pina, S., Sánchez-García, J., Andrés, A., & Heredia, A. (2020). Understanding the role of food matrix on the digestibility of dairy products under elderly gastrointestinal conditions. *Food Research International*, *137*, 109454. Available from https://doi.org/10.1016/j.foodres.2020.109454.

Huang, W., Yu, K., Kang, M., Wang, Q., Liao, W., Liang, P., Liu, G., Cao, Y., & Miao, J. (2022). Identification and functional analysis of three novel osteogenic peptides isolated from tilapia scale collagen hydrolysate. *Food Research International*, *162*, 111993. Available from https://doi.org/10.1016/j.foodres.2022.111993.

Ichikawa, K., Sakuma, S., Yoshihara, A., Miyazaki, H., Funayama, S., Ito, K., & Igarashi, A. (2011). Relationships between the amount of saliva and medications in elderly individuals. *Gerodontology*, *28*(2), 116–120. Available from https://doi.org/10.1111/j.1741-2358.2009.00358.x.

Jiang, B., & Mine, Y. (2001). Phosphopeptides derived from hen egg yolk phosvitin: effect of molecular size on the calcium-binding properties. *Bioscience, Biotechnology, and Biochemistry*, *65*(5), 1187–1190. Available from https://doi.org/10.1271/bbb.65.1187.

Jin, Y., Yu, Y., QI, Y., Wang, F., Yan, J., & Zou, H. (2016). Peptide profiling and he bioactive character of yogurt in the simulated gastrointestinal digestion. *Journal of Proteomics*, *141*, 24–46. Available from https://doi.org/10.1016/j.jprot.2016.04.010.

Katayama, S., Corpuz, H. M., & Nakamura, S. (2021). Potential of plant-derived peptides for the improvement of memory and cognitive function. *Peptides*, *142*, 170571. Available from https://doi.org/10.1016/j.peptides.2021.170571.

Katayama, S., & Nakamura, S. (2019). Emerging roles of bioactive peptides on brain health promotion. *International Journal of Food Science and Technology*, *54*(6), 1949–1955. Available from https://doi.org/10.1111/ijfs.14076.

Korhonen, H. J. (2009). Bioactive components in bovine milk. In Y. W. Park (Ed.), *Bioactive components in milk and dairy products*. Ames, IL, USA: Wiley-Blackwell.

Korhonen, H., & Pihlanto, A. (2006). Bioactive peptides: Production and functionality. *International Dairy Journal*, *16*(9), 945–960. Available from https://doi.org/10.1016/j.idairyj.2005.10.012.

Laugier, R., Bernard, J.-P., Berthezene, P., & Dupuy, P. (1991). Changes in pancreatic exocrine secretion with age: pancreatic exocrine secretion does decrease in the elderly. *Digestion*, *50*(3–4), 202–211. Available from https://doi.org/10.1159/000200762.

Levi, C. S., Goldstein, N., Portmann, R., & Lesmes, U. (2017). Emulsion and protein degradation in the elderly: Qualitative insights from a study coupling a dynamic in vitro digestion model with proteomic analyses. *Food Hydrocolloids*, *69*, 393–401. Available from https://doi.org/10.1016/j.foodhyd.2017.02.017.

Levi, C. S., & Lesmes, U. (2014). Bi-compartmental elderly or adult dynamic digestion models applied to interrogate protein digestibility. *Food function*, *5*, 2402–2409. Available from https://doi.org/10.1039/C4FO00478G.

Li, L., Baima, C., Jiang, J., Liu, Z., Wang, J., Chen, X. D., & Wu, P. (2022). In vitro gastric digestion and emptying of tsampa under simulated elderly and young adult digestive conditions using a dynamic stomach system. *Journal of Food Engineering*, *327*, 111054. Available from https://doi.org/10.1016/j.jfoodeng.2022.111054.

Li, S., Tingting, B., Zheng, J., Liu, L., He, G., & Wu, J. (2019). Preparation, bioavailability, and mechanism of emerging activities of ile-pro-pro and val-pro-pro. *Comprehensive Reviews In Food Science And Food Safety*, *18*(4), 1097–1110. Available from https://doi.org/10.1111/1541-4337.12457.

Liu, Y., & Pischetsrieder, M. (2017). Identification and relative quantification of bioactive peptides sequentially released during simulated gastrointestinal digestion of commercial kefir. *Journal of Agricultural and Food Chemistry*, *65*, 1865–1873. Available from https://pubs.acs.org/doi/10.1021/acs.jafc.6b05385.

Madsen, J. L., & Graff, J. (2004). Effects of ageing on gastrointestinal motor function. *Age and Ageing*, *33*(2), 154–159. Available from https://doi.org/10.1093/ageing/afh040.

Madsen, J. L. (1992). Effects of gender, age, and body mass index on gastrointestinal transit times. *Digestive Diseases and Sciences*, *37*, 1548–1553. Available from https://doi.org/10.1007/BF01296501.

Melchior, S., Moretton, M., Alongi, M., Calligaris, S., & Nicoli, M. C. (2023). Comparison of protein in vitro digestibility under adult and elderly conditions: the case study of wheat, pea, rice, and whey proteins. *Food Research International*, *163*, 112147. Available from https://doi.org/10.1016/j.foodres.2022.112147.

Mello, A. M., Paroni, G., Daragjati, J., & Pilotto, A. (2016). Gastrointestinal microbiota and their contribution to healthy aging. *Digestive Diseases*, *34*(3), 194–201. Available from https://doi.org/10.1159/000443350.

Menard, O., Lesmes, U., Shani-Levi, C. S., Araiza Calahorra, A., Lavoisier, A., Morzel, M., Rieder, A., Feron, G., Nebbia, S., Mashiah, L., Andres, A., Bornhorst, G., Carrière, F., Egger, L., Gwala, S., Heredia, A., Kirkhus, B., Macierzanka, A., Portman, R., … Dupont, D. (2023). Static in vitro digestion model adapted to the general older adult population: an INFOGEST international consensus. *Food & Function*, *14*(10), 4569–4582. Available from https://doi.org/10.1039/D3FO00535F.

Minekus, M., Alminger, M., Alvito, P., Bohn, T., Bourlieu, C., Carrière, F., Boutrou, R., Corredig, M., Dupont, D., Dufour, C., Egger, L., Golding, M., Karakaya, S., Kirkhus, B., Le Feunteun, S., Lesmes, U., Macierzanka, A., Mackie, A., Marze, S., … Brodkorb, A. (2014). A standardised static *in vitro* digestion method suitable for food – an international consensus. *Food & Function*, *5*, 1113–1124. Available from https://doi.org/10.1039/C3FO60702J.

Mishellany-Dutour, A., Renaud, J., Peyrom, M.-A., Rimek, F., & Woda, A. (2008). Is the goal of mastication reached in young dentates, aged dentates and aged denture wearers? *British Journal of Nutrition*, *99*(1), 121–128. Available from https://doi.org/10.1017/S0007114507795284.

Moore, J. G., Tweedy, C., Christian, P. E., & Datz, F. L. (1983). Effect of age on gastric emptying of liquid-solid meals in man. *Digestive Diseases and Sciences*, *28*, 121–128. Available from https://doi.org/10.1007/BF01324951.

Mueller, S., Saunier, K., Hanisch, C., Norin, E., Alm, L., Midtvedt, T., Cresci, A., Silvi, S., Orpianesi, C., Verdenelli, M. C., CLavel, T., Koebnick, C., Zunft, H.-J. F., Doré, J., & Blaut, M. (2006). Differences in fecal microbiota in different european study populations in relation to age, gender, and country: a cross-sectional study. *Applied and Environmental Microbiology*, *72*(2), 1027–1033. Available from https://doi.org/10.1128/AEM.72.2.1027-1033.2006.

Nagler, R., & Hershkovich, O. (2005). Relationships between age, drugs, oral sensorial complaints and salivery profile. *Archives of Oral Biology*, *50*(1), 7–16. Available from https://doi.org/10.1016/J.archoralbio.2004.07.012.

Oliveros, E., Patel, H., Kyung, S., Fugar, S., Gouldberg, A., Madan, N., & Williams, K. A. (2020). Hypertension in older adults: Assessment, management, and challenges. *Clinical Cardiology*, *43*(2), 99–107. Available from https://doi.org/10.1002/clc.23303.

Ottman, N., Smidt, H., De Vos, W. M., & Belzer, C. (2012). The function of our microbiota: who is out there and what do they do? *Frontiers in Cellular and Infection Microbiology*, *2*, 104. Available from https://doi.org/10.3389/fcimb.2012.00104.

Park, Y. W. (2009). Overview of bioactive components in milk and dairy products. In Y. W. Park (Ed.), *Bioactive components in milk and dairy products*. Ames, IL, USA: Wiley-Blackwell.

Pepe, G., Somella, E., Ventre, G., Scala, M. C., Adesso, S., Ostacolo, C., Marzocco, S., & Campiglia, P. (2016). Antioxidant peptides released from gastrointestinal digestion of "Stracchino" soft cheese: Characterization, in vitro intestinal protection and bioavailabity. *Journal of Functional Foods*, *26*, 494–505. Available from https://doi.org/10.1016/j.jff.2016.08.021

Peyron, M.-A., Santé-Lhoutellier, V., Dardevet, D., Hennequin, M., Rémond, D., François, O., & Woda, A. (2019). Addressing various challenges related to food bolus and nutrition with the AM^2 mastication simulator. *Food Hydrocolloids*, *97*, 105229. Available from https://doi.org/10.1016/j.foodhyd.2019.105229.

Picariello, G., Mamone, G., Nitride, C., Addeo, F., & Ferranti, P. (2013). Protein digestomics: Integrated platforms to study food-protein digestion and derived functional and active peptides. *TrAC Trends in Analytical Chemistry*, *52*, 120–134. Available from https://doi.org/10.1016/j.trac.2013.08.001.

Plante, A. M., McCarthy, A. L., Lacey, S., & O'Halloran, F. (2021). Investigating the bioactive properties of cheese-fruit combinations following in vitro digestion using an elderly model. *Current Research in Nutrition and Food Science Journal*, *9*(2), 465–478. Available from https://doi.org/10.12944/CRNFSJ.9.2.10.

Plante, A. M., McCarthy, A. L., & O'Halloran, F. (2021). Cheese as a functional food for older adults: comparing the bioactive properties of different cheese matrices following simulated gastrointestinal in vitro digestion. *International Journal of Food Sciences and Nutrition*, *9*(2), 456–469. Available from https://doi.org/10.1080/09637486.2020.1825644.

Rein, M. J., Renouf, M., Cruz-Hernandez, C., Actis-Goretta, L., Thakkar, S. K., & da Silva Pinto, M. (2013). Bioavailability of bioactive food compounds: a challenging journey to bioefficacy. *British Journal of Clinical Pharmacology*, *75*(3), 588–602. Available from https://doi.org/10.1111/j.1365-2125.2012.04425.x.

Rémond, D., Shahar, D. R., Gille, D., Pinto, P., Kachal, J., Peyron, M.-A., Nunes Dos Santos, C., Walther, B., Bordoni, A., Dupont, D., Tomás-Cobos, L., & Vergères, G. (2015). Understanding the gastrointestinal tract of the elderly to develop dietary solutions that prevent malnutrition. *Oncotarget*, *6*, 13858–13898. Available from https://doi.org/10.18632/oncotarget.4030.

Ribes, S., Genot, M., Aubry, L., Talens, P., Vénien, A., Santé-Lhoutellier, V., & Peyron, M.-A. (2023). Oral impairments decrease the nutrient bioaccessibility of bread in the elderly. *Food Hydrocolloids*, *135*, 108202. Available from https://doi.org/10.1016/j.foodhyd.2022.108202.

Rivero-Pino, F. (2023). Bioactive food-derived peptides for functional nutrition: Effect of fortification, processing and storage on peptide stability and bioactivity within food matrices. *Food Chemistry*, *406*, 135046. Available from https://doi.org/10.1016/j.foodchem.2022.135046.

Rodrigues, D. B., Marques, M. C., Hacke, A., Loubet Filho, S., Cazarin, C. B. B., & Mariutti, L. R. B. (2022). Trust your gut: bioavailability and bioaccessibility of dietary compounds. *Current Research in Food Science*, *5*, 228–233. Available from https://doi.org/10.1016/j.crfs.2022.01.002.

Saito, T., Nakamura, T., Kitazawa, H., Kawai, Y., & Itoh, T. (2000). Isolation and structural analysis of antihypertensive peptides that exist naturally in Gouda cheese. *Journal of Dairy Science*, *83*, 1432–1444. Available from https://doi.org/10.3168/jds.S0022-0302(00)75013-2.

Salles, N. (2007). Basic mechanisms of the aging gastrointestinal tract. *Digestive Diseases*, *25*, 112–117. Available from https://doi.org/10.1159/000099474.

Sánchez-Rivera, L., Diezhandino, I., Ángel Gómez-Ruiz, J., María Fresno, J., Miralles, B., & Recio, I. (2014). Peptidomic study of Spanish blue cheese (Valdeón) and changes after simulated gastrointestinal digestion. *Electrophoresis*, *35*(11), 1627–1636. Available from https://doi.org/10.1002/elps.201300510.

Sarmadi, B. H., & Ismail, A. (2010). Antioxidative peptides from food proteins: A review. *Peptides*, *31*, 1949–1956. Available from https://doi.org/10.1016/j.peptides.2010.06.020.

Schmucker, D. L. (2005). Age-related changes in liver structure and function: implications for disease? *Experimental Gerontology*, *40*(8–9), 650–659. Available from https://doi.org/10.1016/j.exger.2005.06.009.

Serra-Prat, M., Mans, E., & Palomera, E. (2013). Gastrointestinal peptides, gastrointestinal motility, and anorexia of aging in frail elderly persons. *Neurogastroenterology & Motility*, *25*, 291–e245. Available from https://doi.org/10.1111/nmo.12055.

Shang, N., Meram, C., Bandara, N., & Wu, J. (2018). Protein and peptides for elderly health. *Advances in Protein Chemistry and Structural Biology*, *112*, 265–308. Available from https://doi.org/10.1016/bs.apcsb.2018.03.003.

Shani-Levi, C., Alvito, P., Andrés, A., Assunção, R., Barberá, R., Blanquet-Diot, S., Bourlieu, C., Brodkorb, A., Cilla, A., Deglaire, A., Denis, S., Dupont, D., Heredia, A., Karakaya, S., Lucia Giosafatto Giosafatto, C. V., Mariniello, L., Martins, C., Ménard, O., Nehir El, S., ... Lesmes, U. (2017). Extending in vitro digestion models to specific human populations: perspectives, practical tools and bio-relevant information. *Trends in Food Science & Technology*, *60*, 52–63. Available from https://doi.org/10.1016/j.tifs.2016.10.017.

Tsuchita, H., Suzuki, T., & Kuwata, T. (2001). The effect of casein phosphopeptides on calcium absorption from calcium-fortified milk in growing rats. *British Journal of Nutrition*, *85*, 5–10. Available from https://doi.org/10.1079/BJN2000206.

Udenigwe, C. C., & Mohan, A. (2014). Mechanisms of food protein-derived antihypertensive peptides other than ACE inhibition. *Journal of Functional Foods*, *8*, 45–52. Available from https://doi.org/10.1016/j.jff.2014.03.002.

UN. (2020). *World population ageing 2020 highlights: Living arrangements of older persons*. http://www.un.org/development/desa/pd/.

Vellas, B., Balas, D., Moreau, J., Bouisson, M., Senegas-Balas, F., Guidet, M., & Ribet, A. (1988). Exocrine pancreatic secretion in the elderly. *International Journal of Pancreatology*, *3*, 497–502. Available from https://doi.org/10.1007/BF02788208.

Volkert, D., Marie Beck, A., Cederholm, T., Cruz-Jentoft, A., Hooper, L., Kiesswetter, E., Maggio, M., Raynaud-Simon, A., Sieber, C., Sobotka, L., van Asselt, D., Wirth, R., & Bischoff, S. C. (2022). ESPEN practical guideline: Clinical nutrition and hydration in geriatrics. *Clinical Nutrition*, *41*(4), 958–989. Available from https://doi.org/10.1016/j.clnu.2022.01.024.

Wang, C., Zhao, F., Bai, Y., Li, C., Xu, X., Kristiansen, K., & Zhou, G. (2022a). In vitro digestion mimicking conditions in young and elderly reveals marked differences between profiles and potential bioactivity of peptides from meat and soy proteins. *Food Research International*, *157*, 111215. Available from https://doi.org/10.1016/j.foodres.2022.111215.

Wang, C., Zhao, F., Bai, Y., Li, C., Xu, X., Kristiansen, K., & Zhou, G. (2022b). Effect of gastrointestinal alterations mimicking elderly conditions on in vitro digestion of meat and soy proteins. *Food Chemistry*, *383*, 132465. Available from https://doi.org/10.1016/j.foodchem.2022.132465.

Ye, H., Tao, X., Zhang, W., Chen, Y., Yu, Q., & Xie, J. (2022). Food-derived bioactive peptides: production, biological activities, opportunities and challenges. *Journal of Future Foods*, 2(4), 294–306. Available from https://doi.org/10.1016/j.jfutfo.2022.08.002.

Zhang, P., Iqbal, S., Deng, R., Duan, X., Han, H., Dong Chen, X., & Wu, P. (2023). Impact of elderly gastrointestinal alterations on gastric emptying and enzymatic hydrolysis of skim milk: An in vitro study using a dynamic stomach system. *Food Chemistry*, *402*, 134365. Available from https://doi.org/10.1016/j.foodchem.2022.134365.

CHAPTER

3

Oral bioaccessibility and bioavailability of protein hydrolysates and bioactive peptides derived from food sources

Kenji Sato

Division of Applied Biosciences, Graduate School of Agriculture, Kyoto University, Kyoto, Japan

3.1 Background

Dietary proteins serve as a source of amino acids. Dietary proteins are first degraded in the digestive tract by endoproteinases, such as pepsin, trypsin, and chymotrypsin, which generate peptides. The resultant peptides are degraded into amino acids and shorter peptides by exopeptidases, such as carboxypeptidases and aminopeptidases, in pancreatic juice, intestinal mucosa, and blood. The resultant amino acids are used by cells for protein synthesis and as energy sources. To prepare food-grade peptides, food proteins are digested by food-grade proteases on an industrial scale (Toldrá et al., 2018). Enzymatic hydrolysates were originally used as a soluble and stable amino acid source and seasoning ingredient.

Since the 1990s it has been demonstrated that ingestion of some food peptides exerts beneficial activity beyond conventional nutritional activity, namely as the source of amino acids (Mine et al., 2010). Early studies demonstrated that some peptides in fish protein hydrolysates and fermented milk inhibit angiotensin-converting enzyme (ACE) (Nakamura et al., 1995; Yokoyama et al., 1992). An endogenous hypertensive peptide hormone, angiotensin II (Ang II), is generated from an inactive precursor peptide, referred to as angiotensin I (Ang I). ACE is involved in the production of Ang II. Thus in vitro inhibitory activity against ACE by food-derived peptides has been associated with potential antihypertensive activity upon ingestion. In vitro activity-guided fractionation using ACE inhibitory activity and high-performance liquid chromatography (HPLC) separation was performed to identify active peptides in foods. Consequently, many peptides with the ACE inhibitory activity have been identified in protein hydrolysates and fermented foods. These peptides have been proposed to be responsible for the

Protein Digestion-Derived Peptides. DOI: https://doi.org/10.1016/B978-0-443-19141-1.00003-0

antihypertensive activity of some protein hydrolysates and fermented foods. The antihypertensive activity of some products containing food protein hydrolysates or fermented foods has been confirmed in double-blind placebo-controlled human clinical trials using subjects with mild hypertension. Based on the results of human trials and identification of peptides with inhibitory activity against ACE in foods, some commercial products have been approved to present health claims on the label by the Japanese Government. In Japan, such products are referred to as Food for Specified Health Use (FOSHU) (Hayashi et al., 2008). Since this great success, various kinds of food-derived peptides have been demonstrated to exert numerous beneficial effects on human health, such as moderation of hyperlipidemia and hyperglycemia and improvement of skin, joint, liver, and mental conditions. In most cases, the active peptides responsible for these beneficial effects have been identified by the in vitro activity-guided fractionation using enzyme reaction, cell culture, and antioxidant assays. Recently bioactive peptides have been deduced using in silico digestion and bioactivity databases. These active peptides are referred to as food-derived bioactive peptides, which is currently the focus of research. Approximately 6000 hits were obtained by entering "food and bioactive peptide" in the PubMed database. Numerous bioactive peptides have also been identified. To exert biological activities upon ingestion of food containing peptides by the proposed mechanisms that were used for identification of the active peptides, the active peptides must reach the target organs of the proposed mechanism. For example, the peptide, which was identified by the ACE inhibitory assay, must be incorporated into blood circulation to exert antihypertensive activity by inhibiting ACE. However, the bioavailability of the proposed active peptides is reportedly very low. Maximal blood levels of the ACE inhibitory peptides in humans after ingestion of food containing these peptides were demonstrated to be approximately 1–2 nM (Foltz et al., 2007; Matsui et al., 2002), which was far less than the concentrations used for the in vitro ACE inhibitory assay (1–100 μM) (Nakamura et al., 1995; Seki et al., 1995). These facts have raised doubts regarding the proposed mechanism and the active peptide. Indeed, some researchers have demonstrated that the peptides identified by the ACE inhibitory assay exert antihypertensive activity through different mechanisms (Hirota et al., 2011; Shobako et al., 2019; Tanaka et al., 2006; Wu, 2020). However, these facts also raise doubts regarding the rationality of in vitro activity-guided fractionation as follows. Other peptides that were not detected by the ACE inhibitory assay might exert stronger antihypertensive activity by different mechanisms than the proposed one or by conversion to the active form for the proposed mechanism during the digestion and absorption processes. Therefore the structure, activities, and content of food-derived peptides and their metabolites in target organs in the body are more important than those in food (Sato, 2020).

In this chapter, recent knowledge of the bioavailability and metabolic fate of food-derived peptides and their association with structure and health-promoting activities upon ingestion of peptides in food have been introduced.

3.2 Susceptibility of peptides against endoproteinase and exopeptidase

Food protein hydrolysates, namely food peptides, are frequently prepared using protease preparations from heat-stable bacteria, such as *Bacillus thermoproteolyticus* and *B. licheniformis*, as digestion can be carried out at 50°C–60°C, which suppresses the growth of most microorganisms. Food-grade bacterial protease preparations, such as thermolysin, and alcalase, are commercially available and are used for the preparation of food peptides (Toldrá et al., 2018). Bacterial proteases have strong endoproteinase activity, but weak exopeptidase activity (Sumantha et al., 2006). In contrast, fungi proteases generally exhibit both endoproteinase and exopeptidase activities (Sumantha et al., 2006). Food-grade fungal protease preparations from *Aspergillus oryzae* and *Aspergillus niger* have been used to produce food peptides. Thus food peptides produced by the fungal protease preparations generally contain higher ratios of free amino acids and short peptides.

In some studies, it has been reported that a considerable number of peptides in food peptide preparations resist mammalian digestive proteases, such as pepsin and trypsin. Based on these facts, it has been suggested that these peptides can survive in mammalian digestive tracts and are incorporated into blood circulation. Pepsin and trypsin are the mammalian endoproteinases. It is quite natural for peptides generated by microbial endoproteinases to resist mammalian endoproteases. However, most peptides in food peptide preparations are degraded by mammalian exopeptidases, such as carboxypeptidase A in pancreatic juice, intestinal mucosa peptidases, and leucine aminopeptidase in the cytosol (Chen et al., 2019; Miyauchi et al., 2022). Indeed, the levels of peptides susceptible to in vitro digestion by carboxypeptidase A and leucine aminiopeptidase in a rice protein hydrolysate did not increase in the lumen of rats after ingestion of the rice protein hydrolysate (Miyauchi et al., 2022). Thus resistance to pepsin and trypsin does not result in stability in the digestive tract and high bioavailability. Recently in silico digestion software has been frequently used to predict peptide structures after the digestion of food proteins by industrial and digestive enzymes (Panyayai et al., 2019). The in silico digestion software is based on the sequence of substrate proteins and substrate specificity of endoproteinases. However, to the best of my knowledge, the software programs do not consider the substrate specificity of exopeptidases. Therefore, most in silico programs cannot predict the structure of stable food-derived peptides in the intestinal lumen, blood, and organs, but it can predict peptide structure in food protein hydrolysate. Thus the conclusion obtained from in silico analysis should be interpreted with caution.

3.3 Food-derived peptides in body

3.3.1 Hydroxyproline-containing peptides

As mentioned above, the blood levels of food-derived peptides, which were identified by activity-guided fractionation without considering bioavailability, were low nM levels. Thus it has been generally assumed that most peptides in food are degraded into amino acids and only negligible amounts of food-derived peptides are incorporated into blood circulation. However, this concept was partially overturned by the findings of the food-derived collagen peptides in human blood at μM levels (Iwai et al., 2005).

Collagen is a major protein compound in the extracellular matrix that has triple helical domains with small nontriple helical domains (Van der Rest & Garrone, 1991). Collagen forms a molecular family consisting of different gene products and is designated by type and Roman numerals (I–XXVIII). Each collagen molecule consists of 3 subunits. Each subunit is designated by α and Arabic and Roman numerals (Van der Rest & Garrone, 1991). The major collagen type in the skin, tendon, and bone is type I collagen, which is designated as $[\alpha 1(I)]_2\alpha 2(I)$. The major type of collagen in the cartilage is type II ($[\alpha 1(II)]_3$). Collagens in tissues are insoluble in bodily fluids. The triple helical structure of collagen collapses by heat treatment in water, and denatured collagen can be extracted from hot water, and is referred to as gelatin. The native collagen molecule can resist most proteases except collagenases. However, gelatin can also be degraded by digestive enzymes and food-grade proteases. The enzymatically digested gelatin is referred to as collagen hydrolysate or collagen peptide and is used as a food ingredient.

Glycyl residues occupy every three residues in the triple helical domain of collagen. All collagens have posttranslational modified amino acids: hydroxyproline (Hyp) and hydroxylysine. Hyp accounts for approximately 10% (mol/mol) of collagen. Many Gly–Pro–Hyp–Gly repeats are found in type I and II collagens (Van der Rest & Garrone, 1991). Hyp is also present in some proteins, but its content is far less than that of collagen. Thus most peptides comprising Hyp can be considered as collagen peptides. The peptide form of Hyp in the blood can be estimated by subtracting free Hyp levels from the total Hyp levels in the HCl hydrolysate of the deproteinized fraction of blood plasma by selective ethanol precipitation (Iwai et al., 2005). After ingestion of 10–20 g of the collagen hydrolysate, approximately 20–70 μM of the peptide form Hyp, namely, collagen peptide, was present in human blood (Iwai et al., 2005; Ohara et al., 2007); the concentration is more than 10,000 times higher than the levels of food-derived peptides identified by in vitro activity-guided fractionation in previous studies. Even 4–7 h after the ingestion of collagen hydrolysate, significantly high levels of the peptide form of Hyp than baseline levels were observed. To identify food-derived collagen peptides in human blood, collagen peptides in the deproteinized fraction of human blood plasma were resolved by reversed-phase high-performance liquid chromatography (RP-HPLC) (Iwai et al., 2005; Ohara et al., 2007). To improve the resolution of hydrophilic collagen peptides by RP-HPLC, precolumn derivatization with phenyl isothiocyanate (PITC) (Shigemura et al., 2011) and 6-aminoquinolyl-*N*-hydroxysuccinimidyl carbamate (AccQ) (Asai et al., 2019) was performed. PITC- and AccQ-peptide derivatives were sequenced by Edman degradation and tandem mass spectrometry (MS/MS), respectively. Using these methods, at least 13 collagen peptides have been identified in the human blood. The estimated collagen peptides in the human blood were chemically synthesized and used for the optimization of multiple reaction monitoring (MRM) conditions for LC–MS/MS (Asai et al., 2019; Taga et al., 2014), which enables the determination of each collagen peptide with high sensitivity. As shown in Table 3.1, Pro–Hyp was most abundantly in the human blood after the ingestion of collagen hydrolysate. Hyp–Gly was the second leading peptide in the human blood, and its content showed larger individual differences compared to Pro–Hyp (Shigemura, Suzuki, et al., 2018). In addition to the collagen peptides listed in Table 3.1, Glu–Hyp, Ile–Hyp, Ser–Hyp, and Glu–Hyp–Gly were quantified in human blood after ingestion of collagen hydrolysate (Taga et al., 2014) (Table 3.1).

Table 3.1 Reported maximum levels of food-derived peptides in human and animal blood plasma upon ingestion.

Target peptide	Subjects	Test samples	Dose	Maximum content	References
Pro–Hyp	Human	CH	13 g	27 μM	Asai et al. (2019)
Hyp–Gly	Human	CH	13 g	3 μM	Asai et al. (2019)
Ala–Hyp–Gly	Human	CH	13 g	3 μM	Asai et al. (2019)
Ala–Hyp	Human	CH	13 g	3 μM	Asai et al. (2019)
Pro–Hyp–Gly	Human	CH	13 g	2 μM	Asai et al. (2019)
Leu–Hyp	Human	CH	13 g	2 μM	Asai et al. (2019)
Phe–Hyp	Human	CH	13 g	2 μM	Asai et al. (2019)
Ser–Hyp–Gly	Human	CH	13 g	1 μM	Asai et al. (2019)
Gly–Pro–Hyp	Human	CH	13 g	1 μM	Asai et al. (2019)
Pro–Gly	Human	EH	10 g	18 μM	Shigemura et al. (2012)
pyroGlu–Pro	Human	CGH	9 g	100 nM	Ejima et al. (2018)
pyroGlu–Leu–Pro	Human	CGH	9 g	8 nM	Ejima et al. (2018)
pyroGlu–Gln–Pro	Human	CGH	9 g	30 nM	Ejima et al. (2018)
Pro–Ala	Human	CGH	9 g	30 nM	Ejima et al. (2018)
pyroGlu–Pro	Human	CGH	9 g	110 nM	Ejima et al. (2018)

(Continued)

Table 3.1 Reported maximum levels of food-derived peptides in human and animal blood plasma upon ingestion. *Continued*

Target peptide	Subjects	Test samples	Dose	Maximum content	References
pyroGlu–Leu–Pro	Human	WGH	9 g	10 nM	Ejima et al. (2018)
pyroGlu–Gln–Pro	Human	WGH	9 g	40 nM	Ejima et al. (2018)
Pro–Gly	Human	WGH	9 g	60 nM	Ejima et al. (2018)
Pro–Gln	Human	WGH	9 g	30 nM	Ejima et al. (2018)
L-β-Asp–Val	Rat	PLH	167 mg/kg	25 nM	Ejima et al. (2018)
D-α-Asp–Val	Rat	PLH	167 mg/kg	10 nM	Ejima et al. (2019)
D-β-Asp–Val	Rat	PLH	167 mg/kg	23 nM	Ejima et al. (2019)
L-β-Asp–Phe	Rat	PLH	167 mg/kg	8 nM	Ejima et al. (2019)
D-α-Asp–Phe	Rat	PLH	167 mg/kg	3 nM	Ejima et al. (2019)
D-β-Asp–Phe	Rat	PLH	167 mg/kg	28 nM	Ejima et al. (2019)
L-β-ASP–Ile	Rat	PLH	167 mg/kg	14 nM	Ejima et al. (2019)
D-α-Asp–Ile/Leu	Rat	PLH	167 mg/kg	28 nM	Ejima et al. (2019)
D-β-Asp–Ile	Rat	PLH	167 mg/kg	34 nM	Ejima et al. (2019)
L-β-Asp–Leu	Rat	PLH	167 mg/kg	25 nM	Ejima et al. (2019)
D-β-Asp–Leu	Rat	PLH	167 mg/kg	38 nM	Ejima et al. (2019)

CH: collagen hydrolysate; CGH: corn gluten hydrolysate; WGH: wheat gluten hydrolysate; PLH: porcine liver protein hydrolysate.

The degree of hydrolysis of the collagen hydrolysates used for ingestion affects the content and composition of food-derived collagen peptides in human blood (Iwasaki et al., 2022). However, no significant increase of Gly–Pro has been observed in human blood after ingestion of collagen hydrolysates (Asai et al., 2019; Iwai et al., 2005; Ohara et al., 2007; Shigemura et al., 2011; Shigemura, Suzuki, et al., 2018; Taga et al., 2014), while many Gly–Pro motifs are present in collagens. Therefore, the difference in the abundance of collagen peptides in the blood depends not only on the frequency of the motif in the parent protein but also on the substrate specificities of endoproteinases and exopeptidases in the gastrointestinal tract and blood. Some studies have indicated that collagen peptides larger than tripeptides can pass through the Caco-2 monolayer (Shimizu et al., 2010a, 2010b), which has been extensively used as an in vitro intestinal absorption model. These authors concluded that collagen peptides larger than tripeptides might pass through the paracellular pathway. In addition, some animal models using cannulation into the portal vein for blood collection also revealed that collagen peptides larger than tripeptides up to 15 residues were present in the portal blood (Osawa et al., 2018). Furthermore, Hanh et al. (2017) demonstrated that significant amounts of the peptides larger than tripeptides consisting of sarcosine (Sar), such as Gly–Sar–Sar–Sar–Sar, which resist exopeptidase digestion, can be absorbed into the blood circulation after oral administration. These facts suggest that collagen peptides larger than tripeptides can be absorbed into the blood circulation via the paracellular pathway. However, no significant increase of collagen peptides larger than tripeptides have been detected in human and animal blood upon oral ingestion of collagen hydrolysates under physiological conditions (Asai et al., 2019; Iwai et al., 2005; Ohara et al., 2007; Shigemura et al., 2011; Shigemura, Suzuki, et al., 2018; Taga et al., 2014), although the collagen hydrolysate consisted of peptides larger peptides than 500 Da. These controversial results can be explained by the following hypothesis. Peptides larger than tripeptides can potentially pass through the intestinal epithelium via the paracellular route. However, most peptides larger than tripeptides are degraded by exopeptidases in the mucosa and blood. Anticoagulants, such as ethylene diamine tetra acetic acid and heparin, are used for cannulation in animals (Osawa et al., 2018). These anticoagulants can inhibit metalloenzymes including exopeptidases and activate endogenous protease inhibitors, which can explain why blood collected through the cannula contained peptides larger than tripeptides. As mentioned above, no significant increase of Gly–Pro in human blood upon ingestion of collagen hydrolysates has been reported, whereas Gly–Pro is good substrate of a peptide transporter 1 (PepT1) on Caco-2 monolayer and passes through Caco-2 monolayer (Rohm et al., 2019). These facts suggest that the Caco-2 monolayer may lack of some exopeptidases in the digestive tract and blood, which can degrade Gly–Pro. Taken together, the Caco-2 monolayer cannot provide information on the bioavailability of each peptide.

3.3.2 Prolyl dipeptides

All food-derived collagen peptides in the blood listed in Table 3.1 consist of Hyp, a posttranslationally modified amino acid. These facts suggest that the presence of posttranslationally modified amino acids in peptides may be important for their bioavailability. However, a dipeptide consisting of unmodified amino acids, Pro–Gly, was detected in the human blood after ingestion of elastin hydrolysate (Table 3.1) (Shigemura et al., 2012), which is another extracellular matrix protein rich in Gly and Pro residues. Very few Pro–Gly motifs are present in the collagen sequence, as Pro residues in the Gly–X–Pro–Gly motif of collagen are hydroxylated and converted to Gly–X–Hyp–Gly. Many Pro–Gly and Gly–Pro repeats are present in elastin. However, no significant increase of Gly–Pro levels has been detected in human blood after ingestion of the elastin hydrolysate as well as after ingestion of collagen hydrolysates (Shigemura et al., 2012). After ingestion of wheat gluten and corn gluten hydrolysates, blood levels of Pro–Gly, Pro–Ala, and Pro–Gln significantly increased in human blood (Table 3.1), whereas no significant increase was observed in the blood levels of dipeptides with Pro at the carboxy terminal (Ejima et al., 2018). These facts indicate that some dipeptides with Pro at the amino terminal can survive and be absorbed into the blood circulation, while dipeptides with Pro at the carboxyl terminal are degraded by exopeptidases during the digestion and absorption processes. In contrast, the levels of Ala–Hyp, Ser–Hyp, Ile–Hyp, and Phe–Hyp were significantly increased in human blood after ingestion of collagen hydrolysates (Asai et al., 2019; Iwai et al., 2005; Ohara et al., 2007; Shigemura et al., 2011; Shigemura, Suzuki, et al., 2018; Taga et al., 2014), whereas those of Ala–Pro, Ser–Pro, Ile–Pro, and Phe–Pro were not significantly increased in human blood after ingestion of noncollagen hydrolysates (Ejima et al., 2018), which potentially contain these peptide motifs. These facts indicate that the hydroxylation of Pro residues affects the bioavailability of peptides.

3.3.3 Peptides modified during processing

Enzymatic hydrolysates of food proteins have been prepared on an industrial scale and are used as functional food ingredients (Toldrá et al., 2018). Numerous short-chain peptides were present in these hydrolysates. However, most peptides were degraded by in vitro digestion with exopeptidases, except for a few exceptions, while they resisted endoproteinases, such as pepsin and trypsin (Chen et al., 2019). Di- and tripeptides consisting of Pro and Hyp resisted the in vitro exopeptidase digestion, as mentioned above. In addition, some modified peptides, such as pyroglutamyl peptides (Ejima et al., 2018; Miyauchi et al., 2022; Sato et al., 1998), aspartic isopeptides, and D-aspartic peptides/isopeptides (Ejima et al., 2019) in some protein hydrolysates have been demonstrated to resist exopeptidase digestion.

The pyroglutamyl (pyroGlu) peptide is nonenzymatically generated from peptides with Gln at the amino terminus, as shown in Fig. 3.1. Heating enhances the generation of pyroGlu peptides (Suzuki et al., 1999). Therefore, significant amounts of pyroGlu peptides have been detected in many food protein hydrolysates prepared on an industrial scale after heat treatment to terminate the enzyme reaction (Chen et al., 2019; Ejima et al., 2018; Miyauchi et al., 2022; Sato et al., 1998). In some cases, it accounted for 15% of the total peptide content (w/w) (Sato et al., 1998). The pyroGlu peptide is slowly generated, even at room temperature (Suzuki et al., 1999). Thus significant amounts of pyroGlu peptides have been observed in some fermented foods, such as Japanese rice wine (*sake*) (Kiyono et al., 2013) and salted fermented soy paste (*miso*) (Shirako et al., 2021).

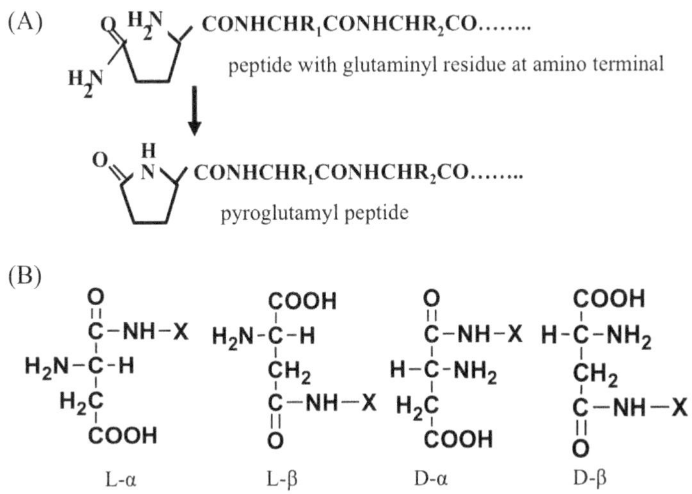

FIGURE 3.1 Structures of pyroglutamyl peptides and racemized and isomerized aspartyl dipeptides.

(A) Formation of pyroglutamyl peptide from peptides with glutaminyl residue at amino terminal. (B) Presence of four isomers of aspartyl dipeptides.

PyroGlu peptides were also found in an exhaustive protease digest of wheat gluten hydrolysate (Sato et al., 1998). To liberate the constituent amino acids, wheat gluten hydrolysate was further digested by pronase E, leucine aminopeptidase, prolidase, and carboxypeptidases A and W. Most amino acids were liberated from the wheat gluten hydrolysate, except for Glu, Gly, and Pro. The nonliberated amino acids were distributed in the pyroGlu peptide, indicating that short-chain pyroGlu peptides resist endoproteinases and exopeptidases. PyroGlu peptides larger than tripeptides are degraded by carboxypeptidases and converted to pyroGlu di- and tripeptides (Chen et al., 2019; Miyauchi et al., 2022). The peptide bond between the carboxy group of pyroGlu and the amino group of the amino acid residue can be cleaved by pyroglutamate aminopeptidase (Higaki-Sato et al., 2003; Kiyono et al., 2013), the activity of which is not very high in the digestive tract (Szewczuk & Kwiatkowska, 1970). Indeed, pyroGlu dipeptides, such as pyroGlu−Gln and pyroGlu−Leu, are stable in crude extracts of the small intestine and blood plasma, while normal peptides, such as Ser−Gln, are rapidly degraded (Miyauchi et al., 2022).

After oral administration of a rice protein hydrolysate, the levels of both hydrophilic and hydrophobic pyroGlu peptides, such as pyroGlu−Gln and pyroGlu−Leu, respectively, significantly increased in the lumen of rats (Miyauchi et al., 2022). In contrast, the levels of only hydrophobic pyroGlu peptides, such as pyroGlu−Leu and pyroGlu−Val, and a pyroGlu peptide with an amino group (pyroGlu−Lys) significantly increased in the blood (Miyauchi et al., 2022). Blood levels of the three pyroGlu peptides were not very high, approximately 10–70 nM, while intestinal tissue contained higher levels of approximately 3–10 μmol/kg. It is believed that di- and tripeptides with amino and imino groups pass via PepT1 on the apical side of the enterocytes. Although the peptide transporter on the basal side of enterocytes remains unknown, it has been believed that food-derived peptides, which resist exopeptidases, pass via a transporter on the basal side of enterocytes, as a significant increase in these peptide levels in blood has been observed. In contrast, pyroGlu–His was shown to be transported via a non-Pept1 transporter on the renal brush border membrane (Skopicki et al., 1998). Thus short-chain pyroGlu peptides can pass via a transporter on the apical side of intestinal cells, but there is no transporter for hydrophilic pyroGlu peptides on the basal side of intestinal cells. Some pyroGlu peptides in the enterocytes may be transported back into the lumen. Hydrophobic pyroGlu peptides (pyroGlu–Ile, pyroGlu–Leu, and pyroGlu–Val) can pass through some transporters on the basal side of enterocytes, but their velocity is not very high. PyroGlu–Lys may pass the same transporter on the basal side to transport other di- and tripeptides with amino groups due to the presence of amino group. In human blood after ingestion of corn and wheat gluten hydrolysates, hydrophobic pyroGlu peptides, pyroGlu–Pro, pyroGlu–Leu–Pro, and pyroGlu–Gln–Pro were significantly increased (Table 3.1). To the best of our knowledge, proteins responsible for the transportation of pyroGlu peptides on the apical and basal sides have not been identified.

Racemized and isomerized aspartyl dipeptides were first detected in a porcine liver protein hydrolysate (Ejima et al., 2019). Asp–Val, Asp–Ile, Asp–Leu, and Asp–Phe were detected in the exopeptidase digest of porcine liver protein hydrolysate. LC–MS/MS analysis revealed the presence of four peaks with the same precursor and product ions but with different retention times for each aspartyl dipeptide. Compared to the synthetic peptides, the four peaks corresponded to isomers of L-α-Asp–X, L-β-Asp–X, D-α-Asp–X, and D-β-Asp–X, as shown in Fig. 3.1. Surprisingly, in the exopeptidase digest of porcine liver protein hydrolysate, the contents of unusual L- and D-aspartyl isopeptides with β peptide bonds were higher than those with α peptide bonds (Ejima et al., 2019). It has been reported that some of the L-aspartyl residues in proteins convert to the D-form and the α-peptide bond between the carboxyl group of the aspartyl residue and the amino group of the adjutant residue and are converted to β-peptide bonds in human tissues during aging (Fujii et al., 2018). The racemization and isomerization of aspartyl residues in proteins are induced via the formation succinimide as a result of the β-carboxyl group of the aspartyl residue and nitrogen in the peptide bond (Fujii et al., 2018). A similar reaction may occur during the processing of porcine liver protein hydrolysate. The presence of racemized and isomerized aspartyl peptides has not been previously reported in other protein hydrolysates. Recently, the author found that these modified peptides were present in Japanese fermented soybean paste (*miso*). The mechanism for underlying the formation of the racemized and isomerized aspartyl peptides in specific protein hydrolysates remains to be elucidated.

After administration of the porcine liver hydrolysate by sonde at 167 mg/kg body weight to rats, only the levels of the unusual forms of the aspartyl peptides (L-β-Asp−X, D-α-Asp−X, and D-β-Asp−X) were significantly increased in the blood (approximately 10−40 nM), while those of the usual form (D-α-Asp−X) did not significantly change (Table 3.1) (Ejima et al., 2019). Thus racemization and isomerization of aspartyl peptides increase bioavailability.

3.4 Biological functions of food-derived peptides in the human body

As summarized in Table 3.1, the levels of collagen-derived di- and tripeptides, dipeptides with Pro residues at the amino terminal, hydrophobic pyroGlu dipeptides, and racemized and/or isomerized aspartyl dipeptides were significantly increased in human and animal blood after oral administration of protein hydrolysates. Hydrophilic pyroGlu peptide levels were significantly increased in the lumen; however, the levels of these peptides did not significantly increase in the blood. The biological functions of the food-derived peptides in the body have also been examined.

Some human clinical trials have demonstrated that the ingestion of collagen hydrolysate shows a variety of beneficial biological functions, such as improvement of the skin condition in elderly women and enhancement of wound healing in patients with pressure ulcer (Wang, 2021). Recent animal studies have also demonstrated that oral administration of the collagen hydrolysate shows other biological functions, such as acceleration of muscle regenerative healing with less scarring after an abdominal wall injury in mice (Jimi et al., 2021) and antidepressant effects in mice (Nogimura et al., 2020). In animal experiments, oral administration of Pro−Hyp, a major food-derived collagen peptide in the human blood, exerted the same effects as those of oral administration of the collagen hydrolysate (Jimi et al., 2021; Nogimura et al., 2020), indicating that Pro−Hyp is responsible for at least some of the biological functions of the collagen hydrolysates. To elucidate the mechanism underlying the beneficial effects of Pro−Hyp, in vitro studies were performed. Pro−Hyp has been demonstrated to specifically initiate the growth of fibroblasts expressing a mesenchymal stem cell marker (p75NTR), which is associated with the improvement of wound healing by ingestion of collagen hydrolysate (Asai et al., 2020), as p75NTR-positive cells accumulate in granulation tissue and play a significant role in wound healing. Furthermore, Pro−Hyp has been demonstrated to suppress the differentiation of chondrocytes into mineralized chondrocytes (Nakatani et al., 2009) and increase the production of hyaluronic acid by fibroblasts (Ohara et al., 2010) and suppress tumor necrosis factor-α-induced inflammatory response in endothelial cells (Kouguchi et al., 2012). These in vitro activities of Pro−Hyp have been linked to the beneficial effects of collagen hydrolysates upon ingestion.

Ingestion of elastin hydrolysate has been demonstrated to improve skin and artery elasticity in human clinical trials (Shiratsuchi et al., 2013). Pro–Gly, a major food-derived elastin peptide in human blood, increased elastin production in cultured fibroblasts (Shigemura et al., 2012). Elastin forms elastic fibers and contributes to tissue elasticity. Thus, it has been suggested that Pro–Gly is at least partially responsible for the beneficial effects of ingestion of the elastin hydrolysate.

PyroGlu–Leu levels significantly increased in the lumen and blood of animals after oral administration of synthetic pyroGlu–Leu (Wada et al., 2013) and rice protein hydrolysate (Miyauchi et al., 2022). Oral administration of small amounts of pyroGlu–Leu (0.1–1.0 mg/kg body weight), which can be obtained by consumption of fermented foods and protein hydrolysates, ameliorates high-fat diet-induced obesity (Shirako et al., 2021) and gut dysbiosis (Shirako et al., 2019). Oral administration of pyroGlu–Leu increases the secretion of rattusin, a member of α-defensin (host antimicrobial peptide), into the lumen (Shirako et al., 2019), which can be associated with the amelioration of gut dysbiosis, as it has been demonstrated that host antimicrobial peptides play significant roles in maintaining proper microbiota (Guo et al., 2017). The improvement of gut microbiota by pyroGlu–Leu could be associated with improvements in many pathological events, including obesity, colitis, and hepatitis, although further studies should be performed to confirm this.

Animal experiments have suggested that the oral administration of porcine liver protein hydrolysate suppresses exercise-induced fatigue (Nakagawasai et al., 2021). Racemized and isomerized Asp–Val, Asp–Ile, Asp–Leu, and Asp–Phe levels were significantly increased in the blood of rats after oral administration of the porcine liver hydrolysate (Ejima et al., 2019). Among these peptides, intraperitoneal injection of L-β-Asp–Leu and D-α-Asp–Phe exerted antifatigue effects at a low concentration (0.03 mg/kg) in the same animal model (Nakagasai et al., 2021). Thus these racemized and isomerized peptides are at least partially responsible for the beneficial activity of the porcine liver protein hydrolysate.

3.5 Conclusion and future prospect

To produce peptides as functional food ingredients, various food proteins have been digested by food-grade proteases and their properties have been studied. Numerous peptides with different lengths and sequences were present in such preparations. It has been assumed that these peptides can pass through the intestinal epithelium via three routes: peptide transporters on the basal and apical sides of enterocytes, the paracellular gap between enterocytes, and endocytosis/exocytosis of intestinal cells (Xu et al., 2019). The author agrees that oligopeptides potentially pass through the intestinal epithelium via these routes. However, most peptides in enzymatic hydrolysates are completely degraded into amino acids by in vitro exopeptidase digestion (Chen et al., 2019; Miyauchi et al., 2022) and in vivo digestion in the rat lumen (Miyauchi et al., 2022). Caco-2 monolayers are extensively used to predict peptide bioavailability. Using this model, some peptides larger than tripeptides were reported to pass through the Caco-2 monolayer, while no significant increase in the peptide levels in blood circulation upon ingestion of protein hydrolysates has been reported. In addition, Gly–Pro is a well-known dipeptide with affinity for PepT1 on Caco-2 cells (Rohm et al., 2019). However, no significant increase in Gly–Pro levels in the human and animal blood after oral administration of food protein hydrolysates containing peptides with the Gly–Pro motifs (Asai et al., 2019; Iwai et al., 2005; Ohara et al., 2007; Shigemura et al., 2011; Shigemura et al., 2012; Shigemura, Suzuki, et al., 2018; Taga et al., 2014) and synthetic Gly–Pro (Rohm et al., 2019) was observed. These facts clearly indicate that the Caco-2 monolayer model cannot be used to predict the bioavailability of food-derived peptides because of the lack of some exopeptidases. In contrast, the Caco-2 monolayer is effective for investigating the route taken by the food-derived peptides in the blood. These facts indicate that the susceptibility of peptides to exopeptidases in the intestinal lumen, enterocytes, and blood, rather than their affinity to transporters, is crucial for the bioavailability of peptides with amino/imino groups.

The presence of Pro residues at the amino terminal and Hyp residues in peptides affect bioavailability via increase of resistance against peptidase digestion. In contrast, pyroGlu peptides, which have no amino groups, remain as intact in the intestinal lumen, but only hydrophobic pyroGlu peptides are absorbed into the blood circulation (Miyauchi et al., 2022). Thus, affinity for transporters, rather than susceptibility to exopeptidases, is crucial for the bioavailability of pyroGlu peptides. Unexpectedly, racemized and isomerized aspartyl peptides were present in the porcine liver hydrolysate (Ejima et al., 2019). These modified peptides showed high bioavailability due to resistance against exopeptidase digestion. The detailed mechanism for such modifications remains to be elucidated. The presence of other modified peptides, such as γ-glutamyl isopeptide (Kuroda & Harada, 2002) and diketopiperazine (Borthwick & Da Costa, 2017), in foods has been reported. These modified peptides display an array of tastes (bitter, astringent, metallic, and umami). However, their bioavailability and other biological functions remain to be elucidated.

A considerable number of food-derived peptides reaching the target organs of the body upon ingestion exert significant in vitro and in vivo biological activities, which have been linked to the biological response upon ingestion of protein hydrolysates and fermented foods. Most peptides in protein hydrolysates are degraded by in vitro exopeptidase digestion (Chen et al., 2019; Miyauchi et al., 2022) and their levels do not significantly increase in the lumen of rats 1 h after the oral administration of protein hydrolysate (Miyauchi et al., 2022). These peptides are rapidly degraded by exopeptidases but are present in the gastrointestinal tract in a short time. The author does not conclude that these labile peptides do not exert any biological functions. These peptides could act as ligands for proteins in the gastrointestinal cells. Alternatively, labile peptides may be converted into other stable and bioactive metabolites. Indeed, Shigemura et al. reported that a diketopiperazine is generated in the human body during digestion and absorption processed (Shigemura, Iwasaki, et al., 2018).

Conventionally, "the active peptides in foods" have been deduced by in vitro assays based on the predicted functions. However, the bioavailability of peptides identified using these techniques is very low in most cases. On the other hand, it has become clear that peptides with special structures are translocated into tissues, even when administered orally. Amounts, structure, and bioactivities of food-derived peptides in body rather than those in foods should be associated with the biological responses upon ingestion of peptides.

References

Asai, T., Takahashi, A., Ito, K., Uetake, T., Matsumura, Y., Ikeda, K., Inagaki, N., Nakata, M., Imanishi, Y., & Sato, K. (2019). Amount of collagen in the meat contained in Japanese daily dishes and the collagen peptide content in human blood after ingestion of cooked fish meat. *Journal of Agricultural and Food Chemistry*, *67*(10), 2831–2838. Available from https://doi.org/10.1021/acs.jafc.8b06896, http://pubs.acs.org/journal/jafcau.

Asai, T. T., Yoshikawa, K., Sawada, K., Fukamizu, K., Koyama, Y., Shigemura, Y., Jimi, S., & Sato, K. (2020). Mouse skin fibroblasts with mesenchymal stem cell marker p75 neurotrophin receptor proliferate in response to prolyl-hydroxyproline. *Journal of Functional Foods*, *66*. Available from https://doi.org/10.1016/j.jff.2020.103792.

Borthwick, A. D., & Da Costa, N. C. (2017). 2,5-Diketopiperazines in food and beverages: Taste and bioactivity. *Critical Reviews in Food Science and Nutrition*, *57*(4), 718–742.

Chen, L., Ejima, A., Gu, R., Lu, J., Cai, M., & Sato, K. (2019). Presence of exopeptidase-resistant and susceptible peptides in a bacterial protease digest of corn gluten. *Journal of Agricultural and Food Chemistry*, *67*(43), 11948–11954. Available from https://doi.org/10.1021/acs.jafc.9b04444.

Ejima, A., Nakamura, M., Suzuki, Y. A., & Sato, K. (2018). Identification of food-derived peptides in human blood after ingestion of corn and wheat gluten hydrolysates. *Journal of Food Bioactives*, *2*. Available from https://doi.org/10.31665/jfb.2018.2145.

Ejima, A., Yamada, K., & Sato, K. (2019). Presence of isomerized aspartic dipeptides in a porcine liver protein hydrolysate and their bioavailability upon ingestion. *Bioactive Compounds in Health and Disease.*, *2*, 155–169.

Foltz, M., Meynen, E. E., Bianco, V., Van Platerink, C., Koning, T. M. M. G., & Kloek, J. (2007). Angiotensin converting enzyme inhibitory peptides from a lactotripeptide-enriched milk beverage are absorbed intact into the circulation. *Journal of Nutrition*, *137*(4), 953–958. Available from https://doi.org/10.1093/jn/137.4.953, http://jn.nutrition.org/cgi/reprint/137/4/953.

Fujii, N., Takata, T., Fujii, N., Aki, K., & Sakaue, H. (2018). D-Amino acids in protein: The mirror of life as a molecular index of aging. *Biochimica et Biophysica Acta - Proteins and Proteomics*, *1866*(7), 840–847. Available from https://doi.org/10.1016/j.bbapap.2018.03.001, http://www.elsevier.com/locate/bbapap.

Guo, X., Li, J., Tang, R., Zhang, G., Zeng, H., Wood, R. J., & Liu, Z. (2017). High fat diet alters gut microbiota and the expression of paneth cell-antimicrobial peptides preceding changes of circulating inflammatory cytokines. *Mediators of Inflammation, 2017*. Available from https://doi.org/10.1155/2017/9474896, http://www.hindawi.com/journals/mi/.

Hanh, V. T., Shen, W., Tanaka, M., Siltari, A., Korpela, R., & Matsui, T. (2017). Effect of aging on the absorption of small peptides in spontaneously hypertensive rats. *Journal of Agricultural and Food Chemistry*, *65*(29), 5935–5943. Available from https://doi.org/10.1021/acs.jafc.7b01727, http://pubs.acs.org/journal/jafcau.

Hayashi, H., Ohno, S., Hashimoto, S., Arai, T., & Suzuki, N. (2008). Use of Foods for Specified Health Use (FOSHU) in treating hypertension. *Japanese Journal of Complementary and Alternative Medicine*, *5*(1), 37–47. Available from https://doi.org/10.1625/jcam.5.37.

Higaki-Sato, N., Sato, K., Esumi, Y., Okumura, T., Yoshikawa, H., Tanaka-Kuwajima, C., Kurata, A., Kotaru, M., Kawabata, M., Nakamura, Y., & Ohtsuki, K. (2003). Isolation and identification of indigestible pyroglutamyl peptides in an enzymatic hydrolysate of wheat gluten prepared on an industrial scale. *Journal of Agricultural and Food Chemistry*, *51*(1), 8–13. Available from https://doi.org/10.1021/jf020528i.

Hirota, T., Nonaka, A., Matsushita, A., Uchida, N., Ohki, K., Asakura, M., & Kitakaze, M. (2011). Milk casein-derived tripeptides, VPP and IPP induced NO production in cultured endothelial cells and endothelium-dependent relaxation of isolated aortic rings. *Heart and Vessels*, *26*(5), 549–556. Available from https://doi.org/10.1007/s00380-010-0096-y, link.springer.de/link/service/journals/00380/index.htm.

Iwai, K., Hasegawa, T., Taguchi, Y., Morimatsu, F., Sato, K., Nakamura, Y., Higashi, A., Kido, Y., Nakabo, Y., & Ohtsuki, K. (2005). Identification of food-derived collagen peptides in human blood after oral ingestion of gelatin hydrolysates. *Journal of Agricultural and Food Chemistry*, *53*(16), 6531–6536. Available from https://doi.org/10.1021/jf050206p.

Iwasaki, Y., Nakatogawa, M., Shimizu, A., Sato, Y., & Shigemura, Y. (2022). Comparison of gelatin and low-molecular weight gelatin hydrolysate ingestion on hydroxyproline (Hyp), Pro-Hyp and Hyp-Gly concentrations in human blood. *Food Chemistry*, *369*. Available from https://doi.org/10.1016/j.foodchem.2021.130869.

Jimi, S., Koizumi, S., Sato, K., Miyazaki, M., & Saparov, A. (2021). Collagen-derived dipeptide ProHyp administration accelerates muscle regenerative healing accompanied by less scarring after wounding on the abdominal wall in mice. *Scientific Reports*, *11*.

Kiyono, T., Hirooka, K., Yamamoto, Y., Kuniishi, S., Ohtsuka, M., Kimura, S., Park, E. Y., Nakamura, Y., & Sato, K. (2013). Identification of pyroglutamyl peptides in Japanese rice wine (sake): Presence of hepatoprotective pyroGlu-Leu. *Journal of Agricultural and Food Chemistry*, *61*(47), 11660–11667. Available from https://doi.org/10.1021/jf404381w.

Kouguchi, T., Ito, A., Iwai, K., Shimizu, M., Takahata, Y., Suzuki, T., Morimatsu, F., & Tanabe, S. (2012). Chicken collagen hydrolysate-derived peptides inhibit tumor necrosis factor-α-induced inflammatory response in endothelial cells. *Food Science and Technology Research*, *18*(5), 667–671. Available from https://doi.org/10.3136/fstr.18.667, https://www.jstage.jst.go.jp/article/fstr/18/5/18_667/_pdf.

Kuroda, M., & Harada, T. (2002). Distribution of γ-glutamyl-β-alanylhistidine isopeptide in the macromolecular fractions of commercial meat extracts and correlation with the color of the macromolecular fractions. *Journal of Agricultural and Food Chemistry*, *50*(7), 2088–2093. Available from https://doi.org/10.1021/jf0111211.

Matsui, T., Tamaya, K., Seki, E., Osajima, K., Matsumoto, K., & Kawasaki, T. (2002). Absorption of Val-Tyr with in vitro angiotensin I-converting enzyme inhibitory activity into the circulating blood system of mild hypertensive subjects. *Biological and Pharmaceutical Bulletin*, *25*(9), 1228–1230. Available from https://doi.org/10.1248/bpb.25.1228.

Mine, Y., Li-Chan, E., & Jiang, B. (2010). *Bioactive proteins and peptides as functional foods and nutraceuticals*. Wiley-Blackwell. Available from http://onlinelibrary.wiley.com/book/10.1002/9780813811048, 10.1002/9780813811048.

Miyauchi, S., Kajiwara, S., & Sato, K. (2022). Metabolic fate of peptides in a rice protein hydrolysate in rat intestine and blood after oral administration. *Journal of Food Bioactives*, *20*, 40–55. Available from https://doi.org/10.31665/JFB.2022.18327.

Nakagawasai, O., Yamada, K., Sakuma, W., Takahashi, K., Odaira, T., Yamagata, R., Nemoto, W., Ejima, A., Sato, K., & Tan-No, K. (2021). A novel dipeptide derived from porcine liver hydrolysate induces recovery from physical fatigue in a mouse model. *Journal of Functional Foods*, *76*. Available from https://doi.org/10.1016/j.jff.2020.104312, http://www.elsevier.com/wps/find/journaldescription.cws_home/717426/description#description.

Nakamura, Y., Yamamoto, N., Sakai, K., Okubo, A., Yamazaki, S., & Takano, T. (1995). Purification and characterization of angiotensin I-converting enzyme inhibitors from sour milk. *Journal of Dairy Science*, *78*(4), 777–783. Available from https://doi.org/10.3168/jds.S0022-0302(95)76689-9.

Nakatani, S., Mano, H., Sampei, C., Shimizu, J., & Wada, M. (2009). Chondroprotective effect of the bioactive peptide prolyl-hydroxyproline in mouse articular cartilage in vitro and in vivo. *Osteoarthritis and Cartilage*, *17*(12), 1620–1627. Available from https://doi.org/10.1016/j.joca.2009.07.001.

Nogimura, D., Mizushige, T., Taga, Y., Nagai, A., Shoji, S., Azuma, N., Kusubata, M., Adachi, S. I., Yoshizawa, F., & Kabuyama, Y. (2020). Prolyl-hydroxyproline, a collagen-derived dipeptide, enhances hippocampal cell proliferation, which leads to antidepressant-like effects in mice. *FASEB Journal*, *34*(4), 5715–5723. Available from https://doi.org/10.1096/fj.201902871R, https://onlinelibrary.wiley.com/journal/15306860.

Ohara, H., Iida, H., Ito, K., Takeuchi, Y., & Nomura, Y. (2010). Effects of Pro-Hyp, a collagen hydrolysate-derived peptide, on hyaluronic acid synthesis using in vitro cultured synovium cells and oral ingestion of collagen hydrolysates in a guinea pig model of osteoarthritis. *Bioscience, Biotechnology, and Biochemistry*, *74*(10), 2096–2099. Available from https://doi.org/10.1271/bbb.100193, http://www.jstage.jst.go.jp/article/bbb/74/10/2096/_pdf.

Ohara, H., Matsumoto, H., Ito, K., Iwai, K., & Sato, K. (2007). Comparison of quantity and structures of hydroxyproline-containing peptides in human blood after oral ingestion of gelatin hydrolysates from different sources. *Journal of Agricultural and Food Chemistry*, *55*(4), 1532–1535. Available from https://doi.org/10.1021/jf062834sJapan, http://pubs.acs.org/doi/pdf/10.1021/jf062834s.

Osawa, Y., Mizushige, T., Jinno, S., Sugihara, F., Inoue, N., Tanaka, H., & Kabuyama, Y. (2018). Absorption and metabolism of orally administered collagen hydrolysates evaluated by the vascularly perfused rat intestine and liver in situ. *Biomedical Research (Japan)*, *39*(1), 1–11. Available from https://doi.org/10.2220/biomedres.39.1, https://www.jstage.jst.go.jp/article/biomedres/39/1/39_1/_pdf/-char/en.

Panyayai, T., Ngamphiw, C., Tongsima, S., Mhuantong, W., Limsripraphan, W., Choowongkomon, K., & Sawatdichaikul. (2019). PeptideDB: A web application for new bioactive peptides from food protein. *Heliyon, 5*.

Rohm, F., Daniel, H., & Spanier, B. (2019). Transport versus hydrolysis: Reassessing intestinal assimilation of di- and tripeptides by LC–MS/MS analysis. *Molecular Nutrition & Food Research, 63*(21). Available from https://doi.org/10.1002/mnfr.201900263.

Sato, K. (2020). Perspective metabolic fate and bioavailability of food-derived peptides: Are normal peptides passed through the intestinal layer to exert biological effects via proposed mechanisms? *Journal of Agricultural and Food Chemistry, 70*, 1461–1466.

Sato, K., Nisimura, R., Suzuki, Y., Motoi, H., Nakamura, Y., Ohtsuki, K., & Kawabata, M. (1998). Occurrence of indigestible pyroglutamyl peptides in an enzymatic hydrolysate of wheat gluten prepared on an industrial scale. *Journal of Agricultural and Food Chemistry, 46*(9), 3403–3405. Available from https://doi.org/10.1021/jf980603i, http://pubs.acs.org/journal/jafcau.

Seki, E., Osajima, K., Matsufuji, H., Matsui, T., & Osajima, Y. (1995). Val-Tyr, an angiotensin I converting enzyme inhibitor from sardines that have resistance to gastrointestinal proteases. *Nippon Nogeikagaku Kaishi, 69*(8), 1013–1020. Available from https://doi.org/10.1271/nogeikagaku1924.69.1013.

Shigemura, Y., Akaba, S., Kawashima, E., Park, E. Y., Nakamura, Y., & Sato, K. (2011). Identification of a novel food-derived collagen peptide, hydroxyprolyl-glycine, in human peripheral blood by pre-column derivatisation with phenyl isothiocyanate. *Food Chemistry, 129*(3), 1019–1024. Available from https://doi.org/10.1016/j.foodchem.2011.05.066.

Shigemura, Y., Iwasaki, Y., Tateno, M., Suzuki, A., Kurokawa, M., Sato, Y., & Sato, K. (2018). A pilot study for the detection of cyclic prolyl-hydroxyproline (Pro-Hyp) in human blood after ingestion of collagen hydrolysate. *Nutrients, 10*(10). Available from https://doi.org/10.3390/nu10101356.

Shigemura, Y., Nakaba, M., Shiratsuchi, E., Suyama, M., Yamada, M., Kiyono, T., Fukamizu, K., Park, E. Y., Nakamura, Y., & Sato, K. (2012). Identification of food-derived elastin peptide, prolyl-glycine (pro-gly), in human blood after ingestion of elastin hydrolysate. *Journal of Agricultural and Food Chemistry, 60*(20), 5128–5133. Available from https://doi.org/10.1021/jf300497p.

Shigemura, Y., Suzuki, A., Kurokawa, M., Sato, Y., & Sato, K. (2018). Changes in composition and content of food-derived peptide in human blood after daily ingestion of collagen hydrolysate for 4 weeks. *Journal of the Science of Food and Agriculture, 98*(5), 1944–1950. Available from https://doi.org/10.1002/jsfa.8677, http://onlinelibrary.wiley.com/journal/10.1002/(ISSN)1097-0010.

Shimizu, K., Sato, M., Zhang, Y., Kouguchi, T., Takahata, Y., Morimatsu, F., & Shimizu, M. (2010a). Molecular size of collagen peptide reverses the permeability of Caco-2 cells. *Bioscience, Biotechnology, and Biochemistry, 74*(5), 1123–1125. Available from https://doi.org/10.1271/bbb.100015, http://www.jstage.jst.go.jp/article/bbb/74/5/1123/_pdf.

Shimizu, K., Sato, M., Zhang, Y., Kouguchi, T., Takahata, Y., Morimatsu, F., & Shimizu, M. (2010b). The bioavailable octapeptide Gly-Ala-Hyp-Gly-Leu-Hyp-Gly-Pro stimulates nitric oxide synthesis in vascular endothelial cells. *Journal of Agricultural and Food Chemistry, 58*(11), 6960–6965. Available from https://doi.org/10.1021/jf100388w.

Shirako, S., Kojima, Y., Hasegawa, T., Yoshikawa, T., Matsumura, Y., Ikeda, K., Inagaki, N., & Sato, K. (2021). Identification of short-chain pyroglutamyl peptides in Japanese salted fermented soy paste (miso) and their anti-obesity effect. *Journal of Food Bioactives*, *12*, 129–139.

Shirako, S., Kojima, Y., Tomari, N., Nakamura, Y., Matsumura, Y., Ikeda, K., Inagaki, N., & Sato, K. (2019). Pyroglutamyl leucine, a peptide in fermented foods, attenuates dysbiosis by increasing host antimicrobial peptide. *npj Science of Food*, *3*(1). Available from https://doi.org/10.1038/s41538-019-0050-z, https://www.nature.com/npjscifood/about.

Shiratsuchi, E., Nakaba, M., Shigemura, Y., Yamada, M., & Sato, K. (2013). *Fish-elastin hydrolysate: Development and impact on the skin and blood vessels* (pp. 467–486). Wiley. Available from 10.1002/9781118375082.ch23.

Shobako, N., Ishikado, A., Ogawa, Y., Sono, Y., Kusakari, T., Suwa, M., Matsumoto, M., & Ohinata, K. (2019). Vasorelaxant and antihypertensive effects that are dependent on the endothelial NO system exhibited by rice bran-derived tripeptide. *Journal of Agricultural and Food Chemistry*, *67*(5), 1437–1442. Available from https://doi.org/10.1021/acs.jafc.8b06341, http://pubs.acs.org/journal/jafcau.

Skopicki, H. A., Fisher, K., Zikos, D., Flouret, G., Bloch, R., Kubillus, S., & Peterson, D. R. (1998). Carrier-mediated transport of pyroglutamylhistidine in renal brush border membrane vesicles. *The American Journal of Physiology*, *255*.

Sumantha, A., Larroche, C., & Pandey, A. (2006). Microbiology and industrial biotechnology of food-grade proteases: A perspective. *Food Technology and Biotechnology*, *44*(2), 211–220. Available from http://public.carnet.hr/ftbrfd/44-211.pdf.

Suzuki, Y., Motoi, H., & Sato, K. (1999). Quantitative analysis of pyroglutamic acid in peptides. *Journal of Agricultural and Food Chemistry*, *47*(8), 3248–3251. Available from https://doi.org/10.1021/jf990003z.

Szewczuk, A., & Kwiatkowska, J. (1970). Pyrrolidonyl peptidase in animal, plant and human tissues: Occurrence and some properties of the enzyme. *European Journal of Biochemistry*, *15*(1), 92–96. Available from https://doi.org/10.1111/j.1432-1033.1970.tb00980.x.

Taga, Y., Kusubata, M., Ogawa-Goto, K., & Hattori, S. (2014). Highly accurate quantification of hydroxyproline-containing peptides in blood using a protease digest of stable isotope-labeled collagen. *Journal of Agricultural and Food Chemistry*, *62*(50), 12096–12102. Available from https://doi.org/10.1021/jf5039597, http://pubs.acs.org/journal/jafcau.

Tanaka, M., Matsui, T., Ushida, Y., & Matsumoto, K. (2006). Vasodilating effect of di-peptides in thoracic aortas from spontaneously hypertensive rats. *Bioscience, Biotechnology, and Biochemistry*, *70*(9), 2292–2295. Available from https://doi.org/10.1271/bbb.60119, http://www.jstage.jst.go.jp/article/bbb/70/9/2292/_pdf.

Toldrá, F., Reig, M., Aristoy, M. C., & Mora, L. (2018). Generation of bioactive peptides during food processing. *Food Chemistry*, *267*, 395–404.

Van der Rest, M., & Garrone, R. (1991). Collagen family of proteins. *FASEB Journal*, *5*(13), 2814–2823. Available from https://doi.org/10.1096/fasebj.5.13.1916105.

Wada, S., Sato, K., Ohta, R., Wada, E., Bou, Y., Fujiwara, M., Kiyono, T., Park, E. Y., Aoi, W., Takagi, T., Naito, Y., & Yoshikawa, T. (2013). Ingestion of low dose pyroglutamyl leucine improves dextran sulfate sodium-induced colitis and intestinal microbiota in mice. *Journal of Agricultural and Food Chemistry*, *61*(37), 8807–8813. Available from https://doi.org/10.1021/jf402515a.

Wang, H. (2021). A review of the effects of collagen treatment in clinical studies. *Polymers, 13*(22). Available from https://doi.org/10.3390/polym13223868.

Wu, J. (2020). A novel angiotensin converting enzyme 2 (ACE2) activating peptide: A reflection of 10 years of research on a small peptide Ile-Arg-Trp (IRW). *Journal of Agricultural and Food Chemistry, 68*(49), 14402–14408. Available from https://doi.org/10.1021/acs.jafc.0c05544.

Xu, Q., Yan, X., Zhang, Y., & Wu, J. (2019). Current understanding of transport and bioavailability of bioactive peptides derived from dairy proteins: A review. *International Journal of Food Science and Technology, 54*(6), 1930–1941. Available from https://doi.org/10.1111/ijfs.14055, http://onlinelibrary.wiley.com/journal/10.1111/(ISSN)1365-2621.

Yokoyama, K., Chiba, H., & Yoshikawa, M. (1992). Peptide inhibitors for angiotensin I-converting enzyme from thermolysin digest of dried bonito. *Bioscience, Biotechnology, and Biochemistry, 56*(10), 1541–1545. Available from https://doi.org/10.1271/bbb.56.1541.

CHAPTER

4 Dynamic simulation of food protein digestion and bioactive peptide release

Begoña Bartolome and M. Victoria Moreno-Arribas

CSIC, Spanish National Research Council, CIAL-Institute of Food Science and Research, Madrid, Spain

4.1 Introduction

Bioactive peptides are fragments of specific proteins that exhibit biological activities, including antihypertensive, antioxidant, antiinflammatory, immunomodulatory, hypolipemic, antidiabetic, anticancer, and antiadhesive (Xu et al., 2019). They are composed of 2–20 amino acid residues with a molecular weight of less than 6000 Da and are inactive when encrypted in the parental protein (Nielsen et al., 2018). Of special interest are the bioactive peptides that are released from food proteins during their passage through the digestive tract as they exert health-promoting properties (Fernandez-Tome & Hernandez-Ledesma, 2020). Food-derived peptides can promote local effects through interaction with digestive receptors and cells, or they can lead to systemic effects after being absorbed and reaching target tissues in an active form (Miner-Williams et al., 2014). Therefore studying the transformations that food proteins undergo in the gastrointestinal tract is essential for elucidating peptide release and its potential bioactivity and impact on human health.

Simulation tools or models are used to address, from a scientific perspective, fundamental factors, such as oral and gastrointestinal digestion, the absorption of nutrients or other compounds, and the bidirectional interaction of food with the colonic microbiota, as well as the mechanisms involved in these factors. Clinical studies have been considered the ideal approach for nutritional research because of their representativeness. However, they are conditioned by their observational nature and by the difficulty of differentiating the study effect from other lifestyle-related variables. In addition, clinical studies have technical, ethical, and economic limitations and are linked to low reproducibility due to interindividual variations (Dupont et al., 2019; Mackie et al., 2020). Animal models have also been widely used; however, they generally involve the death of the animal or surgery to access its gastrointestinal tract for cannula implantation, which has a significant impact on its quality of life. Furthermore, these models, although fundamental in some fields, such as toxicology, have often been questioned in food research due to their differences from human digestion (Brodkorb et al., 2019). As a result,

Protein Digestion-Derived Peptides. DOI: https://doi.org/10.1016/B978-0-443-19141-1.00004-2

there has been a growing interest in developing in vitro and in silico digestion models that can robustly and accurately reproduce the human gastrointestinal tract and its functions.

The simplest in vitro models are static systems consisting of independent bioreactors that simulate oral, stomach, intestinal, or colonic conditions. They have been widely used, especially for studying the influence of different digestion parameters, food structure, component interactions, etc. on nutrient digestibility (Alegría et al., 2015). However, since digestion is a dynamic process, the limitations of these static models (e.g., no flow between the bioreactors, constant concentrations of enzymes and bile, substrate depletion for the microbiota, etc.) may reduce the accuracy of the extrapolation of the results (Dupont et al., 2019). In this sense, dynamic models correct the restrictions of the static models by including physicochemical and mechanical processes and temporal changes in luminal conditions as occur in vivo (Dupont et al., 2019). Dynamic models simulate digestion by introducing characteristics, such as the time variable and the flow of digestates between digestion stages (e.g., gastric or intestinal emptying). In addition, they include variations in the pH of each digestion stage and sequential addition of digestive enzymes, simulating the gradual segregation of these enzymes in the physiological process (Dupont et al., 2019).

This chapter recompiles the related physiological facts, developments, and applications of gastrointestinal digestion dynamic simulators for the study of food protein digestion and peptide release. Firstly, the different stages of the digestion process (oral, gastric, intestinal, and colonic) and their physiological conditions, which are the basis of the in vitro digestion models, are described (Section 4.2). After introducing the fundamental physiological processes, the main static, semi-dynamic, and dynamic models that have been developed to mimic partly or completely the gastrointestinal digestion process are recompiled (Section 4.3). Thereafter, applications of dynamic simulators to monitor protein digestion and peptide release from different sources and food matrices are detailed, with a summary of the operating conditions set for protein studies and the peptide outcomes (Section 4.4). Recapitulation and future perspectives regarding the usefulness of dynamic models in the research of protein digestion-derived peptides are included in Section 4.5.

4.2 Physiology of human digestion

The process of human digestion is a physicochemical, enzymatic and mechanical process that transforms ingested food to supply the organism. This process is divided into four main stages that occur consecutively and serially: the oral stage, which occurs in the mouth; the gastric stage, which takes place in the stomach; the intestinal stage, which occurs along the small intestine; and the colonic stage, mediated by the colonic microbiota, which takes place in the large intestine or colon.

Although the oral cavity is one of the most accessible parts of the anatomy of the human digestive tract and is particularly relevant to the mechanical and structural degradation of food during oral digestion, it is one of the least understood processes. The complexity of the variables affecting oral digestion, which include all the factors involved in mastication, salivary secretion (both in terms of production levels and composition) and bolus formation, implies significant interindividual variability, which in turn is conditioned by age, dentition, musculature and oral microbiota, but also depends on the nature of the food and drink ingested, as well as intake patterns. In general, despite the importance of this stage in gastrointestinal digestion, its simulation is oversimplified or even suppressed (Peyron & Woda, 2016). For example, harmonized static gastrointestinal digestion protocols recommend simulating the oral stage of digestion by grinding the solid food or product to be studied to a 1:1 dilution or pasty texture and adding the digesting enzymes (Brodkorb et al., 2019). In this way, the simulation avoids the impact of the physiological chewing process and the possible interaction of the product with the oral microbiota. On the other hand, there are several chewing simulators that are mainly mechanical and supplemented with in silico models, but which, however, ignore the effect of the oral microbiota or use salivary enzyme ratios and times that are not representative of digestion (Morell et al., 2014; Peyron & Woda, 2016), which is also inadequate when using a full digestion approach. Some simulation models even include ex vivo oral processing (Mandalari et al., 2014), which is more representative, but has limitations when studying products whose food safety is not guaranteed and depends on the presence of volunteers to perform the simulation. Therefore the simulation of oral digestion remains a challenge for the time being.

Gastric digestion is relatively compartmentalized. In the upper part of the stomach, called the "fundus," the pH is relatively high (within the gastric environment) and the food bolus is temporarily stored until it is moved by peristaltic movements into the body of the stomach, where mixing of the food and gastric juices takes place. In the distal part of the stomach, which includes the antrum and the pylorus, peristaltic movements are more intense and these peristaltic forces perform a sieving effect on the particles prior to gastric emptying (Boland, 2016). The result is a solid-liquid system in which the lower zone reduces the particle size to about 2–3 mm, while in the upper zone, gastric juices are remixed with particles too large to pass through the pyloric valve, which is returned to the body of the stomach by a retro-propulsed flow until they reach the appropriate size (Thomas, 2006). The mixing and fragmentation of stomach contents is aided by peristaltic movements of the stomach walls, the speed and pressure of which vary depending on the physical properties of the gastric contents and the area of the stomach (Bornhorst & Singh, 2014). These movements begin in the body of the stomach, where little agitation occurs and phase separation may even take place, with a lipid layer over the rest of the gastric contents (Mackie et al., 2020) and intensify at the antrum and pylorus, allowing the gastric contents to pass through the pyloric sphincter into the duodenum.

In addition to mechanical degradation, the acid pH and the enzymatic action of the gastric juice enable the initiation of the protein and lipid digestion process. The acidic pH of the stomach is due to the high concentration of hydrogen ions secreted by the parietal cells of the stomach together with chloride ions (generally described as HCl) (Isackson & Ashley, 2014). This low-pH environment favors the degradation of stomach contents and protects the gastrointestinal tract from microorganisms susceptible to this acidic environment that may have been introduced along with the food (Howell et al., 2010; Lombardo et al., 2010). With regard to the distribution of pH in the gastric environment and its evolution throughout digestion, data collected in in vivo studies confirm that gradient starting from a pH in the fasting state and after the arrival of the food bolus with a higher pH, there is a progressive drop in pH during digestion, which can be found in estimated average pH values of between 4 and 5.5 when the average digestion time is reached, reducing until reaching pH values of 2 when digestion has elapsed to 95% (Sams et al., 2016) and reaching lower values at the completion of gastric emptying. This process, however, is subject to a high interindividual variability and follows a different evolution depending on the food ingested (Sams et al., 2016).

Enzymatic digestion in the gastric stage proceeds sequentially following the lowering of the pH of the stomach contents. In the first part, carbohydrate digestion takes place. Gastric lipase is activated and reaches its optimum at pH values of between 4 and 5.4 (Mackie et al., 2020), allowing 10%–25% of the triglyceride content present in the stomach to be hydrolyzed (Sams et al., 2016). In addition, it is important to bear in mind that human gastric lipase is unique in having a high specificity for the sn-3 binding sites of triglycerides, producing diglycerides and free fatty acids, which differentiates it from other gastric lipases of non-human origin (Sams et al., 2016), a factor of particular relevance in the in vitro simulation of lipid digestion. Similarly, when the pH of the stomach contents reaches values below 5.5, protein digestion begins. Gastric pepsin, whose optimal degradation pH is between 1.8 and 4.2, is secreted by the parietal cells of the stomach in its zymogen form (pepsinogen), is activated by gastric HCl, and hydrolyzes food proteins to produce peptides and/or amino acids.

The study of gastric emptying is crucial in the digestion process. It starts simultaneously with digestion (Clegg & Shafat, 2010) and increases when the stomach contents are sufficiently degraded (Boland, 2016). The liquid phase, semisolid phase, and smaller particles (less than 2–3 mm) pass through the pyloric valve into the duodenum (Mulet-Cabero et al., 2020). Emptying is the result of the balance between the net effect of the breakthrough forces within the stomach and the resistance of the stomach-duodenal narrowing (pyloric valve). The main driving force is based on the motility of the stomach wall, which has two main phases of movement: the feeding phase, with continuous contractions, and the basal phase, comprising resting phases and strong contractions (Schulze, 2006). Stomach contraction frequency is approximately 2.5–3 contractions per minute and is performed at an estimated linear velocity of

1.5–3 mm/s, taking about one minute to complete the fundus-pylorus course (Bornhorst & Singh, 2014). In terms of stomach contents, liquids, digestible solids, and indigestible solids are emptied differently. Liquids and particles with a size of 1–2 mm are emptied during the digestion phase, while larger or indigestible solids are emptied during the contractions of the basal phase (Hellström et al., 2006).

The gastric emptying process also varies depending on the properties of the ingested food and its quantity, so that intakes with more viscous content, a larger particle size, a higher caloric density, or a high lipid content lead to a slower rate of passage of gastric contents into the duodenum (Mackie et al., 2020). In addition to food-dependent factors and biological factors concerning the individual (such as age or gender, or hormonal regulation), some environmental influences, such as eating food at different temperatures, influence gastric emptying (Kong & Singh, 2008). In general, a deeper understanding of the influence of food properties, gastric motility, and hormonal responses to ingestion is needed to fully explain gastric emptying and, as a consequence, to accurately simulate it in vitro (Li et al., 2020).

The main function of intestinal digestion is the transformation of food components into absorbable compounds that can pass through the intestinal epithelium and enter the systemic circulation. The small intestine can be divided anatomically into three regions: the duodenum, the jejunum, and the ileum. In the duodenum, the mixing of stomach contents (chyme) with pancreatic juices and bile secretions takes place, while along the jejunum and the ileum, enzymatic digestion and absorption of nutrients mainly occurs (Boland, 2016).

When the contents coming from the stomach reach the duodenum they undergo a change in luminal physicochemical conditions—in particular, an increase in pH from gastric values to pH 6.8–9.0 in the duodenum, due to the high buffering effect of pancreatic juice (Bohn, 2014). In addition, pancreatic secretions include pancreatin, which consists of a conglomerate of enzymes secreted by the pancreas as zymogens. These, through the action of pancreatic enteropeptidase, are transformed into their active forms, which include trypsin, chymotrypsin, elastase, carboxypeptidases, pancreatic α-amylase, lipase, cholesterolase, phospholipases, etc. Bile acids or bile salts from the liver, which are essential for lipid emulsification, also reach the duodenum via the common bile duct (Boland, 2016). The chyme that reaches the duodenum is a complex mixture of water and a wide variety of compounds from the diet, including, depending on intake, carbohydrates, proteins and peptides, lipids, minerals, vitamins, and other compounds, such as polyphenols, terpenes, or other xenobiotics. The protein content of the chyme can be made up of proteins, peptides of different sizes, and free amino acids, depending on the intake and the previous degradation of the protein content during the gastric stage.

At the intestinal level, the digestion process continues through the action of proteases in the pancreatic juice. These enzymes are secreted in pro-enzymatic forms (zymogens), which include both endopeptidases (such as trypsin, chymotrypsin, collagenase, and elastase) and exopeptidases (carboxypeptidases A and B) (Campbell et al., 2019). The enzymatic action is triggered by enterokinase or enteropeptidase secreted by duodenal enterocytes and transforms trypsinogen into trypsin, which triggers a chain activation of the remaining pancreatic zymogens (Campbell et al., 2019). Digestion of the generated oligopeptides is completed in the brush border of enterocytes, which have peptidases (aminopeptidases, dipeptidases, and tripeptidases) that hydrolyze the oligopeptides to release free amino acids, dipeptides, or tripeptides absorbable by enterocytes.

Finally, the large intestine begins at the ileocaecal valve and ends at the anus. It consists of three main regions, the ascending, transverse, and descending colon, plus the sigmoid colon and the rectum, where feces accumulate before being expelled through the anal sphincter. The main function of the colon is the reabsorption of water and electrolytes from the luminal contents, which supports the body's water and electrolyte balance, although other compounds, such as metabolites from the colonic microbiota and some minerals and vitamins, may also be reabsorbed. As discussed in the previous sections, although the human microbiota is distributed throughout the body, it is particularly concentrated in the colon, where it is highly functional and diverse, giving colonic fermentation important roles in overall digestion and health maintenance. The gastrointestinal epithelium is covered by a protective mucosal layer (Van den Abbeele et al., 2012), which, at the colonic level, is crucial in reference to the attachment of colonic microbiota and their interaction with the host, through the so-called "host-microbiota interaction." In addition, the colonic region has physiological characteristics that are suitable for maintaining the microbiota, such as anaerobiosis, a temperature suitable for the growth of mesophilic microorganisms, and the availability of nutrients, which allows the microbiota to accumulate and develop, especially in this region of the body.

Although the functions of this microbiota are highly conserved among individuals, the composition of each person's microbiota is characterized by a specific combination of bacterial species due to interindividual and intraindividual variations throughout life (Rinninella et al., 2019). The dominant gut microbial phyla are *Firmicutes*, *Bacteroidetes*, *Actinobacteria*, *Proteobacteria*, *Fusobacteria*, and *Verrucomicrobia*, with *Firmicutes* and *Bacteroidetes* accounting for 90% of the gut microbiota. The phylum *Bacteroidetes* is mainly constituted by *Bacteroides fragilis*, *Bacteroides ovatus*, and *Bacteroides caccae*, while the phylum *Firmicutes* is essentially represented by species belonging to the *Clostridia* of groups XIVa and IV (*Ruminococcus* and *Faecalibacterium prausnitzii*). There are other subdominant groups (*Enterococcus*, *Lactobacillus*, *Bifidobacterium*, *Peptostreptococcus*, *Escherichia*, etc.) or transient microorganisms (yeasts, etc.) that complete the composition of the microbiota (Rajilić-Stojanović et al., 2013). Although each digestion stage (oral, gastric, intestinal, and colonic stage) shows particular physiological conditions, it is probably the colonic digestion stage whose physiological environment is more complex and difficult to simulate in vitro.

4.3 Gastrointestinal digestion models

In vitro systems for simulating human digestion try to mimic physiological conditions, considering pH, digestion time, and digestive enzymes, among other factors. They have been widely applied in recent decades for research related to food and drug digestion, and as a preliminary step for the evaluation of compounds whose food safety is not guaranteed, for example, novel food ingredients or components. However, the design and implementation of fully representative in vitro simulation systems are biologically and technologically challenging, given the inherent complexity of the human gastrointestinal tract in terms of anatomical structures, enzymatic specificity, and hormonal regulation, as well as complex gastrointestinal motility (Li et al., 2020). Furthermore, it should be noted that a model, by definition, is a simplified representation of a process. There are currently several validated and sufficiently representative digestion models that can be classified into static, semidynamic and dynamic models, which allow the study of the most relevant factors of the transformations undergone by food (or other components ingested with it) in the digestive tract to be approached from different scientific perspectives. These models can also be complemented by other in vitro and in vivo studies that simulate relevant physiological processes in the context of digestion and provide a comprehensive view of the process and its implications for health.

4.3.1 Static models

In general, static models of human digestion consist of test vessels, such as tubes or flasks that act as bioreactors, simulating digestion conditions without considering the time variable in terms of fluid transport (Mulet-Cabero et al., 2020). They are often the most simplified models but generally reproduce changes in pH, temperature, and total digestion time and include the main digestive enzymes in the simulation (Minekus et al., 2014). These models are reproducible and allow a wide range of variables to be controlled, so they can often be used to run a large number of simulations in parallel as a preliminary step to other studies. They are also faster and cheaper than in vivo studies and do not carry their ethical restrictions. These characteristics make static simulation models interesting tools for hypothesizing changes in the gastrointestinal tract and for establishing the digestibility and bioaccessibility of nutrients and phytochemicals, as well as pharmaceuticals (Minekus et al., 2014). On the other hand, limitations of the static digestion models can derive from the following characteristics: (1) no gradual exchange of fluids between digestion steps; (2) no simulation of gradual variations in pH or other variables during digestion steps; and (3) constant ratios of enzymes and electrolytes to the product to be digested.

Within these static models, there are several proposals that consider different conditions for each stage of digestion as they have been developed by research teams that approached the characteristics represented in the model according to the needs of their field of study (Biehler & Bohn, 2010; Miller et al., 1981; Versantvoort et al., 2005; Yu & Yang, 2019). However, in recent years, a harmonized food-driven model of gastrointestinal digestion has been proposed. This static model, created and published by the INFOGEST network, was presented in 2014 (Minekus et al., 2014).

Given the wide variety of digestion models described so far, this consensus method aimed to unify and produce reproducible and comparable data across studies in order to deepen the understanding of food digestion and its functionality. Recently, and after several validations and comparisons with in vivo and in vitro digestion results (Egger et al., 2016; Egger et al., 2017), the protocol was revised and updated in Brodkorb et al. (2019), where more details and clarifications on the method are included, in order to make it more reproducible and facilitate comparison of the results. This digestion model has been shown to be useful for simulating the end points of each stage of gastrointestinal digestion, where it has shown good correlations with in vivo data (Brodkorb et al., 2019). However, it is not suitable for kinetic studies or studies that include the time variable. Moreover, for simulating the whole digestive tract, the INFOGEST model would require further inclusion of the colonic fermentation stage, a key step in the study of dietary fiber or phytochemicals, such as polyphenols, whose functionality depends largely on their interaction with colonic microbiota.

Static in vitro models of the colonic stage generally consists in placing the food in a bioreactor together with fecal microbiota. Incubations can be carried out with individual or pooled human fecal inocula (a homogenized mixture of several volunteers to reduce the interindividual variability associated with the colonic microbiota of individuals), although mixed cultures of microorganisms representative of the colonic stage can also be used. Incubation conditions are generally 37°C in an anaerobic environment and for short periods of time, usually less than 72 h. This time is limited, as static systems without fluid exchange are saturable, so the availability of nutrients is limited and the end products of microbial metabolism accumulate. Static colonic fermentation models allow assessment of both microbial metabolism and modulation of the microbiota, in the framework of the bidirectional interaction between microbiota and dietary substrates. The use of these static models is applied to the study of multiple dietary components, such as novel food ingredients, bioactive compounds, and potential prebiotics (Gil-Sánchez et al., 2017; Pompei et al., 2008; Rodríguez-Saavedra et al., 2023; Sánchez-Patán et al., 2015; Tamargo, Cueva, Silva, et al., 2022) with special interest in the formation of bioactive metabolites or the assessment of the selective response of microbial populations. They also allow the potential impact of probiotic strains or commensal and pathogenic species on the colonic community to be assessed (Payne et al., 2012; Vitali et al., 2012) as well as the combined effects of different food ingredients (Tamargo, Cueva, Taladrid, et al., 2022; Tamargo et al., 2023). Other applications for in vitro models of colonic fermentation are those studies that would be limited in vivo by including new products, compounds with possibly unknown effects or for which a potential toxic effect on colonic microbiota is to be ruled out (Payne et al., 2012), for example, nanomaterials (Cueva et al., 2019) or microplastics (Tamargo, Cueva, Silva, et al., 2022).

However, despite being a versatile alternative, there are currently no harmonized models available for colonic fermentations, as is the case for gastrointestinal digestion. Moreover, these colonic fermentation models should be seen as preliminary studies given their limitations and the physicochemical variations of the colonic environment during fermentation, which may affect the composition and functionality of the starting microbial community and therefore limit their representativeness (Gil-Sánchez et al., 2018).

4.3.2 Semidynamic models

Semidynamic simulation models are characterized by the inclusion of at least one of the dynamic features of the gastrointestinal tract (Mackie et al., 2020). These models are an intermediate tool between dynamic and static models. They generally require a simple infrastructure, consisting of specific but accessible equipment or systems, which facilitates their use in many laboratories. In addition, they make it possible to work with small volumes, which reduces the consumption of enzymes and other reagents compared to dynamic models. Although semidynamic simulations are carried out individually, which requires more study time than static models, they are a more physiologically relevant approach given the dynamic nature of digestion (Mackie et al., 2020).

Among other factors, pH is perhaps the first one considered for dynamization during digestion for two main reasons. On the one hand, in vivo digestion studies show that the arrival of the food bolus in the stomach produces a rapid increase in gastric pH, which gradually decreases due to the action of HCl during the digestion period. This gradual change has a considerable impact on the digestion kinetics of macronutrients, especially proteins and lipid emulsions. On the other hand, it is technically feasible to simulate this dynamic feature of digestion thanks to the availability of pH-stat devices or automatic titration systems, which are tools for automatic control of pH and fluid addition, so as to simulate the gastric pH gradient observed in vivo and the gradual secretion of gastric fluids.

Among others, there are some complex and specific models, such as the one proposed by (van den Braak et al., 2013). A compilation of the most commonly used semidynamic models with their main features can be found in (Mackie et al., 2020).

A harmonized semidynamic protocol for the simulation of digestion in the upper gastrointestinal tract has recently been proposed. The selected parameters of this model are based on in vivo data and previous experience with the static harmonized model proposed by the INFOGEST action (Brodkorb et al., 2019). The standardized protocol proposed by (Mulet-Cabero et al., 2020) attempts to combine the characteristics of existing semidynamic simulation models with the most up-to-date scientific evidence, so that a tool can be obtained that allows results to be compared between research groups and that can be used in a wide range of laboratories, as it requires simple and accessible equipment (Mackie et al., 2020).

In general terms, semidynamic models have been used to simulate conditions in the upper gastrointestinal tract. They feature some static stages and others in which they reproduce variables with dynamic characteristics. They usually suppress the oral stage or include it in a simplified and static form, but most simulate the gastric stage in detail, considering as a dynamic variable the evolution of the pH of the chyme during digestion and including the addition of acid and gastric enzyme content in a controlled manner (Mackie et al., 2020). This stage is usually followed by a static intestinal stage that simulates the passage of gastric contents into the duodenum.

4.3.3 Dynamic simulation models

Dynamic models of human digestion are relatively more complex and involve a higher economic investment than static or semidynamic models, as well as a higher maintenance level and operation by specialized technicians (Brodkorb et al., 2019). However, most dynamic models have been shown to be highly representative of the digestion of different classes of food or food ingredients. Moreover, their increased complexity allows them to adapt their simulation conditions depending on the study purposes and/or to represent the characteristics of different population groups, such as the elderly, children, or people with gastrointestinal dysfunctions (Dupont et al., 2019).

In dynamic simulation models, the biochemical environment of the gastrointestinal tract can vary continuously during digestion and physical parameters evolve over time. Furthermore, in most cases, dynamic models include different digestive environments for each stage of digestion and can function as continuous systems (Dupont et al., 2019). Fig. 4.1 depicts the three regions of the digestive tract most commonly simulated in the dynamic models: the stomach, the small intestine, and the large intestine; the latter is further divided into ascending, transverse, and descending colon. One of the main advantages of these dynamic models being able to operate continuously is the opportunity to study medium-term effects (chronic intakes) or include temporal sequences in the intakes—for example, by introducing intakes of a probiotic followed by an extract rich in phenolic compounds to consider the accumulative effects (Gil-Sánchez et al., 2020).

Globally, there are different dynamic models that allow distinct scientific approaches and they have been increasing in number in recent years (Dupont et al., 2019; Mackie et al., 2020; Silva et al., 2020; Tamargo, Cueva, Taladrid, et al., 2022; Tamargo et al., 2023). These models simulate different parameters of the gastrointestinal tract, which means that each of them has specific characteristics and different utilities depending on their design.

There are dynamic models that represent only one stage of digestion. For example, the Human Gastric Simulator (Kong & Singh, 2010) and the Dynamic Gastric Model (Mason et al., 2016; Wickham et al., 2012) represent only the

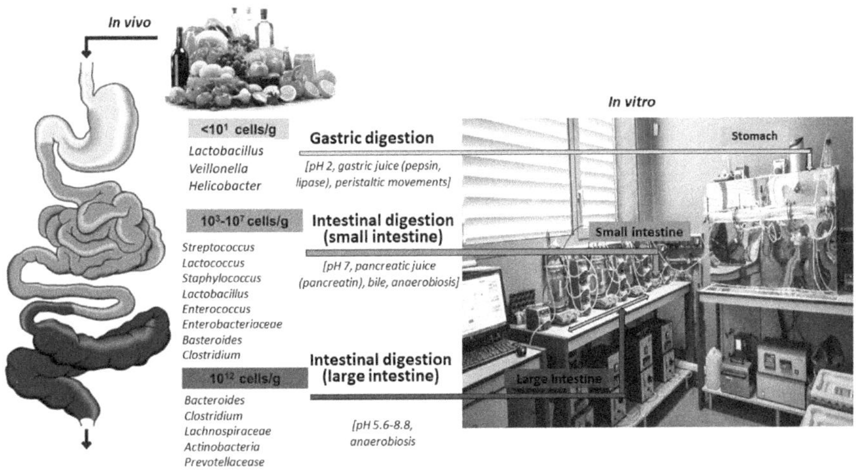

FIGURE 4.1

Main physiological facts of the human digestion to be simulated by dynamic models; as an example, in vivo–in vitro correspondence in the Simgi model (Barroso et al., 2015; Cueva et al., 2015).

gastric stage and are focused on the study of particle degradation during this stage of digestion. Other recently described dynamic simulators of the gastric stage are the Dynamic In Vitro Human Stomach (DIVHS) system, which represents the shape of the stomach, although it does not have gastric pH control (Wang et al., 2019), the Gastric Simulation Model (GSM) (Li et al., 2020), the In Vitro Mechanical Gastric (IMG) system (Barros et al., 2016), and the Artificial Gastric Digestive System (AGDS) (Liu et al., 2019).

The Artificial Colon (ARCOL) model is a monocompartmental model of colonic fermentation that dynamically represents the conditions of the human colon (Blanquet-Diot et al., 2012). This model consists of a bioreactor equipped with several types of probes to control fermentation conditions. The equipment is inoculated with a fecal solution, previously suspended in phosphate buffer and filtered. The inside of the fermenter contains a nutrient culture medium supplemented with ileal effluent. During the simulation, the fermentation medium is sequentially introduced and withdrawn, which gives it its dynamic character. Agitation, pH, and temperature are kept constant throughout the process and, by means of a dialysis system consisting of hollow fiber membranes (30 kDa limit), an optimal concentration of electrolytes in the medium is established. Previous studies have validated the composition and metabolic activity of the colonic microbiota stabilized in the model (Blanquet-Diot et al., 2012; Cordonnier et al., 2015; Fournier et al., 2022; Gérard-Champod et al., 2010; Thévenot et al., 2015).

Bicompartmental dynamic models have also been described, which generally represent the upper gastrointestinal tract, such as the ESIN (Guerra et al., 2016) or the one developed by Lesmes, which is capable of adapting to the characteristics of different population groups, including children, adults, and the elderly (Levi & Lesmes, 2014; Levi et al., 2017). Another bi-compartment dynamic model was the DIDGI (Ménard et al., 2014), but this later evolved into a tri-compartment model, from the stomach to the duodenum/jejunum and then to the ileum (Théron et al., 2018).

Among the multicompartmental models, the TIM (TNO gastroIntestinal Model) (Minekus et al., 1995) stands out, presenting different variants of the original model developed by TNO in 1995 depending on the applications to which they have been oriented. The TIM-1 model represents the upper gastrointestinal tract, the TIM-2 model, which can be coupled to TIM-1, represents the colon (Aguirre et al., 2015), and new derivative models such as TIM-agc (Bellmann et al., 2016), Tiny-TIM (Verwei et al., 2016), and TIM-carbo (Bellmann et al., 2010) have been developed to improve the representativeness of the model in different aspects. Moreover, the TIM-pediatric (Havenaar et al., 2013) was an ad hocsystem specifically designed for pharmaceutical purposes that tried to simulate the conditions of the infant gastrointestinal tract. A further development of this model was the MIDA system whose global aim was studying the digestibility of nutrients, such as starch (Passannanti et al., 2017). The main advantage of TIM models is their ability to simulate peristaltic movements, pressure and absorption of nutrients and water in the small intestine regions, as well as to representatively simulate gastric emptying, which follows the equation described by (Elashoff et al., 1982) from data collected in vivo. These models are adapted to study the gastrointestinal behavior of a wide range of foods and pharmaceuticals. Each compartment consists of a glass jacket and flexible inner walls to reproduce peristaltic movements through pressure changes, and the different reactors are connected by peristaltic valves to allow the flow of chyme. The TIM model system also has an automatic system that records variables and controls the gradual decrease of pH during gastric digestion according to previously defined parameters, while the pH of the small intestine is maintained at fixed values of 6.5, 6.8, and 7.2 for the duodenum, jejunum, and ileum, respectively (Abrahamse et al., 2022; Minekus, 2015).

The digestion simulation in these models might begin with external processing of the food, which is ground and mixed with artificial saliva containing electrolytes and α-amylase. The feed is introduced into the gastric region where it is mixed with gastric secretion, consisting of electrolytes, pepsin and a gastric lipase of fungal origin (Minekus, 2015). The pH of this stage follows a predefined decreasing curve that is controlled by pH electrodes and the addition of HCl. After gastric emptying through the pyloric valve, duodenal secretions consisting of electrolytes, fresh porcine bile and pancreatin are added (Minekus, 2015). Intestinal digestion proceeds through the different regions of the small intestine and the digestion products can be separated according to whether they are hydrophilic or lipophilic (Minekus, 2015). The separation system for compounds with a lipophilic character consists of a filter (50 nm) that allows the micelles to pass through but retains the nonabsorbable phase. This system is a scaled-down version of TIM-1, which allows for a more realistic simulation of the ratio between ingestion volume and study product, which is of particular importance in the case of pharmaceutical formulations (Minekus, 2015). The absorption system is attached to the end of the intestinal stage (or jejunum and ileum in the case of TIM-1) and consists of a series of dialysis membranes with a pore size of 10 kDa, which operate continuously, separating the dialyzed (absorbable) fraction from the retained (nonabsorbable) fraction (Minekus, 2015). The TIM-agc (TIM-advanced gastric compartment) version of the model presents a gastric stage distribution that is more physiologically representative of in vivo digestion conditions.

In TIM-agc, gastric motility is simulated by flexible compartmentalized membranes so that the gastric contents maintain their fluid dynamic distribution more realistically (Bellmann et al., 2016; Minekus, 2015), which is especially evident for high-fat intakes.

One of the most advanced dynamic simulation models and a pioneer in complete gastrointestinal simulation is the Simulator of the Human Intestinal Microbial Ecosystem or SHIME (Molly et al., 1993), a system that simulates the gastrointestinal tract from the stomach to the distal colon, divided into five stages: gastric, intestinal and the three main colonic regions, the ascending, transverse and descending colon. This simulator has several versions, including the Twin-SHIME, the Triple-SHIME and the Quad-SHIME, which allow parallel operation in duplicate, triplicate and quadruplicate, respectively (Dupont et al., 2019). In addition, the model has diversified widely and allows the inclusion and/or coupling of different devices that complement it. This is the case for the M-SHIME, which includes a module for interaction with the intestinal mucosa, with which it is possible to represent mucosal colonization by incorporating mucin-coated devices that function as interaction microcosms (Van den Abbeele et al., 2012). Furthermore, the combination of SHIME and the HMITM module makes it possible to simulate the host-microbiota interaction at the level of the intestinal epithelium, so that the impact of certain treatments on the intestinal wall and the influence of the microbiota on its modulation can be assessed (Dupont et al., 2019). On the other hand, SHIME has been adapted and validated to represent more specifically the gastrointestinal tract of different population groups, such as the elderly and children, and situations of intestinal dysbiosis such as irritable bowel syndrome or *Clostridium difficile* infection (PathoGutTM model), and even to simulate the digestion of other mammals, such as pigs, dogs, and cats (Dupont et al., 2019).

Designing and realizing an in vitro digestion system idea that is able to fully reproduce what happens in the human gastrointestinal tract is a challenge. The inherent complexity of the human digestive system, which involves complex gastrointestinal motility, different anatomical structures, numerous enzymes and specific hormones crucial for food digestion, makes it challenging to simulate the process in vitro in a representative and robust way (Li et al., 2020; Singh et al., 2022).

The Simgi dynamic gastrointestinal simulator is a computer-controlled model of the human gastrointestinal tract that allows the conditions of the gastrointestinal tract to be simulated continuously and individually. It is composed of five compartments that represent the conditions of the stomach, the small intestine and the three main regions of the colon: the ascending, transverse and descending colon, including the microbiota of these areas (Barroso et al., 2015; Cueva et al., 2015). Since the first studies conducted on Simgi and aimed at evaluating colonic microbiota implantation (Barroso et al., 2015) and colonic microbial metabolism (Cueva et al., 2015), the platform has confirmed the suitability of the model with consistency between Simgi results at the metabolic level and those obtained in vivo (Cueva et al., 2015; Muñoz-González et al., 2013). Subsequently, new challenges have been posed with the aim of extending the platform's performance and adjusting the model to different applications in the field of health and food safety (Cueva et al., 2019; Tamargo, Cueva, Silva, et al., 2022; Tamargo, Cueva, Taladrid, et al., 2022).

In conclusion, and although the development of dynamic models is relatively recent, a number of dynamic models are now available to the scientific community. They derived from first prototypes and are still undergoing continuous innovation in order to bring them ever closer to human digestion.

4.4 Simulation of digestion of food proteins using dynamic models

In vivo, and although some transformation of protein structures may occur at the level of the mouth, protein digestion is initiated in the stomach by the acidic environment and by the action of pepsins. Digestion continues in the intestinal lumen through the action of pancreatic endopeptidases, mainly trypsin, elastase, and chymotrypsin together with exopeptidases. It all results in the release of peptides of two to six amino acid residues as well as free amino acids, which are absorbed. Peptides are susceptible to extensive hydrolysis by sequential gastric, pancreatic, and small intestinal brush border membrane (BBM) peptidases. Subsequently, undigested and/or unabsorbed peptides in the small intestine enter the large intestine or colon, where they are metabolized by the intestinal microbiota. A body of evidence proves that dietary peptides that survive luminal digestion can exert local and systemic activities that promote health benefits (Fernandez-Tome & Hernandez-Ledesma, 2020), which makes the characterization and tracking of the "digestome" of food proteins particularly interesting (Picariello et al., 2013).

Among other aims, dynamic simulators of human digestion are seen as investigational tools for understanding the physiological role of food components and its association with diseases (Dixit et al., 2023). In relation to proteins, dynamic simulators have been applied to study the technological and physiochemical characteristics of food proteins as well as the functional properties of both/either proteins and/or their digestion-derived peptides. But, in view of the aim of the chapter, this section only refers to those applications of well-known dynamic simulators (Section 3) concerning protein digestion targeting the release and further study of the derived peptides (Table 4.1). Studies have been revised highlighting the type of food/proteins utilized, targeted populations, operating parameters, and peptide outcomes. With regard to operating parameters, special attention has been given to the conditions characterizing dynamic operation, that is, pH adjustment, continuous addition of digestive fluids and enzymes, and gastric emptying.

Among monocompartmental dynamic models, the HGS (Human Gastric Simulator) (Kong & Singh, 2010) was applied to the characterization of gluten-derived immunogenic peptides in bread (Jayawardana et al., 2022) (Table 4.1). Simulators derived from the DIVHS system (Wang et al., 2019) were used in the study of different technological parameters affecting the release of bioactive peptides from milk proteins (Sun et al., 2023; Zhang et al., 2023) (Table 4.1). Of particular interest was the adaption of the DHSI-IV to the physiological digestion conditions of preschool children (2–5 years) by modifying the volumes of milk intake and simulated digestive fluids (Zhang et al., 2023).

In relation to bi- and tri-compartmental dynamic models, the DIDGI system (Ménard et al., 2014) is probably the simulator for which most applications for assessing the release of peptides from food proteins are reported. Although it was initially developed for simulating the infant gastrointestinal tract, the system has been expanded to other targeted populations and the general population. Moreover, the initial system, which was composed of two compartments (stomach and small intestine), was modified to split the second compartment into duodenum and jejunum + ileum steps. Both bi- and tri-compartmental dynamic models have been applied to digestion simulations of food proteins (Table 4.1). Operation parameters slightly differed among studies, as they were adapted not only to the targeted population but also to the type of food studied; for example, no gastric lipase was added when the objective was focused on protein digestion. Thus the model has been applied to food proteins from both animal (milk, cheese, meat, etc.) and plant (faba bean, pea, soya, etc.) sources, which confirms the model's feasibility. Protein digestives obtained by the DIDGI system have allowed valuable peptide outcomes to be reached according to the evolution of the techniques for peptide analysis and, in particular, for bioactive peptide identification. This is the case for a recent study conducted by (Chauvet et al., 2023) that identified bioactive peptides in DIDGI digestives from different infant formulas, highlighting the importance of the structure and composition (whey proteins and caseins) of the ingredients selected for the infant formulation as it might lead to physiological consequences for the infant.

The in vitro digestion (IVD) model developed by Shani-(Levi & Lesmes, 2014) aimed to facilitate age-tailoring of the digestive fate of liquid formulations. In their first publication, the authors described the design of a two mini-compartment model (composed of a gastric chamber and a duodenal bioreactor) and its application to the digestion of α-lactalbumin, β-lactoglobulin and

Table 4.1 Studies assessing peptide release from food proteins by using in vitro dynamic digestion models.

Model	Food/ingredient	Targeted population	Particular operation parameters	Peptide outcomes	References
Monocompartmental dynamic model					
Human gastric simulator (HGS) (Kong & Singh, 2010)	Bread (gluten)	Not defined	• Inflows: HCl (0.2 N); SGF + pepsin (2000 U/mL) • Emptying after 150 min of gastric digestion	• Free amino groups • Residual R5 gluten epitopes • LC/MS-MS: untargeted peptides and targeted immunogenic peptides	Jayawardana et al. (2022)
In vitro dynamic human stomach system (DHS-IV and DHSI-IV),[a] based on the dynamic in vitro human stomach (DIVHS; Wang et al., 2019)	Milk protein sterilized with UHT or C-UHT	Not defined	• Inflows (DHS-IV): SGF + pepsin (2000 U/mL) + lipase (60 U/mL) + $CaCl_2$ (0.15 mM/L) • Emptying after 2 h of gastric digestion	• LC/MS-MS: untargeted and targeted peptides • % of bioactive peptides (immunomodulatory, antioxidant, antimicrobial, ACE-inhibitory and others, according to the BIOPEP database)	Sun et al. (2023)
	Fermented milks with different casein-to-whey protein ratios	Preschool children	• Inflows (DHSI-IV): SGF + pepsin (4000 U/mL); SIF • Emptying after 1 and 2 h of gastric and intestinal digestion	• LC/MS-MS: untargeted and targeted peptides • Bioactive peptides (antioxidant, ACE-inhibitory opioid, antimicrobial, etc., according to the MBPDB database)	Zhang et al. (2023)

Bi- and tri-compartmental dynamic models					
Digesteur Dynamique Gastro-Intestinale (DIDGI) (Ménard et al., 2014) (2 compartments)	An infant formula	Infants	• Gastric inflows: HCl (0.5 M); pepsin (1250 U/mL); lipase (60 U/mL) • Intestinal inflows: Na_2CO_3 (0.1 M); porcine bile (1%); pancreatin (10%) • Emptying after 210 min of gastric and intestinal digestion	• Residual immunoreactivity by ELISA	Ménard et al. (2014)
	Skimmed milk powder	Adults	• Gastric inflows: HCl (1 M); SGF + pepsin (2000 U/mL) • Intestinal inflows: $NaHCO_3$ (1 M); SIF + pancreatin (100 U of trypsin/mL) + bile (10 mM) • Emptying after 90 min of gastric digestion and 180 min of intestinal digestion	• LC/MS-MS: peptide patterns and targeted sequences	Egger et al. (2019)
	Plant-protein-based infant food	Infants	• Gastric inflows: SGF + pepsin • Intestinal inflows: SIF + pancreatin + bile • Emptying after 78 min of gastric digestion and 200 min of intestinal digestion	• Degree of protein hydrolysis as an indicator of peptide bond cleavages • Size distribution of soluble nitrogen fraction	Le Roux et al. (2020)

(Continued)

Table 4.1 Studies assessing peptide release from food proteins by using in vitro dynamic digestion models. *Continued*

Model	Food/ingredient	Targeted population	Particular operation parameters	Peptide outcomes	References
	Raclette-type cheeses	Adults	• Gastric inflows: HCl (1 M); SGF + pepsin (2000 U/mL) • Intestinal inflows: $NaHCO_3$ (1 M); SIF + pancreatin (100 U of trypsin/mL) + bile (10 mM) • Emptying after 120 min of gastric digestion and 240 min of intestinal digestion	• LC/MS-MS: peptide patterns and identification of sequences with bioactive function	Egger et al. (2021)
	Treated and nontreated human milk	4-week-old infants	• Gastric inflows: HCl (1 M); SGF + pepsin (120 U/mL) + lipase (8.6 U/mL) • Intestinal inflows: $NaHCO_3$ (1 M); SIF + pancreatin (59 U/mL of trypsin) + bile (1.6 mM) • Emptying after 36 min of gastric digestion and 200 min of intestinal digestion	• LC/MS-MS: peptide patterns and identification of sequences with bioactive function	Giribaldi et al. (2022)
	Protein ingredients of infant formulas	4-week-old infants	• Gastric inflows: HCl (1 M); SGF + pepsin (268 U/mL) + rabbit gastric extract (19 U/mL of lipase)	• LC/MS-MS: identification of bioactive peptides (immunomodulatory, antioxidant,	Chauvet et al. (2023)

			• Intestinal inflows: $NaHCO_3$ (1 M); SIF + pancreatin (16 U/mL of trypsin) + bile (3.1 mM) • Emptying after 80 min of gastric digestion and 100 min of intestinal digestion	antimicrobial, ACE-inhibitory and others, according to the BIOPEP database)	
Digesteur Dynamique Gastro-Intestinale (DIDGI) (Ménard et al., 2014) (3 compartments)	Cooked meat digested with and without nitrites	Adults	• Gastric inflows: HCl; SGF + pepsin + lipase • Intestinal inflows: $NaHCO_3$; pancreatin • Emptying after 120 min of gastric digestion, after 150 min in the duodenum/jejunum and after 210 min in the ileum	• N-terminal α-amino groups of peptides • LC/MS-MS: untargeted and targeted peptides	Théron et al. (2018)
	Pork frankfurters	Adults and elderly	• Gastric inflows: HCl; SGF + pepsin + lipase • Emptying after 95 min of gastric digestion	• LC/MS-MS: untargeted and targeted peptides	Peyron et al. (2021)
	Soya milk and tofu	Not Defined	• Gastric inflows: HCl (0.2 M); NaOH (1 M); SSF; SGF + pepsin • Intestinal inflows: HCl (0.2 M); NaOH (1 M); pancreatin	• Degree of protein hydrolysis as an indicator of peptide bond cleavages • Size distribution of soluble nitrogen fraction	Reynaud et al. (2021)

(Continued)

Table 4.1 Studies assessing peptide release from food proteins by using in vitro dynamic digestion models. *Continued*

Model	Food/ingredient	Targeted population	Particular operation parameters	Peptide outcomes	References
In vitro digestion (IVD) model (Levi & Lesmes, 2014; Levi et al., 2017)	α-Lactalbumin, β-lactoglobulin, and lactoferrin	Adults and elderly	• Gastric inflows: · Gastric inflows: HCl; NH_4HCO_3; SGF + pepsin • Duodenal inflows: HCl; NH_4HCO_3; SDF + pancreatic proteases + bile • Emptying after 2–3 h of gastric digestion and after 120–180 min of duodenal digestion	• Peptide profiles by SDS-PAGE	Levi and Lesmes (2014); Levi et al. (2017)
	An enteral formula, α-lactalbumin, β-lactoglobulin, and lactoferrin	Adults and elderly	• Gastric inflows: · Gastric inflows: HCl; NH_4HCO_3; SGF + pepsin + lipase • Duodenal inflows: HCl; NH4HCO3; SDF + pancreatic proteases + lipase + α-amylase + bile • Emptying after 2–3 h of gastric digestion and after 120–180 min of duodenal digestion	• Peptide profiles by SDS-PAGE • LC/MS-MS: untargeted and targeted peptides • In silico PeptideRanker predictive software	Levi et al. (2017)

	Chocolate milk drink	Adults and children	• Gastric inflows: · Gastric inflows: HCl (0.5 M); NH_4HCO_3 (0.5 M); SGF + pepsin (2000 U/mL for adults and 1760 for children) • Duodenal inflows: HCl (0.5 M); NH_4HCO_3 (0.5 M); SDF + trypsin (100 U/mL), chymotrypsin (50 U/mL and 45 U/mL for children), bile (10 mM) • Emptying after 2 h of gastric digestion and after 60 min (adults) or 90 min (children) of duodenal digestion	• LC/MS-MS: identification of bioactive peptides according to the MBPDP database • In silico PeptideRanker predictive software	David, Klaiman, et al. (2020)
	Whey proteins treated with carrageenan	Adults	• Gastric inflows: · Gastric inflows: HCl (0.5 M); NH_4HCO_3 (0.5 M); SGF + pepsin (2000 U/mL) • Duodenal inflows: HCl (0.5 M); NH_4HCO_3 (0.5 M); SDF + trypsin (100 U/mL), chymotrypsin (50 U/mL), bile (10 mM) • Emptying after 120 min of gastric digestion and after 60 min (adults) of duodenal digestion	• LC/MS-MS: identification of bioactive peptides according to the MBPDP database	David, Wojciechowska, et al. (2020)

(Continued)

Table 4.1 Studies assessing peptide release from food proteins by using in vitro dynamic digestion models. *Continued*

Model	Food/ingredient	Targeted population	Particular operation parameters	Peptide outcomes	References
	α-Lactalbumin, β-lactoglobulin, lactoferrin, and whey protein concentrate	Men versus women	• Gastric inflows: HCl (0.2 M); NaOH (0.2 M m); SGF + pepsin • Duodenal inflows: HCl (0.2 M); NaOH (0.2 M); SDF + pancreatic proteases + bile • Emptying after 120–180 min of gastric digestion and after 120–180 min of duodenal digestion	• LC/MS-MS: identification of bioactive peptides according to the MBPDP • In silico PeptideRanker predictive software	Lajterer et al. (2022)
Multicompartment models					
TIM/TIM-1 (TNO gastrointestinal model) (Minekus et al., 1995)	Bread (gluten), a fast food menu	Adults	• Gastric inflows: gastric juice + pepsin pepsin (500 U/mL) • Duodenum inflows: $NaHCO_3$ (1.5 g/L) + pancreatic juice + bile • Jejunum and ileum inflows: $NaHCO_3$ (1.5 g/L) • Emptying after 2.5 h of gastric digestion and after 1.5 h of whole intestinal digestion	• Immunogenic peptides from gliadins and glutenins assessed by monoclonal antibody-based competition assays, Western blot analysis, and proliferation T-cell assays	Mitea et al. (2008)

	Meat	Adults and elderly	• Gastric inflows: HCl (1 M); gastric juice + pepsin pepsin • Duodenum inflows: $NaHCO_3$ (1 M) + pancreatic juice + bile • Jejunum and ileum inflows: $NaHCO_3$ (1 M) • Emptying after 5 h of gastric digestion and intestinal digestion	• LC/MS-MS: untargeted and targeted peptides	Denis et al. (2016)
	Plant proteins, from pea, soy, and rice	Adults	• Gastric inflows: HCL (1 M); gastric juice + pepsin pepsin • Duodenum inflows: $NaHCO_3$ (1 M) + pancreatic juice + bile • Jejunum and ileum inflows: $NaHCO_3$ (1 M) • Emptying after 6 h of gastric digestion and intestinal digestion	• α-amino nitrogen (AAN) values as a measure of free amino acids and short peptides	Keller et al. (2017)
	Whey protein and casein preparations	Not defined	• Gastric inflows: HCL (1 M); SGF + pepsin • Duodenum inflows: $NaHCO_3$ (1 M) + pancreatic juice + bile • Jejunum and ileum inflows: NaHCO3 (1 M) • Emptying after 5 h of gastric digestion and intestinal digestion	• Free amino groups • Peptide characterization by size exclusion chromatography	Abrahamse et al. (2022)

(Continued)

Table 4.1 Studies assessing peptide release from food proteins by using in vitro dynamic digestion models. *Continued*

Model	Food/ingredient	Targeted population	Particular operation parameters	Peptide outcomes	References
Dynamic Gastrointestinal Simulator (Simgi) (Barroso et al., 2015; Cueva et al., 2015)	Whey protein concentrate		• Gastric inflows: HCl; NaOH; SGF + pepsin (1250 U/mL) • Emptying after 90 min of gastric digestion	• LC/MS-MS: identification of peptides	Miralles et al. (2018)
	Whey protein concentrate Whey proteins codigested and noncodigested with food polyphenols		• Gastric inflows: SGF and pepsin (2000 U/mL) • Intestinal inflows: Simulated pancreatic juice consisting of Oxgall Dehydrated Fresh bile (6 g/L) and pancreatin from porcine pancreas powder (9.2 U/mL) acute intake (80 mL) of each studied food model	• Protein and peptide profiles by SDS-PAGE. LC/MS for polyphenols bioaccessibility study in the presence of proteins during digestion • Reduction of α-lactalbumin gastric degradation	Tamargo, Cueva, Silva, et al. (2022); Tamargo, Cueva, Taladrid, et al. (2022)

SSF, *simulated saliva fluid (Minekus et al., 2014);* SGF, *simulated gastric fluid;* SIF, *simulated intestinal fluid (Minekus et al., 2014);* SDF, *simulated duodenal fluid (Lajterer et al., 2022);* LC/MS-MS, *liquid chromatography-tandem mass spectrometry;* SDS-PAGE, *sodium dodecyl sulfate polyacrylamide gel electrophoresis;* MBPDP, *Milk Bioactive Peptide Database.*
[a]*Commercialized by Xiao Dong Pro-health Instrumentation Co Ltd. (Suzhou, China).*

lactoferrin (Table 4.1). The same model was operated under two different conditions trying to simulate the digestive tract of adults and the elderly (>65 years old). In practice, the elderly model was adjusted to account for the distinct gastric mixing, gastric pH gradient, gastric emptying, duodenal pH and mixing, pancreatic and bile secretion as well as the divergence in biochemical composition of the luminal content (Levi & Lesmes, 2014; Levi et al., 2017). At a later date, the model was applied to different protein matrices, including an enteral formula (Levi et al., 2017), a chocolate milk drink (David, Klaiman, et al., 2020), and whey proteins treated with carrageenan (David, Wojciechowska, et al., 2020). In these studies, more advanced peptide analytical techniques (e.g., LC/MS-MS coupled with MBPDP database) were used for identifying peptides as well as for elucidating their bioactivity (e.g., PeptideRanker software) (Table 4.1). Just recently, the same group, using the IVD model, carried out a pioneering experiment in an attempt to elucidate the sex differences in digestive proteolysis (Lajterer et al., 2022). Application of the model to whey protein samples and proteomic analyses of gastric effluents highlights possible differences in the proteolysis of whey proteins that might lead to implications for consumer health (Lajterer et al., 2022).

Among multicompartmental models, the TIM (Minekus et al., 1995) enabled the monitoring of food protein digestion in the stomach and in the different parts of the small intestine (stomach, duodenum, jejunum, and ileum), even mimicking ileal efflux and further nutrient absorption in the jejunal and ileal compartments (Table 4.1). The model also permitted comparison of food-derived peptide release between adult and elderly populations by modifying values for the secretion of pepsin and lipase in the gastric compartment, and pancreatic juice and bile in the duodenal compartment (Denis et al., 2016). Another multicompartmental model, Simgi (Barroso et al., 2015; Cueva et al., 2015), was applied to the study of protein digestion-derived peptides by using only the stomach compartment of the system (Table 4.1). The study of Miralles et al. (2018) on whey proteins reveals that peptide release using a dynamic stomach was consistent with that generated using the static INFOGEST protocol. However, the kinetic behavior of the peptide profile obtained under dynamic conditions better emphasized the effect of the sequential pepsin addition, peristaltic shaking, and gastric emptying on protein digestibility (Miralles et al., 2018). On the other hand, the co-digestion of whey proteins with wine polyphenols affected both polyphenol and food protein digestion and bioaccessibility, as well as their colonic metabolism. In particular, co-digestion of wine with proteins selectively retarded/limited whey protein degradation during digestion and also reduced the proteolytic activity of the protein matrix food model (Tamargo, Cueva, Silva, et al., 2022; Tamargo, Cueva, Taladrid, et al., 2022) (Table 4.1).

All these protein-related studies refer to dynamic simulators covering the stomach and small intestine steps. Another similarity among studies is that food and protein samples that feed the simulators were in liquid form. In the case of solid foods, such as meat, it was necessary to carry out a previous disintegration of the food matrix using mechanical devices, such as the Ultra Turrax.

Moreover, some studies included an oral digestion phase before the gastric simulation. In some studies, oral digestion is carried out in vitro by asking volunteers to chew the initial food and to expectorate the bolus that is subjected to gastric digestion (Jayawardana et al., 2022). In the majority of the studies, oral digestion was simulated by homogenizing the food with artificial saliva (Keller et al., 2017; Mitea et al., 2008) or simulated salivary fluid (SSF) (David, Klaiman, et al., 2020; David, Wojciechowska, et al., 2020; Lajterer et al., 2022; Reynaud et al., 2021; Egger et al., 2021) whose composition differs between adult and elderly individuals (Levi et al., 2017). In other studies, masticator apparatuses are used that simulate this first digestive tract stage (Théron et al., 2018; Peyron et al., 2021).

It is also inferred that dynamic models can be applied to particular populations (children, adults, elderly) by adapting operation conditions to their specific physiological conditions, as also described by (Lee et al., 2023) in a recent review of static and dynamic models simulating protein digestion. On the other hand, some studies validated the dynamic model of protein digestion with parallel animal experiments. (Ménard et al., 2014) compared the evolution of caseins and β-lactoglobulin in the stomach and intestine of 28-day-old piglets with the contents of the corresponding DIDGI compartments after the same digestion time. As determined by ELISA, results showed similar hydrolysis kinetics for both proteins during in vitro and in vivo digestion. In a further study using advanced LC/MS-MS techniques for identifying peptides, Egger et al. (2019) found similar peptide generation from a skimmed milk powder in vitro using the DIDGI simulator and in an animal trial with pigs under similar feeding conditions. (Reynaud et al., 2021), also using the same simulator, found that proteolysis patterns were largely consistent between models at a later stage of digestion, when the stomach was almost empty. Differences were found in individual parameters in the duodenal phase, but in the ileal phase, similar values for nitrogen digestibility in both in vivo and in vitro models were obtained (Reynaud et al., 2021).

Therefore although dynamic digestion models have not yet been widely used in studies on bioactive peptides, possibly because of their recent developments, they are undoubtedly an advanced and accurate approach to studying food protein digestion and peptide release. Coupled with appropriate omic techniques, such as peptidomcs, digestion models could provide a valuable approach to assessing potential bioactivities of food and food components (Silva et al., 2020).

4.5 Conclusions and future perspectives

Although the construction of the first dynamic simulators of the intestinal tract is relatively recent, in the last few years, they have undergone spectacular development through improvement of their design and process control to bring them closer to the physiological conditions of the human digestive tract. In contrast to static digestion models, about which the INFOGEST network has published standardized protocols, no such consensus has yet emerged for dynamic systems of digestion, which remains a future challenge for the scientific community. The study of protein digestion and the subsequent release of peptides has been mostly addressed through the use of static models of the upper digestive tract. However, as dynamic systems have become widely accessible, more and more studies are using these models as an experimental approach to address research on food-derived bioactive peptides. In fact, most of the dynamic models that have been developed −and which are described in Section 4.3—have been applied to the study of food protein digestion and further peptide release, as reviewed in Section 4.4. Moreover, we noticed quite a similarity in the protocols for protein digestion (i.e., pH, simulated fluids, enzyme concentrations, times, etc.) by dynamic simulators, although each model showed its design particularities. Our review also shows the great adaptability of these dynamic models in simulating physiological conditions of different populations (i.e., children, adults, and the elderly), an aspect that is expected to be of great interest in future research. Investigations of bioactive peptides by using dynamic simulators have been focused on major food proteins (dairy, meat, and plant proteins), but it is expected that they will be applied to novel protein sources, such as algae, insects, and fungus, among others. Identifying food-derived bioactive peptides by using dynamic digestion models will be improved by their coupling with appropriate omic techniques, such as peptidomics. Therefore all of this confirms the utility of dynamic digestion models for studying the transformations that food proteins undergo in the gastrointestinal tract, which is essential for elucidating peptide release and its potential bioactivity and impact on human health.

References

Abrahamse, E., Thomassen, G. G. M., Renes, I. B., Wierenga, P. A., & Hettinga, K. A. (2022). Assessment of milk protein digestion kinetics: Effects of denaturation by heat and protein type used. *Food and Function*, *13*(10), 5715−5729. Available from https://doi.org/10.1039/d2fo00542e, http://pubs.rsc.org/en/journals/journal/fo.

Aguirre, M., Eck, A., Koenen, M. E., Savelkoul, P. H. M., Budding, A. E., & Venema, K. (2015). Evaluation of an optimal preparation of human standardized fecal inocula for in vitro fermentation studies. *Journal of Microbiological Methods*, *117*, 78−84. Available from https://doi.org/10.1016/j.mimet.2015.07.019.

Alegría, A., Garcia-Llatas, G., & Cilla, A. (2015). *Static digestion models: General introduction. The impact of food bioactives on health: In vitro and ex vivo models* (pp. 3−12). Spain: Springer International Publishing. Available from https://doi.org/10.1007/978-3-319-16104-4_1.

Barroso, E., Cueva, C., Peláez, C., Martínez-Cuesta, M. C., & Requena, T. (2015). Development of human colonic microbiota in the computer-controlled dynamic SIMulator of the GastroIntestinal tract SIMGI. *LWT − Food Science and Technology*, *61*(2), 283−289. Available from https://doi.org/10.1016/j.lwt.2014.12.014.

Barros, L., Retamal, C., Torres, H., Zúñiga, R. N., & Troncoso, E. (2016). Development of an in vitro mechanical gastric system (IMGS) with realistic peristalsis to assess lipid digestibility. *Food Research International*, *90*, 216–225. Available from https://doi.org/10.1016/j.foodres.2016.10.049.

Bellmann, S., Lelieveld, J., Gorissen, T., Minekus, M., & Havenaar, R. (2016). Development of an advanced in vitro model of the stomach and its evaluation versus human gastric physiology. *Food Research International*, *88*, 191–198. Available from https://doi.org/10.1016/j.foodres.2016.01.030.

Bellmann, S., Minekus, M., Zeijdner, E., Verwei, M., Sanders, P., Basten, W., & Havenaar, R. (2010). *TIM-Carbo: A rapid, cost-efficient and reliable in vitro method for glycemic response after carbohydrate ingestion dietary fibre: New frontiers for food and health* (pp. 467–474). Netherlands: Wageningen Academic Publishers. Available from http://www.wageningenacademic.com/Default.asp ?pageid = 58&docid = 16&artdetail = fibre 2010, https://doi.org/10.3920/978-90-8686-692-2.

Biehler, E., & Bohn, T. (2010). Methods for assessing aspects of carotenoid bioavailability. *Current Nutrition & Food Science*, *6*(1), 44–69. Available from https://doi.org/10.2174/157340110790909545.

Blanquet-Diot, S., Denis, S., Chalancon, S., Chaira, F., Cardot, J. M., & Alric, M. (2012). Use of artificial digestive systems to investigate the biopharmaceutical factors influencing the survival of probiotic yeast during gastrointestinal transit in humans. *Pharmaceutical Research*, *29*(6), 1444–1453. Available from https://doi.org/10.1007/s11095-011-0620-5.

Bohn, T. (2014). Dietary factors affecting polyphenol bioavailability. *Nutrition Reviews*, *72*(7), 429–452. Available from https://doi.org/10.1111/nure.12114, http://onlinelibrary.wiley.com/journal/10.1111/(ISSN)1753-4887.

Boland, M. (2016). Human digestion – A processing perspective. *Journal of the Science of Food and Agriculture*, *96*(7), 2275–2283. Available from https://doi.org/10.1002/jsfa.7601.

Bornhorst, G. M., & Singh, R. P. (2014). Gastric digestion in vivo and in vitro: How the structural aspects of food influence the digestion process. *Annual Review of Food Science and Technology*, *5*(1), 111–132. Available from https://doi.org/10.1146/annurev-food-030713-092346.

Brodkorb, A., Egger, L., Alminger, M., Alvito, P., Assunção, R., Ballance, S., Bohn, T., Bourlieu-Lacanal, C., Boutrou, R., Carrière, F., Clemente, A., Corredig, M., Dupont, D., Dufour, C., Edwards, C., Golding, M., Karakaya, S., Kirkhus, B., Le Feunteun, S., … Recio, I. (2019). INFOGEST static in vitro simulation of gastrointestinal food digestion. *Nature Protocols*, *14*(4), 991–1014. Available from https://doi.org/10.1038/s41596-018-0119-1.

Campbell, J., Berry, J., & Liang, Y. (2019). *Anatomy and physiology of the small intestine. Shackelford's surgery of the alimentary tract: 2 volume set* (pp. 817–841). Elsevier. Available from https://www.sciencedirect.com/book/9780323402323, https://doi.org/10.1016/B978-0-323-40232-3.00071-6.

Chauvet, L., Ménard, O., Le Gouar, Y., Henry, G., Jardin, J., Hennetier, M., Croguennec, T., Van Audenhaege, M., Dupont, D., Lemaire, M., Le Huërou-Luron, I., & Deglaire, A. (2023). Protein ingredient quality of infant formulas impacts their structure and kinetics of proteolysis under in vitro dynamic digestion. *Food Research International*, *169*. Available from https://doi.org/10.1016/j.foodres.2023.112883.

Clegg, M. E., & Shafat, A. (2010). Procedures in the 13C octanoic acid breath test for measurement of gastric emptying: Analysis using BlandAltman methods. *Scandinavian Journal of Gastroenterology*, *45*(7–8), 852–861. Available from https://doi.org/10.3109/00365521.2010.483740.

Cordonnier, C., Thévenot, J., Etienne-Mesmin, L., Denis, S., Alric, M., Livrelli, V., & Blanquet-Diot, S. (2015). Dynamic in vitro models of the human gastrointestinal tract as relevant tools to assess the survival of probiotic strains and their interactions with gut microbiota. *Microorganisms*, *3*(4), 725–745. Available from https://doi.org/10.3390/microorganisms3040725.

Cueva, C., Gil-Sánchez, I., Tamargo, A., Miralles, B., Crespo, J., Bartolomé, B., & Moreno-Arribas, M. V. (2019). Gastrointestinal digestion of food-use silver nanoparticles in the dynamic SIMulator of the GastroIntestinal tract (simgi®). Impact on human gut microbiota. *Food and Chemical Toxicology*, *132*. Available from https://doi.org/10.1016/j.fct.2019.110657.

Cueva, C., Jiménez-Girón, A., Muñoz-González, I., Esteban-Fernández, A., Gil-Sánchez, I., Dueñas, M., Martín-Álvarez, P. J., Pozo-Bayón, M. A., Bartolomé, B., & Moreno-Arribas, M. V. (2015). Application of a new Dynamic Gastrointestinal Simulator (SIMGI) to study the impact of red wine in colonic metabolism. *Food Research International*, *72*, 149–159. Available from https://doi.org/10.1016/j.foodres.2015.03.003, http://www.elsevier.com/inca/publications/store/4/2/2/9/7/0.

David, S., Klaiman, M. M., Shpigelman, A., & Lesmes, U. (2020). Addition of anionic polysaccharide stabilizers modulates in vitro digestive proteolysis of a chocolate milk drink in adults and children. *Foods*, *9*(9). Available from https://doi.org/10.3390/foods9091253.

David, S., Wojciechowska, A., Portmann, R., Shpigelman, A., & Lesmes, U. (2020). The impact of food-grade carrageenans and consumer age on the in vitro proteolysis of whey proteins. *Food Research International*, *130*. Available from https://doi.org/10.1016/j.foodres.2019.108964.

Denis, S., Sayd, T., Georges, A., Chambon, C., Chalancon, S., Santé-Lhoutellier, V., & Blanquet-Diot, S. (2016). Digestion of cooked meat proteins is slightly affected by age as assessed using the dynamic gastrointestinal TIM model and mass spectrometry. *Food and Function*, *7*(6), 2682–2691. Available from https://doi.org/10.1039/c6fo00120c, http://www.rsc.org/Publishing/Journals/FO/about.asp.

Dixit, Y., Kanojiya, K., Bhingardeve, N., Ahire, J. J., & Saroj, D. (2023). In vitro human gastrointestinal tract simulation systems: A panoramic review. *Probiotics and Antimicrobial Proteins*. Available from https://doi.org/10.1007/s12602-023-10052-y, https://www.springer.com/journal/12602.

Dupont, D., Alric, M., Blanquet-Diot, S., Bornhorst, G., Cueva, C., Deglaire, A., Denis, S., Ferrua, M., Havenaar, R., Lelieveld, J., Mackie, A. R., Marzorati, M., Menard, O., Minekus, M., Miralles, B., Recio, I., & Van den Abbeele, P. (2019). Can dynamic in vitro digestion systems mimic the physiological reality? *Critical Reviews in Food Science and Nutrition*, *59*(10), 1546–1562. Available from https://doi.org/10.1080/10408398.2017.1421900, https://www.tandfonline.com/loi/bfsn20.

Egger, L., Ménard, O., Abbühl, L., Duerr, D., Stoffers, H., Berthoud, H., Meola, M., Badertscher, R., Blaser, C., Dupont, D., & Portmann, R. (2021). Higher microbial diversity in raw than in pasteurized milk Raclette-type cheese enhances peptide and metabolite diversity after in vitro digestion. *Food Chemistry*, *340*. Available from https://doi.org/10.1016/j.foodchem.2020.128154.

Egger, L., Ménard, O., Baumann, C., Duerr, D., Schlegel, P., Stoll, P., Vergères, G., Dupont, D., & Portmann, R. (2019). Digestion of milk proteins: Comparing static and dynamic in vitro digestion systems with in vivo data. *Food Research International, 118*, 32–39. Available from https://doi.org/10.1016/j.foodres.2017.12.049.

Egger, L., Ménard, O., Delgado-Andrade, C., Alvito, P., Assunção, R., Balance, S., Barberá, R., Brodkorb, A., Cattenoz, T., Clemente, A., Comi, I., Dupont, D., Garcia-Llatas, G., Lagarda, M. J., Le Feunteun, S., JanssenDuijghuijsen, L., Karakaya, S., Lesmes, U., Mackie, A. R., ... Portmann, R. (2016). The harmonized INFOGEST in vitro digestion method: From knowledge to action. *Food Research International, 88*, 217–225. Available from https://doi.org/10.1016/j.foodres.2015.12.006.

Egger, L., Schlegel, P., Baumann, C., Stoffers, H., Guggisberg, D., Brügger, C., Dürr, D., Stoll, P., Vergères, G., & Portmann, R. (2017). Physiological comparability of the harmonized INFOGEST in vitro digestion method to in vivo pig digestion. *Food Research International, 102*, 567–574. Available from https://doi.org/10.1016/j.foodres.2017.09.047.

Elashoff, J. D., Reedy, T. J., & Meyer, J. H. (1982). Analysis of gastric emptying data. *Gastroenterology, 83*(6), 1306–1312. Available from https://doi.org/10.1016/S0016-5085(82)80145-5.

Fernandez-Tome, S., & Hernandez-Ledesma, B. (2020). Gastrointestinal digestion of food proteins under the effects of released bioactive peptides on digestive health. *Molecular Nutrition and Food Research, 64*, 21. Available from https://doi.org/10.1002/mnfr.202000401.

Fournier, E., Denis, S., Dominicis, A., Van de Wiele, T., Alric, M., Mercier-Bonin, M., Etienne-Mesmin, L., & Blanquet-Diot, S. (2022). A child is not an adult: Development of a new in vitro model of the toddler colon. *Applied Microbiology and Biotechnology, 106*(21), 7315–7336. Available from https://doi.org/10.1007/s00253-022-12199-0, https://www.springer.com/journal/253.

Gil-Sánchez, I., Ayuda-Durán, B., González-Manzano, S., Santos-Buelga, C., Cueva, C., Martín-Cabrejas, M. A., Sanz-Buenhombre, M., Guadarrama, A., Moreno-Arribas, M. V., & Bartolomé, B. (2017). Chemical characterization and in vitro colonic fermentation of grape pomace extracts. *Journal of the Science of Food and Agriculture, 97* (10), 3433–3444. Available from https://doi.org/10.1002/jsfa.8197, http://onlinelibrary.wiley.com/journal/10.1002/(ISSN)1097-0010.

Gil-Sánchez, I., Cueva, C., Sanz-Buenhombre, M., Guadarrama, A., Moreno-Arribas, M. V., & Bartolomé, B. (2018). Dynamic gastrointestinal digestion of grape pomace extracts: Bioaccessible phenolic metabolites and impact on human gut microbiota. *Journal of Food Composition and Analysis, 68*, 41–52. Available from https://doi.org/10.1016/j.jfca.2017.05.005, http://www.elsevier.com/inca/publications/store/6/2/2/8/7/8/index.htt.

Gil-Sánchez, I., Cueva, C., Tamargo, A., Quintela, J. C., de la Fuente, E., Walker, A. W., Moreno-Arribas, M. V., & Bartolomé, B. (2020). Application of the dynamic gastrointestinal simulator (simgi®) to assess the impact of probiotic supplementation in the metabolism of grape polyphenols. *Food Research International, 129*. Available from https://doi.org/10.1016/j.foodres.2019.108790, http://www.elsevier.com/inca/publications/store/4/2/2/9/7/0.

Giribaldi, M., Nebbia, S., Briard-Bion, V., Jardin, J., Peila, C., Coscia, A., Dupont, D., Cavallarin, L., & Deglaire, A. (2022). Simulated dynamic digestion reveals different peptide releases from human milk processed by means of holder or high temperature-short time pasteurization. *Food Chemistry*, *369*. Available from https://doi.org/10.1016/j.foodchem.2021.130998.

Guerra, A., Denis, S., le Goff, O., Sicardi, V., François, O., Yao, A. F., Garrait, G., Manzi, A. P., Beyssac, E., Alric, M., & Blanquet-Diot, S. (2016). Development and validation of a new dynamic computer-controlled model of the human stomach and small intestine. *Biotechnology and Bioengineering*, *113*(6), 1325–1335. Available from https://doi.org/10.1002/bit.25890, http://www.interscience.wiley.com/jpages/0006-3592.

Gérard-Champod, M., Blanquet-Diot, S., Cardot, J. M., Bravo, D., & Alric, M. (2010). Development and validation of a continuous in vitro system reproducing some biotic and abiotic factors of the veal calf intestine. *Applied and Environmental Microbiology*, *76*(16), 5592–5600. Available from https://doi.org/10.1128/AEM.00524-10France, http://aem.asm.org/cgi/reprint/76/16/5592.

Havenaar, R., Anneveld, B., Hanff, L. M., De Wildt, S. N., De Koning, B. A. E., Mooij, M. G., Lelieveld, J. P. A., & Minekus, M. (2013). In vitro gastrointestinal model (TIM) with predictive power, even for infants and children? *International Journal of Pharmaceutics*, *457*(1), 327–332. Available from https://doi.org/10.1016/j.ijpharm.2013.07.053.

Hellström, P. M., Grybäck, P., & Jacobsson, H. (2006). The physiology of gastric emptying. *Best Practice and Research: Clinical Anaesthesiology*, *20*(3), 397–407. Available from https://doi.org/10.1016/j.bpa.2006.02.002.

Howell, M. D., Novack, V., Grgurich, P., Soulliard, D., Novack, L., Pencina, M., & Talmor, D. (2010). Iatrogenic gastric acid suppression and the risk of nosocomial Clostridium difficile infection. *Archives of Internal Medicine*, *170*(9), 784–790. Available from https://doi.org/10.1001/archinternmed.2010.89, http://archinte.ama-assn.org/cgi/reprint/170/9/784, United States.

Isackson, H., & Ashley, C. C. (2014). Secretory functions of the gastrointestinal tract. *Surgery (United Kingdom)*, *32*(8), 396–403. Available from https://doi.org/10.1016/j.mpsur.2014.05.007, http://www.elsevier.com/wps/find/journaldescription.cws_home/709612/description#description.

Jayawardana, I. A., Boland, M. J., Loo, T. S., McNabb, W. C., & Montoya, C. A. (2022). Rapid proteolysis of gluten-derived immunogenic peptides in bread by actinidin in a combined in vivo and in vitro oro-gastrointestinal digestion model. *Food and Function*, *13*(10), 5654–5666. Available from https://doi.org/10.1039/d1fo03740d, http://pubs.rsc.org/en/journals/journal/fo.

Keller, D., van Dinter, R., Cash, H., Farmer, S., & Venema, K. (2017). Bacillus coagulans GBI-30, 6086 increases plant protein digestion in a dynamic, computer-controlled in vitro model of the small intestine (TIM-1). *Beneficial Microbes*, *8*(3), 491–496. Available from https://doi.org/10.3920/BM2016.0196, http://www.wageningenacademic.com/doi/pdf/10.3920/BM2016.0196.

Kong, F., & Singh, R. P. (2008). Disintegration of solid foods in human stomach. *Journal of Food Science*, *73*(5), R67–R80. Available from https://doi.org/10.1111/j.1750-3841.2008.00766.x.

Kong, F., & Singh, R. P. (2010). A human gastric simulator (HGS) to study food digestion in human stomach. *Journal of Food Science*, *75*(9), E627–E635. Available from https://doi.org/10.1111/j.1750-3841.2010.01856.x.

Lajterer, C., Levi, C. S., & Lesmes, U. (2022). An in vitro digestion model accounting for sex differences in gastro-intestinal functions and its application to study differential protein digestibility. *Food Hydrocolloids*, *132*. Available from https://doi.org/10.1016/j.foodhyd.2022.107850.

Lee, S., Jo, K., Jeong, H. G., Choi, Y. S., Yong, H. I., & Jung, S. (2023). Understanding protein digestion in infants and the elderly: Current in vitro digestion models. *Critical Reviews in Food Science and Nutrition*, *63*(7), 975–992. Available from https://doi.org/10.1080/10408398.2021.1957765, https://www.tandfonline.com/loi/bfsn20.

Le Roux, L., Ménard, O., Chacon, R., Dupont, D., Jeantet, R., Deglaire, A., & Nau, F. (2020). Are faba bean and pea proteins potential whey protein substitutes in infant formulas? an in vitro dynamic digestion approach. *Foods*, *9*(3). Available from https://doi.org/10.3390/foods9030362.

Levi, C. S., Goldstein, N., Portmann, R., & Lesmes, U. (2017). Emulsion and protein degradation in the elderly: Qualitative insights from a study coupling a dynamic in vitro digestion model with proteomic analyses. *Food Hydrocolloids*, *69*, 393–401. Available from https://doi.org/10.1016/j.foodhyd.2017.02.017.

Levi, C. S., & Lesmes, U. (2014). Bi-compartmental elderly or adult dynamic digestion models applied to interrogate protein digestibility. *Food and Function*, *5*(10), 2402–2409. Available from https://doi.org/10.1039/c4fo00478g.

Liu, W., Fu, D., Zhang, X., Chai, J., Tian, S., & Han, J. (2019). Development and validation of a new artificial gastric digestive system. *Food Research International*, *122*, 183–190. Available from https://doi.org/10.1016/j.foodres.2019.04.015, http://www.elsevier.com/inca/publications/store/4/2/2/9/7/0.

Li, C., Yu, W., Wu, P., & Chen, X. D. (2020). Current in vitro digestion systems for understanding food digestion in human upper gastrointestinal tract. *Trends in Food Science and Technology*, *96*, 114–126. Available from https://doi.org/10.1016/j.tifs.2019.12.015, http://www.elsevier.com/wps/find/journaldescription.cws_home/601278/description#description.

Lombardo, L., Foti, M., Ruggia, O., & Chiecchio, A. (2010). Increased incidence of small intestinal bacterial overgrowth during proton pump inhibitor therapy. *Clinical Gastroenterology and Hepatology*, *8*(6), 504–508. Available from https://doi.org/10.1016/j.cgh.2009.12.022.

Mackie, A., Mulet-Cabero, A. I., & Torcello-Gomez, A. (2020). Simulating human digestion: Developing our knowledge to create healthier and more sustainable foods. *Food and Function*, *11*(11), 9397–9431. Available from https://doi.org/10.1039/d0fo01981j, http://pubs.rsc.org/en/journals/journal/fo.

Mandalari, G., Grundy, M. M. L., Grassby, T., Parker, M. L., Cross, K. L., Chessa, S., Bisignano, C., Barreca, D., Bellocco, E., Laganà, G., Butterworth, P. J., Faulks, R. M., Wilde, P. J., Ellis, P. R., & Waldron, K. W. (2014). The effects of processing and mastication on almond lipid bioaccessibility using novel methods of in vitro digestion modelling and micro-structural analysis. *British Journal of Nutrition*, *112*(9), 1521–1529. Available from https://doi.org/10.1017/S0007114514002414, http://journals.cambridge.org/BJN.

Mason, L. M., Chessa, S., Huatan, H., Storey, D. E., Gupta, P., Burley, J., & Melia, C. D. (2016). Use of the dynamic gastric model as a tool for investigating fed and fasted sensitivities of low polymer content hydrophilic matrix formulations. *International Journal of Pharmaceutics*, *510*(1), 210–220. Available from https://doi.org/10.1016/j.ijpharm.2016.06.034.

Miller, D. D., Schricker, B. R., Rasmussen, R. R., & Van Campen, D. (1981). An in vitro method for estimation of iron availability from meals. *American Journal of Clinical Nutrition*, *34*(10), 2248–2256. Available from https://doi.org/10.1093/ajcn/34.10.2248.

Minekus, M. (2015). *The TNO gastro-intestinal model (TIM). The impact of food bioactives on health: In vitro and ex vivo models* (pp. 37–46). Netherlands: Springer International Publishing. Available from http://doi.org/10.1007/978-3-319-16104-4.

Minekus, M., Alminger, M., Alvito, P., Ballance, S., Bohn, T., Bourlieu, C., Carrière, F., Boutrou, R., Corredig, M., Dupont, D., Dufour, C., Egger, L., Golding, M., Karakaya, S., Kirkhus, B., Le Feunteun, S., Lesmes, U., MacIerzanka, A., MacKie, A., ... Brodkorb, A. (2014). A standardised static in vitro digestion method suitable for food-an international consensus. *Food and Function*, *5*(6), 1113–1124. Available from https://doi.org/10.1039/c3fo60702j, http://www.rsc.org/Publishing/Journals/FO/about.asp.

Minekus, M., Marteau, P., Havenaar, R., & Huis in't Veld, J. H. J. (1995). A multicompartmental dynamic computer-controlled model simulating the stomach and small intestine. *Alternatives to Laboratory Animals*, *23*(2), 197–209. Available from https://doi.org/10.1177/026119299502300205.

Miner-Williams, W. M., Stevens, B. R., & Moughan, P. J. (2014). Are intact peptides absorbed from the healthy gut in the adult human? *Nutrition Research Reviews*, *27*(2), 308–329. Available from https://doi.org/10.1017/S0954422414000225, http://journals.cambridge.org/NRR.

Miralles, B., del Barrio, R., Cueva, C., Recio, I., & Amigo, L. (2018). Dynamic gastric digestion of a commercial whey protein concentrate. *Journal of the Science of Food and Agriculture*, *98*, 1873–1879. Available from https://doi.org/10.1002/jsfa.8668.

Mitea, C., Havenaar, R., Drijfhout, J. W., Edens, L., Dekking, L., Koning, F., & Dekking, E. H. A. (2008). Efficient degradation of gluten by a prolyl endoprotease in a gastrointestinal model: Implications for coeliac disease. *Gut*, *57*(1), 25–32. Available from https://doi.org/10.1136/gut.2006.111609.

Molly, K., Vande Woestyne, M., & Verstraete, W. (1993). Development of a 5-step multi-chamber reactor as a simulation of the human intestinal microbial ecosystem. *Applied Microbiology and Biotechnology*, *39*(2), 254–258. Available from https://doi.org/10.1007/bf00228615.

Morell, P., Hernando, I., & Fiszman, S. M. (2014). Understanding the relevance of in-mouth food processing. A review of invitro techniques. *Trends in Food Science and Technology*, *35*(1), 18–31. Available from https://doi.org/10.1016/j.tifs.2013.10.005.

Mulet-Cabero, A. I., Egger, L., Portmann, R., Ménard, O., Marze, S., Minekus, M., Le Feunteun, S., Sarkar, A., Grundy, M. M. L., Carrière, F., Golding, M., Dupont, D., Recio, I., Brodkorb, A., & Mackie, A. (2020). A standardised semi-dynamic: In vitro digestion method suitable for food-an international consensus. *Food and Function*, *11* (2), 1702–1720. Available from https://doi.org/10.1039/c9fo01293a, http://pubs.rsc.org/en/journals/journal/fo.

Muñoz-González, I., Jiménez-Girón, A., Martín-Álvarez, P. J., Bartolomé, B., & Moreno-Arribas, M. V. (2013). Profiling of microbial-derived phenolic metabolites in human feces after moderate red wine intake. *Journal of Agricultural and Food Chemistry*, *61* (39), 9470–9479. Available from https://doi.org/10.1021/jf4025135.

Ménard, O., Cattenoz, T., Guillemin, H., Souchon, I., Deglaire, A., Dupont, D., & Picque, D. (2014). Validation of a new in vitro dynamic system to simulate infant digestion. *Food Chemistry*, *145*, 1039–1045. Available from https://doi.org/10.1016/j.foodchem.2013.09.036.

Nielsen, S. D., Beverly, R. L., Underwood, M. A., & Dallas, D. C. (2018). Release of functional peptides from mother's milk and fortifier proteins in the premature infant stomach. *PLoS One*, *13*(11). Available from https://doi.org/10.1371/journal.pone.0208204, https://journals.plos.org/plosone/article/file?id = 10.1371/journal.pone.0208204&type = printable.

Passannanti, F., Nigro, F., Gallo, M., Tornatore, F., Frasso, A., Saccone, G., Budelli, A., Barone, M. V., Nigro, R., & Passi, A. G. (2017). In vitro dynamic model simulating the digestive tract of 6-month-old infants. *PLoS One*, *12*(12). Available from https://doi.org/10.1371/journal.pone.0189807.

Payne, A. N., Zihler, A., Chassard, C., & Lacroix, C. (2012). Advances and perspectives in in vitro human gut fermentation modeling. *Trends in Biotechnology*, *30*(1), 17–25. Available from https://doi.org/10.1016/j.tibtech.2011.06.011.

Peyron, M.-A., Sayd, T., Sicard, J., Mirade, P.-S., Pinguet, J., Chambon, C., & Santé-Lhoutellier, V. (2021). Deciphering the protein digestion of meat products for the elderly by in vitro food oral processing and gastric dynamic digestion, peptidome analysis and modeling. *Food & Function*, *12*(16), 7283–7297. Available from https://doi.org/10.1039/d1fo00969a.

Peyron, M. A., & Woda, A. (2016). An update about artificial mastication. *Current Opinion in Food Science*, *9*, 21–28. Available from https://doi.org/10.1016/j.cofs.2016.03.006, http://www.journals.elsevier.com/current-opinion-in-food-science/.

Picariello, G., Mamone, G., Nitride, C., Addeo, F., & Ferranti, P. (2013). Protein digestomics: Integrated platforms to study food-protein digestion and derived functional and active peptides. *TrAC Trends in Analytical Chemistry*, *52*, 120–134. Available from https://doi.org/10.1016/j.trac.2013.08.001.

Pompei, A., Cordisco, L., Raimondi, S., Amaretti, A., Pagnoni, U. M., Matteuzzi, D., & Rossi, M. (2008). In vitro comparison of the prebiotic effects of two inulin-type fructans. *Anaerobe*, *14*(5), 280–286. Available from https://doi.org/10.1016/j.anaerobe.2008.07.002.

Rajilić-Stojanović, M., Heilig, H. G. H. J., Tims, S., Zoetendal, E. G., & De Vos, W. M. (2013). Long-term monitoring of the human intestinal microbiota composition. *Environmental Microbiology*, *15*(4), 1146–1159. Available from https://doi.org/10.1111/1462-2920.12023.

Reynaud, Y., Couvent, A., Manach, A., Forest, D., Lopez, M., Picque, D., Souchon, I., Rémond, D., & Dupont, D. (2021). Food-dependent set-up of the DiDGI® dynamic in vitro system: Correlation with the porcine model for protein digestion of soya-based food. *Food Chemistry*, *341*. Available from https://doi.org/10.1016/j.foodchem.2020.128276.

Rinninella, E., Raoul, P., Cintoni, M., Franceschi, F., Miggiano, G. A. D., Gasbarrini, A., & Mele, M. C. (2019). What is the healthy gut microbiota composition? A changing ecosystem across age, environment, diet, and diseases. *Microorganisms*, *7*(1). Available from https://doi.org/10.3390/microorganisms7010014, https://www.mdpi.com/2076-2607/7/1/14/pdf.

Rodríguez-Saavedra, M., González de Llano, D., & Moreno-Arribas, M. V. (2023). Beer spoilage lactic acid bacteria from craft brewery microbiota: Microbiological quality and food safety. *Food Research International*, *138*.

Sams, L., Paume, J., Giallo, J., & Carrière, F. (2016). Relevant pH and lipase for in vitro models of gastric digestion. *Food and Function*, *7*(1), 30–45. Available from https://doi.org/10.1039/c5fo00930h, http://www.rsc.org/Publishing/Journals/FO/about.asp.

Schulze, K. (2006). Imaging and modelling of digestion in the stomach and the duodenum. *Neurogastroenterology and Motility*, *18*(3), 172–183. Available from https://doi.org/10.1111/j.1365-2982.2006.00759.x.

Silva, M., Rincon, L., de Llano, D. G., Moreno-Arribas, V. M., & Bartolome, B. (2020). *Omic approaches coupled to gastrointestinal dynamic modelling to assess food bioactivity comprehensive foodomics* (pp. 516–525). Elsevier. Available from https://www.sciencedirect.com/referencework/9780128163962/comprehensive-foodomics, https://doi.org/10.1016/B978-0-08-100596-5.22886-2.

Singh, R., Zogg, H., Ghoshal, U. C., & Ro, S. (2022). Current treatment options and therapeutic insights for gastrointestinal dysmotility and functional gastrointestinal disorders. *Frontiers in Pharmacology*, *13*. Available from https://doi.org/10.3389/fphar.2022.808195, http://www.frontiersin.org/Pharmacology.

Sun, Y., Wang, R., Li, Q., & Ma, Y. (2023). Influence of storage time on protein composition and simulated digestion of UHT milk and centrifugation presterilized UHT milk in vitro. *Journal of Dairy Science*, *106*(5), 3109–3122. Available from https://doi.org/10.3168/jds.2022-22602.

Sánchez-Patán, F., Barroso, E., Van De Wiele, T., Jiménez-Girón, A., Martín-Alvarez, P. J., Moreno-Arribas, M. V., Martínez-Cuesta, M. C., Peláez, C., Requena, T., & Bartolomé, B. (2015). Comparative in vitro fermentations of cranberry and grape seed polyphenols with colonic microbiota. *Food Chemistry*, *183*, 273–282. Available from https://doi.org/10.1016/j.foodchem.2015.03.061.

Tamargo, A., Cueva, C., Silva, M., Molinero, N., Miralles, B., Bartolomé, B., & Moreno-Arribas, M. V. (2022). Gastrointestinal co-digestion of wine polyphenols with glucose/whey proteins affects their bioaccessibility and impact on colonic microbiota. *Food Research International*, *155*. Available from https://doi.org/10.1016/j.foodres.2022.111010, http://www.elsevier.com/inca/publications/store/4/2/2/9/7/0.

Tamargo, A., Cueva, C., Taladrid, D., Khoo, C., Moreno-Arribas, M. V., Bartolomé, B., & González de Llano, D. (2022). Simulated gastrointestinal digestion of cranberry polyphenols under dynamic conditions. Impact on antiadhesive activity against uropathogenic bacteria. *Food Chemistry*, *368*, 130871. Available from https://doi.org/10.1016/j.foodchem.2021.130871.

Tamargo, A., de Llano, D. G., Cueva, C., del Hierro, J. N., Martin, D., Molinero, N., Bartolomé, B., & Victoria Moreno-Arribas, M. (2023). Deciphering the interactions between lipids and red wine polyphenols through the gastrointestinal tract. *Food Research International*, *165*. Available from https://doi.org/10.1016/j.foodres.2023.112524.

Thomas, A. (2006). Gut motility, sphincters and reflex control. *Anaesthesia and Intensive Care Medicine*, *7*(2), 57–58. Available from https://doi.org/10.1383/anes.2006.7.2.57.

Théron, L., Chambon, C., Sayd, T., De La Pomélie, D., Santé-Lhoutellier, V., & Gatellier, P. (2018). To what extent does the nitrosation of meat proteins influence their digestibility? *Food Research International*, *113*, 175–182. Available from https://doi.org/10.1016/j.foodres.2018.06.071.

Thévenot, J., Cordonnier, C., Rougeron, A., Le Goff, O., Nguyen, H. T. T., Denis, S., Alric, M., Livrelli, V., & Blanquet-Diot, S. (2015). Enterohemorrhagic Escherichia coli infection has donor-dependent effect on human gut microbiota and may be antagonized by probiotic yeast during interaction with Peyer's patches. *Applied Microbiology and Biotechnology*, *99*(21), 9097–9110. Available from https://doi.org/10.1007/s00253-015-6704-0.

Van den Abbeele, P., Roos, S., Eeckhaut, V., Mackenzie, D. A., Derde, M., Verstraete, W., Marzorati, M., Possemiers, S., Vanhoecke, B., Van Immerseel, F., & Van de Wiele, T. (2012). Incorporating a mucosal environment in a dynamic gut model results in a more representative colonization by lactobacilli. *Microbial Biotechnology*, *5*(1), 106–115. Available from https://doi.org/10.1111/j.1751-7915.2011.00308.x.

van den Braak, C. C. M., Klebach, M., Abrahamse, E., Minor, M., Hofman, Z., Knol, J., & Ludwig, T. (2013). A novel protein mixture containing vegetable proteins renders enteral nutrition products non-coagulating after invitro gastric digestion. *Clinical Nutrition*, *32*(5), 765–771. Available from https://doi.org/10.1016/j.clnu.2012.11.016.

Versantvoort, C. H. M., Oomen, A. G., Van De Kamp, E., Rompelberg, C. J. M., & Sips, A. J. A. M. (2005). Applicability of an in vitro digestion model in assessing the bioaccessibility of mycotoxins from food. *Food and Chemical Toxicology*, *43*(1), 31–40. Available from https://doi.org/10.1016/j.fct.2004.08.007.

Verwei, M., Minekus, M., Zeijdner, E., Schilderink, R., & Havenaar, R. (2016). Evaluation of two dynamic in vitro models simulating fasted and fed state conditions in the upper gastrointestinal tract (TIM-1 and tiny-TIM) for investigating the bioaccessibility of pharmaceutical compounds from oral dosage forms. *International Journal of Pharmaceutics*, *498*(1–2), 178–186. Available from https://doi.org/10.1016/j.ijpharm.2015.11.048, http://www.elsevier.com/locate/ijpharm.

Vitali, B., Ndagijimana, M., Maccaferri, S., Biagi, E., Guerzoni, M. E., & Brigidi, P. (2012). An in vitro evaluation of the effect of probiotics and prebiotics on the metabolic profile of human microbiota. *Anaerobe*, *18*(4), 386–391. Available from https://doi.org/10.1016/j.anaerobe.2012.04.014.

Wang, J., Wu, P., Liu, M., Liao, Z., Wang, Y., Dong, Z., & Chen, X. D. (2019). An advanced near real dynamic: In vitro human stomach system to study gastric digestion and emptying of beef stew and cooked rice. *Food and Function*, *10*(5), 2914–2925. Available from https://doi.org/10.1039/c8fo02586j, http://pubs.rsc.org/en/journals/journal/fo.

Wickham, M. J. S., Faulks, R. M., Mann, J., & Mandalari, G. (2012). The design, operation, and application of a dynamic gastric model. *Dissolution Technologies*, *19*(3), 15–22. Available from https://doi.org/10.14227/DT190312P15.

Xu, Q., Hong, H., Wu, J., & Yan, X. (2019). Bioavailability of bioactive peptides derived from food proteins across the intestinal epithelial membrane: A review. *Trends in Food Science & Technology*, *86*, 399–411. Available from https://doi.org/10.1016/j.tifs.2019.02.050.

Yu, Y. Q., & Yang, J. Y. (2019). Oral bioaccessibility and health risk assessment of vanadium(IV) and vanadium(V) in a vanadium titanomagnetite mining region by a whole digestive system in-vitro method (WDSM). *Chemosphere*, *215*, 294–304. Available from https://doi.org/10.1016/j.chemosphere.2018.10.042.

Zhang, H., Duan, S., Yu, Y., Wu, R., Wang, J., Chen, X. D., Szeto, I. M. Y., Wu, P., & Jin, Y. (2023). Impact of casein-to-whey protein ratio on gastric emptying, proteolysis, and peptidome profile of fermented milk during in vitro dynamic gastrointestinal digestion in preschool children. *Food Chemistry*, *405*. Available from https://doi.org/10.1016/j.foodchem.2022.134840.

CHAPTER

5

Assessing the bioactivity of dietary peptides and their resistance to digestion: a combined approach using in silico, in vitro, and in vivo analyses

Luis Jiménez Munoz and Milena Corredig

Department of Food Science, CiFood Center, Aarhus University, Skejby, Denmark

5.1 Introduction

Predictive models of world population growth have caused humanity to worry about its potential burden on food systems. Thus efforts have focused on how to increase efficiencies in food chains, to alleviate the increasing pressure on our planetary boundaries (Gerten et al., 2020). Among the myriad of strategies, the gradual transition from meat to plant-based consumption supported by a coordinated action from governmental, societal, health, and environmental institutions is considered as the most holistic and effective approach (Graça et al., 2019). Plant-derived proteins from faba beans, peas, soy, lupin and potatoes are regarded as promising sources due to their adequate nutritional composition (high content of fiber and minerals) (Ložnjak Švarc et al., 2022), high protein quality (Sá et al., 2020), and low environmental impact (Lynch et al., 2018). Other alternative sources, such as microalgae, seaweed, insects, or microbial proteins obtained through genetically edited microorganisms (e.g., recombinant leghemoglobin by *Pichia pastoris*, bovine β-lactoglobulin and ovoalbumin by *Trichoderma reesei* currently used as ingredients by companies such as Impossible Foods, Perfect Day, and Onego Bio) (Chai et al., 2022), could also be included, but due to their novelty, supply chains and extraction challenges, more research efforts are required before a consistent production of foods containing these functional proteins is reached.

Depending on the desired purity and functionality, these protein-rich ingredients are obtained through a variety of different methods: At an industrial scale, plant-based ingredients with a protein content of ~85% (dry matter) often require the use of significant amounts of water during fractionation, as well as energy resources (Schutyser & van der Goot, 2011). Moreover, isolations and

Protein Digestion-Derived Peptides. DOI: https://doi.org/10.1016/B978-0-443-19141-1.00005-4

purifications often include processes with pH shifts or aggregation for efficient physical separations and this may reduce the functional properties of the protein (Dumoulin et al., 2021). Therefore more attention should be put in what is the extent of purification truly required in a particular food formulation. Indeed, when a high level of purity is not required, dry fractionation processes may be applied. For example, for low-oil flours it is possible to produce ingredients with a ~60% protein and ~20% fiber content (D.M) using milling and classification. However, due to these mild processing conditions, while the native structure of the protein bodies may be somewhat maintained, antinutritional factors may also be present, as well as off flavor precursors (De Angelis et al., 2021), thus further processing may be required to hinder their digestibility and appeal. Dry fractionation methods are more energy and water efficient. In addition, as the native structure of the protein is preserved, usually the functionality is higher in comparison to other more purified ingredients (Assatory et al., 2019).

Furthermore, the extraction processes not only influence protein structure in a different manner, but also the composition, that is, the presence of different protein groups. It has been previously demonstrated that applying alkali extraction, followed by isoelectric precipitation (AE-IEP) to legumes (e.g., pea, faba bean, lupin) causes a concentration of globulins in the precipitate fraction, while albumins remain in the suspended fraction (Jiménez-Munoz et al., 2023; Rubio et al., 2014; Shrestha et al., 2021; Vogelsang et al., 2020). In conclusion, the comparison between ingredients of different processing history leads to differences in purity, concentration of antinutritional factors and functionality, and it raises questions on how extraction technologies may affect their nutritional quality (in terms of composition) as well as their digestibility, which does not only relate to the dietary amino acids availability but also the release of peptides with potential health benefits.

Most bioactive peptides are encoded in protein sequences and released through enzymatic processes (e.g., use of commercially available microbial enzymes, enzymatic digestion). However, some can be chemically synthesized and a small number of them are found in their natural sources unmodified (Sánchez & Vázquez, 2017). These peptides affect human health by acting on various systems, including digestive, endocrine, cardiovascular, immune, and nervous. Among the various potential benefits, they include anticancer, antihypertensive, antiinflammatory, antiobesity, antioxidative, antimicrobial, immunomodulatory, and neuroprotective effects (Zaky et al., 2022).

For instance, antihypertensive peptides restrict the production of angiotensin II, a hormone that raises blood pressure, by inhibiting the angiotensin-converting enzyme (ACE) (Sánchez & Vázquez, 2017). Conversely, antidiabetic peptides lead to improved glucose regulation and reduced risk of type II diabetes by inhibiting dipeptidyl peptidase (DPP) III (DPP-III) and DPP IV activity, enzymes involved in the degradation of hormones, such as glucagon-like-peptide-1 (GLP-1) and glucose-dependent insulinotropic peptide (GIP) (Yan et al., 2019). In addition, bioactive peptides also have the potential to be a valuable alternative to traditional opioid analgesics based on their opioid activity (Tyagi et al., 2020).

Opioid peptides have been shown to have pain-relieving properties and a lower risk of addiction compared to synthetic opioids (Mora et al., 2015). Although much is known about certain food-sourced peptides, more research is urgently needed to fully understand the benefits and potential risks of the peptides encoded in the less-known sources of proteins currently proposed as more sustainable alternatives. The research in this area is still in its early stages.

Despite the identified peptide sequences in certain proteins, degradation may occur due to their susceptibility to digestive enzymes (Majura et al., 2022), which might induce their loss of bioactivity. In addition, allergenicity and toxicity of peptides must also be considered, since demonstrating a considerate exposure risk level would be a sufficient reason to stop their further development as therapeutic agents (Duan et al., 2021). Thus discovery and development of food-derived therapeutic peptides remains challenging. Researchers around the globe have resorted to insightful tools like in silico and in vitro/ex vivo to determine the bioaccessibility, bioavailability, allergenicity, and specific bioactivity of food-derived peptides (from animal, dairy, marine, plant, microbial sources) released and absorbed during intestinal digestion (Jakubczyk et al., 2020). Despite providing valuable information that could suggest the existence of in vivo bioactivity, by themselves these methods pose some disadvantages: current in silico models do not take into account the effect of food structure on the mechanisms of digestion and peptide release (Le Feunteun et al., 2020) and in vitro models are limited by not considering the effect of brush border membranes, which could result in a misrepresentation of the peptide stability during in vivo conditions (Barati et al., 2020). In essence, proven in silico *or* in vitro peptide bioactivity does not necessarily guarantee in vivo efficacy. Nevertheless, exploratory studies using in silico and in vitro methods are required to ensure a robust methodology that may lead to the discovery of new peptides with bioactivity. Furthermore, it is also worth investigating how the inclusion of other components in the food matrix, interactions and changes in structure induced by further thermal or nonthermal (e.g., ultrasound, shear, high-pressure, pulsed electric field) processing conditions plausibly will affect protein digestion, peptide absorption, and activity (Ohanenye et al., 2022).

The application of unmethodical peptide screening approaches is often a limitation to progress in the field of peptide discovery. Empirical methods are laborious. In these processes, hydrolysates from commercial proteases are analyzed for in vitro bioactivity. In this screening, there is little consideration of their resistance to gastrointestinal breakdown. Furthermore, commercial enzymes are used due to their affordability, but this comes at the expenses of their lack of purity and presence of side activities. It is therefore necessary to raise the awareness to the application of computational methods for predicting peptide-beneficial health effects, to set up a systematic approach toward a more successful screening (Olsen et al., 2020).

In this chapter, we examine the underlying mechanisms of plant-based protein digestibility and their potential for peptide bioactivity, as affected by processing conditions, using in silico approaches, together with in vitro and in vivo data available in the literature.

5.1.1 An integrated approach for discovering peptides' bioactivity

Selected proteins were used as model systems to demonstrate how an integrated approach could be used to discover bioactivity. Potato (patatin and kunitz trypsin inhibitor), pea (legumin), spinach (RuBisCO), and wheat (alpha-amylase inhibitor, high and low molecular weight glutenin subunits and α/β-gliadin MM1) were analyzed using a predictive in silico approach. The protein sequences were obtained from the Protein Data Bank (https://www.rcsb.org/) and sequences for wheat-derived peptides were taken from the protein sequence database UniProt (https://uniprot.org/). The chosen proteins were subjected to in silico proteolysis carried out using BIOPEP database (https://biochemia.uwm.edu.pl/en/biopep-uwm-2/) (Minkiewicz et al., 2019) using the "enzyme action" function. Pepsin (pH >2), trypsin and chymotrypsin C were selected to simulate gastrointestinal digestion. The predicted degree of hydrolysis was reported in Table 5.1 and generated peptides from the in silico gastrointestinal digestion were evaluated for their biological activity using "search for active fragments." Then, peptides with predicted bioactivity (ACE inhibitory activity, [DPP]-III/IV inhibitory activity, renin inhibitory activity, antioxidant activity and osteoblast stimulating activity) were compared to PeptideRanker (http://distilldeep.ucd.ie/PeptideRanker/), a predictive model that uses a neural network (N-to-1) machine learning algorithm to calculate the probability of bioactivity based on various physicochemical properties, such as amino acid composition, charge, and hydrophobicity (Mooney et al., 2012). The scores ranged from 0 to 1.0, but, for the purpose of this case study, only peptides with values above 0.5 are reported. Potential antioxidant properties of the peptides, including free radical scavenging (FRS) and metal-chelating activity (MCA) scores were predicted using AnOxPePred platform (https://services.healthtech.dtu.dk/services/AnOxPePred-1.0/) and only peptides with FRS or MCA values above 0.5 were reported. Among protein used in the present study, no resulting peptides showed MCA values above 0.5. AllergenFP (http://ddg-pharmfac.net/AllergenFP/index.html) (Dimitrov et al., 2014) was used to predict peptide allergenicity, and Tanimoto similarity coefficient scores are reported in Table 5.1. Peptide toxicity was predicted in using ToxinPred tool (http://www.imtech.res.in/raghava/toxinpred/index.html) (Gupta et al., 2013), none of the peptides studied were toxic.

In the present work, to demonstrate how to complement the in silico data, available reports on peptide bioactivity from in vitro and in vivo studies for potato, pea, spinach, and wheat were then reviewed. The experimental data needs indeed to be compared with the bioactivity predicted in silico. Such integrated approach is an example of how to evaluate potential health benefits of less known proteins. Furthermore, candidate peptides with a bioactive potential (e.g., ACE, DPP-III/IV inhibitory activity) in at least two approaches (in silico, in vitro, and/or in vivo) were then subject of further analysis, for example, using the ligand-based target approach that SwissTargetPrediction offers (swisstargetprediction.ch) (Daina et al., 2019), using *Homo sapiens* as the selected species. For the previous

Table 5.1 Examples from selected plant-based proteins containing peptides with high bioactivity scores.

In silico approach						
Source	**Protein**	**Protein sequence**	**DH%**	**Bioactive peptides**	**Bioactive probability**	**Allergenicity (Tanimoto similarity index)**
Potato	Patatin	4PK9	83.4%	– **ACE Inhibitor: IF, VHW** – **DPP IV Inhibitor: IP, SP**	IF (0.95) VHW (0.59) IP (0.58) SP (0.51)	Probable nonallergens (between 0.84 and 0.86)
	Kunitz Inhibitor	5FZZ	76.3%	– **ACE Inhibitor: IF** – **DPP IV Inhibitor: IW, VF, IP, SP**	IF (0.95) IW (0.94) VF (0.82) IP (0.58) SP (0.51)	
	Profilin	Q38M47_SOLTU	81.5%	– **ACE inhibitor: VW, IG**	VW (0.80) IG (0.50)	**Entire protein sequence: probable allergen (1.0)** **Peptides: VW, IG. Probable nonallergens (0.84–0.86)**
Pea	Legumin	3KSC (pepsin, trypsin and chymotrypsin)	84.6%	– **ACE Inhibitor: IG, IF** – **DPP IV Inhibitor: VF**	IG (0.50) IF (0.95) VF (0.82)	Probable nonallergens (0.86)
		3KSC (hydrolyzed by thermolysin)	29.7%	– **ACE Inhibitor: LRW, VK, VP, AG, AR, LQ, LN, YN, LR, LP** – **DPP IV Inhibitor: LP, AG, AS, AT, FN, LN, LT, VK, VS, YN** – **Renin Inhibitor: LR** – **Antioxidative peptide: LK** – **Stimulating Osteoblasts: LRW**	LRW (0.95) FN (0.5) LP (0.8) LR (0.57) AG (0.55)	Probable nonallergens (0.82–0.87) (Gu et al., 2011)

(Continued)

Table 5.1 Examples from selected plant-based proteins containing peptides with high bioactivity scores. *Continued*

In silico approach						
Source	**Protein**	**Protein sequence**	**DH%**	**Bioactive peptides**	**Bioactive probability**	**Allergenicity (Tanimoto similarity index)**
Spinach (Spinacia oleracea)	RuBisCO	1AUS_1 1AUS_2	82.8%	– **ACE Inhibitor: IF, VF, VW, IP** – **DPP IV Inhibitor: HF, SW, VF, VW, IP, SP** – **DPP III Inhibitor: HP** – **Antioxidative peptide: VW** – **Alpha-glucosidase Inhibitor: VW**	HF (0.95) IF (0.95) SW (0.93) VF (0.81) VW (0.80) IP (0.58) HP (0.57) SP (0.51)	Probable nonallergens (0.84–0.87)
Wheat (Triticum aestivum)	Alpha-amylase inhibitor	P01083	75.6%	– **ACE-Inhibitor: VF, IP** – **DPP-IV Inhibitor: SW, VF, IP**	SW (0.93) VF (0.81) IP (0.58)	**Entire protein sequence: probable allergen (1.0)** **Peptides: SHN, VCY: probable allergens (0.83–0.84)**
		Q4W0V7	77.9%	– **DPP-III Inhibitor: SM**	SM (0.63)	**Entire protein sequence: probable allergen (1.0)** **Peptides: ISE: probable allergen (0.83)**
	Low molecular weight Glutenin subunit	B2Y2Q6	81.0%	– **ACE-Inhibitor: IF, SF, VF, IP** – **DPP-IV Inhibitor: SF, VF, IP, SP** – **Renin Inhibitor: SF**	IF (0.95) SF (0.94) VF (0.81) IP (0.58) SP (0.51)	**Entire protein sequence: probable allergen (0.9)** **Peptides: CSP, SSCH, SHHQ (probable allergens: 0.82–0.84)** **IF, SF, VF, IP, SP (probable nonallergens: 0.84–0.87)**

High molecular weight Glutenin subunit 12			– **ACE-Inhibitor: IF, SF, IP, HP** – **DPP-IV Inhibitor: SF, IP, HP, SP** – **Renin Inhibitor: SF** – **DPP-III Inhibitor: HP**	IF (0.95) SF (0.95) IP (0.58) HP (0.57) SP (0.51)	**Entire protein sequence: probable allergen (1.0)** **Peptides: SSL (Probable allergen: 0.84)** **IF, SF, IP, HP, SP (Probable nonallergens: 0.84–0.87)**
Alpha/beta-gliadin MM1	P18573	80.0%	– **ACE-Inhibitor: IF, SF** – **DPP-IV Inhibitor: SF** – **Renin Inhibitor: SF**	IF (0.95) SF (0.95) IP (0.58)	**Entire protein sequence: probable allergen (1.0)** **Peptides: IF, SF, IP Probable nonallergens (0.84–0.87)**

Peptides from spinach Rubisco, pea-legumin, potato-kunitz inhibitors, and patatin can be found in the primary structure of the protein subunits in the Protein Data Bank (https://www.rcsb.org/). Sequences for profilin and wheat-derived peptides can be found in protein sequence database UniProt (https://www.uniprot.org/). Only peptides with a bioactive probability above 0.5 were taken into account. Degree of Hydrolysis (DH%) calculations were made in BIOPEP–UWM database by considering pepsin (pH > 2), trypsin, and Chymotrypsin C as the acting enzymes. Bioactive probability was obtained by using PeptideRanker (N-to-1 neural network) and Allergenicity index was obtained using AllergenFP v1.0

Legumin Data from Gu, Y., Majumder, K., & Wu, J. (2011). QSAR-aided in silico approach in evaluation of food proteins as precursors of ACE inhibitory peptides. Food Research International, *44(8), 2465–2474. https://doi.org/10.1016/j.foodres.2011.01.051.*

analysis, the amino acid sequences were converted to Simplified Molecular Input Line Entry Specification (SMILES) using BIOPEP (https://biochemia.uwm.edu.pl/en/biopep-uwm-2/) (Minkiewicz et al., 2019).

5.1.2 Predicting bioactivities after gastrointestinal transit

Despite the exhaustive study of plant-based protein ingredients for food, much is still unknown about the health effects of bioactive peptides encoded within the configurations of precursor proteins. This is of particular importance due to the need for a green transition to more sustainable diets. These difficulties are partly due to the difficulty in underpinning the mechanisms based on the low recovery of bioavailable peptides, generated from their oral delivery through foods (Kęska & Stadnik, 2021). Indeed, during gastrointestinal transit, proteins undergo extensive enzymatic breakdown under acidic pH in the stomach, and later on by the action of intestinal enzymes, furthermore, their bioefficacy is linked to their ability to pass, intact, through the intestinal barrier (Gallego et al., 2016).

To gain a deeper understanding of the fate of peptides after digestive processes and intestinal absorption, as well as their impact, researchers have utilized in vitro, in silico, and in vivo models as complementary tools. These models enable the study of digestion in a controlled environment, providing a mechanistic approach to identify the underlying factors that influence the digestion of the proteins within their matrix. These models are useful for understanding how different components are absorbed and metabolized by the body, as well as how potential interactions between different substances that may affect digestion.

The ability to predict the fate of a particular component in the body is extremely helpful for designing healthier foods, by assessing the potential effects of various molecules and their metabolites on human health. Moreover, new interventions may be developed through food for enhancing gut health and the general well-being of the individual. This is why computational methods for prediction of beneficial health effects now go hand in hand with experimental observations in vitro, and in vivo.

Fig. 5.1 depicts the process flow and the relationships between in silico, in vitro, and in vivo data used to study the digestion and absorption of protein derivatives. First, the in silico approach allows for a simulation of the fate of the substance, in other words, leading to the prediction of the breakdown and absorption of peptides and amino acids in the body. Furthermore, the computer modeling will also aid in the evaluation of the effects of various factors on digestion, such as pH and digestive enzyme activity (Gu et al., 2011). This type of analysis can be used as a tool to accelerate the process of discovery of potentially encrypted bioactive peptides in the proteins under investigation. For example, after in silico proteolysis, the identified peptides can be virtually screened for bioactivity with BIOPEP-UWMTM and PeptideRanker database tools. Moreover, to ensure the safety of the studied peptides, the resultant sequences can be further evaluated for their toxicity (e.g., using ToxinPred). When bioactivity (e.g., ACE inhibitory activity) is

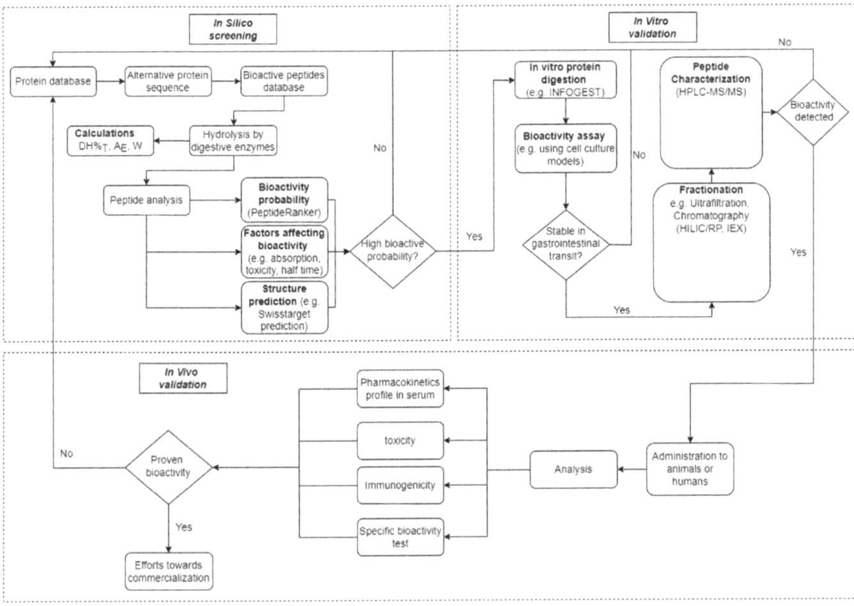

FIGURE 5.1 Flowchart of an integrated approach for discovering bioactivity in peptides derived from proteins by enzymatic action.

The processes include approaches used in silico, in vitro, and in vivo.

predicted in silico, in vitro experiments are recommended in the next step of the process. During the in vitro analysis the digestion is studied in a laboratory setting, using isolated digestive enzymes (pepsin, trypsin, chymotrypsin) or cells (CaCo-2 and HT-29 cells) (Lu et al., 2019). For complex matrices, more relevant in vitro digestion system, which have been harmonized to be comparable among laboratories, would be recommended (Brodkorb et al., 2019).

By carrying out in vitro digestion, researchers can gain a deeper understanding of the mechanisms of protein digestion and the effects that the structure of the matrix, and of the protein involved will have on the release of the various peptides and their distribution. These studies would also provide a platform for the development and evaluation of innovative approaches to optimize their bioefficacy. The use of simulated gastrointestinal conditions, allows for testing of peptide resistance, whereas cell absorption studies allow for the determination of transport mechanisms across the cell monolayer and the bioavailability of the peptides (Xie et al., 2014). To purify peptide fractions that allow for further bioactivity testing, methods, such as ultrafiltration, ion exchange, gel filtration chromatography (separation based on their size, shape, and charge) (Yust et al., 2003), reversed phase chromatography (separation based on hydrophibicity) (Conlon, 2007) can be used. Furthermore, as an alternative to these methods, F-moc (fluorenylmethoxycarbonyl) solid phase peptide synthesis (SPPS) can be utilized in peptide synthesis to complement the in vitro and in vivo methods. In F-moc SPPS, the synthetic peptide is formed by adding individual amino acids one by one to a growing peptide chain that is anchored to a solid support (Behrendt et al., 2016). In this process, the N-terminal amino acid residue is protected with a F-moc group to ensure the stability and reactivity of the growing peptide chain during synthesis (Mäde et al., 2014).

A combination of studies involving synthetic peptides with in vitro digestion coupled with in vitro absorption using cell culture models may provide valuable insight that could lead to a detailed mechanistic understanding of the relationship between peptide design, stability, and bioactivity. These purified peptides can be evaluated for efficacy and safety when used in isolation. After the synthesis or purification step, the specific bioactive potential of each peptide fraction can be further tested, to validate previous in silico predictions.

Experimentally, the entire peptidome of in vitro digests from a protein source can be analyzed by high performance liquid chromatography (HPLC) coupled to tandem mass spectrometry (Fernández-Tomé & Hernández-Ledesma, 2020). This allows for a holistic view at the effect of the matrix as well as potential changes due, for example, to processing conditions. A positive identification of potentially bioactive peptides, also supported by the in silico analysis strengthens then the argument for further in vivo testing. While in vitro and in silico studies aid in a mechanistic understanding of the potential bioactive properties, the study of digestion in living organisms allows researchers to understand the effectiveness of the interventions, for example, for therapeutic or protective use. In vivo studies also validate the results of in silico and in vitro studies, and evaluate potential interactions occurring during digestion, and their physiological effects. For example, in vitro *models* may be used to study the effects of different drugs or dietary supplements on digestive enzymes, while in vivo models are used to evaluate the clinical potential of peptides as therapeutic agents. The therapeutic effect of nutraceuticals can only be achieved once their physicochemical and pharmacokinetic properties have been fully understood and precisely controlled (Ouberai et al., 2017). Overall, the integrative use of in silico, in vitro, and in vivo digestion models has the potential to be a game-changer in the field of digestive health, leading to a wealth of new insights and breakthroughs. The increased sophistication and relevance of these models is providing new tools for a comprehensive approach to understanding the complexities of digestion and will pave the way for the development of more effective and targeted strategies for improving digestive well-being.

5.1.3 A comparison of in silico and in vitro digestion

Gastrointestinal enzymes (pepsin, trypsin, and chymotrypsin) were utilized for in silico digestion analysis (using BIOPEP, Minkiewicz et al., 2019) of potato, pea, spinach, and wheat proteins. The degree of hydrolysis by combined action of the selected proteases on various proteins (patatin and kunitz trypsin inhibitor, legumin, RuBisCO, alpha-amylase inhibitor, high and low molecular weight glutenin subunits) is reported in Table 5.2. Fig. 5.2 shows the comparison for both gastric

Table 5.2 Summary of bioactive peptides found in vitro for different potato, pea, spinach and wheat proteins, diverse processing conditions used to obtain said peptides in vitro and main finding regarding the specific bioactivity found.

In vitro approach					
Source	**Protein**	**Processing conditions**	**Bioactive peptides**	**Bioactivity evaluation method and main findings**	**References**
Potato	Patatin	Peptide synthesis	WG, PRY	Peptide PRY was more potent as ACE and renin-inhibitory peptide than WG. This was determined by intrinsic florescence intensity and Molecular docking data provided evidence of structural mechanisms	(Fu et al., 2017)
	Sporamin	High hydrostatic pressure and alcalase	HDSASGQY, YYMVSA, HDSESGQY, YYIVS, RYYD	Antioxidant activity of the peptides was measured by Oxygen radical absorbance capacity (ORAC). Highest value was 117.3 μg trolox equivalent/mL. - Potential natural antioxidant.	(Zhang & Mu, 2017)
	Patatin and tuberinin	Dehydration and subsequent in vitro digestion	3–10 kDa 1–3 kDa <1 kDa	The peptides fraction from dehydrated potatoes promoted cell migration in the IEC-6 monolayer and showed antiinflammatory as well as antioxidant activity.	(Basilicata et al., 2019)

(Continued)

Table 5.2 Summary of bioactive peptides found in vitro for different potato, pea, spinach and wheat proteins, diverse processing conditions used to obtain said peptides in vitro and main finding regarding the specific bioactivity found. *Continued*

In vitro approach					
Source	**Protein**	**Processing conditions**	**Bioactive peptides**	**Bioactivity evaluation method and main findings**	**References**
Pea	Legumin	Peptide synthesis	LRW	Mouse preosteoblast MC3T3-E1 cells were treated with LRW (10–50 μM). The administration of the tri-peptide had therapeutic potential by promoting osteoblastic bone formation. Thus showing potential to prevent osteoporosis.	(Arora et al., 2020)
				In the in vitro experiment, the tri-peptide exhibited antiinflammatory effect in Angiotensin II-stressed vascular smooth muscle cells from rats (A7r5). Thus showing potential to prevent hypertension.	(Wang et al., 2020)
	Globulins	In vitro digestion followed by IEX Chromatography to purify peptides.	GGSGNY, DLKLP, GSSDNR, MRDLK, HNTPSR	Four peptides resulting from in vitro hydrolysis by gastrointestinal enzymes exhibited potential antihypertensive properties (IC_{50} = 0.073 mg/mL). ESI-Ms/MS was used to identify ACE-Inhibitory peptides.	(Jakubczyk & Baraniak, 2014)

Spinach	RuBisCO	Peptide synthesis and in vitro enzymatic digestion.	YPLDL, YPLDLF	RuBisCO-derived bioactive peptides were obtained by in vitro digestion using gastro-intestinal enzymes. In the guinea pig ileum assay, the IC_{50} was 1110 μM for YLDL and 748 μM for YPLDLF. Thus Rubiscolin-5/-6 peptides had selective δ-opioid agonist activity.	(Yang et al., 2001)
	Rubiscolin-5 and 6	Solid-phase peptide synthesis		Effect of rubiscolins on μ-, δ- and κ-opioid receptors were evaluated using HEK293 cells. Rubisco peptides showed partial agonistic effects on μ and δ opioid receptors.	(Karasawa et al., 2021)
Wheat	Gluten	Hydrolysis by alcalase and Protex from *Bacillus lichenformis*	LY, PY, LVS, YQ, APSY, RGGY	Five bioactive peptides derived from wheat gluten hydrolysate with potent ACE inhibitory activity (77% inhibition rate at a 2.5 mg/mL, IC_{50} = 0.68) and antioxidant activity (6–11 mmol Trolox equivalent/g sample) were identified.	(Liu et al., 2021)
	gliadin peptide (33-mer)	Hydrolysis by pepsin and pancreatic enzymes.	LQLQPFPQPQLPYPQPQLPYPQPQLPYPQPQPF	Proteolytic resistance of the 33-mer peptide fragment to gastrointestinal digestion was shown in vitro through simulated gastrointestinal digestion and in vivo by monitoring peptides and proteins in human feces.	(Comino et al., 2012)

Data are available from literature.

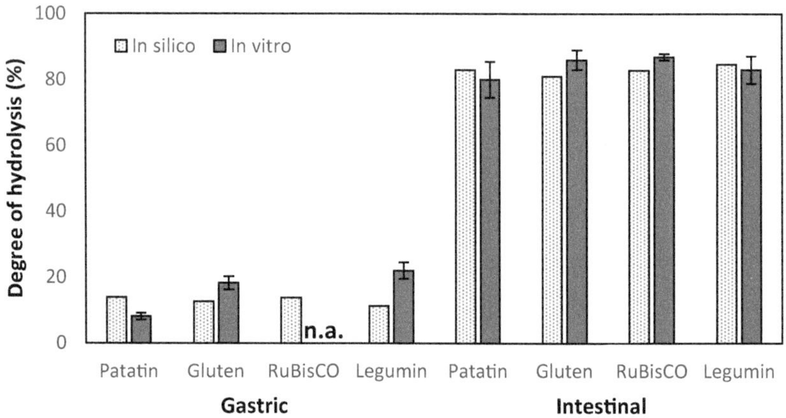

FIGURE 5.2 Comparison of degree of hydrolysis values obtained through in silico approach and in vitro data available in the literature for potato patatin, gluten proteins, RuBisCO, and pea legumin.

In silico degree of hydrolysis was estimated using BIOPEP database, using for the gastric stage, pepsin (pH 1.3; EC 3.4.23.1), and for intestinal conditions, enzymes: pepsin (pH >2; 3.4.23.1), trypsin (EC 3.4.21.4) and chymotrypsin C (EC 3.4.21.2). There were no reported values for gastric digestibility of RuBisCO, which is represented in the graph as "n.a." (nonavailable data).

Data from Qiu, C., Sun, W., Cui, C., Zhao, M. (2013). Effect of citric acid deamidation on in vitro digestibility and antioxidant properties of wheat gluten. Food Chemistry, 141*(3), 2772–2778, doi:10.1016/j. foodchem.2013.05.072; Hansen, M., Andersen, C. A., Jensen, P. R., Hobley, T. J. (2022). Scale-up of alfalfa (Medicago sativa) protein recovery using screw presses.* Foods, 11*(20), http://www.mdpi.com/journal/ foods. doi:10.3390/foods11203229; Jiménez-Munoz, L., Tsochatzis, E. D., & Corredig, M. (2022). Impact of the structural modifications of potato protein in the digestibility process under semi-dynamic simulated human gastrointestinal in vitro system. Nutrients, 14(12). https://doi.org/10.3390/nu14122505. Jiménez-Munoz, L., Torp Nielsen, M., Roman, L., Corredig, M. (2023). Variation of in vitro digestibility of pea protein powder dispersions from commercially available sources.* Food Chemistry, 401, *doi:10.1016/j. foodchem.2022.134178.*

and intestinal digestion stages of in silico and in vitro values for patatin, high molecular weight glutenin subunit, RuBisCO and legumin. In vitro degree of hydrolysis values for these proteins were obtained from literature, it should be noted that the ingredients containing these proteins were not completely pure, as they contained other minor proteins. Specifically, the ingredients were a potato protein isolate rich in patatin fraction, deamidated wheat gluten, alfalfa (*Medicago sativa*) protein concentrate rich in RuBisCO, and a pea protein isolate rich in globulin fraction. Similar values where found when comparing in silico with reported experimental values from in vitro gastrointestinal digestion experiments. The similarities shown in Fig. 5.2 indicate the robustness of the in silico digestion to predict protein digestibility. However, the in silico model is still limited, as it does not take into account the influence of processing conditions, matrix components, and complexity (Le Feunteun et al., 2020). This is also the case for many in vitro reports. Values for the digestibility of RuBisCO (from spinach, alfalfa, microalgae, and seaweed) at the gastric stage, in vitro are not yet reported in the literature, thus not shown in Fig. 5.2.

The intestinal in silico and in vitro digestibilities for patatin were 83% and 80.0% ± 1.0 (Jiménez-Munoz et al., 2022), for gluten 81% and 86.0% ± 3.0% (Qiu et al., 2013), RuBisCO 83% and 87.4% ± 0.9% (Hansen et al., 2022), respectively. For pea legumin, it was 85%, while for a pea protein isolate it was 83% ± 2.5% (Jiménez-Munoz et al., 2023). There were no available data in the literature focusing on the in vitro digestibility of legumin; thus digestibility of all protein groups present in peas (vicilin, convicilin, legumin, albumins) was used instead to allow the comparison.

Different types of potato proteins (patatin, kunitz inhibitor, and profilin), legumin from pea, and RuBisCO from spinach and wheat (alpha-amylase inhibitor, low Mw and high Mw glutenin subunits) were further analyzed in silico for their bioactivity and allergenicity indices. Potato patatin showed the highest predicted DH% (83.4%) while kunitz inhibitor (PKI) had the lowest values (76.3%), due to its known serine protease-inhibitory activity (Komarnytsky et al., 2011). Interestingly, in spite of the low Tanimoto similarity index, suggesting low allergenicity of patatin and its digestion-derived peptides, this protein has previously been reported as a major allergen (Seppälä et al., 1999). Allergic responses in the adult population to raw potatoes have been previously documented through contact with the oral cavity and the pharynx, while an allergenic response to heat-treated potatoes has only been reported for children (Przekora et al., 2020). As a thermostable protein, patatin may retain its allergenic properties (Przekora et al., 2020). In addition, aspartyl protease inhibitor, cystatin, serine protease inhibitor, profilin, and glucanase also have allergenic properties in potatoes. As an example, the in silico analysis of potato profilin is shown in Table 5.1. The entire protein sequence of the intact protein has the highest value (1.0) of allergenicity index. In contrast, peptides derived from profilin digestion, such as VW and IG, are likely not allergens (0.84–0.86). Furthermore, the three potato proteins studied in silico showed a potential ACE-inhibitory activity, with patatin and kunitz inhibitor also showing DPP IV-inhibitory activity. The peptides IF and VW reported the highest probability to show ACE-Inhibitory activity with 0.95 and 0.80, respectively. Peptide IW showed the highest probability (0.94) to possess DPP IV-inhibitory activity, while VHW showed potential antioxidant activity, as its FRS values were high (0.54).

It is important to critically evaluate the methodologies used for the in vitro evaluations. In the case of potato protein for example, when the in silico studies summarized above were contrasted with in vitro studies, the exact peptides did not match. The discrepancies may derive from the fact that in most of the in vitro studies reported, potato proteins were not subjected to digestion by gastrointestinal enzymes, but by alcalase, or using synthetic peptide sequences. Only the study by (Basilicata et al., 2019) focused on in vitro gastrointestinal simulation, but instead of screening specific peptides, it evaluated the range of molecular weight (Mw) in which they were found. Despite these discrepancies, patatin-derived peptides are consistently showing ACE inhibitory activity in both in silico and in vitro studies (Fu et al., 2017), in particular the peptide IF (Tsai et al., 2020). Furthermore, it is predicted an FRS antioxidant activity for peptides generated in silico (VHW, 0.54) and in vitro (WG, PRY; 0.51, 0.50). The results therefore strongly suggest potato proteins contain peptides resistant to digestive enzymes, which could potentially yield protective effects against hypertension.

Pea legumin has also been extensively studied. It is known that thermolysin causes extensive hydrolysis (29.7%) before digestion (Gu et al., 2011). The tripeptide Leu-Arg-Trp (LRW) produced by this process was predicted to have potent osteoblast stimulation and ACE-inhibitory activity (0.95). Further in vitro studies (Arora et al., 2020), using the synthesized peptide LRW, demonstrated its therapeutic potential to prevent osteoporosis as it promoted osteoblastic bone formation. In addition, Wang et al. (2020) showed that the same LRW tripeptide could potentially aid in the prevention of hypertension, as the peptide had an anti-inflammatory effect in Angiotensin II-stressed vascular smooth rat muscle cells. However, when further tested in vivo *(Data shown in* Table 5.3) (Fan et al., 2022), the oral administration of the LRW was not able to significantly reduce blood pressure in rats. It was hypothesized that the reason for the discrepancy between in silico, in vitro, and in vivo results could be that the tri-peptide was not resistant to hydrolysis by gastrointestinal enzymes and could not reach the bloodstream intact (measured using a CaCo-2 cell absorption model). Thus to ensure a therapeutic effect of such nonresistant peptides, further protection strategies, such as encapsulation are required. The in silico analysis identified the tripeptide LRW sequence within the structure of 11S/legumin. After digestion, however, the hydrolysis of legumin was reported to be 84.6%, and LRW was no longer identified, due to its digestion. This example shows the importance of not only proving that a peptide is bioactive as well as its mechanism but also demonstrating its resistance to digestive enzymes.

Another important functional land-derived protein, RuBisCO was studied as an example of its potential to encode bioactive peptides. The in silico hydrolysis of RuBisCO from spinach showed about 82.8% hydrolysis. The remaining resistant peptides were predicted to have a high probability for FRS activity (HP, 0.54), ACE-inhibitory (IF; 0.95), DPP IV-inhibitory (HF, SW; 0.95, 0.93), antioxidative and α-glucosidase inhibitory activity (VW; 0.81). The elevated in silico probabilities suggest the potential of these RuBisCO peptides as therapeutic agents. In addition, the RuBisCO peptides resulting from digestion showed no allergenicity in silico.

Table 5.3 Summary of bioactive peptides found in vivo for different potato, pea, spinach and wheat proteins, diverse processing conditions used to obtain said peptides in vitro and main finding regarding the specific bioactivity found.

In vivo studies					
Source	**Protein**	**Processing conditions**	**Bioactive peptide**	**Bioactivity evaluation method and main findings**	**References**
Potato	Potato protein hydrolysate	Alcalase hydrolysis	DIKTNKPVIF	Evaluation using western blot. Oral supplementation and intraperitoneal injection of 10 mM potato peptide to the mice enhanced lipid metabolism and showed antiinflammatory activity.	Dumeus et al. (2018)
	Potato protein hydroly-sate	Alcalase hydrolysis	IF	Evaluation cell death/survival using western blot, apoptosis measured by TUNEL assay. Potato-derived peptide-IF (10 mg/kg body weight) showed ACE inhibitory peptide that reduced hypertension in hypertensive rats while regulating oxidative stress associated to hypertension.	Tsai et al. (2020)
Pea	Legumin	Peptide synthesis.	LRW	Oral administration of LRW (15 mg/kg body weight/day during 12 days) did not reduce significantly blood pressure in rats. Possibly, LRW could not reach the bloodstream intact due to low gastrointestinal instability (measured using CaCo-2 cells). Thus LRW failed to reduce blood pressure in vivo.	Fan et al. (2022)
Spinach	RuBisCO	Peptide synthesis (F-moc and purified by RP-HPLC)	MRWRD, MRW, LRIPVA, IAYKPAG	From peptides obtained by pepsin-pancreatin digestion, MRW and LRIPVA showed a potent antihypertensive effect after oral administration in hypertensive rats, with an ACE inhibition (IC_{50} value of 0.6 and 0.38 μM, respectively).	Yang et al. (2003)
			YPLDLF	Oral (100 mg/kg) and intraperitoneal (10 mg/kg) administration of rubiscolin-6 had an anxiolytic	Hirata et al. (2007)

(Continued)

Table 5.3 Summary of bioactive peptides found in vivo for different potato, pea, spinach and wheat proteins, diverse processing conditions used to obtain said peptides in vitro and main finding regarding the specific bioactivity found. *Continued*

In vivo studies					
Source	**Protein**	**Processing conditions**	**Bioactive peptide**	**Bioactivity evaluation method and main findings**	**References**
				effect. Results from this study suggest that anxiolytic activity of rubiscolin-6 is mediated by δ-opioid receptor.	
			SYLPPLTT, SYLPPLT, YHIEPV	The 3 orally administered peptides to mice showed significant anxiety-reducing effects. Contrarily, there was no effect after intraperitoneal administration. This suggest that peptides acted directly on the gastrointestinal tract. Moreover, absorption from the gastrointestinal tract may not be required for anxiolytic effects.	Kimura et al. (2018)
			YHIEPV	Orally administered YHIEPV (0.3 mg/kg/day) reduced body weight and food intake in mice with dietary obesity by increasing neural leptin responsiveness.	Kaneko et al. (2022)
Wheat	Gluten	Hydrolysate (alcalase/neutrase) further hydrolyzed by pancreatin and leucine-aminopeptidase.	PA, PG, PQ, ZP, ZLP, ZQP	Six peptides had a significant increase in human plasma (between 10–100 nM), after ingestion of wheat gluten hydrolysate. Thus it was concluded that these peptides remained intact after digestion by intestinal enzymes and are absorbed in the blood.	Ejima et al. (2018)

Data are available from literature.

In an in vitro study (Yang et al., 2001), two naturally occurring peptide fragments (rubiscolin-5 and -6) produced by pepsin and pancreatin digestion showed opioid activity. The concentration of the enzymatically released rubiscolin-5 peptide (YPLDL) was high in the digest, suggesting that oral delivery of spinach RuBisCO leads to high bioactivity. Recently (Karasawa et al., 2021) the opioid receptors response to Rubisculin-5 and -6 were tested using human embryonic kidney (HEK293) cell lines, to confirm their agonistic effects. It was concluded that these peptides could be used as selective δ-opioid receptors that can be orally administered for treating pain with lesser side effects than currently available opioid analgesics. Diverse in vivo studies (Hirata et al., 2007; Kimura et al., 2018) validated the anxiety-reducing effect of rubiscolin peptides, which were mediated by δ-opioid receptors. In these studies, it was also established that these peptides acted directly on the gastrointestinal tissue, and that absorption through the intestinal barrier may not be required for anxiolytic effects.

Another plant-derived protein group which has been studied in detail is that from wheat. The degree of hydrolysis of low and high Mw glutenin subunits have been shown to be 81% and 87.5%, respectively, during in silico analysis. Moreover, the hydrolysis values of the two alpha-amylase inhibitor and the gliadin derived from wheat were $\leq$ 80%. The full protein sequence of both, low and high, Mw glutenins has shown to have high potential allergenicity (0.9 and 1.0, respectively). Glutenin-derived peptides (CSP, SSCH, SHHQ, and SSL) were also predicted as potential allergens (0.82–0.84). A similar in silico prediction was found for the full sequence of α-amylase inhibitors (highest probability score: 1.0) and the digestion-derived peptides (SHN, VCY, and ISE) (0.83–0.84). Commonly found in cereal sources, the consumption of amylase and trypsin inhibitors has been associated to intestinal responses associated with nonceliac and celiac disease, as well as with food and respiratory allergies (Geisslitz et al., 2022). Meanwhile, the proteolytic resistant 33-mer gliadin-peptide (LQLQPFPQPQLPYPQPQLPYPQPQLPYPQPQPF) has been described as the major immunogenic sequence responsible for triggering the adaptive immune response in celiac disease (Schalk et al., 2017). The proteolytic resistance to digestive enzymes by gliadin from wheat gluten (33% of the peptide fragment remained intact) has been demonstrated to be due to the high content of proline and glutamine (Gutiérrez et al., 2017). Generally, gastrointestinal stability is considered to be influential in the allergenic potential of proteins (Fu, 2002; Rao et al., 2018; Verhoeckx et al., 2019). However, small peptide fragments derived from the initial protein structure may also possess allergenic properties (Bøgh et al., 2009; Hartmann et al., 2007; Moreno, 2007). In addition, other digestion-derived peptides from gliadin, glutenin subunits, and α-amylase inhibitors, such as SF (0.94), VW (0.81) SW (0.93), and IF (0.95), have been shown to have ACE and/or DPP-IV inhibitory activity using in silico analysis. In vitro generated peptides, such as LY, PY, and YQ, from hydrolyzed wheat gluten, all show an FRS activity (scores >0.50). The seemingly contradictive results between bioactivity and potential negative side effects highlight the importance of considering the toxicity and allergenicity of peptide sources when screening for bioactivity. This comprehensive and holistic approach is necessary to ensure the development of functional and safe food products. This is particularly important when studying proteins that are less established in our diet but suggested as potential new, more sustainable ingredients.

5.1.4 Predicting the bioactivity mechanisms of peptides in the human body

It is possible to further analyze the potential targets in the human body, once the analyses described above show their potential bioactivity and stability to digestion. Using the examples summarized in Table 5.2, with demonstrated bioactivity with at least two models (i.e., in silico and in vitro), an additional prediction platform (SwissTargetPrediction) was applied. Table 5.4 shows the probability of these peptides to bind to a particular binding site. Furthermore, for those with the highest binding probability, their binding to specific protein is shown in Table 5.2 Peptides that showed validated bioactivity in at least two models were the following: IF from patatin (in silico and in vivo), LRW (in vitro and in vivo) from pea legumin and the RuBisCO Spinach-derived peptide YPLDLF (in vitro and in vivo).

As shown in Table 5.4, the patatin IF peptide exhibited a 40% probability of binding to proteases, more specifically, a high binding ACE (49%) and Calpain 1 (53%) probability (data shown in Table 5.5). While the effect of patatin peptides on Calpain 1 has only been shown in silico and is yet to be confirmed, the ACE result is in agreement with the present in silico analysis, as well as with the in vitro (Pihlanto et al., 2008) and in vivo data (Tsai et al., 2020) are presented in Table 5.2 for IF. In addition, IF peptide has been demonstrated to protect cardiac tissue and reduce cardiac hypertrophy and fibrosis in spontaneously hypertensive rats (Daroit & Brandelli, 2021).

This analysis clearly demonstrates how in silico studies can aid in the development of healthy foods from less studied proteins. For example, the data presented in this study supports potatoes as promising ingredients for the elaboration of functional foods with probable and favorable effects on cardiovascular health. Nevertheless, additional research is still needed to corroborate the bioactive peptide and examine the antihypertensive effect of the nutraceutical product in humans. LRW from peas showed a 33% probability of binding to C3a anaphylatoxin chemotactic receptor, a family G-protein-coupled receptor, involved in inflammatory processes by mediating smooth muscle contraction and increasing vascular permeability (Crass et al., 1996). It is important to highlight that LRW was shown to be a nonresistant peptide; thus therapeutic effects reported in the literature are unlikely unless a successful encapsulation strategy is applied. In addition, RuBisCO peptide YPLDLF showed a 40% binding probability to family A/G-protein-coupled receptors and a high probability (58%) to bind to δ-opioid receptor, but only a 22% probability to bind to μ-opioid receptors. These results were validated in vitro and in vivo (Hirata et al., 2007; Karasawa et al., 2021), in which RuBisCO peptides -5/-6 showed partial activation and anxiolytic effects on μ- and δ-opioid receptors.

Table 5.4 Binding probability of three bioactive peptides (IF, LRW, YPLDLF) derived from digestion of plant protein sources to target proteins in the human body as predicted by in silico analysis using SwissTargetPrediction tool.

<table>
<tr><th>Peptide structure</th><th>Binding probability of bioactive peptide to target receptors</th></tr>
<tr><td>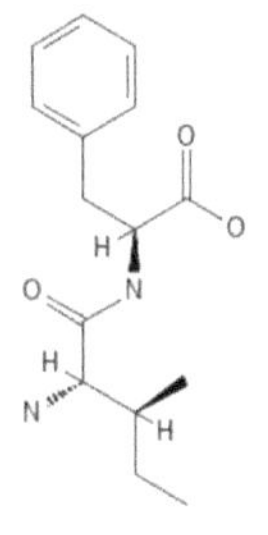
IF</td><td>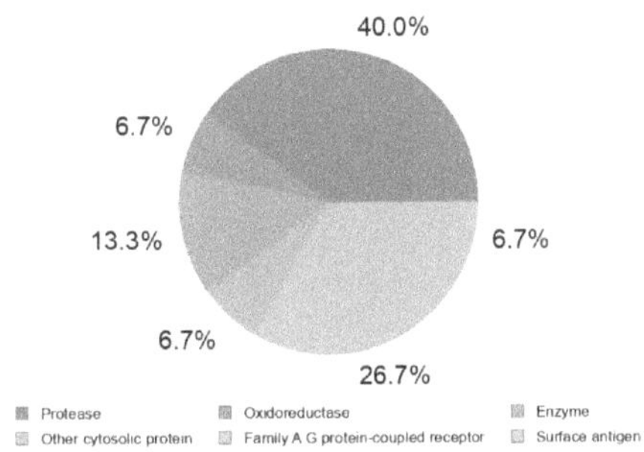
</td></tr>
<tr><td>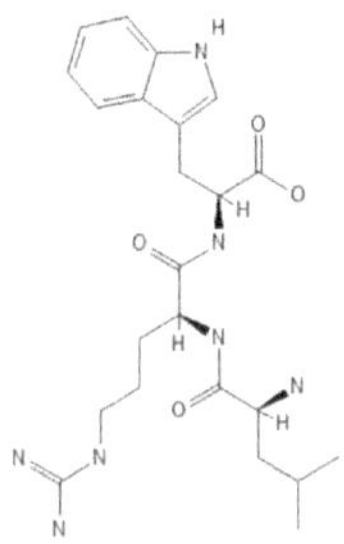
LRW</td><td>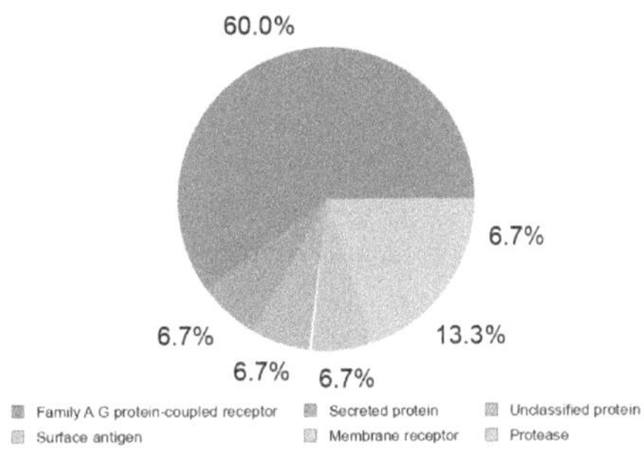
</td></tr>
<tr><td>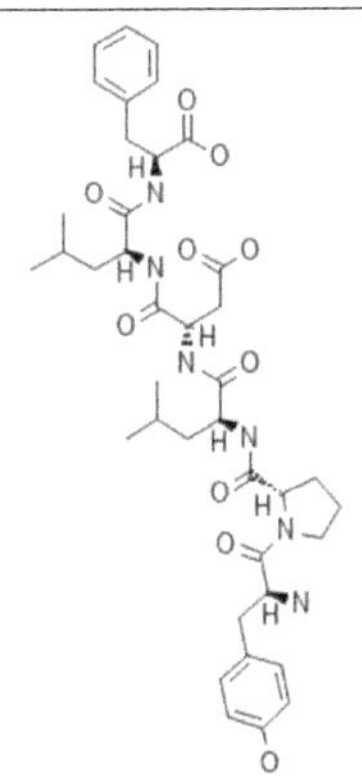
YPLDLF</td><td>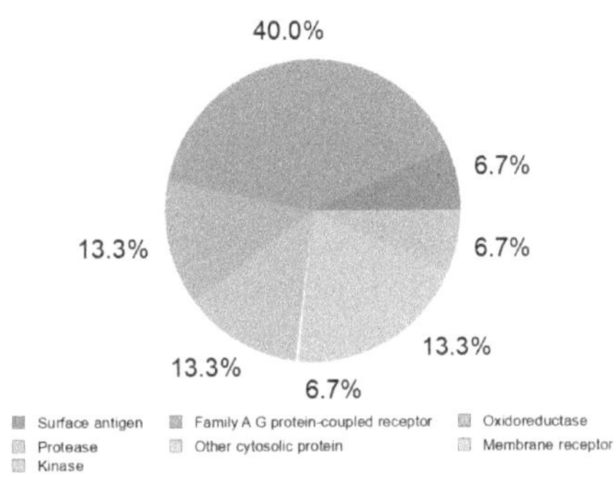
</td></tr>
</table>

Table 5.5 Predicted binding probabilities of three bioactive plant-derived peptides (IF, LRW, YPLDLF) to human protein receptors, as determined by in silico analysis.

SEQUENCE	Protein 1	Protein 2	Protein 3
IF	Calpain 1 (0.53)	ACE (0.49)	Cyclooxygenase-2 (0.17)
LRW	C3a-ACR (0.33)	NTR 1 (0.24)	MCR4 (0.18)
YPLDLF	HLA class I histocompatibility antigen A-3 (0.45)	Mu opioid receptor (by homology) (0.28)	Delta opioid receptor (by homology) (0.28)

ACE, *Angiotensin-converting enzyme;* C3a-ACR, *anaphylatoxin chemotactic receptor;* CO2, *cyclooxygenase-2;* NTR1, *neurotensin receptor 1;* MCR4; *melanocortin receptor 4;* HLA1-HCA3; *HLA class I histocompatibility antigen A-3;* MOR; *Mu opioid receptor;* DOR; *delta opioid receptor.*

5.2 Conclusions

Advances with in silico and in vitro models allow for better screening of dietary proteins and their potential bioactivity, and in the future, they will aid in discovery of more peptide-based therapeutic approaches to be delivered orally. This study brought some examples of peptides first evaluated from in silico gastrointestinal digestion of different plant-based proteins from potato, pea, spinach and wheat that were profiled based on their potential bioactivity, allergenicity, toxicity, degree of hydrolysis and target properties. The use of multiple tools, such as BIOPEP, PeptideRanker, AnOxPePred, AllergenFP, ToxinPred, and SwissTargetPrediction, to evaluate these peptides properties provided an example of a comprehensive and systematic approach to peptide screening and prediction. A rigorous comparison of the in silico results with in vitro and then, in vivo studies would allow for evaluating more closely the effectiveness of in silico models as an analytical tool for peptide screening and prediction. The focus on potato, pea, spinach, and wheat proteins, which are commonly consumed and have been shown to contain bioactive peptides, is also a relevant and important topic. This book chapter presents a compilation of strong evidence, as reported by in silico, in vitro, and in vivo studies, suggesting that the patatin fraction derived from potatoes contains bioactive peptides (IF) with ACE inhibitory activity which would resist during digestion. A target analysis of the potato-derived peptide IF suggests high binding probability (40%) to proteases as targeted proteins in the human body, more specifically, Calpain-1, angiotensin-converting enzyme, and cyclooxygenase-2.

Overall, the integrative use of in silico, in vitro, and in vivo digestion models has the potential to be a game-changer in the field of digestive health, leading to a wealth of new insights and breakthroughs. This is of particular importance in food science, where the current green transition to more sustainable diets is presenting to consumer food products containing protein sources not fully understood.

Furthermore, the understanding of the effect of processing on the modification of digestion kinetics and peptide resistance is still at its infancy. It is therefore important to conclude that with the improved sophistication and relevance of these in silico and in vitro models it is now possible to carry out a comprehensive approach to understanding the complexities of digestion which will pave the way for the development of not only more sustainable, but also healthier foods.

References

Arora, H., Shang, N., Bhullar, K. S., & Wu, J. (2020). Pea protein-derived tripeptide LRW shows osteoblastic activity on MC3T3-E1 cells: Via the activation of the Akt/Runx2 pathway. *Food and Function*, *11*(8), 7197–7207. Available from https://doi.org/10.1039/d0fo00497a, http://pubs.rsc.org/en/journals/journal/fo.

Assatory, A., Vitelli, M., Rajabzadeh, A. R., & Legge, R. L. (2019). Dry fractionation methods for plant protein, starch and fiber enrichment: A review. *Trends in Food Science and Technology*, *86*, 340–351. Available from https://doi.org/10.1016/j.tifs.2019.02.006, http://www.elsevier.com/wps/find/journaldescription.cws_home/601278/description#description.

Barati, M., Javanmardi, F., Mousavi Jazayeri, S. M. H., Jabbari, M., Rahmani, J., Barati, F., Nickho, H., Davoodi, S. H., Roshanravan, N., & Mousavi Khaneghah, A. (2020). Techniques, perspectives, and challenges of bioactive peptide generation: A comprehensive systematic review. *Comprehensive Reviews in Food Science and Food Safety*, *19*(4), 1488–1520. Available from https://doi.org/10.1111/1541-4337.12578, http://onlinelibrary.wiley.com/journal/10.1111/(ISSN)1541-4337.

Basilicata, M. G., Pepe, G., Rapa, S. F., Merciai, F., Ostacolo, C., Manfra, M., Sarno, V. D., Autore, G., de Vita, D., Marzocco, S., & Campiglia, P. (2019). Anti-inflammatory and antioxidant properties of dehydrated potato-derived bioactive compounds in intestinal cells. *International Journal of Molecular Sciences*, *20*(23). Available from https://doi.org/10.3390/ijms20236087, https://www.mdpi.com/1422-0067/20/23/6087/pdf.

Behrendt, R., White, P., & Offer, J. (2016). Advances in Fmoc solid-phase peptide synthesis. *Journal of Peptide Science*, *22*(1), 4–27. Available from https://doi.org/10.1002/psc.2836.

Brodkorb, A., Egger, L., Alminger, M., Alvito, P., Assunção, R., Ballance, S., Bohn, T., Bourlieu-Lacanal, C., Boutrou, R., Carrière, F., Clemente, A., Corredig, M., Dupont, D., Dufour, C., Edwards, C., Golding, M., Karakaya, S., Kirkhus, B., Le Feunteun, S., … Recio, I. (2019). INFOGEST static in vitro simulation of gastrointestinal food digestion. *Nature Protocols*, *14*(4), 991–1014. Available from https://doi.org/10.1038/s41596-018-0119-1.

Bøgh, K. L., Kroghsbo, S., Dahl, L., Rigby, N. M., Barkholt, V., Mills, E. N. C., & Madsen, C. B. (2009). Digested Ara h 1 has sensitizing capacity in Brown Norway rats. *Clinical & Experimental Allergy*, *39*(10), 1611–1621. Available from https://doi.org/10.1111/j.1365-2222.2009.03333.x.

Chai, K. F., Ng, K. R., Samarasiri, M., & Chen, W. N. (2022). Precision fermentation to advance fungal food fermentations. *Current Opinion in Food Science*, *47*100881. Available from https://doi.org/10.1016/j.cofs.2022.100881.

Comino, I., Real, A., Vivas, S., Síglez, M. Á., Caminero, A., Nistal, E., Casqueiro, J., Rodríguez-Herrera, A., Cebolla, A., & Sousa, C. (2012). Monitoring of gluten-free diet compliance in celiac patients by assessment of gliadin 33-mer equivalent epitopes in feces. *The American Journal of Clinical Nutrition*, *95*(3), 670–677. Available from https://doi.org/10.3945/ajcn.111.026708.

Conlon, J. M. (2007). Purification of naturally occurring peptides by reversed-phase HPLC. *Nature Protocols*, *2*(1), 191–197. Available from https://doi.org/10.1038/nprot.2006.437.

Crass, T., Raffetseder, U., Martin, U., Grove, M., Klos, A., Köhl, J., & Bautsch, W. (1996). Expression cloning of the human C3a anaphylatoxin receptor (C3aR) from differentiated U-937 cells. *European Journal of Immunology*, *26*(8), 1944–1950. Available from https://doi.org/10.1002/eji.1830260840, http://onlinelibrary.wiley.com/journal/10.1002/(ISSN)1521-4141.

Daina, A., Michielin, O., & Zoete, V. (2019). SwissTargetPrediction: Updated data and new features for efficient prediction of protein targets of small molecules. *Nucleic Acids Research*, *47*(1), W357–W3664. Available from https://doi.org/10.1093/nar/gkz382, https://academic.oup.com/nar/issue.

Daroit, D. J., & Brandelli, A. (2021). In vivo bioactivities of food protein-derived peptides – A current review. *Current Opinion in Food Science*, *39*, 120–129. Available from https://doi.org/10.1016/j.cofs.2021.01.002, http://www.journals.elsevier.com/current-opinion-in-food-science/.

Dimitrov, I., Naneva, L., Doytchinova, I., & Bangov, I. (2014). AllergenFP: Allergenicity prediction by descriptor fingerprints. *Bioinformatics (Oxford, England)*, *30*(6), 846–851. Available from https://doi.org/10.1093/bioinformatics/btt619.

Duan, X., Zhang, M., & Chen, F. (2021). Prediction and analysis of antimicrobial peptides from rapeseed protein using in silico approach. *Journal of Food Biochemistry*, *45*(4), 1–12. Available from https://doi.org/10.1111/jfbc.13598.

De Angelis, D., Pasqualone, A., Allegretta, I., Porfido, C., Terzano, R., Squeo, G., & Summo, C. (2021). Antinutritional factors, mineral composition and functional properties of dry fractionated flours as influenced by the type of pulse. *Heliyon*, *7*(2). Available from https://doi.org/10.1016/j.heliyon.2021.e06177.

Dumeus, S., Shibu, M. A., Lin, W. T., Wang, M. F., Lai, C. H., Shen, C. Y., Lin, Y. M., Viswanadha, V. P., Kuo, W. W., & Huang, C. Y. (2018). Bioactive peptide improves diet-induced hepatic fat deposition and hepatocyte proinflammatory response in SAMP8 ageing mice. *Cellular Physiology and Biochemistry*, *48*(5), 1942–1952. Available from https://doi.org/10.1159/000492518, http://www.karger.ch/journals/cpb/cpb_jh.htm.

Dumoulin, L., Jacquet, N., Malumba, P., Richel, A., & Blecker, C. (2021). Dry and wet fractionation of plant proteins: How a hybrid process increases yield and impacts nutritional value of faba beans proteins. *Innovative Food Science & Emerging Technologies*, *72*. Available from https://doi.org/10.1016/j.ifset.2021.102747.

Ejima, A., Nakamura, M., Suzuki, Y. A., & Sato, K. (2018). Identification of food-derived peptides in human blood after ingestion of corn and wheat gluten hydrolysates. *Journal of Food Bioactives*, *2*. Available from https://doi.org/10.31665/jfb.2018.2145.

Fan, H., Wu, K., & Wu, J. (2022). Pea-derived tripeptide LRW fails to reduce blood pressure in spontaneously hypertensive rats due to its low gastrointestinal stability and transepithelial permeability. *Food Bioscience*, *49*(May)101964. Available from https://doi.org/10.1016/j.fbio.2022.101964.

Fernández-Tomé, S., & Hernández-Ledesma, B. (2020). Gastrointestinal digestion of food proteins under the effects of released bioactive peptides on digestive health. *Molecular Nutrition and Food Research*, *64*(21), 1–42. Available from https://doi.org/10.1002/mnfr.202000401.

Fu, T. (2002). Digestion stability as a criterion for protein allergenicity assessment. *Annals of the New York Academy of Sciences*, *964*, 99–110. Available from https://doi.org/10.1111/j.1749-6632.2002.tb04135.x.

Fu, Y., Alashi, A. M., Young, J. F., Therkildsen, M., & Aluko, R. E. (2017). Enzyme inhibition kinetics and molecular interactions of patatin peptides with angiotensin I-converting enzyme and renin. *International Journal of Biological Macromolecules*, *101*, 207–213. Available from https://doi.org/10.1016/j.ijbiomac.2017.03.054, http://www.elsevier.com/locate/ijbiomac.

Gallego, M., Grootaert, C., Mora, L., Aristoy, M. C., Van Camp, J., & Toldrá, F. (2016). Transepithelial transport of dry-cured ham peptides with ACE inhibitory activity through a Caco-cell monolayer. *Journal of Functional Foods*, *21*, 388–395. Available from https://doi.org/10.1016/j.jff.2015.11.046, http://www.elsevier.com/wps/find/journaldescription.cws_home/717426/description#description.

Geisslitz, S., Weegels, P., Shewry, P., Zevallos, V., Masci, S., Sorrells, M., Gregorini, A., Colomba, M., Jonkers, D., Huang, X., De Giorgio, R., Caio, G. P., D'Amico, S., Larré, C., & Brouns, F. (2022). Wheat amylase/trypsin inhibitors (ATIs): Occurrence, function and health aspects. *European Journal of Nutrition*, *61*(6), 2873–2880. Available from https://doi.org/10.1007/s00394-022-02841-y, https://www.springer.com/journal/394.

Gerten, D., Heck, V., Jägermeyr, J., Bodirsky, B. L., Fetzer, I., Jalava, M., Kummu, M., Lucht, W., Rockström, J., Schaphoff, S., & Schellnhuber, H. J. (2020). Feeding ten billion people is possible within four terrestrial planetary boundaries. *Nature Sustainability*, *3*(3), 200–208. Available from https://doi.org/10.1038/s41893-019-0465-1, http://www.nature.com/natsustain/.

Graça, J., Godinho, C. A., & Truninger, M. (2019). Reducing meat consumption and following plant-based diets: Current evidence and future directions to inform integrated transitions. *Trends in Food Science and Technology*, *91*(July), 380–390. Available from https://doi.org/10.1016/j.tifs.2019.07.046.

Gupta, S., Kapoor, P., Chaudhary, K., Gautam, A., Kumar, R., & Raghava, G. P. S. (2013). In silico approach for predicting toxicity of peptides and proteins. *PLoS One*, *8*(9). Available from https://doi.org/10.1371/journal.pone.0073957India, http://www.plosone.org/article/fetchObjectAttachment.action;jsessionid = BFD4330D2451F927CB9068BBF3DEFC6B?uri = info%3Adoi%2F10.1371%2Fjournal.pone.0073957&representation = PDF.

Gutiérrez, S., Pérez-Andrés, J., Martínez-Blanco, H., Ferrero, M. A., Vaquero, L., Vivas, S., Casqueiro, J., & Rodríguez-Aparicio, L. B. (2017). The human digestive tract has proteases capable of gluten hydrolysis. *Molecular Metabolism*, *6*(7), 693–702. Available from https://doi.org/10.1016/j.molmet.2017.05.008, http://www.elsevier.com/wps/find/journaldescription.cws_home/728618/description.

Gu, Y., Majumder, K., & Wu, J. (2011). QSAR-aided in silico approach in evaluation of food proteins as precursors of ACE inhibitory peptides. *Food Research International*, *44*(8), 2465–2474. Available from https://doi.org/10.1016/j.foodres.2011.01.051.

Hansen, M., Andersen, C. A., Jensen, P. R., & Hobley, T. J. (2022). Scale-up of alfalfa (Medicago sativa) protein recovery using screw presses. *Foods*, *11*(20). Available from https://doi.org/10.3390/foods11203229, http://www.mdpi.com/journal/foods.

Hartmann, R., Wal, J. M., Bernard, H., & Pentzien, A. K. (2007). Cytotoxic and allergenic potential of bioactive proteins and peptides. *Current Pharmaceutical Design*, *13*(9), 897–920. Available from https://doi.org/10.2174/138161207780414232, http://www.ingentaconnect.com/content/ben/cpd/2007/00000013/00000009/art00004, Germany.

Hirata, H., Sonoda, S., Agui, S., Yoshida, M., Ohinata, K., & Yoshikawa, M. (2007). Rubiscolin-6, a δ opioid peptide derived from spinach Rubisco, has anxiolytic effect via activating σ1 and dopamine D1 receptors. *Peptides*, *28*(10), 1998–2003. Available from https://doi.org/10.1016/j.peptides.2007.07.024.

Jakubczyk, A., & Baraniak, B. (2014). Angiotensin I converting enzyme inhibitory peptides obtained after in vitro hydrolysis of pea (Pisum sativum var. Bajka) globulins. *BioMed Research International*, *2014*. Available from https://doi.org/10.1155/2014/438459.

Jakubczyk, A., Karas, M., Rybczynska-Tkaczyk, K., Zielinska, E., & Zielinski, D. (2020). Current trends of bioactive peptides - New sources and therapeutic effect. *Foods*, *9*(7). Available from https://doi.org/10.3390/foods9070846, https://www.mdpi.com/2304-8158/9/7/846.

Jiménez-Munoz, L., Torp Nielsen, M., Roman, L., & Corredig, M. (2023). Variation of in vitro digestibility of pea protein powder dispersions from commercially available sources. *Food Chemistry*, *401*. Available from https://doi.org/10.1016/j.foodchem.2022.134178.

Jiménez-Munoz, L., Tsochatzis, E. D., & Corredig, M. (2022). Impact of the structural modifications of potato protein in the digestibility process under semi-dynamic simulated human gastrointestinal in vitro system. *Nutrients*, *14*(12). Available from https://doi.org/10.3390/nu14122505.

Kaneko, K., Takekuma, Y., Goto, T., & Ohinata, K. (2022). An orally active plant Rubisco-derived peptide increases neuronal leptin responsiveness. *Scientific Reports*, *12*(1). Available from https://doi.org/10.1038/s41598-022-12595-6.

Karasawa, Y., Miyano, K., Fujii, H., Mizuguchi, T., Kuroda, Y., Nonaka, M., Komatsu, A., Ohshima, K., Yamaguchi, M., Yamaguchi, K., Iseki, M., Uezono, Y., & Hayashida, M. (2021). In vitro analyses of spinach-derived opioid peptides, rubiscolins: Receptor selectivity and intracellular activities through G protein- and β-arrestin-mediated pathways. *Molecules (Basel, Switzerland)*, *26*(19). Available from https://doi.org/10.3390/molecules26196079.

Kimura, S., Uchida, T., Tokuyama, Y., Hosokawa, M., Nakato, J., Kurabayashi, A., Sato, M., Suzuki, H., Kanamoto, R., & Ohinata, K. (2018). Identification of Rubisco anxiolytic-like peptides (rALPs) by comprehensive analysis of spinach green leaf protein digest. *Biochemical and Biophysical Research Communications*, *505*(4), 1050–1056. Available from https://doi.org/10.1016/j.bbrc.2018.09.195.

Komarnytsky, S., Cook, A., & Raskin, I. (2011). Potato protease inhibitors inhibit food intake and increase circulating cholecystokinin levels by a trypsin-dependent mechanism. *International Journal of Obesity*, *35*(2), 236–243. Available from https://doi.org/10.1038/ijo.2010.192.

Kęska, P., & Stadnik, J. (2021). Potential DPP IV inhibitory peptides from dry-cured pork loins after hydrolysis: An in vitro and in silico study. *Current Issues in Molecular Biology*, *43*(3), 1335–1349. Available from https://doi.org/10.3390/cimb43030095.

Le Feunteun, S., Mackie, A. R., & Dupont, D. (2020). In silico trials of food digestion and absorption: How far are we? *Current Opinion in Food Science*, *31*, 121–125. Available from https://doi.org/10.1016/j.cofs.2020.04.006, http://www.journals.elsevier.com/current-opinion-in-food-science/.

Liu, W. Y., Zhang, J. T., Miyakawa, T., Li, G. M., Gu, R. Z., & Tanokura, M. (2021). Antioxidant properties and inhibition of angiotensin-converting enzyme by highly active peptides from wheat gluten. *Scientific Reports*, *11*(1). Available from https://doi.org/10.1038/s41598-021-84820-7, http://www.nature.com/srep/index.html.

Ložnjak Švarc, P., Jensen, M. B., Langwagen, M., Poulsen, A., Trolle, E., & Jakobsen, J. (2022). Nutrient content in plant-based protein products intended for food composition databases. *Journal of Food Composition and Analysis*, *106*. Available from https://doi.org/10.1016/j.jfca.2021.104332, http://www.elsevier.com/inca/publications/store/6/2/2/8/7/8/index.htt.

Lu, X., Li, C., & Huang, Q. (2019). Combining in vitro digestion model with cell culture model: Assessment of encapsulation and delivery of curcumin in milled starch particle stabilized Pickering emulsions. *International Journal of Biological Macromolecules*, *139*, 917–924. Available from https://doi.org/10.1016/j.ijbiomac.2019.08.078.

Lynch, H., Johnston, C., & Wharton, C. (2018). Plant-based diets: Considerations for environmental impact, protein quality, and exercise performance. *Nutrients*, *10*(12), 1–16. Available from https://doi.org/10.3390/nu10121841.

Majura, J. J., Cao, W., Chen, Z., Htwe, K. K., Li, W., Du, R., Zhang, P., Zheng, H., & Gao, J. (2022). The current research status and strategies employed to modify food-derived bioactive peptides. *Frontiers in Nutrition*, *9*. Available from https://doi.org/10.3389/fnut.2022.950823, Journal.frontiersin.org/journal/nutrition.

Minkiewicz, P., Iwaniak, A., & Darewicz, M. (2019). BIOPEP-UWM database of bioactive peptides: Current opportunities. *International Journal of Molecular Sciences*, *20*(23). Available from https://doi.org/10.3390/ijms20235978.

Mooney, C., Haslam, N. J., Pollastri, G., & Shields, D. C. (2012). Towards the improved discovery and design of functional peptides: Common features of diverse classes permit generalized prediction of bioactivity. *PLoS One*, *7*(10). Available from https://doi.org/10.1371/journal.pone.0045012, http://www.plosone.org/article/fetchObjectAttachment.action?uri = info%3Adoi%2F10.1371%2Fjournal.pone.0045012&representation = PDF, Ireland.

Mora, L., Aristoy, M. C., & Toldrá, F. (2015). Bioactive peptides in foods. *Encyclopedia of Food and Health*, 395–400. Available from http://www.sciencedirect.com/science/book/9780123849533, 10.1016/B978-0-12-384947-2.00067-2.

Moreno, F. J. (2007). Gastrointestinal digestion of food allergens: Effect on their allergenicity. *Biomedicine and Pharmacotherapy*, *61*(1), 50–60. Available from https://doi.org/10.1016/j.biopha.2006.10.005.

Mäde, V., Els-Heindl, S., & Beck-sickinger, A. G. (2014). Automated solid-phase peptide synthesis to obtain therapeutic peptides. *Beilsetein Journal of Organic Chemistry*, *10*, 1197–1212. Available from https://doi.org/10.3762/bjoc.10.118.

Ohanenye, I. C., Ekezie, F. G. C., Sarteshnizi, R. A., Boachie, R. T., Emenike, C. U., Sun, X., Nwachukwu, I. D., & Udenigwe, C. C. (2022). Legume seed protein digestibility as influenced by traditional and emerging physical processing technologies. *Foods*, *11*(15). Available from https://doi.org/10.3390/foods11152299, http://www.mdpi.com/journal/foods.

Olsen, T. H., Yesiltas, B., Marin, F. I., Pertseva, M., García-Moreno, P. J., Gregersen, S., Overgaard, M. T., Jacobsen, C., Lund, O., Hansen, E. B., & Marcatili, P. (2020). AnOxPePred: Using deep learning for the prediction of antioxidative properties of peptides. *Scientific Reports*, *10*(1). Available from https://doi.org/10.1038/s41598-020-78319 w, http://www.nature.com/srep/index.html.

Ouberai, M. M., Dos Santos, A. L. G., Kinna, S., Madalli, S., Hornigold, D. C., Baker, D., Naylor, J., Sheldrake, L., Corkill, D. J., Hood, J., Vicini, P., Uddin, S., Bishop, S., Varley, P. G., & Welland, M. E. (2017). Controlling the bioactivity of a peptide hormone in vivo by reversible self-assembly. *Nature Communications*, *8*(1). Available from https://doi.org/10.1038/s41467-017-01114-1, http://www.nature.com/ncomms/index.html.

Pihlanto, A., Akkanen, S., & Korhonen, H. J. (2008). ACE-inhibitory and antioxidant properties of potato (Solanum tuberosum). *Food Chemistry*, *109*(1), 104–112. Available from https://doi.org/10.1016/j.foodchem.2007.12.023.

Przekora, J., Wawrzyniak, A., Bujnowska, A., Rustecka, A., & Kalicki, B. (2020). Severe atopic dermatitis in a boy with potato allergy. *Pediatria i Medycyna Rodzinna*, *16*(3), 301–305. Available from https://doi.org/10.15557/pimr.2020.0055.

Qiu, C., Sun, W., Cui, C., & Zhao, M. (2013). Effect of citric acid deamidation on in vitro digestibility and antioxidant properties of wheat gluten. *Food Chemistry*, *141*(3), 2772–2778. Available from https://doi.org/10.1016/j.foodchem.2013.05.072.

Rao, H., Tian, Y., Fu, W., & Xue, W. (2018). In vitro digestibility and immunoreactivity of thermally processed peanut. *Food and Agricultural Immunology*, *29*(1), 989–1001. Available from https://doi.org/10.1080/09540105.2018.1499710.

Rubio, L. A., Pérez, A., Ruiz, R., Guzmán, M. A., Aranda-Olmedo, I., & Clemente, A. (2014). Characterization of pea (Pisum sativum) seed protein fractions. *Journal of the Science of Food and Agriculture*, *94*(2), 280–287. Available from https://doi.org/10.1002/jsfa.6250.

Schalk, K., Lang, C., Wieser, H., Koehler, P., & Scherf, K. A. (2017). Quantitation of the immunodominant 33-mer peptide from α-gliadin in wheat flours by liquid chromatography tandem mass spectrometry. *Scientific Reports*, *7*. Available from https://doi.org/10.1038/srep45092, http://www.nature.com/srep/index.html.

Schutyser, M. A. I., & van der Goot, A. J. (2011). The potential of dry fractionation processes for sustainable plant protein production. *Trends in Food Science and Technology*, *22*(4), 154–164. Available from https://doi.org/10.1016/j.tifs.2010.11.006.

Seppälä, U., Alenius, H., Turjanmaa, K., Reunala, T., Palosuo, T., & Kalkkinen, N. (1999). Identification of patatin as a novel allergen for children with positive skin prick test responses to raw potato. *Journal of Allergy and Clinical Immunology*, *103*(1), 165–171. Available from https://doi.org/10.1016/s0091-6749(99)70541-5.

Shrestha, S., Hag, L. V. T., Haritos, V. S., & Dhital, S. (2021). Lupin proteins: Structure, isolation and application. *Trends in Food Science and Technology*, *116*, 928–939. Available from https://doi.org/10.1016/j.tifs.2021.08.035, http://www.elsevier.com/wps/find/journaldescription.cws_home/601278/description#description.

Sánchez, A., & Vázquez, A. (2017). Bioactive peptides: A review. *Food Quality and Safety*, *1*(1), 29–46. Available from https://doi.org/10.1093/fqsafe/fyx006.

Sá, A. G. A., Moreno, Y. M. F., & Carciofi, B. A. M. (2020). Plant proteins as high-quality nutritional source for human diet. *Trends in Food Science and Technology*, *97* (January), 170–184. Available from https://doi.org/10.1016/j.tifs.2020.01.011.

Tsai, B. C. K., Hsieh, D. J. Y., Lin, W. T., Tamilselvi, S., Day, C. H., Ho, T. J., Chang, R. L., Viswanadha, V. P., Kuo, C. H., & Huang, C. Y. (2020). Functional potato bioactive peptide intensifies Nrf2-dependent antioxidant defense against renal damage in hypertensive rats. *Food Research International*, *129*. Available from https://doi.org/10.1016/j.foodres.2019.108862, http://www.elsevier.com/inca/publications/store/4/2/2/9/7/0.

Tyagi, A., Daliri, E. B. M., Ofosu, F. K., Yeon, S. J., & Oh, D. H. (2020). Food-derived opioid peptides in human health: A review. *International Journal of Molecular Sciences*, *21*(22), 1–26. Available from https://doi.org/10.3390/ijms21228825, https://www.mdpi.com/1422-0067/21/22/8813/pdf.

Verhoeckx, K., Bøgh, K. L., Dupont, D., Egger, L., Gadermaier, G., Larré, C., Mackie, A., Menard, O., Adel-Patient, K., Picariello, G., Portmann, R., Smit, J., Turner, P., Untersmayr, E., & Epstein, M. M. (2019). The relevance of a digestibility evaluation in the allergenicity risk assessment of novel proteins. Opinion of a joint initiative of COST action ImpARAS and COST action INFOGEST. *Food and Chemical Toxicology*, *129*, 405–423. Available from https://doi.org/10.1016/j.fct.2019.04.052, http://www.elsevier.com/locate/foodchemtox.

Vogelsang, M. D., Petersen, I. L., Joehnke, M. S., Sørensen, J. C., Bez, J., Detzel, A., Busch, M., Krueger, M., Mahony, J. A. O., Arendt, E. K., & Zannini, E. (2020). Comparison of faba bean protein ingredients environmental performance. *Foods*, *9*.

Wang, X., Bhullar, K. S., Fan, H., Liao, W., Qiao, Y., Su, D., & Wu, J. (2020). Regulatory effects of a pea-derived peptide Leu-Arg-Trp (LRW) on dysfunction of rat aortic vascular smooth muscle cells against angiotensin II stimulation. *Journal of Agricultural and Food Chemistry*, *68*(13), 3947–3953. Available from https://doi.org/10.1021/acs.jafc.0c00028, http://pubs.acs.org/journal/jafcau.

Xie, N., Liu, S., Wang, C., & Li, B. (2014). Stability of casein antioxidant peptide fractions during in vitro digestion/Caco-2 cell model: Characteristics of the resistant peptides. *European Food Research and Technology*, *239*(4), 577–586. Available from https://doi.org/10.1007/s00217-014-2253-5.

Yang, Y., Marczak, E. D., Yokoo, M., Usui, H., & Yoshikawa, M. (2003). Isolation and antihypertensive effect of angiotensin I-converting enzyme (ACE) inhibitory peptides from spinach Rubisco. *Journal of Agricultural and Food Chemistry*, *51*(17), 4897–4902. Available from https://doi.org/10.1021/jf026186y.

Yang, S., Yunden, J., Sonoda, S., Doyama, N., Lipkowski, A. W., Kawamura, Y., & Yoshikawa, M. (2001). Rubiscolin, a δ selective opioid peptide derived from plant Rubisco. *FEBS Letters*, *509*(2), 213–217. Available from https://doi.org/10.1016/S0014-5793(01)03042-3.

Yan, J., Zhao, J., Yang, R., & Zhao, W. (2019). Bioactive peptides with antidiabetic properties: A review. *International Journal of Food Science & Technology*, *54*(6), 1909–1919. Available from https://doi.org/10.1111/ijfs.14090.

Yust, M. M., Pedroche, J., Girón-Calle, J., Alaiz, M., Millán, F., & Vioque, J. (2003). Production of ace inhibitory peptides by digestion of chickpea legumin with alcalase. *Food Chemistry*, *81*(3), 363–369. Available from https://doi.org/10.1016/S0308-8146(02)00431-4.

Zaky, A. A., Simal-Gandara, J., Eun, J. B., Shim, J. H., & Abd El-Aty, A. M. (2022). Bioactivities, applications, safety, and health benefits of bioactive peptides from food and by-products: A review. *Frontiers in Nutrition*, *8*. Available from https://doi.org/10.3389/fnut.2021.815640, journal.frontiersin.org/journal/nutrition.

Zhang, M., & Mu, T. H. (2017). Identification and characterization of antioxidant peptides from sweet potato protein hydrolysates by Alcalase under high hydrostatic pressure. *Innovative Food Science and Emerging Technologies*, *43*(July), 92–101. Available from https://doi.org/10.1016/j.ifset.2017.08.001.

CHAPTER

6 Bioactive peptides from milk and dairy proteins: models of digestion and intestinal barrier

Carlotta Giromini[1,2], Davide Lanzoni[1], Raffaella Rebucci[1] and Antonella Baldi[1]

[1]*Department of Veterinary Medicine and Animal Sciences (DIVAS), University of Milan, Lodi, Italy*

[2]*CRC, Innovation for Well-Being and Environment, Università degli Studi di Milano, Milano, Italy*

6.1 Introduction

6.1.1 Bioactivity and bioavailability

Milk and dairy products contain many bioactive components that have been reported to have health benefits, including minerals, proteins, and/or their bioactive peptides. Bioactive peptides are generally defined as short chains of amino acid (AA) monomers that are mostly connected with peptide bonds and have beneficial effects on human health. Bioactivity represents all events related to how the bioactive compound reaches the target tissue, interacts with biomolecules and how it is metabolized and transformed to induce a physiologic response. Bovine milk, which is widely consumed around the world, contains approximately 32–34 g/L of protein, of which 80% (w/w) is casein and 20% (w/w) is whey protein. There are three main groups of milk proteins that participate in the formation of peptides: caseins (α, β, and k-casein), whey proteins (β-lactoglobulin, α-lactalbumin, lactoferrin, immunoglobulins, serum albumin, glycomacropeptides, enzymes and growth factors), and proteins of milk fat globule including mucin-1 and xanthine dehydrogenase/oxidase. Whey protein is also rich in branched-chain AAs (such as leucine, isoleucine, and valine, while casein contains more histidine, methionine, phenylalanine, proline, serine, tyrosine, and valine (Fekete et al., 2013).

Milk peptides and AAs are released after gastrointestinal digestion and absorbed through intestinal barrier to be effective in their nutritional and extranutritional function. Protein digestion begins in the stomach and is favored by acidic pH conditions (pH 1–3) and a transit time ranging from 30 minutes to 3 hours. Here, gastric juice unfolds the protein structure allowing gastric enzymes

Protein Digestion-Derived Peptides. DOI: https://doi.org/10.1016/B978-0-443-19141-1.00006-6

to access it and break it down into peptides (Giromini, Cheli, et al., 2019). However, the behaviour between milk proteins is different. At the gastric level, casein proteins are largely degraded, as reported by (Giromini, Cheli, et al., 2019) and (Boland & Singh, 2019), whereas whey proteins, due to their globular structure, are more resistant to gastrointestinal hydrolysis. Notably, β-lactoglobulin can also interact with phosphatidylcholine in the gastric mucosa without being affected by gastric and duodenal digestion (Giromini, Cheli, et al., 2019; Giromini, Lovegrove, et al., 2019). Digestion subsequently continues in the small intestine by enzymes, such as trypsin, carboxypeptidase, and aminopeptidase. Here, the decomposed peptides (di- and tri-peptides) can be transported by specific transporters or hydrolyzed into simple AAs at the level of intestinal epithelial cells and then reach the whole body via the bloodstream (Giromini, Cheli, et al., 2019; Shimizu, 2004). Unabsorbed polypeptides, on the other hand, reach the large intestine, the site of fermentation and water reabsorption with a transit time of more than 12 hours in adults (Guerra et al., 2012).

Bioavailability and bioactivity of a peptide is under the dependence of gastrointestinal parameters, food processing and food matrix complexity, which, altogether, affect the efficiency of the digestion and absorption processes. Milk peptides are bioavailable when they are stable during transport across the gut mucosa, resistant to brush border peptidases and they reach the target site in an active form through body vessels. The absorption and bioavailability steps represent key points in milk peptide bioactivity at the target tissue (Giromini et al., 2020).

Bioavailability is usually determined by in vivo assays in which the individual variability, physiological state, dose, and copresence of other food components are considered. Bioavailability includes the concept of bioactivity (Etcheverry et al., 2012). Bioactivity represents all events linked to how the bioactive compound reaches the target tissue, interacts with biomolecules and how it is metabolized and transformed to induce a physiologic response (Giromini, Cheli, et al., 2019).

Di and tri-peptides and AAs from ingested proteins are easily absorbed by intestinal epithelial cells; however, orally administered long-chain bioactive peptides may have difficulty in crossing the intestinal epithelial cell barrier and lose the ability to act in vivo.

A bioactive peptide to exert its activity after oral administration must remain intact and active after intestinal transport and brush border peptidase contact.

Natural antioxidants and bioactive components derived from different food sources have significant roles in maintaining the body functions, including prevention of different pathological conditions (Giromini & Cavalleri, 2022). Milk peptides may exert their effects on health through different bioactive functions. In contrast with pharmacological treatments, milk proteins have not been shown to cause any side effects, making them promising and safe to consume by healthy individuals and by those with a variety of other diseases (Fekete et al., 2016).

In the food industry, new extraction and isolation techniques have been developed and used to obtain bioactive peptides with specific activities and improved bioactivity as bioavailability, which depend on the degree of hydrolysis during isolation process. In this sense, the commercialization of bioactive peptide from milk and dairy source requires both in vitro/ex vivo and in vivo clinical trials.

Clinical trials represent the gold standard models to reproduce the digestive process and study bioactive components efficacy in humans (Marcano et al., 2015). However, the analysis using these models is complex, very expensive, variable, and unethical (Albenzio et al., 2017; Lucas-González et al., 2018). For this reason, research is investigating alternative methods. Among these, ex vivo models are possible candidates, although still little studied in the literature. This method consists of simulating the digestive process in the laboratory using gastric and intestinal fluids taken from humans and/or animals without the use of commercial digestive enzymes (Lanzoni et al., 2024). These represent the link between in vivo and in vitro models. They make it possible to reproduce the physiological process in vivo at a lower cost and by reducing analysis time. However, their development requires special attention. As reported by Ulleberg (2011), it is necessary to form gastric and intestinal pools to avoid variability and to monitor enzyme activity and stability, particularly after the storage period (typically −20°C). In fact, as argued by De Gara et al. (1986) and Muller and Ghale (1982), enzymes, such as pepsin, lipase, and trypsin, are subject to a loss of their activity, thus affecting the digestive process. To overcome the criticalities of previous models and to assess the health-promoting role of bioactive peptides in milk and dairy products, attention has increasingly focused on in vitro digestion models. These models, in addition to being faster and cheaper, can elucidate the mode of action of nutrients mechanistically with a significant reduction in the use of laboratory animals, following the principle of replacement, reduction and refinement of animal studies (Giromini, Cheli, et al., 2019). Furthermore, an attempt to model the in vitro digestion of milk proteins can determine the bioaccessibility of peptides, the effect of their chemical form and, the synergistic effect of all peptides produced during the digestion process (Muller & Ghale, 1982).

Considering the antihypertensive and ACE-inhibitory effects of milk proteins observed in vivo, they are mediated by intestinal absorption of serum and casein bioactive compounds and cell wall receptors (Fekete et al., 2016) . In fact, it has been demonstrated that the potency of blood pressure reduction via ACE-inhibitory activity greatly depends on the extent of intact peptides permeability (Sangsawad et al., 2018). It is well established that casein and whey have differential effects on gastric emptying and kinetics of digestion and absorption. Even though casein proteins have a structure that makes them sensitive to enzymatic hydrolysis, they clot in the stomach due to the low pH and are digested more slowly, which results in a prolonged AA release. In contrast, intact whey (which is acid soluble) or hydrolyzed whey are absorbed more rapidly, with a faster AA release and half-life (Boirie et al., 1997). As a result of their different inherent AA compositions leading to distinct absorption and kinetic behavior, they may also have differential effects on health.

Short chain peptides are absorbed intact via different intestinal transporters, such as carrier mediated, paracellular diffusion, and induce systemic bioactive effects. Wuerzner et al. (2009) reported that valine-proline-proline and isoleucine-proline-proline were poorly absorbed and rapidly eliminated, precluding ACE-inhibitory activity effect. Therefore, the phenomenon of ACE-inhibitory activity induced by some peptides in vitro does not always translate into significant effect in human studies (Fekete et al., 2016). Giromini et al. (2017) demonstrated how the release and inhibitory activity of ACE-1 compounds from dairy and plant in predigested protein fraction, gastric and pancreatic digestion fractions are influenced by the digestive process The same samples were tested in a human intervention trial, showing promising effect at blood pressure level and improvement in vascular reactivity (Fekete et al., 2016), even though the authors failed to detect significant reduction of circulating ACE-inhibitory activity in the intervention study. Information on the in vivo kinetics of bioactive peptides derived from milk proteins is very inconsistent, yet an essential factor for understanding the action of milk proteins in the human body. For this, further studies are required to substantiate the importance of ACEi or alternative mechanisms of action.

6.2 From proteins to peptides: methods to obtain bioactive peptides and amino acids

The digestive process is a complex combination of versatile physicochemical events that guide the intake of food, its disintegration into particles of appropriate size, absorption, transport to related organs, and elimination of the remaining waste products (Sensoy, 2021). More precisely, digestion includes two main simultaneous events: mechanical and enzymatic transformation. The first begins in the mouth and ends in the stomach, involving the transformation of food into small particles. The second implicates the action of multiple enzymes that have the task of breaking down macro- into micro-molecules, which are subsequently absorbed into the bloodstream. This phase begins in the mouth and ends in the intestine (Lucas-González et al., 2018). Together, these processes play a key role, as the digested nutrients are used by the body for growth, cellular repair, and energy development (Lucas-González et al., 2018). The close link between nutrition and health has driven the food industry, supported by scientific research, to understand in detail the results of each ingested nutrient (Manach et al., 2005; Sensoy, 2021). Indeed, foods contain components that can have beneficial or adverse effects on human health. Therefore it is essential to understand the individual mechanisms that may influence their release to achieve health benefits (Sensoy, 2021).

In particular, milk and dairy products are one of the most important sources of protein due to their high AA profile and the presence of bioactive peptides. The latter benefit human physiological health through the cardiovascular, skeletal, nervous, immune, and especially gastrointestinal systems (Choi et al., 2012; Giromini & Givens, 2022). Most bioactive peptides are inactive in the original protein sequence and are only activated following the process of hydrolysis, which can occur at different stages, such as secretion, storage, and processing of milk by enzymes naturally present in the milk itself or digestion by digestive enzymes (Albenzio et al., 2017). As reported by Giromini, Cheli, et al. (2019) and Giromini, Lovegrove, et al. (2019), the most important health benefit obtained from milk proteins and dairy products is obtained following gastrointestinal digestion and the subsequent release (Giromini, Lovegrove, et al., 2019) of peptides. Therefore, to assess the role of these bioactive peptides, scientific research has relied on the design and study of digestive models capable of accurately simulating the complex physicochemical and physiological events that occur during the entire digestive process, as reported in Fig. 6.1.

6.2.1 Milk and dairy protein digestion using ex vivo models

The digestion process can be simulated by means of ex vivo digestion models. As shown in Table 6.1, there are few studies in the literature that focus on the milk protein digestion using ex vivo models.

Predicting the nutritional quality of food products and their nutrients requires a better understanding of their digestibility. For this reason, the digestibility of the individual proteins and peptides that compose milk is increasingly being investigated in the literature. In the study conducted by (Devle et al., 2014), the authors wanted to investigate not only protein digestibility, but also how this was affected by the presence of lipids within homogenized whole and skimmed milk samples.

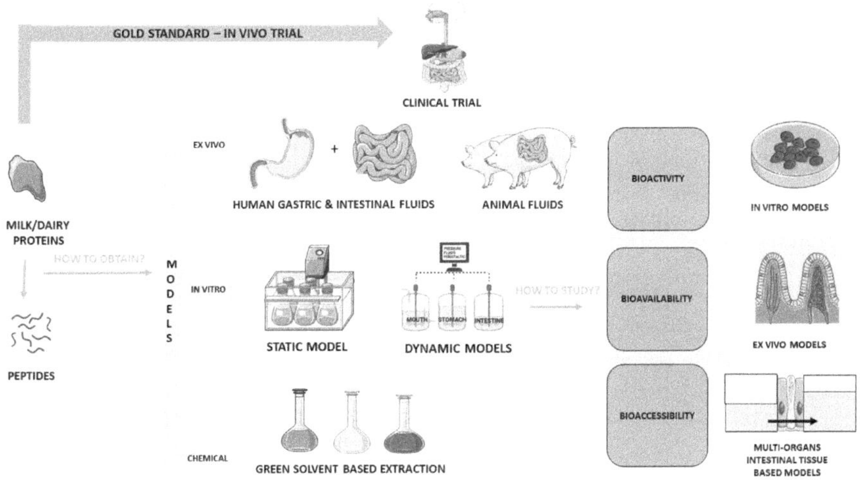

FIGURE 6.1 Available models to assess dairy proteins efficacy.

Overview of available models to assess dairy proteins efficacy.

The figure was partly generated using Servier Medical Art, provided by Servier, licensed under a creative Commons Attribution 3.0 unported license.

Table 6.1 Milk protein digestion using ex vivo models (from 2011 to).

Matrix	Digestion conditions	Proteins/ peptides	References
Bovine milk	GASTRIC (40 mins, 37 C, reduction of pH to 2.5); INTESTINAL (120 mins, 37°C, pH 7).	Casein β-lactoglobulin	Devle et al. (2014)
Bovine milk	GASTRIC (60 mins, 37°C, pH 3); INTESTINAL (120 mins, 37°C, pH 7).	Casein β-lactoglobulin Lactoalbumin	Islam et al. (2017)
Bovine milk	GASTRIC (40 mins, 37°C, reduction of pH to 2.5); INTESTINAL (120 mins, 37°C, pH 7).	Casein (β, αs1, κ, αs2) β-lactoglobulin α-albumin	**Islam et al. (2015)**
Bovine milk	GASTRIC (120 mins, 37°C, pH 3); INTESTINAL (120 mins, 37°C, pH 7).	β-Casein β-casomorphin	Asledottir et al. (2018)
Buffalo milk	GASTRIC (40 mins, 37°C, reduction of pH to 2.5); INTESTINAL (120 mins, 37°C, pH 7).	β-Casein β-lactoglobulin α-Lactoalbumin	**Ulleberg et al. (2011); Ulleberg (2011)**
Buffalo milk	GASTRIC (40 mins, 37°C, reduction of pH to 2.5); INTESTINAL (120 mins, 37°C, pH 7).	Casein (β, αs, k) β-lactoglobulin α-Lactoalbumin	Islam et al. (2016)
Buffalo and Bovine milk	GASTRIC (40 mins, 37°C, reduction of pH to 2.5); INTESTINAL (120 mins, 37°C, pH 7).	Casein (β, αs) β-lactoglobulin	Alegría et al. (2015)

As reported by the authors, caseins tend to be degraded during the gastric phase (after only 40 minutes) in both samples, following the physiological situation. Lactoglobulins, in particular β-, on the other hand, show a different behaviour. In whole milk, its protein structure is unaltered even after duodenal digestion (160 minutes), whereas in skimmed milk, it is digested within 60 minutes of duodenal digestion (Devle et al., 2014). This difference is due to the influence of the fatty acids released during digestion of whole milk. In fact, as reported by several authors in the literature (Loch et al., 2011; Narayan & Berliner, 1998; Ragona et al., 2000), lactoglobulin is able to bind fatty acids in the hydrophobic protein pocket, forming a bond that could stabilize the protein during digestion, hindering subsequent binding with bile acids. These results are confirmed in the study conducted by Islam et al. (2015, 2017). More specifically, in the latter, caseins were fully digested after only 40 minutes of the gastric phase. All the peptides formed were characterized by the presence of proline, known as a helix-breaker in the protein structure, preventing proteolytic enzymes from gaining access to the hydrophobic sequence for further proteolysis (Islam et al., 2015). Differently, some α-lactalbumins and lactoglobulins, particularly β-, were found to be intact 120 minutes after duodenal digestion, confirming previous findings (Devle et al., 2014; Islam et al., 2015). As can be guessed, caseins have a structure that makes them more sensitive to enzymatic hydrolysis. However, they are considered slow proteins due to their greater retention in the stomach than whey proteins (fast proteins). The greater retention in the stomach results in greater persistence of casein-derived AAs in plasma than whey proteins (Giromini, Cheli, et al., 2019). In most cases, this does not represent a health problem. However, β-casein, when digested, leads to the formation of β-casomorphin. This peptide is negatively involved in cardiovascular problems, diabetes, and abdominal pain, in addition to be able to bind to opioid receptors located on epithelial cell of the intestine and inhibit gastrointestinal functions causing a decrease in gastrointestinal motility (Asledottir et al., 2018). For this reason, Asledottir et al. (2018) wanted to investigate, using ex vivo digestion protocols, which genetic variant of β-casein was most degraded and led to the formation of β-casomorphin. As reported in their study, although all genetic variants led to the formation of this peptide, variant F was the most degraded, reaching maximum digestion at 120 minutes duodenal minutes. The studies reported so far have only been conducted on cow's milk. Different results were obtained following ex vivo digestion on buffalo skimmed milk. As shown by Islam et al. (2015), protein digestion on buffalo milk protein showed a complete degradation of caseins within 40 minutes of gastric digestion, confirming previous reports. However, in contrast to studies conducted on cow's milk, α-lactalbumin and β-lactoglobulin, although they resisted gastric digestion, were readily digested after 5 minutes of duodenal digestion (Islam et al., 2015). This difference in results can be explained in several ways: (1) β-lactoglobulins from cow's milk have a different AA composition compared to buffalo milk; (2) possible changes in tertiary structure; (3) changes in surface hydrophobicity due to the substitution of some AAs. All these mechanisms could alter the rate of protein degradation (Islam et al., 2015; Ulleberg, 2011). Furthermore, as reported by Islam et al. (2015), another limiting factor could be the concentration of bile salts within the intestinal fluid. As suggested by the same authors, the concentration required to lead to complete degradation must be above 2.0 mM, which is sufficient for it to bind to the protein structure causing its destabilization and further duodenal digestion (Islam et al., 2015). These findings are confirmed in a further study conducted by Islam et al. (2016) and (Islam et al., 2015).

In light of the above, the ex vivo methods proved to be reliable predictive models. The results obtained showed consistent trends that suggest they can be used to simulate the physiological process of digesting matrices, such as milk to assess protein degradation and identify released peptides.

6.2.2 Milk and dairy protein digestion using in vitro digestion models

In vitro models can be classified into static and dynamic. Static models were developed with the aim of reproducing the physiological conditions (temperature, agitation, pH, enzymes, and chemical composition) and sequence of events of each phase of the digestive process (oral, gastric, and intestinal), in which the food matrix remains stationary in a single bioreactor (Albenzio et al., 2017; Alegría et al., 2015; Fernández-García et al., 2009) Although, these models do not allow the simulation of parameters, such as mixing, hydration, and peristaltic force, they are characterized by high simplicity, low cost, and reproducibility (Fernández-García et al., 2009). Several in vitro static digestion protocols exist in the literature, making it difficult to compare results. For this reason, (Minekus et al., 2014) developed a harmonized model, improving health properties of food by sharing knowledge on the digestive process (INFOGEST), with the aim of simulating the in vitro physiological process of food and increasing its reproducibility between different laboratories and countries. The characteristics of INFOGEST protocol are shown in Fig. 6.2. The INFOGEST protocol was developed and validated on milk powder samples, and the relative bioaccessibility of milk proteins was used to assess its final suitability.

As reported by (Bohn et al., 2018), the INFOGEST protocol used to digest milk and dairy products samples, in contrast to other in vitro digestion methods, resulted in a high hydrolysis potential and a more homogeneous protein pattern with no detectable intact protein. In particular, in a study conducted by (Egger et al., 2016), digested milk powder samples showed that casein proteins were completely hydrolyzed in the stomach by the action of the enzyme pepsin, unlike whey proteins (β-lactoglobulin), where complete digestion only occurred in the intestinal phase (Egger et al., 2016). These data not only confirm what has been reported so far but are also confirmed in further literature studies. More precisely, Bohn et al. (2018) demonstrated how the milk powder used to validate the INFOGEST method, when administered to pigs, underwent the same degree of hydrolysis and bioaccessibility both gastric and intestinal. Egger et al. (2019) wanted to compare the INFOGEST method with dynamic protocols and in vivo trials conducted in pig models. As reported by the authors, the two in vitro methods showed similar kinetic behaviour. In particular, although caseins were degraded at different times (within 30 minutes, static protocol and within 60 minutes, dynamic protocol) they led to the formation of the same peptides, detectable already in the first gastric phase but decreasing during the digestive process. In parallel, β-lactoglobulins, resisting gastric digestion, led to the formation of peptides detectable only in the intestinal phase. More precisely, as the authors pointed out, these peptide patterns were highly comparable with those obtained at gastric and intestinal (jejunum) levels in pigs (Egger et al., 2017; Egger et al., 2019). Following the harmonization of the static digestion protocol, multiple studies were conducted using the INFOGEST method with the aim of assessing the bioaccessibility and bioactivity of proteins and peptides from milk

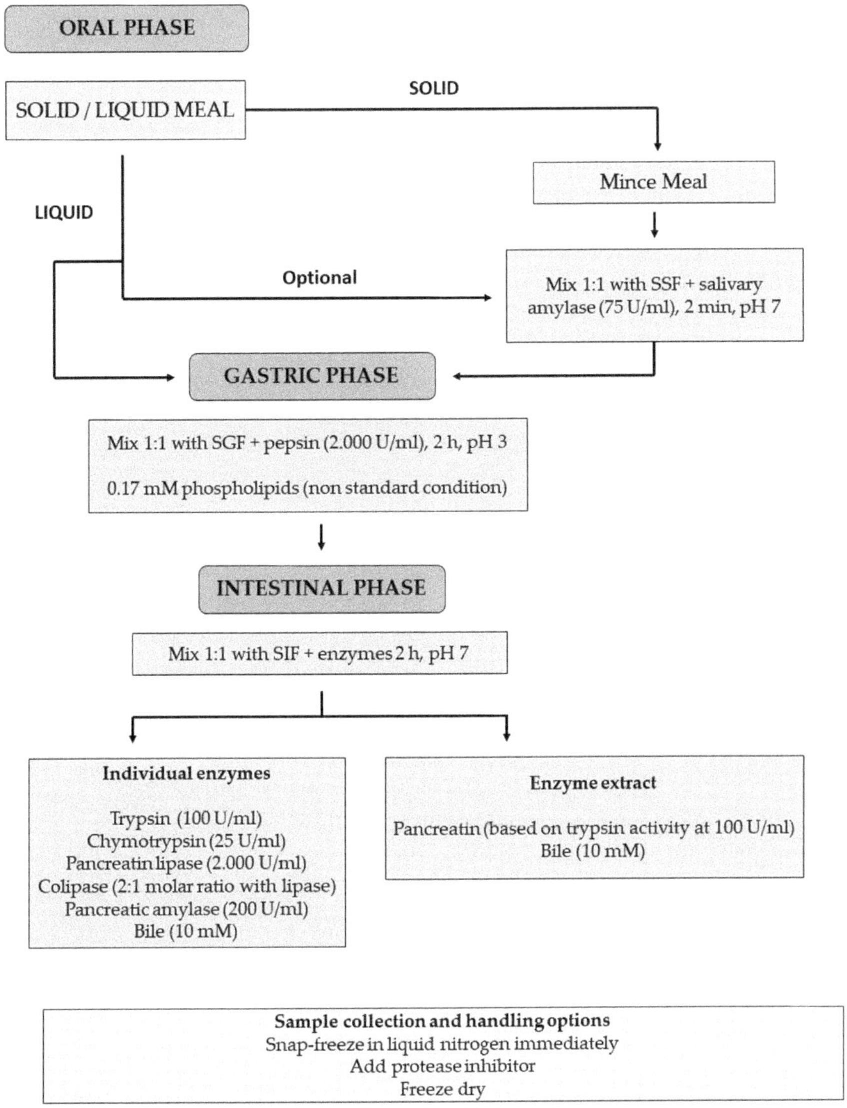

FIGURE 6.2 In vitro digestion method by INFOGEST.

Flow diagram of simulated in vitro digestion method, developed by INFOGEST. SSF, SGF, and SIF are Simulated Salivary Fluid, Simulated Gastric Fluid and Simulated Intestinal Fluid, respectively. Enzyme activities are in units per mL of final digestion mixture at each corresponding digestion phase.

From Minekus, M., Alminger, M., Alvito, P., Ballance, S., Bohn, T., Bourlieu, C., Carrière, F., Boutrou, R., Corredig, M., Dupont, D., Dufour, C., Egger, L., Golding, M., Karakaya, S., Kirkhus, B., Le Feunteun, S., Lesmes, U., Maclerzanka, A., MacKie, A., ... Brodkorb, A. (2014). A standardised static in vitro digestion method suitable for food-an international consensus. Food and Function, 5*(6), 1113–1124. https://doi.org/10.1039/c3fo60702j.*

and dairy products. (Giromini et al., 2017), following in vitro digestion of caseins and whey proteins, observed an increase in ACE-1 inhibitory activity. Bioactivity has also been confirmed in human studies, in which these proteins, when administered orally, led to a reduction in blood pressure and an improvement in cardiovascular biomarkers in adults with prehypertension and mild hypertension (Fekete et al., 2016, 2018). An interesting study was conducted by (De Noni et al., 2015). The authors wanted to study, following in vitro digestion of different types of cheeses, the release of β-casomorphin (5−7) widely investigated in the literature for their potential biological activity. Cheeses, as suggested by the authors, are favorable products for the release of these peptides, as proteolysis of β-caseins occurs during the cheese production and ripening process, although little information is still available on this subject. However, as the authors point out, although β-casomorphin 7 peptide patterns have been identified in the intestine, their concentration is so low that they cannot result in any relevant physiological activity (De Noni et al., 2015).

In conclusion, static models allow standardization of results by combining simplicity and low cost of analysis. However, in spite of their potential and wide applicability, none of the static models can reproduce the complex dynamic environment of the gut (Giromini, Cheli, et al., 2019; Giromini, Lovegrove, et al., 2019).

Dynamic methods were developed to meet the need for a more physiological model able of simulating not only enzymatic hydrolysis but also mechanical processes, which are fundamental in the digestive process. Indeed, they can reproduce changes in pH, enzyme secretions, peristaltic forces, and microbial fermentation continuously (Liu et al., 2019). Such models allow any food matrix to be realistically processed, enabling direct comparison with the results of in vivo studies. They were developed to allow the study of parameters, such as bioaccessibility of nutrients, effect of food structure, and interaction between different nutrients (Dupont et al., 2019). They are divided into mono-compartmental and multicompartmental. The best known of a mono-compartmental model is the dynamic gastric model (DGM), characterized by a body and an antrum. Here, gastric emptying is regulated by a valve that allows small particles to leave the stomach, while larger particles are refluxed into the upper chamber for further digestion (Wickham et al., 2009). Such models are less complex than multicompartmental models. The latter make it possible to faithfully reproduce each stage of the digestive process with realistic transit times (Guerra et al., 2012) An example is the TIM-1 model consisting of stomach and small intestine. This method allows the integration of all key parameters of human digestion, such as temperature, pH kinetics (gastric and intestinal), transit times, mixing, peristaltic transposition, sequential addition of digestive secretions, and passive absorption of water and small molecules (Guerra et al., 2012)

With regard to the evaluation of the digestion of milk proteins and derivatives, dynamic models are attracting increasing attention, as shown in Table 6.2.

Dynamic models have been developed to mimic digestion in adult subjects. As reported by Giromini, Cheli, et al. (2019), Giromini, Lovegrove, et al. (2019), this

Table 6.2 Dynamic model complexity and application to milk and dairy digestion.

System	Model set-up	Application	References
Dynamic gastric model	Mono-compartmental (gastric digestion)	Milk protein digestion-proteomic approach	Zhang et al. (2014)
Human gastric simulator	Mono-compartmental (gastric digestion)	Whey protein hydrolysis	Guo et al. (2014)
TIM-1	Multicompartmental (gastric and small intestinal digestion)	Digestion of whey protein extract	Nabil et al. (2011)
Simulator of the gastrointestinal tract (Simgi)	Multicompartmental (stomach, small intestinal and colonic microbiota)	Digestion of whey protein concentrate	Miralles et al. (2018)
DIDGI system	Multicompartmental (gastric and small intestinal digestion)	Peptide mapping, Milk proteolysis, Intestinal bioaccessibility of AA	de Oliveira et al. (2016); Meddah et al. (2001); Sánchez-Rivera et al. (2015)
Simulator of the human intestinal microbial ecosystem (SHIME)	Multicompartmental (gastric, small, and large intestinal digestion)	Whey retentate effect on the microbial community	(Nebbia et al., 2020)
DIDGI system	Multicompartmental (gastric and small intestinal digestion)	Digestion of milk proteins in skim milk powder	Egger et al. (2019)
DIDGI system	Multicompartmental (gastric and small intestinal digestion)	Digestion of milk to study pasteurization effect	Wang et al. (2018)
Human gastric simulator	Mono-compartmental (gastric digestion)	Digestion of dairy proteins to study the physicochemical behavior	Li et al. (2022)
Human gastric simulator	Mono-compartmental (gastric digestion)	Digestion of goat milk to study the gastric digestion behavior	Wang et al. (2022)
Human gastric simulator	Mono-compartmental (gastric digestion)	Digestion of plant-based milk and bovine milk	Roy et al. (2021)
Human gastric simulator	Mono-compartmental (gastric digestion)	Digestion of goat, sheep and bovine milk to study physical characteristics	Zhang et al. (2023)

(Continued)

Table 6.2 Dynamic model complexity and application to milk and dairy digestion. *Continued*

System	Model set-up	Application	References
Dynamicin vitrohuman stomach (DIVHS)	Multicompartmental (gastric and small intestinal digestion)	Evaluation of gastric emptying and enzymatic hydrolysis of skim milk	de Oliveira et al. (2016)
Human gastric simulator	Mono-compartmental (gastric digestion)	Evaluation of heat treatment and homogenization of gastric behaviors of protein	de Oliveira et al. (2016)

Adapted from Giromini, C., Cheli, F., Rebucci, R. & Baldi, A. (2019). Invited review: Dairy proteins and bioactive peptides: Modeling digestion and the intestinal barrier. Journal of Dairy Science, 102*(2), 929–942. https://doi.org/10.3168/jds.2018-15163.*

physiological process differs between children, the elderly, and pathological patients (inflammatory intestinal disease), not allowing a valid comparison. However, these in vitro methods, as previously reported, represent a valid solution to study the mechanisms of nutrient digestion, absorption, and bioaccessibility. Therefore the harmonization of a dynamic digestion protocol represents a future goal of nutritional research. In general, in the study of matrices, such as milk and dairy products, they have proved useful in studying their digestion and bioaccessibility of AAs, with high predictive power (Giromini, Cheli, et al., 2019; Giromini, Lovegrove, et al., 2019; Miralles et al., 2018). To date, their lack of standardization and high cost limit their widespread application in research laboratories.

6.2.3 Chemical hydrolysis

Bioactive peptides can be generated using different approaches as chemical, enzymatic, and fermentation methods. In chemical hydrolysis, a solvent is used to hydrolyze the proteins. The application of organic solvents, such as hexane, acetone, petroleum ether, and ethanol, in food and dairy industry is diminishing mainly due to food safety concerns (Tan et al., 2021). The milder conditions of enzyme-catalyzed hydrolysis are less damaging to the proteins substrate and are and ideal way to generate food-grade protein hydrolysates. Among researchers, food scientist, and technologist, there is a shift toward the use of "green solvents," along with green technologies. The term "green" is used to describe solvents that are in accordance to the Environmental, Safety and Health (ESH) characteristics; mainly a green solvent is one that: (1) originate from a renewable biomass resource (water based, vegetable oil, wood, fruits, etc.), (2) environment friendly, in reference to greenhouse effect and environmental pollution, (3) biodegradable, and (4) low toxicity levels (Li et al., 2019).

These are attracting interest as they can be applied for the extraction of active substances, including protein hydrolysis. One of these is the subcritical water extraction. Subcritical water extraction technology uses water as an extraction solvent maintained at temperature between 100°C and 374°C and pressure below its critical point. In subcritical state, water is able to acquire unique properties, including change in hydrogen bonding structure and increase in ionic product (Kw). In turn, it has the ability to act as either acid or basic catalyst. The subcritical water extraction is characterized by low process cost, mild operating condition, and environmental sustainability. Water is a cheap and nontoxic solvent, the obtained extracts are nontoxic and environmentally sustainable, and it can be applied to both low-polar and nonpolar molecules. On the other side, the limits of the technique are mainly related to the risk of unexpected reactions, and degradation of temperature-sensitive compounds. To date, the studies available are focused on the application of the subcritical extraction in plant-based compounds, such as soybean and rice seafood. To the best of our knowledge, the only study considering a dairy protein (β-casein) was conducted by Powell et al. (2016). The authors compare the technique with trypsin treatment observing that, under mild subcritical water conditions, phosphorylation was retained on the hydrolyzed product.

Other examples of green extraction procedures are available in the literature and applied to dairy products. In the study conducted by Kamal et al. (2023), ethyl acetate, ethanol, isopropanol, *n*-heptane, and cyclopentyl methyl ether were selected as the green solvents for the possible substitution of conventional solvents namely, hexane and ethyl ether. The manuscript assessed the sustainable solvent recovery of valuable protein from dairy and nondairy expired waste, the extraction efficiency and quality of extracted protein using conventional and green solvents and the commercial valorization potential of recycled dairy proteins. The results highlighted that green solvent depicted a higher protein recovery than conventional. However, the overall protein recovery was low. Further studies are emerging, being of paramount importance to assess the suitability of greener methods to obtain high-quality peptides and AAs from dairy products.

6.3 In vitro, ex vivo, and multiorgans intestinal tissue-based models

In the last few years, new approaches combining in vitro gastrointestinal digestion and intestinal cell models have been optimized and applied for estimation of the bioavailability of peptides due to lack of reproducibility of cell-free systems (Amigo & Hernandez-Ledesma, 2020). This chapter will describe in vitro, ex vivo, and multiorgans intestinal tissue-based models with benefits and limitation related to the bioavailability, bioactivity, and bioaccessibility of peptides derived dairy digested products.

6.3.1 In vitro intestinal models

An ideal model to mimic the intestinal barrier should contain all the gut epithelial cell types and these cells should have the ability to be cultured for the defined assay time without losing their in vivo characteristics (Ghiselli et al., 2021). Also, the intestinal biochemical environment should be considered and include the epithelium/immune system or enteric nervous system crosstalk.

In vitro intestinal models allow to create well-controlled and repeatable conditions with some limitations related to the models considered since they do not include all the representative cells of the intestinal barrier. Primary cells are rarely used in nutritional research due to scarce reproducibility of the results and short life span. The main experimental limit is that primary cells are able to survive only few days in cell culture. The American Type Culture Collection offers a wide-ranging array of commercially available human intestinal cell lines, of which Caco-2 and HT-29 cells represent the most commonly used (Jochems et al., 2018). Caco-2 cell line represents the gold standard well-established model to be applied in food science and nutrition studies related to nutrition absorption, transport, and metabolism in gut (Maznah, 1999). Caco-2 cell line is derived from human colon adenocarcinoma and during routine culture in vitro, cells undergo to spontaneous morphological and biochemical differentiation. In this way, Caco-2 cells will be similar to intestinal cells and acquire some specific characteristics, such as microvilli structure, brush border on the surface of cells, tight junctions between cells, secretion of hydrolases, and synthesis of carrier transport systems for sugar, amino acids, and drugs (Schrek & Melzig, 2018; Xu et al., 2019). In general, Caco-2 cells begin to differentiate on 7 days after standard culturing and differentiation completes on about 21 days. Undifferentiated Caco-2 cells obtained after short-term culturing are generally used to study the anticancer or antioxidant activity of active ingredients, while the differentiated Caco-2 monolayers are useful to simulate the intestinal tract in vitro due to its morphology, expression of marker enzymes and permeability characteristics which similar to those of human small intestine (Ding et al., 2021). Recently the differentiated Caco-2 cell model has been significantly developed due to the advancement of various technologies, which not only overcomes the limits of the traditional model but also further increases its application. To better mimic gut barrier, permeable membranes represent the most common supports used to replicate the intestinal barrier configuration. This system grants access to both the apical and basolateral compartments as a model to predict intestinal adsorption of peptide. Cells create the intestinal barrier, growing as a monolayer on a permeable membrane (i.e., Transwell filters), and with this approach, it is possible to mimic the in vivo intestinal barrier physiology and functionality (Yeste et al., 2018). Studies with Caco-2 models have been used to predict which bioactive peptides are resistant to brush border enzymes and are adsorbed from the apical to the basolateral compartments. Lacroix et al. (2017) demonstrated that DPP-IV inhibitory peptides from milk may be adsorbed intact through the transepithelial cells and exert their biological effect. On the contrary, other peptides (lactoferrin β-derived peptide and milk-derived peptides VLPVQK) are susceptible to brush border peptidases, and new sequences are further adsorbed

(Fernández-Musoles et al., 2013). Caco-2 cells are also characterized by the absence of a protective mucus layer that, sometimes, is compensated by the co-culture of these cells with different mucus-secreting globet cells. The human colon carcinoma HT-29 cells, besides being a model to study the different aspects of the biology of human cancer, received interest in food digestion and bioactivity studies. Goblet cells have the crucial role of secreting mucins and can be mimicked in vitro using mucus-secreting cell line HT29-MTX-E12 (Goto & Kiyono, 2012). The HT29-MTX-E12 cells were extensively used to study the role of bioactive dairy peptides in mucus production and gut functionality (Giromini et al., 2015; Martínez-Maqueda et al., 2013). The literature of in vitro dairy protein studies shows that the Caco-2 cell line is mainly used in absorption or transport studies, whereas the HT29-MTX cell line is mainly involved in mucus-modulation studies. Among, porcine intestinal epithelial cell lines, IPEC-J2 and IPEC-1 cell lines are good candidate models to study the antibacterial activities of dairy peptides at the gastrointestinal level (Brosnahan & Brown, 2012). IPEC-J2 is a morphologically and functionally more differentiated cell line isolated from the jejunum of a neonatal unsuckled piglet in comparison to IPEC-1.

The simultaneous use of cell lines derived from different tissues allows to track the fate of whey proteins through the upper gastrointestinal tract, their uptake across the intestinal barrier and then assess the physiological impact to downstream target cells (Corrochano et al., 2019). Simulated gastrointestinal digestion released a selection of whey peptides some of which were transported across a Caco-2/HT-29 intestinal barrier, inhibited free radical formation in muscle and liver cells. Milk has been shown to contain peptides that deliver important signals to the brains of neonates and infants. To reach the brain, milk peptides have to resist proteolytic degradation in the gastrointestinal tract, cross the gastrointestinal barrier and later cross the highly selective blood–brain barrier (BBB). (Christensen et al., 2020) used a coupled intestinal and BBB model to study the transport of milk-derived peptides. Bottani et al. (2020) used combined in vitro intestinal model (Caco-2/HT29 co-culture) in combination with Saos-2 cells, an in vitro model of human osteoblast-like cells, to evaluate the potential role of infant biscuits formulated with bovine milk proteins in promoting bone development. Although still a simulation, this strategy creates new possibilities for investigating and understanding the mechanism of nutrient absorption.

In addition to the above-mentioned studies, Table 6.3 reports a summary of the recent in vitro intestinal cell lines used to investigate dairy protein and peptide absorption and bioactivity.

The intestinal compartment is the most investigated target of nutrients digestion and absorption studies. However, the single-tissue approach is far from the in vivo physiology as the process of digestion and intestinal absorption is modulated by cross-interaction between different tissues. In this direction the development of dynamic multiorgan approach, where different organs communicate through an exchange of liquids (e.g., cell culture medium) may represent an increasing of the physiological relevance of the protocol. A dynamic in vitro model of the intestinal epithelium has been used for the investigation of P-glycoprotein activity (Costa et al., 2020). The advantages of the model are related with the presence of a

Table 6.3 Recent in vitro intestinal cell lines used for the study of dairy protein and peptides.

Cells	Proteins/peptides considered	Study outcome	References
Caco-2, colon epithelial enterocyte like, human origin	Peptides derived from in vitro digestion of human milk	specific milk peptides may be able to reach the bloodstream and exert bioactivity.	(Liang et al., 2022)
	Peptides obtained from milk casein hydrolyzed with neutrase	cholesterol-lowering activity	(Jiang et al., 2020)
	Milk protein-derived peptides RHPHPHLSFM, ARHPHPHLSF, HPHPHLSFM, VPYPQR, YVPR,VLPVPEK	Promote antioxidant activity (in vitro ABTS[a] and DPPH scavenging assays, crocin bleaching test;7 TbOOH[b]-treated cells)	(Tonolo et al., 2019)T
	Casein derived peptides NTVPAKSCQAQPTTMc, QGPIVLNPWDQVKR, APSFSDIPNPIGSENSE, ARHPHPHLSFM, AVPYPQR, NPYVPR, KVLPVPEK	Promote antioxidant activity (In vitro ABTS[a] scavenging assays and crocin bleaching test; H_2O_2- and TbOOH[b]-treated cells);	(Tonolo et al., 2020) (Tonolo et al., 2018)
	Casein derived peptides LLY	Promote antioxidant activity (In vitro ABTS[a] scavenging assays and oxygen radical absorbance capacity (ORAC) assay; H_2O_2-treated cells)	(Sowmya et al., 2018)
	Buffalo milk protein, buffalo mozzarella cheese hydrolysate, major sequence of CKYVCTCKM	Stimulated differentiation, modulated adherent junction formation and counteracted the TNF-α inflammation via regulating NF-κB pathway in differentiated Caco-2 cells; counteracted the DNBS[c]-induced increase in intestinal permeability and restored the organization of the colonic adherent junction	(Tenore et al., 2015); (Tenore et al., 2019)
	Bovine lactoferrin	Increase TEER[d]	(Anderson et al., 2017)
	Casein Hydrolisate	Transport of intact peptides: via paracellular route [high-molecular weight peptides (1,600–500 Da)]/via PepT1 (low-molecular weight peptides, <500 Da)	(Wang & Li, 2018)

T84, colon, epithelia, human, small intestinal, epithelial like, rat origin	whey protein and casein hydrolysates	migration and proliferation activities in a cell model for wound healing	(Purup et al., 2019)
IEC-6, colon, epithelia, human, small intestinal, epithelial like, rat origin	"Stracchino" soft-cheese, β-casein peptides, EAMAPK AVPYPQ	Promote antioxidant acitivity in H_2O_2-treated cells	(Pepe et al., 2016)
HT29-MTX, colon Epithelial goblet like, human origin	In vitro digested casein protein (CNP)	beneficial potential of CNP at the intestinal goblet cell level.	Giromini, Cheli, et al. (2019); Giromini, Lovegrove et al. (2019)
	Whey and casein proteins digestates	Cell proliferation modulation	(Giromini et al., 2016)
	β-Casofensin	Protects intestinal epithelial cells against apoptosis	(Bessette et al., 2016)
Caco-2 + HT29-MTX	Commercial whey products (CWP) whey protein isolate (WPI), galacto-oligosaccharide-whey protein concentrate (GOS-W), and glycomacropeptide (GMP)	Glycomacropeptide enhanced activity in downregulating bacterial endotoxin lipopolysaccharides (LPS) '- induced inflammation in vitro	(Arbizu et al., 2020)
HCT-116, colon adenocarcinoma cell line	Water-soluble peptides (WSPs) extracts of Cheddar cheeses made with buffalo and cow milk	Anti-proliferative activity	(Rafik et al., 2020)
	Goat milk-derived bioactive peptides	anticarcinogenic effect	(Cakir & Tunali-Akbay, 2021)

DPPH, 2,2-diphenyl-1-picrylhydrazyl. Examples of application implemented with Giromini, Cheli, et al. (2019).

[a]*ABTS, 2,2-Azino-bis-(3-ethyl-benzthiazoline-6-sulfonic acid).*

[b]*TbOOH t-butyl hydroperoxide.*

[c]*DNBS, 2,4-dinitrobenzene sulfonic acid.*

[d]*TEER, Transepithelial electrical resistance.*

From Giromini, C., Cheli, F., Rebucci, R., & Baldi, A. (2019). Invited review: Dairy proteins and bioactive peptides: Modeling digestion and the intestinal barrier. Journal of Dairy Science, *102(2), 929–942. https://doi.org/10.3168/jds.2018-15163.*

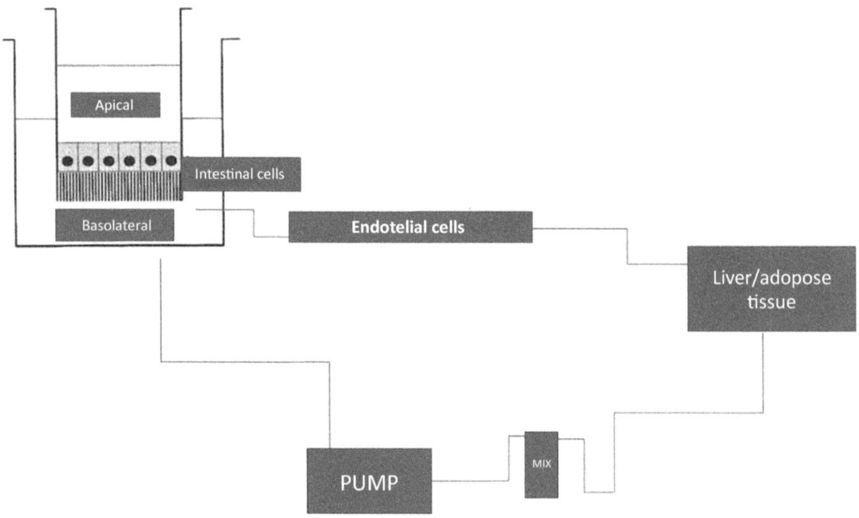

FIGURE 6.3 Dynamic multiorgan system set-up.

Graphical representation of dynamic multiorgan system set-up.

dynamic environment, where the medium flows as blood in human circulation (Fig. 6.3). Indeed, the flow of the medium improve the cells viability and metabolism, inducing a mechano-transduction effect on the surface of the tissues. This strategy allows to evaluate complex scenario characterized by interconnection between different tissues in cross-modulation from liquid exchange.

To the best of our knowledge, there are no studies available using the fluidic chambers to test the effect of dairy/milk proteins, and although dynamic multiorgan systems provide an approximation of the human reality they might be a valid tool to increase the predictivity of in vitro models leading to obtain results with higher predictive power.

6.3.2 Ex vivo models

As underlined before, the major limit of the principal in vitro models is that the intestinal epithelium also involves goblet cells, endocrine cells, M cells and mucus, which are not considered in the described cell monolayer models. Ex vivo models, using intact intestinal tissue segments from an animal, provides a better representation of the in vivo situation, since it includes the morphological and physiological features of the intestinal wall, including the multicellular assembly and presence of the mucus layer allowing the simulation of the possible further hydrolysis of some peptides by aminopeptidases located on the brush border membranes (Mulet-Cabero et al., 2020). Thus ex vivo models mainly represented by everted sac and Ussing-chamber have been proposed to cover these aspects using intestinal tissues with the great advantage to have specie specific models due to the fact that tissue from various animals may be used depending on the availability and purpose of the research. For human nutrition, intestinal tissue derived from pigs is the most considered since they are considered the best non-primate model, due to the fact that pigs have a large single-stomach and a comparable gut physiology to humans (Roura et al., 2016).

The everted sac model represents a fast and inexpensive model to study the intestinal transport of food components (Álvarez-Olguín et al., 2022). The model consists of a small piece on intestine flushed with a specific buffer with the serose part inside the sac and the mucosa directly in contact with the buffer, filled with oxygenated buffer and placed in a backer containing the peptides to be tested. Xue et al. (2021) demonstrated the bioactive peptide release and the absorption tracking of casein in the gastrointestinal digestion of rats using an everted gut sac model.

In a Ussing chamber, biopsy from intestinal animal tissue segment are placed between the apical and the basolateral chamber with the epithelia side exposed to the mucosal (apical) compartments and the lamina propria (serosal) in the basolateral compartment. The Ussing chamber, used to measure transport across epithelial membranes, takes its name from Danish biologist Hans Ussing in the 1950s. The system was established to understand the phenomenon of active Na + transport as a first step for the development of models of transepithelial transport, including the discovery of the Na + /K + ATPase pump (Skou & Esmann, 1992). The Ussing system offers an ex vivo measurement of permeability with fluorescent probes and/or electrophysiological measurements. The system design could accommodate multiple chambers which allow to increase the replicates' number. The chambers support both epithelial tissue and cell monolayers in such a way that each side of the membrane is isolated and faces a separate chamber-half. The potential difference across epithelial tissue can be measured using Ag/AgCl electrodes and transepithelial resistance (TEER) can be calculated to measure gut integrity. Subsequently, low TEER values are correlated with an increased permeability. Short circuit current (Isc) is also often used during electrophysiological measurements. Isc refers to the current that is required to nullify tissue potential difference. It is a sum of all ionic currents across the epithelium and can be calculated from potential difference and TEER values (Thomson et al., 2019). The Ussing chamber system has been not largely used to investigate the behaviour, transport, and bioactivities of milk and dairy proteins at intestinal barrier level. The study of (Bruno et al., 2017) have applied the Ussing chamber system to study the bioactivity of β-casofensin in the modulation of the intestinal barrier function using jejunum segments of adult rats.

Simulated gastric digestion (SGD) coupled to the Ussing chamber model may represent a new approach to study bioavailability of food components. Ozorio et al. (2020) studied the activity of brush border enzymes on undigested whey protein hydrolysate and on simulated intestinal digested whey hydrolysate and the absorption of resulting peptides through the proximal jejunum of a 7-week old piglet setup in an Ussing chamber model. Absorbed peptides were identified by mass spectrometry confirming an evident hydrolysis activity from brush border piglet proximal jejunum cells. Mulet-Cabero et al. (2020), set-up a protein model system (8% total protein) with caseins and whey mixtures digested differently using a semidynamic model, a static intestinal model, and an ex vivo absorption Ussing chambers model demonstrating that caseins delayed but whey proteins accelerated bioaccessibility and absorption. There are still a few studies investigating the absorption of AAs by Ussing chamber but the results are difficult to compare since the experimental setups are different, including the source of the animal tissue (rat, mice) and tissue size (Chen et al., 2020). Literature search revealed only one study (Awati et al., 2009) using the Ussing chamber model with porcine cells to investigate the permeation of AAs from whey proteins, and no such research on the dietary absorption of whey oligopeptides was evident.

It can be concluded that the current trends on coupling SGD with cell-based models offer the opportunity to better evaluate the bioavailability of food peptides. To be considered that only few studies determined peptide bioactivity after absorption evaluation and the models described need to be fully explored with bioactive milk and dairy peptides.

6.4 Conclusion

Bioactive peptides showed potential role as nutraceuticals and food supplements. They have not been shown to cause any side effects, making them promising and safe to consume by healthy individuals as prevention treatment against several diseases. Long-term clinical trials are the gold standard to assess their efficacy. However, in vitro and ex vivo systems to study digestion/extraction, absorption, and bioactivity must be developed and standardized rendering them applicable from laboratory scale to industrial scale. The development of novel approaches and models to assess the bioactivity of milk and dairy proteins will increase the predictive power of the in vitro available models. In vitro digestion systems can be combined with cell-based models or ex vivo tissues and maintained in static or dynamic conditions. The recent developments of non-invasive methods to follow these digestive parameters and to assess the bioactivity of milk and dairy proteins open new avenues and may help in a near future to fill the scientific and technological gaps. The ongoing development of models that are fully validated will enlarge even more their range of applications in food and pharmaceutical fields and boost the development of new target products in food, nutrition, and health.

References

Albenzio, M., Santillo, A., Caroprese, M., Malva, A. D., & Marino, R. (2017). Bioactive peptides in animal food products. *Foods*, *6*(5), 1–14. Available from https://doi.org/10.3390/foods6050035, https://www.mdpi.com/2304-8158/6/5/35.

Alegría, A., Garcia-Llatas, G., & Cilla, A. (2015). *Static digestion models: General introduction. The impact of food bioactives on health: In vitro and ex vivo models* (pp. 3–12). Spain: Springer International Publishing. Available from http://doi.org/10.1007/978-3-319-16104-4, 10.1007/978-3-319-16104-4_1.

Álvarez-Olguín, M. A., Beltrán-Barrientos, L. M., Hernandez-Mendoza, A., Hernandez-Mendoza, A. F., & Vallejo-Cordoba, B. (2022). Current trends and perspectives on bioaccessibility and bioavailability of food bioactive peptides: In vitro and ex vivo studies. *Journal of the Science of Food and Agriculture*, *102*(15), 6824–6834.

Amigo, L., & Hernandez-Ledesma, B. (2020). Current evidence on the bioavailability of food bioactive peptides. *Molecules*, *25*(19), 4479.

Anderson, R. C., Basset, S. A., Haggarty, N. W., Gopal, P. K., Armstrong, K. M., & Roy, N. C. (2017). Early-lactation, but not mid-lactation, bovine lactoferrin preparation increases epithelial barrier integrity of Caco-2 cell layers. *Journal of Dairy Science*, *100*(2), 886–891.

Arbizu, S., Chew, B., Mertens-Talcott, S. U., & Noratto, G. (2020). Commercial whey products promote intestinal barrier function with glycomacropeptide enhanced activity in downregulating bacterial endotoxin lipopolysaccharides (LPS)-induced inflammation in vitro. *Food & Function*, *11*(7), 5842–5852.

Asledottir, T., Le, T. T., Poulsen, N. A., Devold, T. G., Larsen, L. B., & Vegarud, G. E. (2018). Release of β-casomorphin-7 from bovine milk of different β-casein variants after ex vivo gastrointestinal digestion. *International Dairy Journal*, *81*, 8–11. Available from https://doi.org/10.1016/j.idairyj.2017.12.014, http://www.elsevier.com/inca/publications/store/4/0/5/8/6/0/index.htt.

Awati, A., Rutherfurd, S. M., Plugge, W., Reynolds, G. W., Marrant, H., Kies, A. K., & Moughan, P. J. (2009). Ussing chamber results for amino acid absorption of protein hydrolysates in porcine jejunum must be corrected for endogenous protein. *Journal of the Science of Food and Agriculture*, *89*(11), 1857–1861.

Bessette, C., Benoit, B., Sekkal, S., Bruno, J., Estienne, M., Léonil, J., Ferrier, L., Théodorou, V., & Plaisancié, P. (2016). Protective effects of β-casofensin, a bioactive peptide from bovine β-casein, against indomethacin-induced intestinal lesions in rats. *Molecular Nutrition & Food Research*, *60*(4), 823–833.

Bohn, T., Carriere, F., Day, L., Deglaire, A., Egger, L., Freitas, D., Golding, M., Le Feunteun, S., Macierzanka, A., Menard, O., Miralles, B., Moscovici, A., Portmann, R., Recio, I., Rémond, D., Santé-Lhoutelier, V., Wooster, T. J., Lesmes, U., Mackie, A. R., & Dupont, D. (2018). Correlation between in vitro and in vivo data on food digestion. What can we predict with static in vitro digestion models? *Critical Reviews in Food Science and Nutrition*, *58*(13), 2239–2261. Available from https://doi.org/10.1080/10408398.2017.1315362, http://www.tandf.co.uk/journals/titles/10408398.asp.

Boirie, Y., Dangin, M., Gachon, P., Vasson, M.-P., Maubois, J.-L., & Beaufrère, B. (1997). Slow and fast dietary proteins differently modulate postprandial protein accretion. *Proceedings of the National Academy of Sciences*, *94*(26), 14930–14935. Available from https://doi.org/10.1073/pnas.94.26.14930.

Boland, M., & Singh, H. (2019). *Milk proteins: From expression to food Milk Proteins: From Expression to Food* (pp. 1–764). New Zealand: Elsevier. Available from https://www.elsevier.com/books/milk-proteins/boland/978-0-12-815251-5.

Bottani, M., Cattaneo, S., Pica, V., Stuknytė, M., De Noni, I., & Ferraretto, A. (2020). In vitro antioxidant properties of digests of hydrolyzed casein and caseinophosphopeptide preparations in cell models of human intestine and osteoblasts. *Journal of Functional Foods*, *64*, 103673

Brosnahan, A. J., & Brown, D. R. (2012). Porcine IPEC-J2 intestinal epithelial cells in microbiological investigations. *Veterinary Microbiology*, *156*(3-4), 229–237.

Bruno, J., Nicolas, A., Pesenti, S., Schwarz, J., Simon, J. L., Léonil, J., & Plaisancié, P. (2017). Variants of β-casofensin, a bioactive milk peptide, differently modulate the intestinal barrier: In vivo and ex vivo studies in rats. *Journal of Dairy Science*, *100*(5), 3360–3372.

Cakir, B., & Tunali-Akbay, T. (2021). Potential anticarcinogenic effect of goat milk-derived bioactive peptides on HCT-116 human colorectal carcinoma cell line. *Analytical Biochemistry*, *622*, 114166.

Chen, Y., Dinges, M. M., Green, A., Cramer, S. E., Larive, C. K., & Lytle, C. (2020). Absorptive transport of amino acids by the rat colon. *American Journal of Physiology-Gastrointestinal and Liver Physiology*, *318*(1), G189–G202.

Choi, J., Sabikhi, L., Hassan, A., & Anand, S. (2012). Bioactive peptides in dairy products. *International Journal of Dairy Technology*, *65*(1), 1–12. Available from https://doi.org/10.1111/j.1471-0307.2011.00725.x.

Christensen, B., Toth, A. E., Nielsen, S. S. E., Scavenius, C., Petersen, S. V., Enghild, J. J., Rasmussen, J. T., Nielsen, M. S., & Sørensen, E. S. (2020). Transport of a peptide from bovine α_{s1}-casein across models of the intestinal and blood-brain barriers. *Nutrients*, *12*(10), 3157.

Corrochano, A. R., Ferraretto, A., Arranz, E., Stucknyte, M., Bottani, M., O'Connor, P. M., Kelly, P. M., De Noni, I., Buckin, V., & Giblin, L. (2019). Bovine whey peptides transit the intestinal barrier to reduce oxidative stress in muscle cells. *Food Chemistry*, *288*, 306–314.

Costa, J., Almonti, V., Cacopardo, L., Poli, D., Rapposelli, S., & Ahluwalia, A. (2020). A dynamic in-vitro model of the intestinal epithelium for the investigation of P-glycoprotein activity. *Research Square*, Submitted for publication.

De Gara, C. J., Burget, D. W., Sivakumaran, T., & Hunt, R. H. (1986). The effect of temperature and ph on the stability of human pepsin in stored gastric juice: A method to prevent activity loss. *Scandinavian Journal of Gastroenterology*, *21*(6), 650–654. Available from https://doi.org/10.3109/00365528609011096.

De Noni, I., Stuknytė, M., & Cattaneo, S. (2015). Identification of β-casomorphins 3 to 7 in cheeses and in their in vitro gastrointestinal digestates. *LWT - Food Science and Technology*, *63*(1), 550–555. Available from https://doi.org/10.1016/j.lwt.2015.03.036.

de Oliveira, S. C., Deglaire, A., Ménard, O., Bellanger, A., Rousseau, F., Henry, G., Dirson, E., Carrière, F., Dupont, D., & Bourlieu, C. (2016). Holder pasteurization impacts the proteolysis, lipolysis and disintegration of human milk under in vitro dynamic term newborn digestion. *Food Research International*, *88*, 263–275. Available from https://doi.org/10.1016/j.foodres.2015.11.022, http://www.elsevier.com/inca/publications/store/4/2/2/9/7/0.

Devle, H., Ulleberg, E. K., Naess-Andresen, C. F., Rukke, E. O., Vegarud, G., & Ekeberg, D. (2014). Reciprocal interacting effects of proteins and lipids during exvivo digestion of bovine milk. *International Dairy Journal*, *36*(1), 6–13. Available from https://doi.org/10.1016/j.idairyj.2013.11.008.

Ding, X., Hu, X., Chen, Y., Xie, J., Ying, M., Wang, Y., & Yu, Q. (2021). Differentiated Caco-2 cell models in food-intestine interaction study: Current applications and future trends. *Trends in Food Science & Technology*, *107*, 455–465.

Dupont, D., Alric, M., Blanquet-Diot, S., Bornhorst, G., Cueva, C., Deglaire, A., Denis, S., Ferrua, M., Havenaar, R., Lelieveld, J., Mackie, A. R., Marzorati, M., Menard, O., Minekus, M., Miralles, B., Recio, I., & Van den Abbeele, P. (2019). Can dynamic in vitro digestion systems mimic the physiological reality? *Critical Reviews in Food*

Science and Nutrition, *59*(10), 1546–1562. Available from https://doi.org/10.1080/10408398.2017.1421900, https://www.tandfonline.com/loi/bfsn20.

Egger, L., Ménard, O., Baumann, C., Duerr, D., Schlegel, P., Stoll, P., Vergères, G., Dupont, D., & Portmann, R. (2019). Digestion of milk proteins: Comparing static and dynamic in vitro digestion systems with in vivo data. *Food Research International*, *118*, 32–39. Available from https://doi.org/10.1016/j.foodres.2017.12.049, http://www.elsevier.com/inca/publications/store/4/2/2/9/7/0.

Egger, L., Ménard, O., Delgado-Andrade, C., Alvito, P., Assunção, R., Balance, S., Barberá, R., Brodkorb, A., Cattenoz, T., Clemente, A., Comi, I., Dupont, D., Garcia-Llatas, G., Lagarda, M. J., Le Feunteun, S., JanssenDuijghuijsen, L., Karakaya, S., Lesmes, U., Mackie, A. R., ... Portmann, R. (2016). The harmonized INFOGEST in vitro digestion method: From knowledge to action. *Food Research International*, *88*, 217–225. Available from https://doi.org/10.1016/j.foodres.2015.12.006, http://www.elsevier.com/inca/publications/store/4/2/2/9/7/0.

Egger, L., Schlegel, P., Baumann, C., Stoffers, H., Guggisberg, D., Brügger, C., Dürr, D., Stoll, P., Vergères, G., & Portmann, R. (2017). Physiological comparability of the harmonized INFOGEST in vitro digestion method to in vivo pig digestion. *Food Research International*, *102*, 567–574. Available from https://doi.org/10.1016/j.foodres.2017.09.047, http://www.elsevier.com/inca/publications/store/4/2/2/9/7/0.

Etcheverry, P., Grusak, M. A., & Fleige, L. E. (2012). Application of in vitro bioaccessibility and bioavailability methods for calcium, carotenoids, folate, iron. *magnesium, polyphenols, zinc, and vitamins B 6, B 12, D, and E. Frontiers in Physiology*, *3*. Available from https://doi.org/10.3389/fphys.2012.00317, http://www.frontiersin.org/Journal/DownloadFile.ashx?pdf = 1&FileId = 6249&articleId = 31039&Version = 1&ContentTypeId = 21&FileName-= fphys-03-00317.pdf, United States.

Fekete, A. A., Giromini, C., Chatzidiakou, Y., Givens, D. I., & Lovegrove, J. A. (2016). Whey protein lowers blood pressure and improves endothelial function and lipid biomarkers in adults with prehypertension and mild hypertension: Results from the chronic Whey2Go randomized controlled trial1,2. *American Journal of Clinical Nutrition*, *104*(6), 1534–1544. Available from https://doi.org/10.3945/ajcn.116.137919, http://ajcn.nutrition.org/content/104/6/1534.full.pdf.

Fekete, A. A., Givens, D. I., & Lovegrove, J. A. (2013). The impact of milk proteins and peptides on blood pressure and vascular function: A review of evidence from human intervention studies. *Nutrition Research Reviews*, *26*(2), 177–190. Available from https://doi.org/10.1017/S0954422413000139, http://journals.cambridge.org/NRR.

Fernández-García, E., Carvajal-Lérida, I., & Pérez-Gálvez, A. (2009). In vitro bioaccessibility assessment as a prediction tool of nutritional efficiency. *Nutrition Research*, *29*(11), 751–760. Available from https://doi.org/10.1016/j.nutres.2009.09.016.

Fernández-Musoles, R., Salom, J. B., Martínez-Maqueda, D., López-Díez, J. J., Recio, I., & Manzanares, P. (2013). Antihypertensive effects of lactoferrin hydrolyzates: Inhibition of angiotensin-and endothelin-converting enzymes. *Food Chemistry*, *139*(1-4), 994–1000.

Ghiselli, F., Rossi, B., Piva, A., & Grilli, E. (2021). Assessing intestinal health. In vitro and ex vivo gut barrier models of farm animals: Benefits and limitations. *Frontiers in Veterinary Science*, *8*, 723387.

Giromini, C., Baldi, A., Fusi, E., Rebucci, R., & Purup, S. (2015). Effect of growth factors, estradiol 17-β, and short chain fatty acids on the intestinal HT29-MTX cells: Growth factors and SCFAs effects on intestinal E12 cells. *Cell Biology and Toxicology*, *31*, 199–209.

Giromini, C., & Cavalleri, M. (2022). *Bioactive peptides: Cytomodulatory activity. Bioactive peptides from food: Sources, analysis, and functions* (pp. 461–484). CRC Press. Available from http://www.tandfebooks.com/doi/book/9781000551938, 10.1201/9781003106524-29.

Giromini, C., Cheli, F., Rebucci, R., & Baldi, A. (2019). Invited review: Dairy proteins and bioactive peptides: Modeling digestion and the intestinal barrier. *Journal of Dairy Science*, *102*(2), 929–942. Available from https://doi.org/10.3168/jds.2018-15163, http://www.elsevier.com/wps/find/journaldescription.cws_home/721317/description#description.

Giromini, C., Fekete, Á., Givens, D., Baldi, A., & Lovegrove, J. (2017). Short-communication: A comparison of the in vitro angiotensin-1-converting enzyme inhibitory capacity of dairy and plant protein supplements. *Nutrients*, *9*(12), 1352. Available from https://doi.org/10.3390/nu9121352.

Giromini, C., Fekete, Á. A., Pinotti, L., & Baldi, A. (2020). *Milk proteins: Their role in cardiovascular health. Milk and dairy foods: Their functionality in human health and disease* (pp. 145–172). Elsevier. Available from https://www.elsevier.com/books/milk-and-dairy-foods/givens/978-0-12-815603-2, 10.1016/B978-0-12-815603-2.00006-1.

Giromini, C., & Givens, I. (2022). Benefits and risks associated with meat consumption during key life processes and in relation to the risk of chronic diseases. *Foods*, *11*(14), 2063.

Giromini, C., Lovegrove, J. A., Givens, D. I., Rebucci, R., Pinotti, L., Maffioli, E., Tedeschi, G., Sundaram, T. S., & Baldi, A. (2019). In vitro-digested milk proteins: Evaluation of angiotensin-1-converting enzyme inhibitory and antioxidant activities, peptidomic profile, and mucin gene expression in HT29-MTX cells. *Journal of Dairy Science*, *102*(12), 10760–10771. Available from https://doi.org/10.3168/jds.2019-16833, http://www.elsevier.com/wps/find/journaldescription.cws_home/721317/description#description.

Giromini, C, Rebucci, R., Fusi, E., Rossi, L., Saccone, F., & Baldi, A. (2016). Cytotoxicity, apoptosis, DNA damage and methylation in mammary and kidney epithelial cell lines exposed to ochratoxin A. *Cell Biology and Toxicology*, *32*, 249–258.

Goto, Y., & Kiyono, H. (2012). Epithelial barrier: An interface for the cross-communication between gut flora and immune system. *Immunological Reviews*, *245*(1), 147–163.

Guerra, A., Etienne-Mesmin, L., Livrelli, V., Denis, S., Blanquet-Diot, S., & Alric, M. (2012). Relevance and challenges in modeling human gastric and small intestinal digestion. *Trends in Biotechnology*, *30*(11), 591–600. Available from https://doi.org/10.1016/j.tibtech.2012.08.001.

Guo, Q., Ye, A., Lad, M., Dalgleish, D., & Singh, H. (2014). Effect of gel structure on the gastric digestion of whey protein emulsion gels. *Soft Matter*, *10*(8), 1214–1223. Available from https://doi.org/10.1039/c3sm52758a.

Islam, M. A., Devle, H., Comi, I., Ulleberg, E. K., Rukke, E. O., Vegarud, G. E., & Ekeberg, D. (2017). Ex vivo digestion of raw, pasteurised and homogenised milk – Effects on lipolysis and proteolysis. *International Dairy Journal*, *65*, 14–19. Available from https://doi.org/10.1016/j.idairyj.2016.09.008, http://www.elsevier.com/inca/publications/store/4/0/5/8/6/0/index.htt.

Islam, M. A., Ekeberg, D., Rukke, E. O., & Vegarud, G. E. (2015). Ex vivo digestion of milk from Red Chittagong cattle focusing proteolysis and lipolysis. *Asian-Australasian Journal of Animal Sciences*, *28*(4), 559–567. Available from https://doi.org/10.5713/ajas.14.0269, http://www.ajas.info/upload/pdf/ajas-28-4-559.pdf.

Islam, M. A., Ekeberg, D., Rukke, E. O., & Vegarud, G. E. (2016). Ex vivo digestion of proteins and fat in buffalo milk. *International Dairy Journal*, *52*, 82–91. Available from https://doi.org/10.1016/j.idairyj.2015.08.008, http://www.elsevier.com/inca/publications/store/4/0/5/8/6/0/index.htt.

Islam, M. A., Ekeberg, D., Rukke, E. O., & Vegarud, G. E. (2015). Ex vivo digestion of omega-3 enriched buffalo skimmed milk. *Journal of Functional Foods*, *19*, 842–851. Available from https://doi.org/10.1016/j.jff.2014.08.016, http://www.elsevier.com/wps/find/journaldescription.cws_home/717426/description#description.

Liang, N., Kim, B. J., & Dallas, D. C. (2022). Bioavailability of peptides derived from the in vitro digestion of human milk assessed by caco-2 cell monolayers. *Journal of Agricultural and Food Chemistry*, *70*(23), 7077–7084.

Liu, W., Ye, A., Han, F., & Han, J. (2019). Advances and challenges in liposome digestion: Surface interaction, biological fate, and GIT modeling. *Advances in Colloid and Interface Science*, *263*, 52–67. Available from https://doi.org/10.1016/j.cis.2018.11.007.

Jiang, X., Pan, D., Zhang, T., Liu, C., Zhang, J., Su, M., Wu, Z., Zeng, X., Sun, Y., & Guo, Y. (2020). Novel milk casein–derived peptides decrease cholesterol micellar solubility and cholesterol intestinal absorption in caco-2 cells. *Journal of Dairy Science*, *103*(5), 3924–3936.

Jochems, P. G. M., Garssen, J., Van Keulen, A. M., Masereeuw, R., & Jeurink, P. V. (2018). Evaluating human intestinal cell lines for studying dietary protein absorption. *Nutrients*, *10*(3), 322.

Kamal, H., Ali, A., & Le, C. F. (2023). Green solvent processing: Effect of type of solvent on extraction and quality of protein from dairy and non-dairy expired milk products. *Food Chemistry*, *400*, 133988.

Lacroix, I. M., Chen, X. M., Kitts, D. D., & Li-Chan, E. C. (2017). Investigation into the bioavailability of milk protein-derived peptides with dipeptidyl-peptidase IV inhibitory activity using Caco-2 cell monolayers. *Food & Function*, *8*(2), 701–709.

Lanzoni, D., Rebucci, R., Cheli, F., Cavaliere, R., Ghilardi, G., Marchetti, L., Crotti, A., Baldi, A., & Giromini, C. (2024). Functional characterisation of *Euglena gracilis* following growth medium enrichment. *Italian Journal of Animal Science*, *23*(1), 53–64.

Li, W. G., Fan, Z. G., Wu, Y. Y., Jiang, Z. G., & Shi, R. C. (2019). Eco-friendly extraction and physicochemical properties of pectin from jackfruit peel waste with subcritical water. *Journal of the Science of Food and Agriculture*, *99*(12), 5283–5292.

Li, S., Ye, A., Pan, Z., Cui, J., Dave, A., & Singh, H. (2022). Dynamic in vitro gastric digestion behavior of goat milk: Effects of homogenization and heat treatments. *Journal of Dairy Science*, *105*(2), 965–980. Available from https://doi.org/10.3168/jds.2021-20980, http://www.elsevier.com/wps/find/journaldescription.cws_home/721317/description#description.

Loch, J., Polit, A., Górecki, A., Bonarek, P., Kurpiewska, K., Dziedzicka-Wasylewska, M., & Lewiński, K. (2011). Two modes of fatty acid binding to bovine β-lactoglobulin-crystallographic and spectroscopic studies. *Journal of Molecular Recognition*, *24*(2), 341–349. Available from https://doi.org/10.1002/jmr.1084.

Lucas-González, R., Viuda-Martos, M., Pérez-Alvarez, J. A., & Fernández-López, J. (2018). In vitro digestion models suitable for foods: Opportunities for new fields of application and challenges. *Food Research International*, *107*, 423–436. Available from https://doi.org/10.1016/j.foodres.2018.02.055, http://www.elsevier.com/inca/publications/store/4/2/2/9/7/0.

Manach, C., Williamson, G., Morand, C., Scalbert, A., & Rémésy, C. (2005). Bioavailability and bioefficacy of polyphenols in humans. I. Review of 97 bioavailability studies. *The American Journal of Clinical Nutrition*, *81*(1), S230–S242. Available from https://doi.org/10.1093/ajcn/81.1.230s.

Marcano, J., Hernando, I., & Fiszman, S. (2015). Invitro measurements of intragastric rheological properties and their relationships with the potential satiating capacity of cheese pies with konjac glucomannan. *Food Hydrocolloids*, *51*, 16–22. Available from https://doi.org/10.1016/j.foodhyd.2015.04.028.

Martínez-Maqueda, D., Miralles, B., Cruz-Huerta, E., & Recio, I. (2013). Casein hydrolysate and derived peptides stimulate mucin secretion and gene expression in human intestinal cell. *International Dairy Journal*, *32*(1), 13–19.

Maznah, I. (1999). The use of caco-2 cells as an in vitro method to study bioavailability of iron. *Malaysian Journal of Nutrition*, *5*(1), 31–45.

Meddah, A. T. T., Yazourh, A., Desmet, I., Risbourg, B., Verstraete, W., & Romond, M. B. (2001). The regulatory effects of whey retentate from Bifidobacteria fermented milk on the microbiota of the simulator of the human intestinal microbial ecosystem (SHIME). *Journal of Applied Microbiology*, *91*(6), 1110–1117. Available from https://doi.org/10.1046/j.1365-2672.2001.01482.x.

Minekus, M., Alminger, M., Alvito, P., Ballance, S., Bohn, T., Bourlieu, C., Carrière, F., Boutrou, R., Corredig, M., Dupont, D., Dufour, C., Egger, L., Golding, M., Karakaya, S., Kirkhus, B., Le Feunteun, S., Lesmes, U., Maclerzanka, A., MacKie, A., ... Brodkorb, A. (2014). A standardised static in vitro digestion method suitable for food-an international consensus. *Food and Function*, *5*(6), 1113–1124. Available from https://doi.org/10.1039/c3fo60702j, http://www.rsc.org/Publishing/Journals/FO/about.asp.

Miralles, B., del Barrio, R., Cueva, C., Recio, I., & Amigo, L. (2018). Dynamic gastric digestion of a commercial whey protein concentrate. *Journal of the Science of Food and Agriculture*, *98*(5), 1873–1879. Available from https://doi.org/10.1002/jsfa.8668, http://onlinelibrary.wiley.com/journal/10.1002/(ISSN)1097-0010.

Mulet-Cabero, A. I., Torcello-Gómez, A., Saha, S., Mackie, A. R., Wilde, P. J., & Brodkorb, A. (2020). Impact of caseins and whey proteins ratio and lipid content on in vitro digestion and ex vivo absorption. *Food Chemistry*, *319*, 126514.

Muller, D. P. R., & Ghale, G. K. (1982). Stability of pancreatic enzyme activities in duodenal juice after pancreatic stimulation by a test meal or exogenous hormones. *Annals of Clinical Biochemistry*, *19*(2), 89–93. Available from https://doi.org/10.1177/000456328201900205.

Nabil, S., Gauthier, S. F., Drouin, R., Poubelle, P. E., & Pouliot, Y. (2011). In vitro digestion of proteins and growth factors in a bovine whey protein extract as determined using a computer-controlled dynamic gastrointestinal system (TIM-1). *Food Digestion*, *2*(1-3), 13–22. Available from https://doi.org/10.1007/s13228-011-0010-9.

Narayan, M., & Berliner, L. J. (1998). Mapping fatty acid binding to β-lactoglobulin: Ligand binding is restricted by modification of Cys 121. *Protein Science*, *7*(1), 150–157. Available from https://doi.org/10.1002/pro.5560070116, http://onlinelibrary.wiley.com/journal/10.1002/(ISSN)1469-896X.

Nebbia, S., Giribaldi, M., Cavallarin, L., Bertino, E., Coscia, A., Briard-Bion, V., Ossemond, J., Henry, G., Ménard, O., Dupont, D., & Deglaire, A. (2020). Differential impact of holder and high temperature short time pasteurization on the dynamic in vitro digestion of human milk in a preterm newborn model. *Food Chemistry*, *328*, 127126.

Ozorio, L., Mellinger-Silva, C., Cabral, L. M. C., Jardin, J., Boudry, G., & Dupont, D. (2020). The influence of peptidases in intestinal brush border membranes on the absorption of oligopeptides from whey protein hydrolysate: An ex vivo study using an ussing chamber. *Foods*, *9*(10), E1415.

Pepe, G., Sommella, E., Ventre, G., Scala, M. C., Adesso, S., Ostacolo, C., Marzocco, S., Novellino, E., & Campiglia, P. (2016). Antioxidant peptides released from gastrointestinal digestion of "Stracchino" soft cheese: Characterization, in vitro intestinal protection and bioavailability. *Journal of Functional Foods*, *26*, 494–505.

Powell, T., Bowra, S., & Cooper, H. G. (2016). Subcritical water processing of proteins: An alternative to enzymatic digestion? *Analytical Chemistry*, *88*(12), 6425–6432.

Purup, S., Nielsen, S. D., Le, T. T., Bertelsen, S. H., Sørensen, J., & Larsen, L. B. (2019). Wound healing properties of commercial milk hydrolysates in intestinal cells. *International Journal of Peptide Research and Therapeutics*, *25*, 483–491.

Rafik, S., Gulzar, N., Huma, N., Hussain, I., & Murtaza, M. S. (2020). Evaluation of antiproliferative activity of Cheddar cheeses using colon adenocarcinoma (HCT-116) cell line. *International Journal of Dairy Technology*, *73*(1), 255–260.

Ragona, L., Fogolari, F., Zetta, L., Pérez, D. M., Puyol, P., Kruif, K. De, Löhr, F., Rüterjans, H., & Molinari, H. (2000). Bovine β-lactoglobulin: Interaction studies with palmitic acid. *Protein Science*, *9*(7), 1347–1356. Available from https://doi.org/10.1110/ps.9.7.1347, http://onlinelibrary.wiley.com/journal/10.1002/(ISSN)1469-896X.

Roura, E., Koopmans, S. J., Lallès, J. P., Le Huerou-Luron, I., de Jager, N., Shuurman, T., & Val-Laillet, D. (2016). Critical review evaluating the pig as a model for human nutritional physiology. *Nutrition Research Reviews*, *29*(1), 60–90.

Roy, D., Ye, A., Moughan, P. J., & Singh, H. (2021). Structural changes in cow, goat, and sheep skim milk during dynamic in vitro gastric digestion. *Journal of Dairy Science*, *104* (2), 1394–1411. Available from https://doi.org/10.3168/jds.2020-18779, http://www.elsevier.com/wps/find/journaldescription.cws_home/721317/description#description.

Sangsawad, P., Choowongkomon, K., Kitts, D. D., Chen, X. M., Li-Chan, E. C., & Yongsawatdigul, J. (2018). Transepithelial transport and structural changes of chicken angiotensin I-converting enzyme (ACE) inhibitory peptides through Caco-2 cell monolayers. *Journal of Functional Foods*, *45*, 401–408.

Schrek, K., & Melzig, M. F. (2018). Intestinal saturated long-chain fatty acid, glucose and fructose transporters and their inhibition by natural plant extracts in Caco-2 cells. *Molecules*, *23*, 2544.

Sensoy, I. (2021). A review on the food digestion in the digestive tract and the used in vitro models. *Current Research in Food Science*, *4*, 308–319. Available from https://doi.org/10.1016/j.crfs.2021.04.004, https://www.journals.elsevier.com/current-research-in-food-science.

Shimizu, M. (2004). Food-derived peptides and intestinal functions. *BioFactors, 21*(1–4), 43–47. http://onlinelibrary.wiley.com/journal/10.1002/(ISSN)1872-8081. 10.1002/biof.552210109.

Sánchez-Rivera, L., Ménard, O., Recio, I., & Dupont, D. (2015). Peptide mapping during dynamic gastric digestion of heated and unheated skimmed milk powder. *Food Research International*, *77*, 132–139. Available from https://doi.org/10.1016/j.foodres.2015.08.001, http://www.elsevier.com/inca/publications/store/4/2/2/9/7/0.

Skou, J. C., & Esmann, M. (1992). The na, k-atpase. *Journal of Bioenergetics and Biomembranes*, *24*, 249–261.

Sowmya, K., Mala, D., Bhat, M. I., Kumar, N., Bajaj, R. K., Kapila, S., & Kapila, R. (2018). Bio-accessible milk casein derived tripeptide (LLY) mediates overlapping anti-inflammatory and anti-oxidative effects under cellular (Caco-2) and in vivo milieu. *The Journal of Nutritional Biochemistry*, *62*, 167–180.

Tan, B., He, J., Zhang, S., Xu, C., Chen, S., Liu, H., & Li, W. (2021). Insight into anti-corrosion nature of Betel leaves water extracts as the novel and eco-friendly inhibitors. *Journal of Colloid and Interface Science*, *585*, 287–301.

Tenore, C. G., Pagano, E., Lama, S., Vanacore, D., Di maro, S., Maisto, M., Capasso, R., Merlino, F., Borrelli, F., Stiuso, P., & Novellino, E. (2019). Intestinal anti-inflammatory effect of a peptide derived from gastrointestinal digestion of buffalo (*Bubalus bubalis*) mozzarella cheese. *Nutrients*, *11*(3), 610.

Tenore, G. C., Ritieni, A., Campiglia, P., Stiuso, P., Di Maro, S., Sommella, E., Pepe, G., D'Urso, E., & Novellino, E. (2015). Antioxidant peptides from "Mozzarella di Bufala Campana DOP" after simulated gastrointestinal digestion: In vitro intestinal protection, bioavailability, and anti-haemolytic capacity. *Journal of Functional Foods*, *15*, 365–375.

Thomson, A., Smart, K., Somerville, M. S., Lauder, S. N., Appanna, G., Horwood, J., Sunder Raj, L., Srivastava, B., Durai, D., Scurr, M. J., Keita, A. V., Gallimore, A. M., & Godkin, A. (2019). The Ussing chamber system for measuring intestinal permeability in health and disease. *BMC Gastroenterology*, *19*(1), 1–14.

Tonolo, F., Fiorese, F., Moretto, L., Folda, A., Scalcon, V., Grinzato, A., Ferro, S., Arrigoni, G., Bindoli, A., Feller, E., Bellamio, M., Marin, O., & Rigobello, M. P. (2020). Identification of new peptides from fermented milk showing antioxidant properties: Mechanism of action. *Antioxidants*, *9*(2), 117.

Tonolo, F., Folda, A., Scalcon, V., Moretto, L., Marin, O., Bindoli, A., & Rigobello, M. P. (2019). *Activation of Nrf2 by milk-derived bioactive peptides in Caco-2 cells*, . *Free radical biology and medicine*, (1st, p. S54)(139, p. S54). New York, NY, 10169 USA: Elsevier Science.

Tonolo, F., Sandre, M., Ferro, S., Folda, A., Scalcon, V., Scutari, G., Feller, E., Marin, O., Bindoli, A., & Rigobello, M. P. (2018). Milk-derived bioactive peptides protect against oxidative stress in a Caco-2 cell model. *Food & Function*, *9*(2), 1245–1253.

Ulleberg, E. K., Comi, I., Holm, H., Herud, E. B., Jacobsen, M., & Vegarud, G. E. (2011). Human gastrointestinal juices intended for use in in vitro digestion models. *Food Digestion*, *2*(1-3), 52–61. Available from https://doi.org/10.1007/s13228-011-0015-4.

Ulleberg, E. K. (2011). In vitro digestion of caprine whey proteins by human gastrointestinal juices: Effect of whey hydrolysates and peptides on in vitro cell responses. In *PhD thesis*, Ås, Norway: Department of Chemistry, Biotechnology and Food Science (IKBM), Norwegian University of Life Sciences (UMB).

Wang, B., & Li, B. (2018). Charge and hydrophobicity of casein peptides influence transepithelial transport and bioavailability. *Food Chemistry*, *245*, 646–652.

Wang, X., Ye, A., Dave, A., & Singh, H. (2022). Structural changes in oat milk and an oat milk-bovine skim milk blend during dynamic in vitro gastric digestion. *Food Hydrocolloids*, *124*107311. Available from https://doi.org/10.1016/j.foodhyd.2021.107311.

Wang, X., Ye, A., Lin, Q., Han, J., & Singh, H. (2018). Gastric digestion of milk protein ingredients: Study using an in vitro dynamic model. *Journal of Dairy Science*, *101*(8), 6842–6852. Available from https://doi.org/10.3168/jds.2017-14284, http://www.elsevier.com/wps/find/journaldescription.cws_home/721317/description#description.

Wickham, M., Faulks, R., & Mills, C. (2009). In vitro digestion methods for assessing the effect of food structure on allergen breakdown. *Molecular Nutrition and Food Research*, *53*(8), 952–958. Available from https://doi.org/10.1002/mnfr.200800193, http://www3.interscience.wiley.com/cgi-bin/fulltext/122512006/PDFSTART, United Kingdom.

Wuerzner, G., Peyrard, S., Blanchard, A., Lalanne, F., & Azizi, M. (2009). The lactotripeptides isoleucine-proline-proline and valine-proline-proline do not inhibit the N-terminal or C-terminal angiotensin converting enzyme active sites in humans. *Journal of Hypertension*, *27*(7), 1404–1409.

Xu, Q., Hong, H., Wu, J., & Yan, X. (2019). Bioavailability of bioactive peptides derived from food proteins across the intestinal epithelial membrane: A review. *Trends in Food Science & Technology*, *86*, 399–411.

Xue, H., Han, J., He, B., Yi, M., Liu, X., Song, H., & Li, J. (2021). Bioactive peptide release and the absorption tracking of casein in the gastrointestinal digestion of rats. *Food & Function*, *12*(11), 5157–5170.

Yeste, J., Illa, X., Alvarez, M., & Villa, R. (2018). Engineering and monitoring cellular barrier models. *Journal of Biological Engineering*, *12*, 18.

Zhang, Q., Cundiff, J. K., Maria, S. D., McMahon, R. J., Wickham, M. S. J., Faulks, R. M., & Van Tol, E. A. F. (2014). Differential digestion of human milk proteins in a simulated stomach model. *Journal of Proteome Research*, *13*(2), 1055–1064. Available from https://doi.org/10.1021/pr401051u.

Zhang, P., Iqbal, S., Deng, R., Duan, X., Han, H., Dong Chen, X., & Wu, P. (2023). Impact of elderly gastrointestinal alterations on gastric emptying and enzymatic hydrolysis of skim milk: An in vitro study using a dynamic stomach system. *Food Chemistry*, *402*. Available from https://doi.org/10.1016/j.foodchem.2022.134365, http://www.elsevier.com/locate/foodchem.

CHAPTER

7 Bioactive peptides from marine sources after simulated gastrointestinal digestion

Bin Zhang

College of Food and Bioengineering, Henan University of Science and Technology, Luoyang, P.R. China

7.1 Introduction

Marine organisms account for about one-half of the total global biodiversity and are rich sources of structurally diverse components, some of which can be proteinaceous in nature and include proteins, peptides, and amino acids (Harnedy & FitzGerald, 2012). Bioactive peptides, usually ranging from 2 to 20 amino acid residues in size, are encrypted within their parent protein and can be released through (1) enzymatic hydrolysis by digestive enzymes, such as pepsin and trypsin, (2) hydrolysis by microbial proteinases and peptidases during fermentation, or (3) proteolysis by enzymes derived from microorganisms or plants, or a combination of all the above (Harnedy & FitzGerald, 2011).

As one of the hydrolysis techniques, simulated gastrointestinal digestion not only can be used to hydrolyze proteins into peptides, it is also utilized to reveal the possible structural and chemical changes for foods during gastrointestinal digestion. Thus in this chapter, bioactive peptides from marine sources after simulated gastrointestinal digestion have been discussed.

7.2 In vitro gastrointestinal digestion

7.2.1 General process for in vitro gastrointestinal digestion

In recent researches, in vitro gastrointestinal digestion is usually adopted to investigate the digestion process of foods in human body and to reveal their possible nutrient functions, adsorption and transportation mechanisms. In addition, in vitro gastrointestinal digestion has also been used to test the resistance of peptides against gastrointestinal digestion for their potential biological activities intended for food and pharmaceutical designs.

Protein Digestion-Derived Peptides. DOI: https://doi.org/10.1016/B978-0-443-19141-1.00007-8

Although for the foods mainly composed of proteins, the in vitro gastrointestinal digestion process primarily includes the enzymatic digestion that simulates the physiological process occurred in both stomach and intestine, mouth mastication is also considered in some experiments using polytron (Ultra-turrax T18 basic, IKA) for 10 seconds at a speed of 18,000 rpm, in order to mimic the whole process as close as possible (Trigo et al., 2021). On such a basis, specific parameters adopted for in vitro gastrointestinal digestion, such as enzymes, pH, temperature, and duration, may differ to some extent due to the major food compositions and research objectives.

As the most important parameter, the enzymes most frequently used within this process include α-amylase for the simulated oral digestion, pepsin for the simulated gastric digestion, and trypsin, chymotrypsin, pancreatin, and Corolase PP for the simulated intestinal digestion, in which pancreatin is the most commonly used enzyme, accounting for almost a half, which is followed by trypsin alone or together with chymotrypsin, or Colorase PP possessing chymotrypsin, elastase, dipeptidase, tryptic and aminopeptidase activities (Guan et al., 2018).

Meanwhile, some other biological molecules, such as bile salt, and mucin may sometimes be used to reflect the gastrointestinal environment more closely (Hur et al., 2011). Furthermore, in some experiments, instead of using buffers, simulated saliva fluid consisting of 12.50 mM KCl, 14.02 mM KH_2PO_4, 7.99 mM $NaHCO_3$, 4.10 mM NaCl, 0.29 mM $MgCl_2(H_2O)_6$, and simulated gastric fluid consisting of 35.08 mM KCl, 0.63 mM KH_2PO_4, 25.98 mM $NaHCO_3$, 41.00 mM NaCl, 0.59 mM $MgCl_2(H_2O)_6$, and simulated duodenal fluid consisting of 6.91 mM KCl, 0.56 mM KH_2PO_4, 43.00 mM $NaHCO_3$, 32.85 mM NaCl, 0.32 mM $MgCl_2(H_2O)_6$ were used to mimic the oral, gastric and intestinal digestion environment, respectively.

On the other hand, the real environment in stomach is greatly distinguished from that in intestine based on the pH value. For most experiments, the pH adopted for the simulated gastric digestion is selected around pH 2.0 (in the range from pH 1.5 to pH 3.0). While in some experiments, the simulated gastric digestion is conducted stepwisely. For example, for the in vitro gastrointestinal digestion of salmon (*Salmo salar* L.) protein, 2 hours gastric digestion phase has been adopted, including mixing at initial pH for 5 minutes, at pH 5.0 for 10 minutes and at pH 2.5 for the rest of time (Borawska et al., 2016). While for the intestinal digestion, neutral pH, ranging from 6.5 to 8.0 is adopted for most experiments. For the duration of the digestion, it varies from 1 to 4 hours for both gastric and intestinal digestion, and the duration for the gastric phase is usually equal to or less than that for the intestinal digestion.

Although the parameters used for in vitro gastrointestinal digestion vary from experiment to experiment, the temperature is consistently set at the human body temperature, 37°C. Based on the review on the present publications above, a universal and well-recognized procedure should be established for the subsequent researches.

7.2.2 Biological peptides derived from marine sources after in vitro gastrointestinal digestion

Marine organisms have proven to be invaluable sources of peptides with unique structures and diverse bioactivities (Pavlicevic et al., 2020). Such peptides generally occur as inactive sequences encrypted within their parent proteins and could be released via enzymatic hydrolysis, gastrointestinal digestion, or microbial fermentation. Peptides derived from marine sources after in vitro gastrointestinal digestion have been demonstrated to possess a wide variety of bioactivities, such as antioxidant activity, angiotensin-I-converting enzyme (ACE)-inhibitory activity, and other activities, including α-glucosidase inhibitory activity, osteogenesis promoting activity, monoamine oxidase A inhibitory activity, and dipeptidyl peptidase-IV (DPP-IV) inhibitory activity (Table 7.1).

7.2.2.1 Antioxidant peptides

Reactive oxygen species (ROS), a series of chemically reactive products of stepwise reduction of oxygen, can be induced by sunlight, ultraviolet, chemical reactions, and metabolic processes (Wang et al., 2009). The overexpression and residual of ROS in human body can attack membrane lipids, protein, DNA, and other bioactive macromolecules, leading to a wide range of pathological effects, such as DNA damage, carcinogenesis, and ageing-related cellular degeneration or tissue injury (Wang et al., 2009).

Antioxidant peptides are usually obtained through in vitro enzymatic hydrolysis, gastrointestinal digestion, and microbial fermentation, which release the peptides from their precursor proteins as an active form by hydrolysis. The antioxidant activity of the released peptides is partly ascribed to their amino acid compositions, such as hydrophobic amino residues, like leucine, isoleucine, and proline and aromatic amino residues, like tyrosine (Wong et al., 2020).

As one of the frequently used techniques, in vitro gastrointestinal digestion is also used to discovery antioxidant peptides from marine sources, such as fishes, molluscs, and algae.

For example, antioxidant peptides Phe-Ile-Lys-Lys, His-Leu, Ile-Tyr, Pro-His-Leu, Pro-Trp, Val-Pro-Trp were identified in both of the hydrolyzed Salmon (*S. salar* L.) myofibryllar protein (MP) and sarcoplasmic protein (SP) after in vitro gastrointestinal digestion, while an additional antioxidant peptide, Val-Tyr, was also detected in the in vitro hydrolysate of SP (Borawska et al., 2016). Spanish mackerel (*Scomberomorous niphonius*) has been tested for its possibility to produce antioxidant peptides. After hydrolysis using five kinds of enzymes, including pepsin, Neutrase, papain, trypsin, and Alcalase, as well as in vitro gastrointestinal digestion, the hydrolysate prepared using in vitro gastrointestinal digestion exhibited the highest degree of hydrolysis (27.45 ± 1.76%) and 2,2-diphenyl-1-picrylhydrazyl (DPPH) radical scavenging activity (52.58 ± 2.68%) at the concentration of 10 mg protein/mL among the six hydrolysates (Zhao et al., 2019). Through ultrafiltration in combination

Table 7.1 Bioactive peptides released from marine sources after in vitro gastrointestinal digestion.

Sources	Gastric digestion procedure	Intestinal digestion procedure	Type of bioactivity detected	References
Alaska pollock (*Gadus chalcogrammus*) skin collagen	Incubate with pepsin (2.5%, w/w) at pH 2.0 and 37°C for 1.5 hours	Incubate with Corolase PP (1%, w/w) at pH 7.5 and 37°C for 2.5 hours	• Antioxidant activities • ACE inhibitory activity • DPP-IV inhibitory activity	Guo et al. (2015)
Crassostrea gigas tubulin α-1 C Chain	Incubate with pepsin (1%, w/w) at pH 2.5 and 37°C for 2 hours	Incubate with pancreatin (2%, w/w) at pH 8.0 and 37°C for 4 hours	• Osteogenesis promoting activity	Chen, Zhang, et al. (2019)
Hairtail (*Trichiurus japonicus*)	Incubate with pepsin (2400 U/g) at pH 2.0 and 37°C for 4 hours	Incubate with pancreatin (300 U/g) at pH 7.5 and 37°C for 4 hours	• Monoamine oxidase A inhibitory activity	Yang et al. (2020)
Miiuy croaker	Incubate with pepsin (dosage not mentioned) at pH 1.5 for 2 hours	Incubate with trypsin (dosage not mentioned) at pH 7.0 for 2 hours	• ACE inhibitory activity	Zhu et al. (2022)
Salmon (*Salmo salar* L.) protein fraction	Premix for 3 minutes; and Incubate with porcine pepsin solution (15 U/g of protein) at initial pH for 5 minutes, at pH 5.0 for 10 minutes and at pH 2.5 for the rest of time, with a total duration of 2 hours	Incubate with Corolase PP (50 mg/mL) and bile salts (111.6 mg/mL) for 1 hour (both pH and temperature not mentioned)	• Antioxidant activity	Borawska et al. (2016)
Silver warehou (*Seriollela punctata*)	Incubate with pepsin (P7000, porcine	Incubate with pancreatin solution (1 mL/5 g	• Antioxidant activity • ACE inhibitory activity	Manikkam et al. (2016)

(Continued)

Table 7.1 Bioactive peptides released from marine sources after in vitro gastrointestinal digestion. *Continued*

Sources	Gastric digestion procedure	Intestinal digestion procedure	Type of bioactivity detected	References
	stomach mucosa; 1 mL/5 g fillets) at pH 2.0 and 37°C for 2 hours	fillets) at pH 7.5 and 37°C for 2 hours	• Trypsin inhibitory activity	
Skipjack Tuna (*Katsuwonus pelamis*) Head	Incubate with pepsin (1%, w/w) at pH 1.5 and 37.0 ± 2°C for 2 hours	Incubate with trypsin (1%, w/w) at pH 7.0 and 37.0 ± 2°C for 2 hours	• Antioxidant activity	Zhang et al. (2019)
Spanish Mackerel (*Scomberomorous niphonius*) Muscle	Incubate with pepsin (1%, w/w) at pH 1.5 and 37.0 ± 2°C for 2 hours	Incubate with trypsin (1%, w/w) at pH 7.0 and 37.0 ± 2°C for 2 hours	• Antioxidant activity	Zhao et al. (2019)
Tilapia (*Oreochromis niloticus*) skin gelatin	Incubate with pepsin (2.86%, w/w) at pH 2.5 and 37°C for 1 hour	Incubate with pancreatin (4%, w/w) at pH 7.5 and 37°C for 2 hours	• ACE inhibitory activity	Ling et al. (2018)
Nigerian marine molluscs: *Tympanatonus fuscatus* var radula (L.) and *Pachymelania aurita* (M.)	Incubate with pepsin (8%, w/w) at pH 2.1 and 37°C for 1 hour	Incubate with trypsin (4%, w/w) and chymotrypsin (2% w/w) at pH 7.5 and 37°C for 1.5 hours	• Antioxidant activity	Paul et al. (2021)
Sea cucumber (*Isostichopus badionotus*)	Incubate with pepsin (20 mg/g protein) at pH 2.0 and 37°C for 1.5 hours	Incubate with Corolase PP (40 mg/g protein) at pH 7.5 and 37°C for 1.5 hours	• Antioxidant activity • -radical scavenging activities • -iron reducing capacity • ACE-inhibitory activity • Cytotoxic effects against colorectal cancer cells	Perez-Vega et al. (2013)

(Continued)

Table 7.1 Bioactive peptides released from marine sources after in vitro gastrointestinal digestion. *Continued*

Sources	Gastric digestion procedure	Intestinal digestion procedure	Type of bioactivity detected	References
Isochrysis zhanjiangensis	Incubate with pepsin (1%, w/w) at pH 2.5 and 37°C for 2.5 hours	Incubate with trypsin and chymotrypsin (1%, w/w) at 37°C for 2.5 hours (pH not mentioned)	• Antioxidant activity	Chen, Zhang, et al. (2019)
Porphyra haitanensis	Incubate with pepsin (dosage not mentioned) at initial pH 2.0 and 37°C for 2 hours	Incubate with trypsin (dosage not mentioned) at pH 7.0 and 37°C for 4 hours	• Antioxidant activity • α-Glucosidase inhibitory activity	Wen et al. (2020)

with multiple chromatography methods, 4 peptides were identified as Pro-Glu-Leu-Asp-Trp, Trp-Pro-Asp-His-Trp, Phe-Gly-Tyr-Asp-Trp-Trp, and Tyr-Leu-His-Phe-Trp showing higher antioxidant activities, including DPPH radical scavenging activity (EC_{50} of 1.53, 0.70, 0.53, and 0.97 mg/mL, respectively), hydroxyl radical scavenging activity (EC_{50} of 1.12, 0.38, 0.26, and 0.67 mg/mL, respectively), and superoxide anion radical scavenging activity (EC_{50} of 0.85, 0.49, 0.34, and 1.37 mg/mL, respectively). These peptides could inhibit lipid peroxidation and protect plasmid DNA against oxidative damage in a dose-dependent manner, and were stable under a low temperature (<60 °C) and moderate pH environment (pH 5–9), showing a great potential for application in health food products. Antioxidant peptides were also prepared from the skipjack tuna (*Katsuwonus pelamis*) heads using the in vitro gastrointestinal digestion. 6 peptides were purified from the hydrolysate via ultrafiltration and a serial of chromatography techniques, and identified as Val-Glu-Glu (P1), Trp-Met-Phe-Asp-Trp (P2), Asp-Ala-Gly-Pro-Tyr-Gly-Pro-Ile (P3), Trp-Met-Gly-Pro-Tyr (P4), Glu-Arg-Gly-Pro-Leu-Gly-Pro-His (P5) and Glu-Met-Gly-Pro-Ala (P6), among which P2, P4, and P6 showed the highest radical scavenging activities against DPPH (EC_{50} of 0.31, 0.33, and 0.46 mg/mL, respectively), hydroxyl (EC_{50} of 0.30, 0.43, and 0.52 mg/mL, respectively), and superoxide anion (EC_{50} of 0.56, 0.38, and 0.71 mg/mL, respectively) (Zhang et al., 2019). Other investigations also demonstrated that antioxidant peptides can be released from silver warehou (Manikkam et al., 2016) and Alaska pollock skin collagen (Guo et al., 2015) after in vitro gastrointestinal digestion.

Molluscs are another sources for bioactive peptide discovery. Two commonly consumed marine molluscs known as periwinkles in southern Nigeria, that is, *Pachymelania aurita* and *Tympanatonus fuscatus* var *radula* were hydrolyzed by simulated gastrointestinal digestion (Paul et al., 2021). The hydrolysates and their fractions have demonstrated high antioxidant activities, including antilipid peroxidation activity, hydroxyl radical scavenging activity, ferric reducing power, and metal chelation activity, in which the antioxidant activities were more pronounced for the peptides of lower molecular weights. In addition, the in vitro gastrointestinal digest of the cooked sea cucumber has demonstrated antioxidant activity, in which the activity of the digest obtained after the gastric phase was significantly lower that obtained after the intestinal phase, and higher activities also occurred to the peptides of lower molecular weights, indicating a positive correlation between the antioxidant activity and the molecular weight (Perez-Vega et al., 2013).

Marine microalgae also can be used as a source for the development of cosmeceuticals, nutraceuticals, and functional foods. The gastrointestinal hydrolysate of *Isochrysis zhanjiangensis* showed relatively high free radical scavenging activity as compared to its hydrolysates using digestive enzymes, including chymotrypsin, trypsin, and pepsin alone. After purification using sequential chromatography methods, an antioxidant peptide has been identified from the hydrolysate using Q-TOF ESI/MS as Asn-Asp-Ala-Glu-Tyr-Gly-Ile-Cys-Gly-Phe. The identified antioxidant peptide is capable to alleviate ethanol-induced cytotoxicity and inhibit the production of intracellular ROS, indicating that this highly active peptide can probably be developed as an agent against alcohol toxicity (Chen, Zhang, et al., 2019). Antioxidant peptides are also identified from other marine algae, such as *Porphyra haitanensis*. The antioxidant activities of the hydrolysate were confirmed by DPPH radical scavenging assay (ranging from 30.6% at the concentration of 0.115 mg/mL to 57.1% at the concentration of 0.241 mg/mL) and ferric reducing antioxidant power (FRAP) assay (ranging from 0.003 mM Fe^{2+} to 0.149 mM Fe^{2+}). After further resolving by gel filtration chromatography packed with Sephadex G-25, a significantly positive correlation has been observed between the peptide contents and the DPPH radical scavenging rate as well as the FRAP value.

7.2.2.2 ACE-inhibitory peptides

As a key enzymatic factor mediating systemic hypertension, inhibition of ACE can be considered as an efficient strategy for prevention and management of hypertension. In recent days, ACE-inhibitory peptides have been widely investigated and obtained from edible proteins for their antihypertension activity.

Proteins from marine organisms are also an abundant source for discovery of ACE-inhibitory peptides. After successive simulated gastrointestinal digestion and further separation by transport across Caco-2 cell monolayer, ACE-inhibitory peptides were separated by a series of chromatographic steps from the hydrolysate of tilapia skin gelatin, in which two ACE inhibitory peptides, that is, Val-Gly-Leu-Pro-Asn-Ser-Arg and Gln-Ala-Gly-Leu-Ser-Pro-Val-Arg (IC_{50} of 80.90 and 68.35 μM, respectively) were identified. As revealed by molecular docking, both peptides interacted with ACE via hydrogen bonds (Ling et al., 2018). Through in vitro gastrointestinal digestion, novel ACE inhibitory peptides have also been identified from the swim bladder hydrolysate of *miiuy croaker*, including Asp-Glu-Gly-Pro-Glu, Glu-Val-Gly-Ile-Gln, Ser-His-Gly-Glu-Tyr, Gly-Pro-Trp-Gly-Pro-Ala, Gly-Pro-Phe-Gly-Thr-Asp, Ser-Pro-Tyr-Gly-Phe, and Val-Ile-Gly-Pro-Phe, in which Ser-His-Gly-Glu-Tyr and Ser-Pro-Tyr-Gly-Phe exhibited potent ACE inhibitory activity with IC_{50} values ranged from 0.86 to 0.37 mg/mL. The interaction mechanism between both peptides and ACE has also been suggested to involve hydrophobic interaction, electrostatic force and hydrogen bonding via molecular docking, with affinity energy of -8.7 and -9.7 kcal/mol, respectively (Zhu et al., 2022). In addition, ACE inhibitory peptides were also detected from the digest of sea cucumber (Antonio Perez-Vega et al., 2013), silver warehou (Manikkam et al., 2016) and Alaska pollock skin collagen (Guo et al., 2015) after in vitro gastrointestinal digestion.

7.2.2.3 Other bioactive peptides

Besides antioxidant and ACE inhibitory activities, other bioactive peptides were also found from the in vitro gastrointestinal digests of marine proteins. For example, in addition to antioxidant and ACE inhibitory activities, peptides from the digest of sea cucumber showed cytotoxic capacity against colorectal HT -29 cells (Antonio Perez-Vega et al., 2013). In the digest of *Porphyra haitanensis*, α-glucosidase inhibitory activity has been detected, which could be increased after separation by Sephadex-25 gel filtration chromatography, in which the fraction F3 showed the highest inhibitory activity at an observed inhibitory rate of 42.1% (Wen et al., 2020). The osteogenic effect of a novel peptide (Tyr-Arg-Gly-Asp-Val-Val-Pro-Lys) from the digest of *Crassostrea gigas* has been demonstrated in an assay using the synthetic peptide, in which the proliferation of MC3T3-E1 cells was significantly promoted as determined by 3-(4,5-dimethylthiazol-2-yl)-2,5-diphenyltetrazoliumbromide (MTT) method, and confirmed by the increased activity of alkaline phosphatase (ALP). Further investigation suggested that the promotion mechanism of the peptide on MC3T3-E1 could probably be ascribed to the activation of MAPK signaling pathway induced by binding with the peptide (Chen, Xu, et al., 2019). Additionally, monoamine oxidase A (MAO-A) inhibitory peptides has been reported from the in vitro simulated gastrointestinal digest of hairtail. Ultrafiltered fractions ($<$ 3 kDa) from both of the hydrolysates by pepsin and simulated gastrointestinal enzymes exhibited the highest MAO-A inhibitory activities, with IC_{50} values of 0.61 and 2.54 mg/mL, respectively. After screening, 11 high-score peptides were identified, among which Val-Val-Phe-Glu-Val-Phe-Trp showed the highest MAO-A inhibitory activity ($IC_{50} = 0.41$ mM).

7.2.3 Effect of in vitro gastrointestinal digestion on bioactivities of peptides from marine sources

Resistance to gastrointestinal digestion is often required for effective utilization of bioactive peptides in food or nutraceutical applications due to their bioavailability and downstream bioactivity (Shen & Matsui, 2019). Accordingly, in another aspect, in vitro gastrointestinal digestion is also used to test the gastrointestinal stability of the bioactive peptides identified (Table 7.2). In the recent publications, we have found that antioxidant activity can be maintained or improved in almost all instances, whereas ACE-inhibitory activity will be reduced. Generally, it is apparent that small peptides, such as di- and tri-peptides may be less susceptible to gastrointestinal enzymes and retain their activities following in vitro gastrointestinal digestion, while larger peptides have higher probability to be hydrolyzed after digestion. The maintained or improved antioxidant activity indicates that smaller peptides may exert higher activity. While for ACE-inhibitory activity, the digestion will result in breakdown of the identified peptides, which further decrease their ACE-inhibitory activities. The difference between the influence on antioxidant activity and that on ACE inhibitory activity could be attributed to different mechanisms involved, that is, for antioxidant activity, chemical reaction of the peptides is the main factor, whereas for ACE inhibition, the interaction between the peptides and ACE is crucial. The difference can be further interpreted as that ACE inhibitory activity is more closely related the composition and conformation of peptides, rather than the molecular weight. Besides, in vitro gastrointestinal digestion has also been used for evaluation on bioaccessibility, such as those conducted on *Pyropia columbina* (Cian et al., 2015) and *Ulva fenestrate* (Trigo et al., 2021).

7.3 In silico gastrointestinal digestion

In contrast to in vitro gastrointestinal digestion, which demands for time consuming and laborious experiments, in silico digestion in combination with virtual screening for bioactive peptides is attracting and promising (Yuan et al., 2022). Accordingly, investigations on in silico gastrointestinal digestion of proteins from marine sources have been reported in recent decade.

7.3.1 General process for in silico gastrointestinal digestion

Generally, in silico gastrointestinal digestion is consisted of enzymatic cleavage of the known amino acid sequences of the target proteins by pepsin and trypsin, and usually chymotrypsin, or sometimes other proteinases, such as elastase, carboxypeptidase A and B and aminopeptidases, using in silico tools, such as BIOPEP-UWM, ExPASy PeptideCutter, and MS-Digest.

Table 7.2 Effect of in vitro gastrointestinal digestion on the bioactivities of the peptides from marine sources.

Sources	Oral digestion procedure	Gastric digestion procedure	Intestinal digestion procedure	Objectives	Effects	References
Atlantic Mackerel (*Scomber scombrus*)	Phosphate buffer (10 mM) at pH 6.8 and 37°C for 2 minutes	Pepsin (1%, w/w) at pH 1.5–2 and 37°C for 60 minutes	Pancreatin (1%, w/w), trypsin(1%, w/w), and bile extract (35%, w/w) at pH 7.5 and 37°C for 30 minutes, followed by adding intestinal acetone powder solution (1%, w/w) at 37°C for 30 minutes	Gastrointestinal stability on ACE inhibitory activity	• Retained ACE-inhibitory activity	Hokmabadinazhad et al. (2022)
Atlantic salmon (*Salmo salar*) co-product-derived protein	Not applicable	Pepsin (2.5%, w/w) at pH 2.0 and 37°C for 90 minutes	Corolase PP (1%, w/w) at pH 7.0 and 37°C for 150 minutes	Gastrointestinal stability on antidiabetic activity	• Retained GLP-1 secretory and DPP-IV inhibitory activity • Improved insulin secretory activity	Harnedy et al. (2018)
Boarfish (*Capros aper*) protein	Not applicable	Pepsin (2.5%, w/w) at pH 2.0 and 37°C for 1.5 hours	Corolase PP (1%, w/w) at pH 7.0 and 37°C for 2.5 hours	Gastrointestinal stability on antidiabetic activity	• Enhanced GLP-1 release • Higher oral bioavailability	Parthsarathy et al. (2019)
Cobia (Rachycentron canadum) skin	Not applicable	Pepsin (4%) at pH 2.0 and 37°C for 4 hours	Pancreatin (1:25) at neutral pH and 37°C for 4 hours	Gastrointestinal stability on ACE-inhibitory activity	• Reduced ACE-inhibitory activity	Lin et al. (2019)
Cobia (Rachycentron canadum) Liver Protein	Not applicable	Pepsin (25000 U/mL) in simulated gastric fluid at 37°C for 2 hours	Pancreatin solution (800 U/mL) in e simulated intestinal fluid at 37°C for 2 hours	Gastrointestinal stability on antioxidant activity	• Increased scavenge DPPH radical activity	Wang et al. (2020)

Striped weakfish (*Cynoscion guatucupa*)	Not applicable	Pepsin (4%, w/w) at pH 2.0 and 37°C for 2 hours	Pancreatin (4%, w/w) in intestinal fluid at pH 7.5 and 37°C for 4 hours	Gastrointestinal stability on antioxidant activity	• All samples are bioaccessible • Increased 2,2′-azino-bis(3-ethylbenzothiazoline-6-sulfonic acid) (ABTS) and hydroxyl radical scavenging activities	Lima et al. (2019)
Tilapia (*Oreochromis niloticus*) muscle proteins	Not applicable	Pepsin (2.86%, w/w) at pH 2.0 and 37°C for 1 hour	Pancreatin (4%, w/w) and α-chymotrypsin (0.52 U/mg substrate) added with 0.5 mL mixture of sodium taurocholate and sodium glycodeoxycholate (176 mM each) at pH 6.5 and 37°C for 2 hours	Gastrointestinal stability on ACE inhibitory activity	• Reduced ACE inhibitory activity • Enhanced transport across Caco-2 cell monolayer	Toopcham et al. (2017)
Yellowfin Tuna (*Thunnus albacares*) dark meat	Not applicable	Pepsin (4%, w/w) at pH 2.0 and 37°C for 1 hour	Pancreatin (8%, w/w) at pH 7.5 and 37°C for 2 hours	Gastrointestinal stability on antioxidant activity	• Reduced DPPH radical scavenging activity • Slightly increased ABTS radical scavenging activity	Unnikrishnan et al. (2021)
Marine Snail (*Neptunea arthritica cumingii*) meat and visceral mass	Not applicable	Pepsin (8.33%, w/w), at pH 1.0 and 37.0 ± 2°C for 2 hours	Trypsin (8.33%, w/w) at pH 8.3 and 37.0 ± 2°C for 2 hours, followed by chymotrypsin(8.33%, w/w) at pH 8.3 and 37.0 ± 2°C for 2 hours	Gastrointestinal stability on antioxidant and ACE-inhibitory activity	• Maintained DPPH radical scavenging activity • Maintained ACE inhibitory activity	Zhang et al. (2018)

(*Continued*)

Table 7.2 Effect of in vitro gastrointestinal digestion on the bioactivities of the peptides from marine sources. *Continued*

Sources	Oral digestion procedure	Gastric digestion procedure	Intestinal digestion procedure	Objectives	Effects	References
Scallop adductor (*Patinopecten yessoensis*)	Not applicable	Simulated gastric fluid at pH 3.0 and 37°C for 2 hours	Simulated intestinal fluid at pH 7.0 and 37°C for 2 hours	Stability	• Low molecular weight (<4000 Da) SAH–Zn complex remain comparably stable in in vitro gastrointestinal digestion	Sun et al. (2022)
Pyropia columbina	Not applicable	Pepsin digestion mixture for 2 hours	Pancreatin solution for 2 hours	Gastrointestinal stability on ACE-inhibitory and antioxidant activity	• Decreased ACE inhibitory activity • Increased DPPH scavenging activity and copper chelating activity	Cian et al. (2015)
Porphyra dioica	Not applicable	Pepsin (2.5%, w/w) at pH 2.5 and 37°C for 1.5 hours	Corolase PP (1%, w/w) at pH 7.5 and 37°C for 3 hours	Susceptible to hydrolysis with gastrointestinal enzymes	• Increased or maintained oxygen radical absorbance capacity (ORAC) values • Decreased or maintained ACE and DPP-IV inhibitory activities	Cermeno et al. (2019)
Red Seaweed (*Palmaria palmata*)	Simulated salivary fluid and 150 U/mL amylase at pH 7.0 and 37.0°C for 2 minutes	Simulated gastric fluid and pepsin (1000 U/mL) at pH 3.0 and 37.0°C for 2 hours	Simulated duodenal fluid (SDF), pancreatin (100 U/mL) and porcine bile extract (2.4 mg/mL) at pH 7.0 and 37.0°C for 2 hours	Gastrointestinal stability on Renin inhibitory activity	• Tridecapeptide Ile-Arg-Leu-Ile-Ile-Val-Leu-Met-Pro-Leu-Ile-Met-Ala cleaved and activated to the dipeptide Ile-Arg	Fitzgerald et al. (2014)

Rhodophyta (*Gracilariopsis lemaneiformis*)	Not applicable	Pepsin (0.8 mg/mL) at pH 2.0 and 37°C for 4 hours	Pancreatin (10 mg/mL) at pH 7.0 and 37°C for 4 hours	Gastrointestinal stability on ACE-inhibitory activity	• Enhanced GLP-1 release • Higher oral bioavailability	Deng et al. (2018)
Sea lettuce (*Ulva fenestrata*)	Polytron at a speed of 18 000 rpm for 10 seconds and α-amylase solution (1050 U/mL) at 37°C, pH 7.0 for 2 minutes	Pepsin solution at pH 2.0 and 37°C for 1 hour	Not applicable	Bio-accessibility	• Both *Ulva fenestrata* and the protein extracts are as bioavailable as casein	Trigo et al. (2021)
Undaria pinnatifida	Artificial saliva solution	Porcine pepsin (0.04 g pepsin in 0.1 M HCl) at pH 2.0 and 37°C for 1 hour	bile salts glycodeoxycholate, taurodeoxycholate and taurocholate (0.04 g, 0.03 g and 0.04 g in 1 mL saline, respectively) and pancreatin (0.04 g in 500 mL saline) at pH 7.4 and 37°C for 2.5 hours	Gastrointestinal stability on antioxidant activity of glycoprotein	• Decreased antioxidant activity during the gastric phase • Increased antioxidant activity during the duodenal phase	Rafiquzzaman et al. (2013)

After digestion, the resulting peptide sequences are collected and subjected to further analysis. Generally, two in silico tools for screening bioactive peptides are frequently used, that is, PeptideRanker and BIOPEP-UWM. PeptideRanker is an on-line prediction tool that is trained in fivefold cross-validation using a novel N-to-1 Neural Network (N1-NN) (Mooney et al., 2012). PeptideRanker allows users to submit a list of peptide sequences and will predict the probability for these peptide sequences being bioactive with a score in the range from 0 to 1. PeptideRanker can be commonly used for primary screening of bioactive peptides from a large pool of peptides derived from in silico digestion or peptidome. While in BIOPEP-UWM database, the profile of a potential bioactivity screening is based on the idea that the same bioactive fragment, especially a short one (2–3 amino acid residues), cannot be attributed to a given protein, but may be present in many sequences (many from the so-called common subsequences) (Minkiewicz et al., 2019). The BIOPEP-UWM database has recorded a wide range of bioactivities, including ACE inhibitory activity, α-amylase inhibitory activity, α-glucosidase inhibitory activity, antiamnestic activity, antibacterial activity, anticancer activity, antifungal activity, antiinflammatory activity, antioxidative activity, antithrombotic activity, antiviral activity, DPP-IV inhibitory activity, and so on, the details of which can be found in Table 7.2 of the article published by Minkiewicz et al. (Minkiewicz et al., 2019). After searching in BIOPEP-UWM, possible bioactivity will be given to a specific peptide sequence, which is very helpful for subsequent analysis.

Besides bioactivity analysis, other properties for the peptides will also be considered and evaluated in the context of application as cosmeceuticals, nutraceuticals, and functional foods, such as physical and chemical properties, toxicity and adsorption, distribution, metabolism, excretion, and toxicity (ADMET) analysis. All these analyses can be accomplished by in silico tools, such as Peptide solubility calculator, ToxinPred, admetSAR, and SwissADME.

Finally, the molecular mechanism for the screened peptides can be further revealed using molecular docking or molecular dynamics analysis. Molecular docking methodology explores the behavior of small molecules in the binding site of a target protein. Molecular docking programs perform a search algorithm in which the conformation of the ligand is evaluated recursively until the convergence to the minimum energy is reached. Finally, an affinity scoring function, ΔG, is employed to rank the candidate poses as the sum of the electrostatic and van der Waals energies (Pagadala et al., 2017). The lowest ΔG value for each peptide can be compared to give the most potent inhibitor for the receptor, which generally is an enzyme that plays an important role in the physiological process, such as ACE, α-amylase, α-glucosidase, and DPP-IV. At present, more than 60 different docking tools and programs have been developed for both academic and commercial purposes, among which AutoDock Vina, GOLD, and MOE-Dock predict top-ranking poses with best scores (Pagadala et al., 2017).

7.3.2 Biological peptides derived from marine sources after in silico gastrointestinal digestion

Primarily, in silico gastrointestinal digestion can be used to release bioactive peptides from marine resources and also to provide in-depth understanding on the nutrient basis for the proteins from which the peptides are derived (Table 7.3).

For example, the utilization of sarcoplasmic fish proteins as the protein ingredient for humans is limited due to the lack of understanding of their characteristics and nutritional properties. On such a basis, the bioactive peptides released after in silico gastrointestinal digestion were evaluated, in which 183 different peptides were obtained after the intestinal digestion from titin, collagen and troponin T, and composed of proline, glycine, aspartic acid, and glutamic acid. Among these peptides, a total of 16 antimicrobial peptides were predicted (Carrera et al., 2013). In another investigation, the potential of paramyosins from 9 seafoods has been tested for producing ACE and DPP-IV inhibitory peptides after in silico gastrointestinal digestion. Among 2853 fragments released, 26 known and 3 putative ACE-inhibitory peptides, and 53 known and 1 putative DPP-IV inhibitory peptides were discovered using BIOPEP-UWM and SwissTargetPrediction, which indicating that paramyosins can be used as a promising source for developing ACE and DPP-IV inhibitory peptides (Chai et al., 2022). In addition, in silico digestion of oyster (*Crassostrea gigas*) proteins, including myosin heavy chain (MHC), actin, and paramyosin has been investigated. Based on the PeptideRanker score and water solubility, 17 peptides were identified and subjected to ADMET analysis and bioactivity prediction. Among these 17 peptides, 12 peptides were predicted to possess ACE inhibitory activity, and 9 peptides to possess DPP-IV inhibitory activity, in which Met-Met-Asp-Phe and Met-Arg were considered as the most potent ACE and DPP-IV inhibitor, respectively. The binding modes between Met-Met-Asp-Phe and ACE, and Met-Arg and DPP-IV were analyzed by molecular docking, which indicated that oyster could be a promising source for developing bioactive peptides (Yuan et al., 2022). In order to find DPP-IV inhibitory peptides from *Chlorella vulgaris*, 43 protein sequences were subjected to in silico gastrointestinal digested and 468 di- and tri-peptides were obtained, from which 6 peptides, that is, Ala-Ala-Arg, Val-Pro-Ala, Val-Pro-Trp, Ile-Pro-Leu, Ile-Pro-Arg, and Pro-Pro-Leu, were selected for DPP-IV inhibitory assay based on their sequence feature. These peptides were further demonstrated for their inhibitory activity on DPP-IV in mouse serum, indicating that *C. vulgaris* proteins would be a good source for DPP-IV inhibitory peptides (Zhu et al., 2017). DPP-IV inhibitory peptides were also prepared from Largemouth bass (*Micropterus salmoides*) based on in silico gastrointestinal digestion in combination with pharmacophore and 3D-QSAR analysis. Based on the constructed pharmacophore model and 3D-QSAR model, 6 tripeptides, including Val-Ser-Met, Ile-Ser-Trp, Val-Ser-Trp, Ile-Cys-Tyr, Ile-Ser-Asp and Ile-Ser-Glu were selected and subjected to in vitro DPP-IV inhibition assay, in which Ile-Cys-Tyr was detected as the most active tripeptide (IC_{50} value of 0.73 mM) (Wang et al., 2020). Based on

Table 7.3 Bioactive peptides released from marine sources using in silico gastrointestinal digestion.

Sources	Digestion process	Evaluation tools	Bioactivities evaluated	References
Carp (*Cyprinus carpio*) and Atlantic salmon (*Salmo salar*)	• BIOPEP-UWM	• Quantitative parameters in BIOPEP-UWM	• ACE inhibitor • Antioxidant peptide	Borawska-Dziadkiewicz et al. (2021)
European Carp (*Cyprinus carpio* L.)	• BIOPEP-UWM Pepsin (EC 3.4.23.1) (stomach) Trypsin (EC 3.4.21.4) and chymotrypsin (EC 3.4.21.1) (intestine)	• In vitro ACE inhibition assay	• ACE inhibitor (as a counterpart to in vitro and ex vivo digestion)	Darewicz et al. (2017)
Largemouth bass (*Micropterus salmoides*)	• BIOPEP-UWM	• BIOPEP-UWM • AllerTOP • Toxicity • Pharmacophore modeling and virtual screening	• DPP-IV inhibitor	Wang et al. (2020)
Salmon (*Salmo salar*)	• BIOPEP-UWM Pepsin (EC 3.4.23.1) (stomach) Trypsin (EC 3.4.21.4) and chymotrypsin (EC 3.4.21.1) (intestine)	• In vitro ACE inhibition assay	• ACE inhibitor (as a counterpart to in vitro and ex vivo digestion)	Darewicz et al. (2014)
15 commonly consumed fish species in Europe	• Ms-Digest Pepsin (stomach) Trypsin, chymotrypsin, elastase, carboxypeptidase A and B and aminopeptidases (intestine)	• PeptideRanker • CAMP database • BIOPEP-UWM • BioPD	• ACE inhibitor • Antimicrobial peptide	Carrera et al. (2013)
9 sea food paramyosin proteins	• BIOPEP-UWM Pepsin (EC 3.4.23.1) (stomach) Trypsin (EC 3.4.21.4) and chymotrypsin (EC 3.4.21.1) (intestine)	• BIOPEP-UWM • SwissADME • AllerTOP • Toxicity • Molecular Docking • Molecular Dynamics	• ACE-1 inhibitor • DPP-IV inhibitor	Chai et al. (2022)

Crassostrea gigas	• BIOPEP-UWM Pepsin (EC 3.4.23.1) (stomach) Trypsin (EC 3.4.21.4) and chymotrypsin (EC 3.4.21.1) (intestine)	• PeptideRanker • ToxinPred • AdmetSAR2 • BIOPEP-UWM • iDPPIV-SCM • Molecular docking	• ACE inhibitor • DPP-IV inhibitor	Yuan et al. (2022)
Chlorella vulgaris	• BIOPEP-UWM Pepsin (EC 3.4.23.1) (stomach) Trypsin (EC 3.4.21.4) and chymotrypsin (EC 3.4.21.1) (intestine)	• Molecular docking using LibDock	• ACE inhibitor	Xie et al. (2018)
Chlorella vulgaris	• BIOPEP-UWM Pepsin (EC 3.4.23.1) (stomach) Trypsin (EC 3.4.21.4) and chymotrypsin (EC 3.4.21.1) (intestine)	• Type of N-terminal residue analysis	• DPP-IV inhibitor	Zhu et al. (2017)
Tilapia (*Oreochromis niloticus*)	• ExPASy PeptideCutter Pepsin pH 1.3 and pH > 2.0 (EC 3.4.23.1) (stomach) Trypsin (EC 3.4.21.4) and chymotrypsin-low specificity (intestine)	• In vitro antioxidant activities • Intracellular ROS scavenging activities	• Gastrointestinal Stability on antioxidant activity	Zhang et al. (2021)
Green Macroalga *Ulva lactuca*	• ExPASy PeptideCutter Pepsin pH 1.3 and pH > 2.0 (EC 3.4.23.1) (stomach) Trypsin (EC 3.4.21.4) and chymotrypsin (EC 3.4.21.1) (intestine)	• BIOPEP-UWM • AllerTOP • ToxinPred • PeptideRanker	• Gastrointestinal Stability	Garcia-Vaquero et al. (2019)
Nannochloropsis oculata	• ExPASy PeptideCutter Pepsin pH 1.3 and pH > 2.0 (EC 3.4.23.1) (stomach) Trypsin (EC 3.4.21.4) (intestine)	• PeptideRanker • BIOPEP-UWM	• Gastrointestinal Stability on ACE inhibitory activity	Hayes et al. (2022)

above, we do believe that in silico digestion together with bioinformatics analysis can be a promising way to screen out bioactive peptides due to the advantages, such as less labor required, time-saving, and cost-effective. Thus based on in silico digestion and virtual screening, in vitro and in vivo studies, randomized controlled trials, and long-term studies can be performed to further confirm their activities.

Other investigations involving in silico gastrointestinal digestion are intended to use such technique to test the gastrointestinal stability of the bioactive peptides identified in experimental researches, which is crucial, especially when the peptides are designed for oral administration. As compared to in vitro gastrointestinal digestion, in silico gastrointestinal digestion does not need to synthesize the peptides and avoids subsequent steps, including enzymatic digestion and HPLC detection. Due to these benefits and the well-acceptable results, gastrointestinal stability has been tested on the bioactive peptides from green macroalga *Ulva lactuca*, microalga *Nannochloropsis oculate*, and Tilapia using in silico digestion.

7.4 Prospects and challenges

Simulated gastrointestinal digestion, both in vitro and in silico, is highly evaluated in this chapter for the production of marine organism-derived bioactive peptides. In addition, its utilization for gastrointestinal stability test on identified peptides was also explained. Despite a large number of researches conducted using simulated gastrointestinal digestion, the parameters involved vary a lot from experiment to experiment. Since the simulated gastrointestinal digestion is an enzymatic process consisting of several steps, the digestive enzymes selected and their dosages are extremely important to mimic the physiological conditions. Besides, the factors that influence the actual environment of gastrointestinal digestion are complex, which involves pH, ionic strength, type of inorganic salts, etc. Thus a standardized protocol for simulated gastrointestinal digestion is highly demanded to reflect the gastrointestinal process as actual as possible.

On the other hand, in these publications mentioned above, although a series of hydrolysates or specific peptides have been identified with single or multiple bioactivities, these hydrolysates or peptides are still far from application in foods or pharmaceuticals due to their high production cost and lack of confirmation on in vivo efficacy, especially in randomized controlled trials, and long term studies. Thus it should be clearly recognized that there is still a long way for investigation on bioactive peptides, especially those of marine origin.

References

Antonio Perez-Vega, J., Olivera-Castillo, L., Angel Gomez-Ruiz, J., & Hernandez-Ledesma, B. (2013). Release of multifunctional peptides by gastrointestinal digestion of sea cucumber (Isostichopus badionotus). *Journal of Functional Foods*, *5*(2), 869–877. Available from https://doi.org/10.1016/j.jff.2013.01.036.

Borawska-Dziadkiewicz, J., Darewicz, M., & Tarczynska, A. S. (2021). Properties of peptides released from salmon and carp via simulated human-like gastrointestinal digestion described applying quantitative parameters. *PLoS One*, *16*(8). Available from https://doi.org/10.1371/journal.pone.0255969.

Borawska, J., Darewicz, M., Pliszka, M., & Vegarud, G. E. (2016). Antioxidant properties of salmon (Salmo salar L.) protein fraction hydrolysates revealed following their ex vivo digestion and *in vitro* hydrolysis. *Journal of the Science of Food and Agriculture*, *96*(8), 2764–2772. Available from https://doi.org/10.1002/jsfa.7441.

Carrera, M., Canas, B., & Gallardo, J. M. (2013). The sarcoplasmic fish proteome: Pathways, metabolic networks and potential bioactive peptides for nutritional inferences. *Journal of Proteomics*, *78*, 211–220. Available from https://doi.org/10.1016/j.jprot.2012.11.016.

Cermeno, M., Stack, J., Tobin, P. R., O'Keeffe, M. B., Harnedy, P. A., Stengel, D. B., & FitzGerald, R. J. (2019). Peptide identification from a Porphyra dioica protein hydrolysate with antioxidant, angiotensin converting enzyme and dipeptidyl peptidase IV inhibitory activities. *Food & Function*, *10*(6), 3421–3429. Available from https://doi.org/10.1039/c9fo00680j.

Chai, T.-T., Wong, C. C.-C., Sabri, M. Z., & Wong, F.-C. (2022). Seafood paramyosins as sources of anti-angiotensin-converting-enzyme and anti-dipeptidyl-peptidase peptides after gastrointestinal digestion: A cheminformatic investigation. *Molecules (Basel, Switzerland)*, *27*(12). Available from https://doi.org/10.3390/molecules27123864.

Chen, H., Xu, Z., Fan, F., Shi, P., Tu, M., Wang, Z., & Du, M. (2019). Identification and mechanism evaluation of a novel osteogenesis promoting peptide from Tubulin Alpha-1C chain in Crassostrea gigas, [Article; Proceedings Paper]*Food Chemistry*, *272*, 751–757. Available from https://doi.org/10.1016/j.foodchem.2018.07.063.

Chen, M.-F., Zhang, Y. Y., He, M. D., Li, C. Y., Zhou, C. X., Hong, P. Z., & Qian, Z.-J. (2019). Antioxidant peptide purified from enzymatic hydrolysates of isochrysis zhanjiangensis and its protective effect against ethanol induced oxidative stress of HepG2 cells. *Biotechnology and Bioprocess Engineering*, *24*(2), 308–317. Available from https://doi.org/10.1007/s12257-018-0391-5.

Cian, R. E., Garzon, A. G., Ancona, D. B., Guerrero, L. C., & Drago, S. R. (2015). Hydrolyzates from Pyropia columbina seaweed have antiplatelet aggregation, antioxidant and ACE I inhibitory peptides which maintain bioactivity after simulated gastrointestinal digestion. *Lwt-Food Science and Technology*, *64*(2), 881–888. Available from https://doi.org/10.1016/j.lwt.2015.06.043.

Darewicz, M., Borawska-Dziadkiewicz, J., Vegarud, G. E., & Minkiewicz, P. (2017). European carp (Cyprinus carpi L.) protein-derived ex vivo digests and *in vitro* hydrolysates differ in the ACE I inhibitory activity and composition of released ACE inhibitory. *Peptides. Protein and Peptide Letters*, *24*(2), 156–164. Available from https://doi.org/10.2174/0929866523666160816115441.

Darewicz, M., Borawska, J., Vegarud, G. E., Minkiewicz, P., & Iwaniak, A. (2014). Angiotensin I-converting enzyme (ACE) inhibitory activity and ACE inhibitory peptides of salmon (Salmo salar) protein hydrolysates obtained by human and porcine gastrointestinal enzymes. *International Journal of Molecular Sciences*, *15*(8), 14077–14101. Available from https://doi.org/10.3390/ijms150814077.

Deng, Z., Liu, Y., Wang, J., Wu, S., Geng, L., Sui, Z., & Zhang, Q. (2018). Antihypertensive effects of two novel angiotensin I-converting enzyme (ACE) inhibitory peptides from Gracilariopsis lemaneiformis (rhodophyta) in spontaneously hypertensive rats (SHRs. *Marine Drugs*, *16*(9). Available from https://doi.org/10.3390/md16090299.

Fitzgerald, C., Aluko, R. E., Hossain, M., Rai, D. K., & Hayes, M. (2014). Potential of a renin inhibitory peptide from the red seaweed Palmaria palmata as a functional food ingredient following confirmation and characterization of a hypotensive effect in spontaneously hypertensive rats. *Journal of Agricultural and Food Chemistry*, *62*(33), 8352–8356. Available from https://doi.org/10.1021/jf500983n.

Garcia-Vaquero, M., Mora, L., & Hayes, M. (2019). *In vitro* and *in silico* approaches to generating and identifying angiotensin-converting enzyme i inhibitory peptides from green macroalga Ulva lactuca. *Marine Drugs*, *17*(4). Available from https://doi.org/10.3390/md17040204.

Guan, H. N., Diao, X. Q., Jiang, F., Han, J. C., & Kong, B. H. (2018). The enzymatic hydrolysis of soy protein isolate by Corolase PP under high hydrostatic pressure and its effect on bioactivity and characteristics of hydrolysates. *Food Chemistry*, *245*, 89–96. Available from https://doi.org/10.1016/j.foodchem.2017.08.081.

Guo, L., Harnedy, P. A., Zhang, L., Li, B., Zhang, Z., Hou, H., Zhao, Xue, & FitzGerald, R. J. (2015). *In vitro* assessment of the multifunctional bioactive potential of Alaska pollock skin collagen following simulated gastrointestinal digestion. *Journal of the Science of Food and Agriculture*, *95*(7), 1514–1520. Available from https://doi.org/10.1002/jsfa.6854.

Harnedy, P. A., & FitzGerald, R. J. (2011). Bioactive Proteins, Peptides, and Amino Acids from Macroalgae. *Journal of Phycology*, *47*(2), 218–232. Available from https://doi.org/10.1111/j.1529-8817.2011.00969.x.

Harnedy, P. A., & FitzGerald, R. J. (2012). Bioactive peptides from marine processing waste and shellfish: A review. *Journal of Functional Foods*, *4*(1), 6–24. Available from https://doi.org/10.1016/j.jff.2011.09.001.

Harnedy, P. A., Parthsarathy, V., McLaughlin, C. M., O'Keeffe, M. B., Allsopp, P. J., McSorley, E. M., … FitzGerald, R. J. (2018). Atlantic salmon (Salmo salar) co-product-derived protein hydrolysates: A source of antidiabetic peptides. *Food Research International*, *106*, 598–606. Available from https://doi.org/10.1016/j.foodres.2018.01.025.

Hayes, M., Mora, L., & Lucakova, S. (2022). Identification of Bioactive Peptides from Nannochloropsis oculata Using a Combination of Enzymatic Treatment, *in Silico* Analysis and Chemical Synthesis. *Biomolecules*, *12*(12). Available from https://doi.org/10.3390/biom12121806.

Hokmabadinazhad, S. A., Thibodeau, J., Fliss, I., Marette, A., Bazinet, L., & Beaulieu, L. (2022). Identification of gastrointestinal digestion stable antihypertensive fish peptides from atlantic mackerel (Scomber scombrus). *Journal of Medicinal Food*, *25*(10), 952–962. Available from https://doi.org/10.1089/jmf.2021.0185.

Hur, S. J., Lim, B. O., Decker, E. A., & McClements, D. J. (2011). *In vitro* human digestion models for food applications. *Food Chemistry*, *125*(1), 1–12. Available from https://doi.org/10.1016/j.foodchem.2010.08.036.

Lima, K. O., de Quadros, C. d C., da Rocha, M., Jocelino Gomes de Lacerda, J. T., Juliano, M. A., Dias, M., & Prentice, C. (2019). Bioactivity and bioaccessibility of protein hydrolyzates from industrial byproducts of Stripped weakfish (Cynoscion guatucupa). *Lwt-Food Science and Technology*, *111*, 408–413. Available from https://doi.org/10.1016/j.lwt.2019.05.043.

Ling, Yuan, Liping, Sun, & Yongliang, Zhuang (2018). Preparation and identification of novel inhibitory angiotensin-I-converting enzyme peptides from tilapia skin gelatin hydrolysates: Inhibition kinetics and molecular docking. *Food & Function*, *9*(10), 5251–5259. Available from https://doi.org/10.1039/c8fo00569a.

Lin, Y.-H., Chen, C.-A., Tsai, J.-S., & Chen, G.-W. (2019). Preparation and identification of novel antihypertensive peptides from the *in vitro* gastrointestinal digestion of marine cobia skin hydrolysates. *Nutrients*, *11*(6). Available from https://doi.org/10.3390/nu11061351.

Manikkam, V., Mathai, M. L., Street, W. A., Donkor, O. N., & Vasiljevic, T. (2016). Sustainable use of silver warehou (Seriollela punctata): Effects of storage, processing conditions and simulated gastrointestinal digestion on selected in-vitro bioactivities. *Journal of Food Science and Technology-Mysore*, *53*(9), 3574–3582. Available from https://doi.org/10.1007/s13197-016-2338-1.

Minkiewicz, P., Iwaniak, A., & Darewicz, M. (2019). BIOPEP-UWM database of bioactive peptides: Current opportunities. *International Journal of Molecular Sciences*, *20*(23). Available from https://doi.org/10.3390/ijms20235978.

Mooney, Catherine, Haslam, Niall J., Pollastri, Gianluca, & Shields, Denis C. (2012). Towards the improved discovery and design of functional peptides: Common features of diverse classes permit generalized prediction of bioactivity. *PLoS One*, *7*(10). Available from https://doi.org/10.1371/journal.pone.0045012.

Pagadala, N. S., Syed, K., & Tuszynski, J. (2017). Software for molecular docking: A review. *Biophysical reviews*, *9*(2), 91–102. Available from https://doi.org/10.1007/s12551-016-0247-1.

Parthsarathy, V., McLaughlin, C. M., Harnedy, P. A., Allsopp, P. J., Crowe, W., McSorley, E. M., FitzGerald, R. J., & O'Harte, F. P. M. (2019). Boarfish (Capros aper) protein hydrolysate has potent insulinotropic and GLP-1 secretory activity in vitro and acute glucose lowering effects in mice. *International Journal Of Food Science And Technology*, *54*(1), 271–281. Available from https://doi.org/10.1111/ijfs.13975.

Paul, A. A., Osoniyi, O. R., Eghianruwa, Q. A., Oparinde, O. G., & Adesina, A. S. (2021).). Antioxidant activity of the simulated gastro-intestinal digestion hydrolysate of two edible Nigerian marine molluscs: Tympanatonus fuscatus var radula (L.) and Pachymelania aurita (M.). *International Journal of Biochemistry and Molecular Biology*, *12*(1), 8–16.

Pavlicevic, Milica, Maestri, Elena, & Marmiroli, Marta (2020). Marine bioactive peptides—An overview of generation, structure and application with a focus on food sources. *Marine Drugs*, *18*(8). Available from https://doi.org/10.3390/md18080424.

Perez-Vega, Juan Antonio, Olivera-Castillo, Leticia, Gomez-Ruiz, Jose Angel, & Hernandez-Ledesma, Blanca (2013). Release of multifunctional peptides by gastrointestinal digestion of sea cucumber (Isostichopus badionotus). *Journal of Functional Foods*, *5*(2), 869–877. Available from https://doi.org/10.1016/j.jff.2013.01.036.

Rafiquzzaman, S. M., Kim, E.-Y., Kim, Y.-R., Nam, T.-J., & Kong, I.-S. (2013). Antioxidant activity of glycoprotein purified from Undaria pinnatifida measured by an *in vitro* digestion model. *International Journal of Biological Macromolecules*, *62*, 265–272. Available from https://doi.org/10.1016/j.ijbiomac.2013.09.009.

Shen, W., & Matsui, T. (2019). Intestinal absorption of small peptides: A review. *International Journal of Food Science and Technology*, *54*(6), 1942–1948. Available from https://doi.org/10.1111/ijfs.14048.

Sun, J., Liu, X., Wang, Z., Yin, F., Liu, H., Nakamura, Y., ... Zhou, D. (2022). Gastrointestinal digestion and absorption characterization *in vitro* of zinc-chelating hydrolysate from scallop adductor (Patinopecten yessoensis). *Journal of the Science of Food and Agriculture*, *102*(8), 3277–3286. Available from https://doi.org/10.1002/jsfa.11673.

Toopcham, T., Mes, J. J., Wichers, H. J., Roytrakul, S., & Yongsawatdigul, J. (2017). Bioavailability of angiotensin I-converting enzyme (ACE) inhibitory peptides derived from Virgibacillus halodenitrificans SK1-3-7 proteinases hydrolyzed tilapia muscle proteins. *Food Chemistry*, *220*, 190–197. Available from https://doi.org/10.1016/j.foodchem.2016.09.183.

Trigo, J. P., Engstrom, N., Steinhagen, S., Juul, L., Harrysson, H., Toth, G. B., ... Undeland, I. (2021). *In vitro* digestibility and Caco-2 cell bioavailability of sea lettuce (Ulva fenestrata) proteins extracted using pH-shift processing. *Food Chemistry*, *356*. Available from https://doi.org/10.1016/j.foodchem.2021.129683.

Unnikrishnan, P., Kizhakkethil, B. P., George, J. C., Abubacker, Z. A., Ninan, G., & Nagarajarao, R. C. (2021). Antioxidant peptides from dark meat of yellowfin tuna (Thunnus albacares): Process optimization and characterization. *Waste and Biomass Valorization*, *12*(4), 1845–1860. Available from https://doi.org/10.1007/s12649-020-01129-8.

Wang, K., Yang, X., Lou, W., & Zhang, X. (2020). Discovery of dipeptidyl peptidase 4 inhibitory peptides from Largemouth bass (Micropterus salmoides) by a comprehensive approach. *Bioorganic Chemistry*, *105*. Available from https://doi.org/10.1016/j.bioorg.2020.104432.

Wang, Y. G., Zhu, F. R., Fan, Q. L., Han, F. S., & Wang, H. Y. (2009). Purification, characterization and effect of antioxidative dipeptide from salmon protamine hydrolysate: Study on the properties of bioactive peptides. *Amino Acids*, *37*(1), 33–34.

Wen, Lin, Tan, Siwen, Zeng, Limin, Wang, Zhufeng, Ke, Xingyao, Zhang, Zheqing, Tang, Huanwen, Guo, Honghui, & Xia, Enqin (2020). Ultrasound-assisted extraction and in vitro simulated digestion ofPorphyra haitanensisproteins exhibiting antioxidative and alpha-glucosidase inhibitory activity. *Journal of Food Measurement and Characterization*, *14*(6), 3291–3298. Available from https://doi.org/10.1007/s11694-020-00564-4.

Wong, F. C., Xiao, J. B., Wang, S. Y., Ee, K. Y., & Chai, T. T. (2020). Advances on the antioxidant peptides from edible plant sources. *Trends in Food Science & Technology*, *99*, 44–57. Available from https://doi.org/10.1016/j.tifs.2020.02.012.

Xie, Jingli, Chen, Xujun, Wu, Junjie, Zhang, Yanyan, Zhou, Yan, Zhang, Lujia, Tang, Ya-Jie, & Wei, Dongzhi (2018). Antihypertensive Effects, Molecular Docking Study, and Isothermal Titration Calorimetry Assay of Angiotensin I-Converting Enzyme Inhibitory Peptides from Chlorella vulgaris. *Journal of Agricultural and Food Chemistry*, *66*(6), 1359–1368. Available from https://doi.org/10.1021/acs.jafc.7b04294.

Yang, X., Wang, K., Liu, Q., & Zhang, X. (2020). Discovery of monoamine oxidase A inhibitory peptides from hairtail (Trichiurus japonicus) using *in vitro* simulated gastrointestinal digestion and *in silico* studies. *Bioorganic Chemistry*, *101*. Available from https://doi.org/10.1016/j.bioorg.2020.104032.

Yuan, Zhao, Li, Chen, Shuo, Jia, Yishan, Li, Yannan, Chen, Wenjie, Li, & Bin, Zhang (2022). Bioinformatics identification and molecular mechanism of angiotensin-converting enzyme and dipeptidyl peptidase-IV inhibitory peptides from in silico digest of Crassostreagigas. *Journal of Food Measurement and Characterization*, *16*(6), 4540–4550. Available from https://doi.org/10.1007/s11694-022-01548-2.

Zhang, Shan-Shan, Han, Li-Wen, Shi, Yong-Ping, Li, Xiao-Bin, Zhang, Xuan-Ming, Hou, Hai-Rong, Lin, Hou-Wen, & Liu, Ke-Chun (2018). Two novel multi-functional peptides from meat and visceral mass of marine snail Neptunea arthritica cumingii and their activities in vitro and in vivo. *Marine Drugs*, *16*(12). Available from https://doi.org/10.3390/md16120473.

Zhang, X., Noisa, P., & Yongsawatdigul, J. (2021). Identification and characterization of tilapia antioxidant peptides that protect AAPH-induced HepG2 cell oxidative stress. *Journal of Functional Foods*, *86*. Available from https://doi.org/10.1016/j.jff.2021.104662.

Zhang, L., Zhao, G.-X., Zhao, Y.-Q., Qiu, Y.-T., Chi, C.-F., & Wang, B. (2019). Identification and active evaluation of antioxidant peptides from protein hydrolysates of skipjack tuna (Katsuwonus pelamis) head. *Antioxidants*, *8*(8). Available from https://doi.org/10.3390/antiox8080318.

Zhao, G.-X., Yang, X.-R., Wang, Y.-M., Zhao, Y.-Q., Chi, C.-F., & Wang, B. (2019). Antioxidant peptides from the protein hydrolysate of Spanish mackerel (Scomberomorous niphonius) muscle by *in vitro* gastrointestinal digestion and their *in vitro* activities. *Marine Drugs*, *17*(9). Available from https://doi.org/10.3390/md17090531.

Zhu, Q., Chen, X., Wu, J., Zhou, Y., Qian, Y., Fang, M., ... Wei, D. (2017). Dipeptidyl peptidase IV inhibitory peptides from Chlorella vulgaris: *In silico* gastrointestinal hydrolysis and molecular mechanism. *European Food Research and Technology*, *243* (10), 1739–1748. Available from https://doi.org/10.1007/s00217-017-2879-1.

Zhu, W.-Y., Wang, Y.-M., Suo, S.-K., Zheng, S.-L., & Wang, B. (2022). Isolation, identification, molecular docking analysis, and cytoprotection of seven novel angiotensin I-converting enzyme inhibitory peptides from miiuy croaker byproducts-swim bladders. *Frontiers in Marine Science*, *9*. Available from https://doi.org/10.3389/fmars.2022.977234.

CHAPTER

8 Bioactive peptides released from meat and meat products during gastrointestinal digestion

Leticia Mora and Fidel Toldra

Department of Food Science, Institute of Agrochemistry and Food Technology, Spanish National Research Council (CSIC), Paterna, Spain

8.1 Introduction

Meat is a highly significant source of dietary proteins, providing the human body all the essential amino acids required for optimal nutrition. Furthermore, meat is rich in minerals, such as zinc, iron, selenium, iron, and phosphorus, as well as nutrients like niacin, vitamins B12 and B6, choline, and riboflavin. Additionally, certain types of meat are also abundant in vitamin K. On the other hand, the richness of meat and meat products makes them a potential and important source of bioactive peptides (BAP). Thus meat and meat products constitute a good source of nutrients, but even more, they can also influence or modulate the different physiological systems through the generation of BAP (López-Pedrouso et al., 2023).

BAP are hidden within the protein sequences of meat and meat products and can be liberated from alkali, acid, or enzymatic digestion. Main enzymatic ways to obtain BAP are: (1) during the processing of meat or meat products, including curing or fermentation due to the action of endogenous muscular or added microbial enzymes, (2) during gastrointestinal (GI) digestion through the action of salivary, stomachal, intestinal and pancreatic enzymes (Capriotti et al., 2015; Pepe et al., 2016), and (3) due to the action of commercial enzymes to hydrolyze meat or meat-derived products under controlled conditions (Fig. 8.1).

Main key mechanisms that involve GI digestion are related to the action of sequential oral, gastric, and intestinal digestion enzymes, including amylase, pepsin, lipase (gastric and pancreatic), trypsin and chymotrypsin. These enzymes are active in the human body being able to generate numerous BAP that could be affecting different physiological mechanisms, such as circulatory, respiratory, digestive, or nervous systems (Brodkorb et al., 2019). In this sense, in the last decade, there is an increasing interest in studying and understanding the food digestion in humans. However, the use of in vivo trials is complicated due to the

Protein Digestion-Derived Peptides. DOI: https://doi.org/10.1016/B978-0-443-19141-1.00008-X

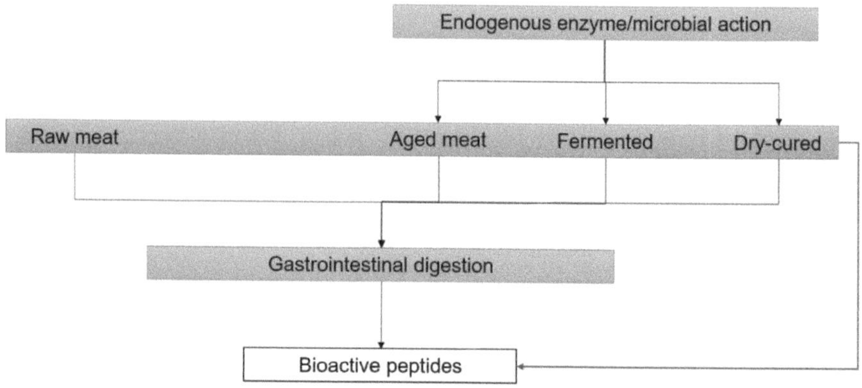

FIGURE 8.1

Generation of BAP. Meat and meat products BAP main ways of generation, that includes the processing due to the action of endogenous muscular or added microbial enzymes and the gastrointestinal digestion through the action of salivary, stomach, intestinal, and pancreatic enzymes.

difficulties in justifying ethical reasons and the expensive price so the use of in vitro models, such as the Infogest 2.0 model, has been widely spread in research approaches to simulate the digestion of different food, ingredients, or even pharmaceutical compounds (Brodkorb et al., 2019). On the other hand, the obtained results after testing in vivo effects of BAP not always are in line with the previous in vitro tests, which is why it is so important to consider the effect of the action happening during digestion and absorption.

Research studies of BAP derived from meat and meat products suffer the challenge of being able to reach the bloodstream and later arrive to the target organ in order to exert its biological activity (Sun et al., 2020). The studies involving the consumption of BAP are especially difficult considering the complications occurred after oral consumption due its inactivation by acids and enzymes participating in the GI digestion and present in the bloodstream, in addition to the challenge of crossing intact the intestinal barrier where frequently occurs the degradation of BAP due to the activity of transepithelial transport cells (Gallego et al., 2016).

Regarding BAP identification, many databases that compile the sequences mentioned in the literature are commonly used and result highly valuable for relating certain bioactive sequences previously included in the database with the sequences identified in the studies (Carrasco-Castilla et al., 2012). Nevertheless, there is very little data regarding the concentration of these BAP present in the processed food, as well as the quantity of peptides bioavailable in the bloodstream following GI digestion. In this sense, small peptides, such as di- and tripeptides, use to be those with the highest probabilities to exert their physiological effect in an effective way but the quantitation of BAP encounter multiple challenges due to their small molecular size, which is between metabolomics and proteomics disciplines, and the complexity of food matrices (Gallego et al., 2019). Consequently, the application of advanced mass spectrometry techniques is crucial for accurately identify the sequence of these small peptides although difficulties, such as cross contaminations, interferences, and the low abundancy of the most interesting sequences, complicate the development of adequate methodologies for their quantitation from difficult matrices like meat and meat products (Mora et al., 2017). Recently, the importance of structural properties of BAP on their resistance during GI digestion has been reported on a systematic review using PRISMA guidelines, concluding that peptides resistant to in vitro GI digestion compared to nonresistant peptides show lower molecular size and hydrophobicity, higher positive net charge at intestinal pH, branched-chain aliphatic N-terminal residues, absence of C-terminal leucine, and higher histidine and proline content at the C-terminal (Ahmed et al., 2022).

8.2 Bioactive peptides released from meat

The most studied bioactivities in BAP derived from raw meat are angiotensin-converting enzyme (ACE)-I inhibitory and antioxidant activities.

ACE plays a crucial role in regulating blood pressure within the renin-angiotensin system, as it is responsible for transforming the inactive decapeptide angiotensin I into the potent vasoconstrictor octapeptide, angiotensin II. This conversion leads to an increase in blood pressure. Additionally, ACE has the ability to deactivate bradykinin. Consequently, peptides that inhibit ACE have the potential to lower blood pressure when administered orally a few hours after intake. These ACE inhibitory peptides achieve this effect by inhibiting ACE, which is typically membrane-bound in vascular endothelial cells.

Raw meat has been widely studied as a source of BAP when digested under controlled conditions of hydrolysis using different commercial proteases under specific hydrolysis conditions (Arihara et al., 2001; Katayama et al., 2003). However, not always the study of the effect of GI digestion is considered. In this sense, pork raw meat has been previously identified as a source of ACE inhibitory peptides after simulated in vitro GI digestion (Escudero et al., 2010).

8.2.1 SNL meat compared to CG meat

Regarding chicken meat, the ACE-inhibitory activity of a breast muscle extract treated with an *Aspergillus protease* and gastric proteases (trypsin, chymotrypsin, and intestinal juice) was tested in spontaneously hypertensive rats (SHRs) and a decrease in blood pressure was observed after oral administration (Saiga et al., 2003). Also, significant differences in ACE-inhibitory activity was observed after in vitro GI digestion of commercial broiler breast meat using different cooking methods (Sangsawad et al., 2017). In duck meat, potential toxicity, resistance against gastrointestinal proteases and physicochemical properties of the synthetized peptides identified after hydrolysis with was determined using in silico analysis. They were stable against digestion, indicating that duck-derived antioxidant peptide could serve as a promising candidate against oxidative stress (Li et al., 2020).

The potential of rabbit meat protein as a source of ACEI-inhibitory peptides after in silico GI digestion have been studied using the enzymes of the GI digestion tract, pepsin, trypsin, and chymotrypsin. The "Enzyme action" tool available in the BIOPEP-UWM database (http://www.uwm.edu.pl/biochemia/index.php/pl/biopep) was used to simulate the enzymatic digestion, and seven novel ACE inhibitory peptides with good absorption, distribution, metabolism, elimination and toxicity (ADMET) properties were finally identified and confirmed in vitro, analyzed in admetSAR (http://lmmd.ecust.edu.cn/admetsar1/) (Chen et al., 2022).

Antioxidant peptides can be categorized into two main groups based on the types of reactions they participate in: (i) The first group involves reactions that rely on hydrogen atom transfer. These peptides work by reducing free radicals, and their efficacy can be assessed using various assays like the oxygen radical absorbance capacity assay (ORAC), total radical trapping antioxidant parameter (TRAP), and β-carotene bleaching assay. (ii) The second group involves reactions based on electron transfer, where the peptides reduce oxidants. These reactions can be measured through assays, such as the 2,2-azino-bis(3-ethylbenzothiazoline-6-sulfonic acid) (ABTS) radical scavenging assay, ferric-reducing antioxidant power assay, and 1-diphenyl-2 picrylhydrazyl (DPPH) radical scavenging activity. Synthetic antioxidant compounds like t-butylhydroquinone (TBHQ), butylated hydroxyanisole (BHA), butylated hydroxytoluene (BHT), and propyl gallate are widely used in several food products. However, similar to the situation with ACE-inhibitory compounds, there is a growing interest in exploring and finding new natural antioxidant peptides as an alternative to synthetic antioxidants to avoid the recent safety concerns associated with the use of synthetic antioxidants in food products. Antioxidant and antihypertensive activities of *Longissimus dorsi* muscle from Suino Nero Lucano (SNL) and modern crossbred (CG) pigs, before and after cooking and in vitro GI digestion have been studied, concluding that in vitro GI digestion positively influenced antioxidative and antihypertensive activity of SNL meat compared to CG meat (Simonetti et al., 2016).

Type-2 diabetes is a significant global health concern, affecting millions of adults worldwide, and closely linked to overnutrition and obesity. Recently BAP have been reported to exert positive effects in preventing and modulating this type of diabetes mainly enhancing insulin sensitivity. The enzymes a-glucosidase and dipeptidyl dipeptidase IV (DPPIV) play essential roles in regulating glucose levels in postprandial plasma, and in diabetes type-2 management, an effective strategy could involve inhibiting both of these enzymes. DPPIV is a serine protease that primarily cleaves X-proline or X-alanine dipeptides from the N-terminus of various substrates. Its inhibition would reduce its action on glucagon-like peptide (GLP-1) and glucose-dependent insulinotropic peptide (GIP), leading to a lower degradation of GLP-1. GLP-1 is an important hormone that promotes insulin secretion and regulates blood sugar levels. By inhibiting DPPIV and preserving GLP-1, the body can maintain better control over glucose levels (Yin et al., 2022).

The potential contribution to generate BAP after consumption of grilled (140°C for 5 minutes) pork, beef, chicken, and turkey meat was studied following in vitro GI digestion following the Infogest protocol., and pork was found to be the best source of dipeptidyl peptidase (DPP)-IV inhibitory peptides whereas chicken meat showed the highest ACE-inhibitory activity (Martini et al., 2019). More recently, in vitro models of GI digestion simulating the conditions of the GI tract of healthy adults and elderly individuals were applied to investigate differences in the digestibility of meat (chicken, beef and pork).

8.3 Bioactive peptides released from meat-derived products

In meat-derived products there is a previous processing step before consumption that includes ageing, curing or fermentation. The ultimate characteristics of processed meats are influenced by numerous factors, being the proteolysis one of the key contributors. This proteolysis occurs as a result of the activity of endogenous enzymes present in products like matured meat and dry-cured ham. Additionally, microbial peptidases also play a role in the proteolysis of dry-fermented meats, where the addition of starters containing *Lactobacillus* is very common and results in the combination of microbial and endogenous proteolysis (Lepper-Blilie et al., 2016; Leroy et al., 2004; Warner et al., 2022).

The identification and characterization of BAP in processed meats also need the use of robust mass spectrometry methodologies. In fact, the most common proteomics approaches are not adequate for peptidomics, therefore alternative strategies need to be considered.

8.3.1 Aged/ripened meat

Ageing stands as the predominant method employed by meat processors to elevate meat texture and palatability (Lepper-Blilie et al., 2016). Usually meat is packed and stored under chilling conditions (between −1 and 4°C) for a period of 2–4 weeks or even up to 12–20 weeks, especially for long distance export markets (Hughes et al., 2015). In this regards, the duration of aging together with the temperature have an important influence on the final quality of meat. Long aging times were reported to improve texture and flavor characteristics although the optimal time for dry aging of beef is 14–21 days and 7–10 days for wet aging at 0–1°C (Khan et al., 2016; Leroy et al., 2004).

The impact of ageing on meat tenderness occurs from the breakdown of structural muscle proteins, such as titin, nebulin, and desmin, by muscle endopeptidases. This phenomenon has been extensively studied and reviewed in order to understand its mechanisms and effects (Warner et al., 2022). Nevertheless, the role of exopeptidases and their influence during meat ageing has received less attention in spite of their crucial role in the generation of small peptides and their impact on the final quality of meat. Also there is little information regarding the effect of GI digestion in ripened meats. Fig. 8.2 supports the evidence that aged beef meat has the potential to be a valuable source of ACE-inhibitory peptides, adding extra value to this product. Based on this, cooking does not have a significant impact on the detected bioactivity, while in vitro GI digestion leads to a significant increase in the ACE-inhibitory activity from the first week. This increase is probably due to the strong generation of small peptides resulting from the activity of gastrointestinal enzymes (Mora et al., 2017).

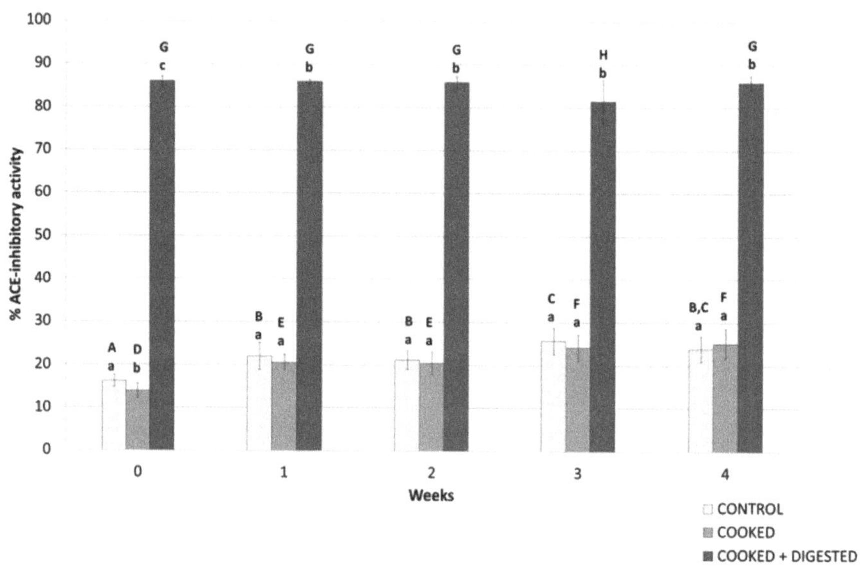

FIGURE 8.2

Aged beef meat bioactivity. ACE-inhibitory activity (%) of control, samples after cooking, and cooking + GI digestion at 0, 1, 2, 3, and 4 weeks of storage ($n = 6$). Differences in letters (a, b, c, ...) indicate significant differences between control and processed samples whereas differences in capital letters (A, B, C, D, ...) indicate significant differences in the same sample between time of storage ($P < .5$).

From Mora, L., Bolumar, T., Heres, A., & Toldrá, F. (2017). Effect of cooking and simulated gastrointestinal digestion on the activity of generated bioactive peptides in aged beef meat. Food and Function, *8(12), 4347–4355. https://doi.org/10.1039/c7fo01148. With the permission of the Royal Society of Chemistry.*

8.3.2 Dry-cured meat products

Dry-curing involves different biochemical reactions that result in the generation of multiple compounds that characterize the final quality of the meat product in terms of color, texture, taste, and flavor. The intense proteolysis occurred by the action endogenous muscle enzymes results in an accumulation of peptides (Mora et al., 2009, 2017; Wang et al., 2020) and free amino acids at the end of the dry-curing process (Sforza et al., 2001; Toldrá et al., 2000), being these products of great importance for its contribution to the final quality of the product. The action of GI enzymes contributes to the generation of even smaller peptides like di- and tripeptides which many of them have been widely described as bioactive in literature. The contribution of gastrointestinal enzymes to the generation of peptides able to modify or influence the different biological systems is very important as the small size of these compounds facilitate them to arrive intact to the target organs to exert their bioactivity.

The scheme represented in Fig. 8.3 shows how the intense proteolysis occurred in the proteins of dry-cured ham after the curing period together with the effect of GI enzymes permit the generation of small peptides that can be transported through the intestinal membrane without modification and reach the bloodstream and organs to exert their physiological action (Toldrá et al., 2020).

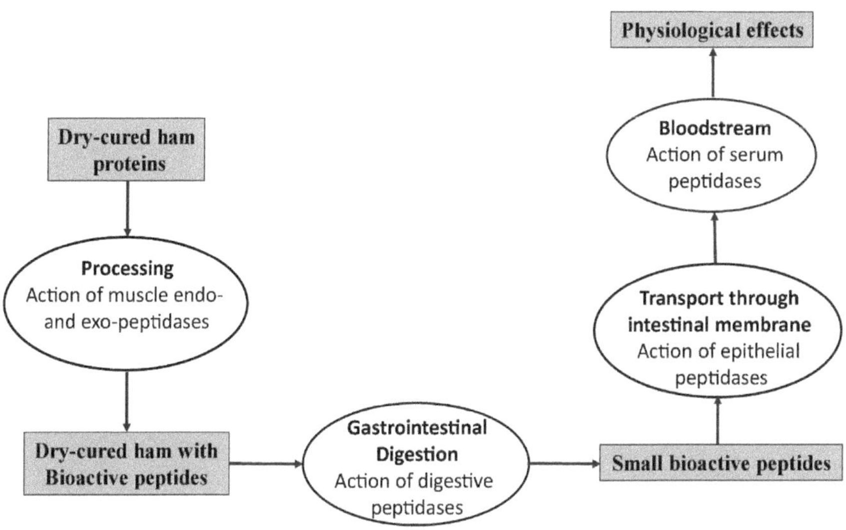

FIGURE 8.3

Scheme about dry-cured ham BAP generation. Scheme of generation of bioactive peptides in dry-cured ham.

From Toldrá, F., Gallego, M., Reig, M., Aristoy, M. C., & Mora, L. (2020). Bioactive peptides generated in the processing of dry-cured ham. Food Chemistry, 321, *126689. https://doi.org/10.1016/j.foodchem.2020.12668. With the permission of Elsevier.*

In a recent study about the impact of in vitro GI digestion on the peptide profiles and their antioxidant activity of dry-cured hams, it was observed that the antioxidant activity of the digested samples decreased when assessed by 2,2-diphenyl-1-picryl-hydrazyl (DPPH) radical scavenging activity and ferric-reducing antioxidant measurement, but improved when analyzed using 2,2′-azino-bis-(3-ethylbenzothiazoline-6-sulfonic) acid radical scavenging methodology (Gallego et al., 2020). Also, chromatographic and mass spectrometry analysis revealed changes in the peptide profiles as shown in Fig. 8.4, evidencing that the characteristics of amino acid composition, length, or hydrophobicity affects the bioactivity and bioavailability of the peptides. The analysis of peptide profiles before and after the simulated GI digestion of dry-cured hams revealed the hydrolysis of proteins, predominantly titin (TTN), that is the largest protein in muscle with primary structural function, by GI enzymes. This characterization highlights the role of GI enzymes in the proteolytic degradation of dry-cured ham proteins as well as prove the importance of peptide characteristics on the antioxidant activity. In this sense, the principal component analysis shown in Fig. 8.5 evidences on the differences between the undigested and digested samples as well as information about the most influential peptides. According to these results, the presence of peptides derived from titin is the main responsible for the observed distribution of the score plot in the analysis. This fact is due to the substantial number of titin-derived peptides identified after simulated GI digestion.

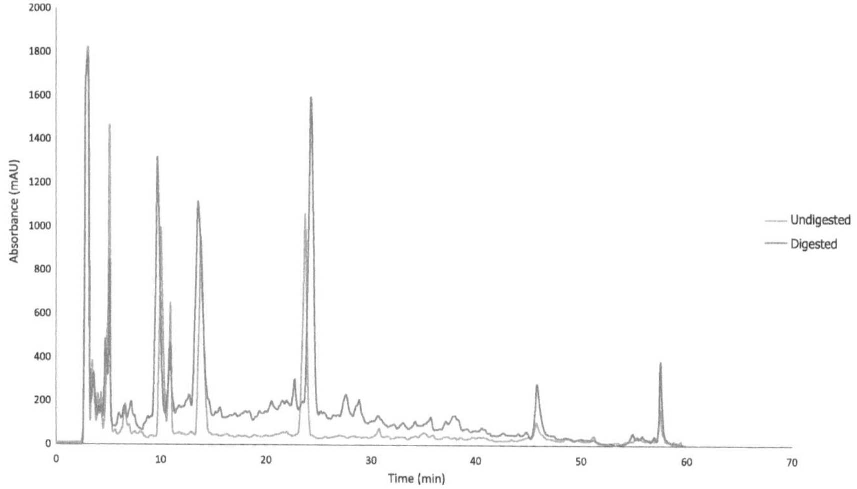

FIGURE 8.4

Dry-cured ham peptides profile. Reversed-phase chromatographic separation of the undigested sample compared to the digested dry-cured ham sample.

From Gallego, M., Mauri, L., Aristoy, M. C., Toldrá, F., & Mora, L. (2020). Antioxidant peptides profile in dry-cured ham as affected by gastrointestinal digestion. Journal of Functional Foods, 69, *103956. https://doi.org/10.1016/j.jff.2020.10395. With the permission of Elsevier.*

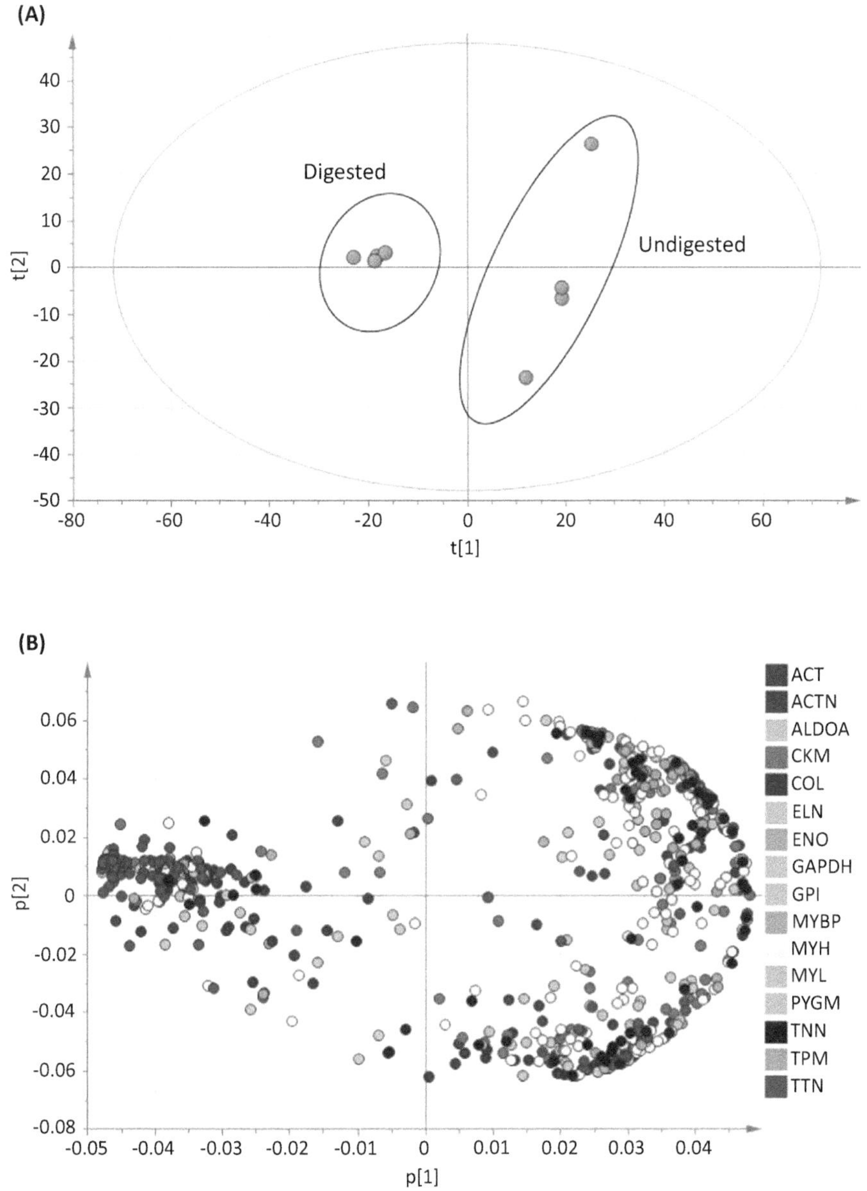

FIGURE 8.5

Dry-cured ham peptidomics statistics. (A) Principal component analysis score plot of main identified proteins to assess the variance among all the peptides generated in the undigested and digested dry-cured ham samples ($n = 4$). (B) Loading plot showing the peptides affecting the score plot distribution and colored by protein of origin.

From Gallego, M., Mauri, L., Aristoy, M. C., Toldrá, F., & Mora, L. (2020). Antioxidant peptides profile in dry-cured ham as affected by gastrointestinal digestion. Journal of Functional Foods, 69, *103956. https://doi.org/10.1016/j.jff.2020.103956. With the permission of Elsevier.*

8.3.3 Fermented meat products

Traditional fermented meat products can be prepared using different formulations, types of raw meat, steps of processing, and microbiological starters. The final quality of these products will depend on the generation of different compounds from the breakdown of proteins, lipids and carbohydrates, in which both endogenous muscle and microbial enzymes participate (Fernández et al., 2000; Ordóñez et al., 1999).

Main characteristics of these types of products is the use of starter cultures to assure the safety of final product as well as its technological properties. Main characteristics of these types of products is the use of starter cultures to assure the safety of final product as well as its technological properties. Lactic acid bacteria and staphylococci are commonly found in European dry-fermented sausages and have been recognized as the most active microorganisms involved in several processes, such as acidification, denitrification, lipolysis, and proteolysis (Flores & Toldrá, 2011). Thus selecting specific strains it is possible to achieve enhanced sensory characteristics and gain technological advantages in the production of dry-fermented sausages.

There are few studies regarding the effect of GI digestion in these products although the observed changes in peptide profile and amino acid contents of three European dry-fermented sausages produced in Spain, Italy, and Belgium, respectively (Fig. 8.6), proved the intense hydrolysis occurred and the extreme generation peptides and free amino acids (Gallego et al., 2016). The ACEI-inhibitory and antioxidant activity of each type of dry-fermented sausage was evaluated in this study after fractionation by size-exclusion chromatography. Spanish and Belgian dry-fermented sausages exhibited ACE inhibition values of approximately 85% at a concentration of 200 mg/mL. In addition, Belgian samples showed the highest levels of DPPH radical-scavenging activity and ferric reducing power capacity when compared to the Spanish sausages. This study, confirms that the combined action of muscle and microbial enzymes would contribute to the generation of BAP showing ACE inhibitory and antioxidant activities. The obtained results improved the knowledge about the intense proteolysis that takes place during the processing of dry-fermented Spanish, Italian, and Belgian sausages as well as evidence their potential as natural sources of BAP.

The enormous amount of peptides generated due to the action of endogenous muscle enzymes and microbial enzymes might be enhanced by the GI enzymes, which could contribute to decrease the final size of the peptides making them more adequate to reach the blood system and later exert its physiological activity.

8.4 In vivo studies

Different studies regarding naturally generated BAP resulted in significant differences between in vitro and in vivo results, which make necessary to confirm the biological activity with further studies.

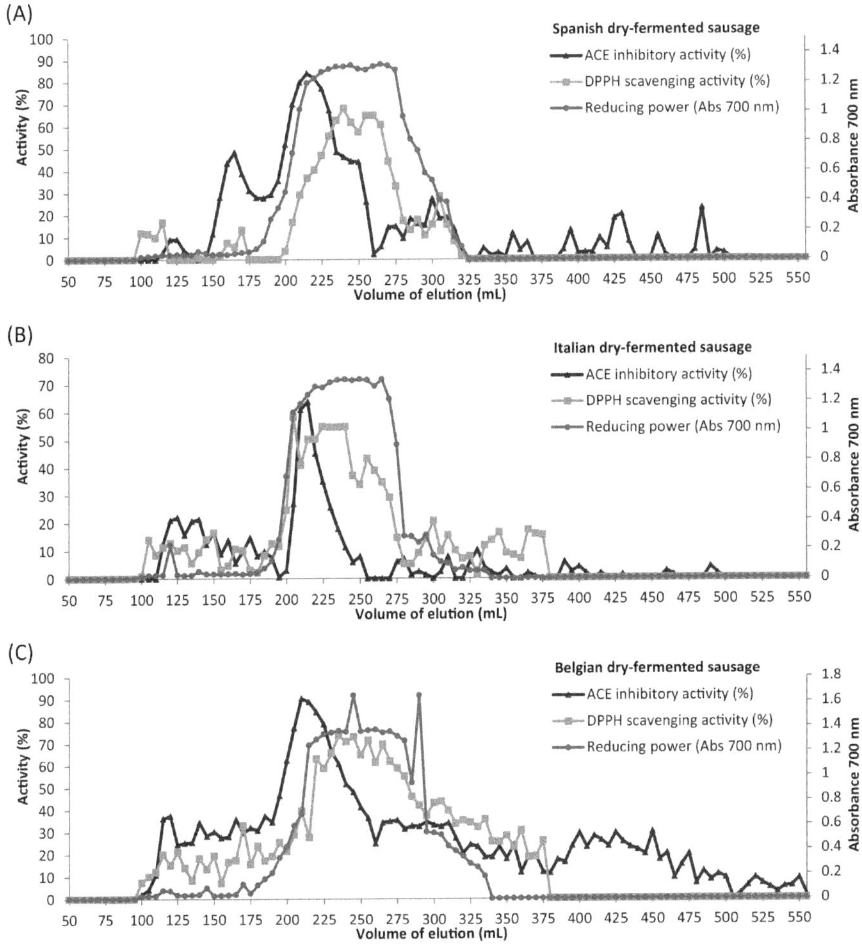

FIGURE 8.6

Dry-fermented sausages bioactive profiles. ACE inhibitory activity and antioxidant activity (assayed by the measurement of DPPH radical-scavenging activity and reducing power) of the fractions collected from size-exclusion chromatography of (A) Spanish dry-fermented sausage, (B) Italian dry-fermented sausage, and (C) Belgian dry-fermented sausage.

From Gallego, M., Mora, L., Escudero, E., & Toldrá, F. (2018). Bioactive peptides and free amino acids profiles in different types of European dry-fermented sausages. International Journal of Food Microbiology, *276, 71–78. https://doi.org/10.1016/j.ijfoodmicro.2018.04.00. With the permission of Elsevier.*

Regarding this, the peptide AAATP was isolated and identified in dry-cured ham resulting in an IC_{50} value of 100 μM for ACEI-inhibitory activity. The activity of this peptide was also confirmed in vivo resulting in the decrease of systolic blood pressure by -25.62 ± 4.5 mm Hg ($P < .05$) in SHR after 8 hours of administration. Results from this study show that Spanish dry-cured ham may represent a source of natural peptides with potential benefit for human health (Escudero et al., 2013).

In order to understand the effect of gut epithelial membrane on the transport of BAP, transport assays through Caco-2 cells were used as a model for small intestinal transport of peptides. In order to understand the effect of gut epithelial membrane on the transport of BAP, transport assays through Caco-2 cells were used as a model for small intestinal transport of peptides. The used Caco-2 cells keep the morphology and role of mature enterocytes, and express peptidases and transporters that may be in contact with the studied peptides affecting their transport, being Caco-2 a model that is frequently used to study the phenomena of absorption in the small intestine (Hidalgo et al., 1989; Yee, 1997). In this sense, Gallego et al. (2016) studied the effect of transport through Caco-2 cells in AAATP peptide and other sequences and evidenced that the transport can result in the degradation of ACE inhibitory peptides into smaller fragments (Fig. 8.7) (Gallego et al., 2016). This occurs mainly because peptides are susceptible to be hydrolyzed by brush border peptidases, resulting in the generation of di- and tri-peptides that are likely to be transported through the intestinal peptide transporter T1 (PepT1). Wang and Li (2017) reported that peptides transported by this active route exhibit higher bioavailability compared to peptides transported via the paracellular route. However, it is also important to consider that the amino acid sequence of short peptides can also impact their bioavailability. The use of cell cultures has contributed to our understanding about the diverse functional properties of antihypertensive peptides. These peptides have resulted to be active by using different metabolic routes, including ACE inhibition, control in the release of nitric oxide and endothelin, as well as reductions in oxidative stress and inflammation (Chakrabarti et al., 2017; Liu et al., 2016).

Regarding human studies, clinical trials are the most accurate way to evaluate the efficacy and health effects of naturally generated BAP from food sources, but the complexity, high cost, need of medical personnel, and time consuming make these assays very difficult to be conducted. Recent studies established a relation between the consumption of bioactive compounds from dry-cured ham and positive health effects observed in a clinical trial with 40 volunteers. Main results showed a modification of systolic blood pressure, total cholesterol, low-density lipoproteins and basal glucose levels after dry-cured ham consumption even though results were not conclusive due to the low number of volunteers (Montoro-García et al., 2017). More recently this study was confirmed with a second clinical trial involving fifty-four volunteers with stage 1 prehypertension concluding that regular dry-cured ham consumption improve systolic and diastolic blood pressure and facilitates the maintenance of metabolic pathways, being positive in the prevention of cardiovascular diseases (Montoro-García et al., 2022).

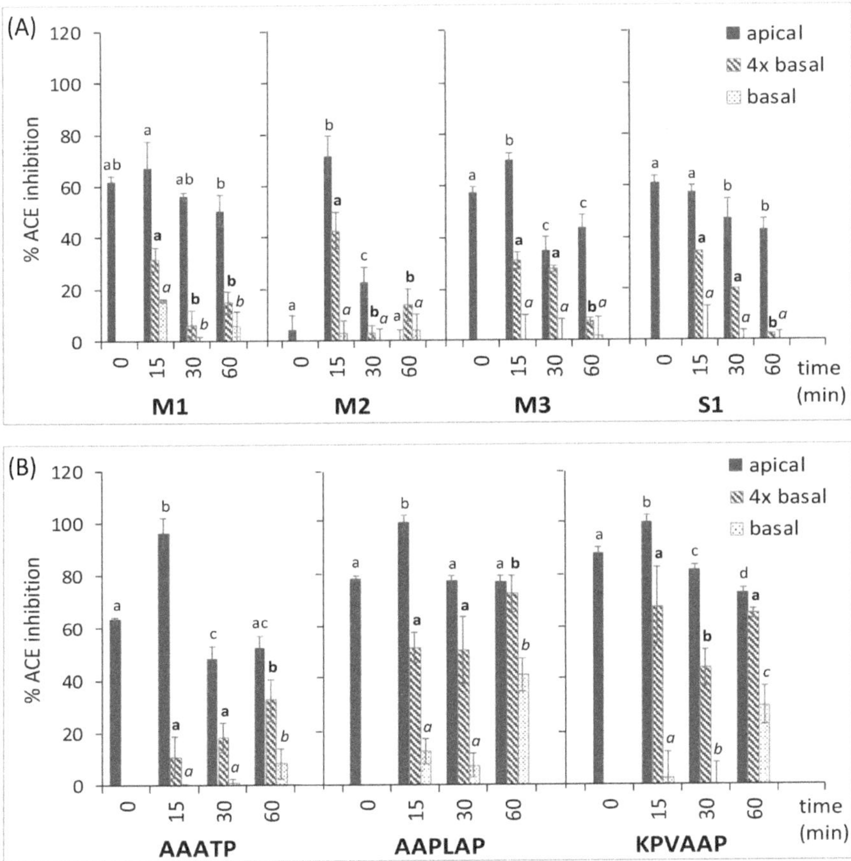

FIGURE 8.7

Caco-2 cells transport. ACE inhibitory activity (%) observed in the apical, basal, and four times concentrated basal compartments obtained from the cellular transport after the application of (A) dry-cured ham extracts and (B) purified peptides. Bar letters indicate significant differences at $P<.05$ ($N=3$) between different times of the cellular transport (0, 15, 30, 60 minutes) in each compartment and sample (a = apical; **a** = basal 4×; *a* = basal).

From Gallego, M., Grootaert, C., Mora, L., Aristoy, M. C., Van Camp, J., & Toldrá, F. (2016). Transepithelial transport of dry-cured ham peptides with ACE inhibitory activity through a Caco-cell monolayer. Journal of Functional Foods, 21, *388–395. https://doi.org/10.1016/j.jff.2015.11.04. With the permission of Elsevier.*

8.5 In silico studies

The advancement in the identification and characterization of the biological activity of naturally generated peptides is extremely dependent on modern mass spectrometry instruments, bioinformatics tools, and updated protein databases as well as on expensive biological test based on ELISA protocols necessary in empirical approaches.

The use of in silico methods together with proteomic approaches can be complementary tools in the identification of peptides from meat protein sources as many steps of traditional empirical approaches can be changed and simplified by using bioinformatics. Also, the use of statistics tools facilitates the selection of best candidates as bioactive sequences, and later bioactivity assays with the synthesized peptides would confirm the hypothesis.

On the other hand, the high difficulty of the study of GI digestion and its products in direct, have resulted in the development of simulation tools that are simple and useful digestion models developed specially to study the effects of food processing on the proteins and evaluate the stability of BAP in the GI tract (Mora et al., 2017; Sayd et al., 2016). These models do not reproduce all the dynamic aspects of the GI process, but show very good correlations with in vivo outcomes even in the case of complex foods that can be subjected to structural changes due to the digestive environment (Bohn et al., 2017).

Regarding main in silico tools that are used in the analysis and prediction of BAP, BIOPEP can be used to simulate the hydrolysis using up to three different

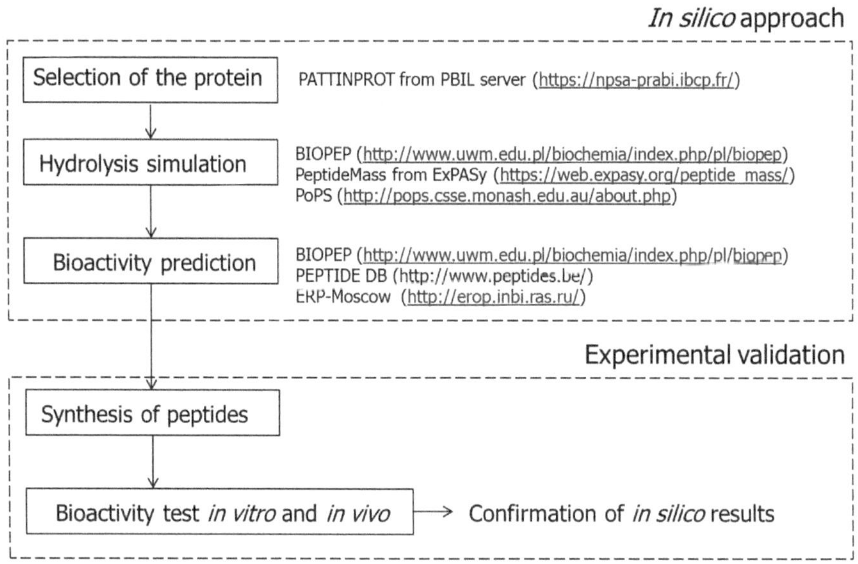

FIGURE 8.8

In silico approach scheme. Main steps of in silico approaches and open access databases for the selection of the protein, hydrolysis simulation and bioactivity prediction.

From Toldrá, F., Gallego, M., Reig, M., Aristoy, M. C., & Mora, L. (2020). Bioactive peptides generated in the processing of dry-cured ham. Food Chemistry, 321, *126689. https://doi.org/10.1016/j.foodchem.2020.12668. With the permission of Elsevier.*

enzymes as well as to predict the potential of a certain sequence to be bioactive. The empirical validation of the results using analytical techniques as well as the bioactivity confirmation in vivo is frequently necessary to confirm the obtained results as indicated in the diagram (Fig. 8.8).

8.6 Conclusions and future perspectives

The influence and responsibility of GI digestion enzymes on the generation of natural BAP from meat and meat products has been widely proved in the last years. Regarding this, it is expected that meat results a highly valuable source of BAP due to its high content in proteins. This fact gives an extra value to the consumption of meat due to the positive health benefits of these compounds that are mainly generated after the GI digestion of the meat-derived food.

On the other hand, also there is an increasing interest in using co- and by-products from meat industry for the production of added-value peptides mainly derived from protein hydrolysates to be used as ingredients in feed, food, pet food, as well as food complements or food supplements. This perspective is opening interesting alternatives that support the circular economy of the sector by decreasing the environmental impact of by-products increasing their economic potential.

References

Ahmed, T., Sun, X., & Udenigwe, C. C. (2022). Role of structural properties of bioactive peptides in their stability during simulated gastrointestinal digestion: A systematic review. *Trends in Food Science and Technology*, *120*, 265–273. Available from https://doi.org/10.1016/j.tifs.2022.01.008, http://www.elsevier.com/wps/find/journaldescription.cws_home/601278/description#description.

Arihara, K., Nakashima, Y., Mukai, T., Ishikawa, S., & Itoh, M. (2001). Peptide inhibitors for angiotensin I-converting enzyme from enzymatic hydrolysates of porcine skeletal muscle proteins. *Meat Science*, *57*(3), 319–324. Available from https://doi.org/10.1016/S0309-1740(00)00108-X.

Bohn, T., Carriere, F., Day, L., Deglaire, A., Egger, L., Freitas, D., et al. (2017). Correlation between in vitro and in vivo data on food digestion. What can we predict with static in vitro digestion models? *Critical Reviews in Food Science and Nutrition*, *58*(13), 2239–2261. Available from http://dx.doi.org/10.1080/10408398.2017.13153621.

Brodkorb, A., Egger, L., Alminger, M., Alvito, P., Assunção, R., Ballance, S., Bohn, T., Bourlieu-Lacanal, C., Boutrou, R., Carrière, F., Clemente, A., Corredig, M., Dupont, D., Dufour, C., Edwards, C., Golding, M., Karakaya, S., Kirkhus, B., Le Feunteun, S., ... Recio, I. (2019). INFOGEST static in vitro simulation of gastrointestinal food digestion. *Nature Protocols*, *14*(4), 991–1014. Available from https://doi.org/10.1038/s41596-018-0119-1, http://www.natureprotocols.com/.

Capriotti, A. L., Caruso, G., Cavaliere, C., Samperi, R., Ventura, S., Zenezini Chiozzi, R., & Laganà, A. (2015). Identification of potential bioactive peptides generated by simulated gastrointestinal digestion of soybean seeds and soy milk proteins. *Journal of Food Composition and Analysis*, *44*, 205–213. Available from https://doi.org/10.1016/j.jfca.2015.08.007, http://www.elsevier.com/inca/publications/store/6/2/2/8/7/8/index.htt.

Carrasco-Castilla, J., Hernández-Álvarez, A. J., Jiménez-Martínez, C., Gutiérrez-López, G. F., & Dávila-Ortiz, G. (2012). Use of proteomics and peptidomics methods in food bioactive peptide science and engineering. *Food Engineering Reviews*, *4*(4), 224–243. Available from https://doi.org/10.1007/s12393-012-9058-8, http://www.springer.com/life + sci/food + science/journal/12393.

Chakrabarti, S., Liao, W., & Davidge, S. T. (2017). Milk-derived tripeptides IPP (Ile-Pro-Pro) and VPP (Val-Pro-Pro) differentially modulate angiotensin II effects on vascular smooth muscle cells. *Journal of Functional Foods*, *30*, 151–158.

Chen, J., Yu, X., Chen, Q., Wu, Q., & He, Q. (2022). Screening and mechanisms of novel angiotensin-I-converting enzyme inhibitory peptides from rabbit meat proteins: A combined in silico and in vitro study. *Food Chemistry*, *370*, 131070. Available from https://doi.org/10.1016/j.foodchem.2021.131070.

Escudero, E., Mora, L., Fraser, P. D., Aristoy, M. C., Arihara, K., & Toldrá, F. (2013). Purification and Identification of antihypertensive peptides in Spanish dry-cured ham. *Journal of Proteomics*, *78*, 499–507. Available from https://doi.org/10.1016/j.jprot.2012.10.019.

Escudero, E., Sentandreu, M. A., Arihara, K., & Toldrá, F. (2010). Angiotensin I-Converting enzyme inhibitory peptides generated from in vitro gastrointestinal digestion of pork meat. *Journal of Agricultural and Food Chemistry*, *58*(5), 2895–2901. Available from https://doi.org/10.1021/jf904204n.

Fernández, M., Ordóñez, J. A., Bruna, J. M., Herranz, B., & de la Hoz, L. (2000). Accelerated ripening of dry-fermented sausages. *Trends in Food Science & Technology*, *11*, 201–209.

Flores, M., & Toldrá, F. (2011). Microbial enzymatic activities for improved fermented meats. *Trends in Food Science & Technology*, *22*(2), 81–90.

Gallego, M., Grootaert, C., Mora, L., Aristoy, M. C., Van Camp, J., & Toldrá, F. (2016). Transepithelial transport of dry-cured ham peptides with ACE inhibitory activity through a Caco cell monolayer. *Journal of Functional Foods*, *21*, 388–395. Available from https://doi.org/10.1016/j.jff.2015.11.046, http://www.elsevier.com/wps/find/journaldescription.cws_home/717426/description#description.

Gallego, M., Mauri, L., Aristoy, M. C., Toldrá, F., & Mora, L. (2020). Antioxidant peptides profile in dry-cured ham as affected by gastrointestinal digestion. *Journal of Functional Foods*, *69*, 103956. Available from https://doi.org/10.1016/j.jff.2020.103956.

Gallego, M., Mora, L., & Toldrá, F. (2019). The relevance of dipeptides and tripeptides in the bioactivity and taste of dry-cured ham. *Food Production, Processing and Nutrition*, *1*(1). Available from https://doi.org/10.1186/s43014-019-0002-7, https://www.springer.com/journal/43014.

Hidalgo, I. J., Raub, T. J., & Borchardt, R. T. (1989). Characterization of the human colon carcinoma cell line (Caco-2) as a model system for intestinal epithelial permeability. *Gastroenterology*, *96*(2), 736–749. Available from https://doi.org/10.1016/S0016-5085(89)80072-1.

Hughes, J. M., McPhail, N. G., Kearney, G., Clarke, F., & Warner, R. D. (2015). Beef longissimus eating quality increases up to 20 weeks of storage and is unrelated to meat colour at carcass grading. *Animal Production Science*, *55*(2), 174. Available from https://doi.org/10.1071/an14304.

Katayama, K., Tomatsu, M., Fuchu, H., Sugiyama, M., Kawahara, S., Yamauchi, K., Kawamura, Y., & Muguruma, M. (2003). Purification and characterization of an angiotensin I-converting enzyme inhibitory peptide derived from porcine troponin C. *Animal Science Journal*, *74*(1), 53–58. Available from https://doi.org/10.1046/j.1344-3941.2003.00086.x.

Khan, M. I., Jung, S., Nam, K. C., & Jo, C. (2016). Postmortem aging of beef with a special reference to the dry aging. *Korean Journal for Food Science of Animal Resources*, *36*(2), 159–169. Available from https://doi.org/10.5851/kosfa.2016.36.2.159, http://ocean.kisti.re.kr/downfile/volume/ksfsar/CSSPBQ/2016/v36n2/CSSPBQ_2016_v36n2_159.pdf.

Lepper-Blilie, A. N., Berg, E. P., Buchanan, D. S., & Berg, P. T. (2016). Effects of post-mortem aging time and type of aging on palatability of low marbled beef loins. *Meat Science*, *112*, 63–68. Available from https://doi.org/10.1016/j.meatsci.2015.10.017, http://www.elsevier.com/locate/meatsci.

Leroy, B., Lambotte, S., Dotreppe, O., Lecocq, H., Istasse, L., & Clinquart, A. (2004). Prediction of technological and organoleptic properties of beef Longissimus thoracis from near-infrared reflectance and transmission spectra. *Meat Science*, *66*(1), 45–54. Available from https://doi.org/10.1016/S0309-1740(03)00002-0, http://www.elsevier.com/locate/meatsci.

Li, T., Shi, C., Zhou, C., Sun, X., Ang, Y., Dong, X., Huang, M., & Zhou, G. (2020). Purification and characterization of novel antioxidant peptides from duck breast protein hydrolysates. *LWT*, *125*, 109215. Available from https://doi.org/10.1016/j.lwt.2020.109215.

Liu, R. L., Ge, X. L., Gao, X. Y., Zhan, H. Y., Shi, T., Su, N., et al. (2016). Two angiotensin-converting enzyme-inhibitory peptides from almond protein and the protective action on vascular endothelial function. *Food & Function*, *7*(9), 3733–3739.

López-Pedrouso, M., Zaky, A. A., Lorenzo, J. M., Camiña, M., & Franco, D. (2023). A review on bioactive peptides derived from meat and by-products: Extraction methods, biological activities, applications and limitations. *Meat Science*, *204*, 109278. Available from https://doi.org/10.1016/j.meatsci.2023.109278.

Martini, S., Conte, A., & Tagliazucchi, D. (2019). Comparative peptidomic profile and bioactivities of cooked beef, pork, chicken and turkey meat after in vitro gastro-intestinal digestion. *Journal of Proteomics*, *208*, 103500. Available from https://doi.org/10.1016/j.jprot.2019.103500.

Montoro-García, S., Velasco-Soria, Á., Mora, L., Carazo-Díaz, C., Prieto-Merino, D., Avellaneda, A., et al. (2022). Beneficial impact of pork dry-cured ham consumption on blood pressure and cardiometabolic markers in individuals with cardiovascular risk. *Nutrients*, *14*, 298. Available from https://doi.org/10.3390/nu14020298.

Montoro-García, S., Zafrilla-Rentero, M. P., Celdrán-de Haro, F. M., Piñero-de Armas, J. J., Toldrá, F., Tejada-Portero, L., et al. (2017). Effects of dry-cured ham rich in bioactive peptides on cardiovascular health: a randomized controlled trial, *Journal of Functional Foods*, 38(A), 160–167. Available from https://doi.org/10.1016/j.jff.2017.09.012.2.

Mora, L., Bolumar, T., Heres, A., & Toldrá, F. (2017). Effect of cooking and simulated gastrointestinal digestion on the activity of generated bioactive peptides in aged beef meat. *Food and Function*, *8*(12), 4347–4355. Available from https://doi.org/10.1039/c7fo01148b, http://pubs.rsc.org/en/journals/journal/fo.

Mora, L., Sentandreu, M. A., Fraser, P. D., Toldrá, F., & Bramley, P. M. (2009). Oligopeptides arising from the degradation of creatine kinase in spanish dry-cured ham. *Journal of Agricultural and Food Chemistry*, *57*(19), 8982–8988. Available from https://doi.org/10.1021/jf901573t, http://pubs.acs.org/doi/pdfplus/10.1021/jf901573t, United Kingdom.

Ordóñez, J. A., Hierro, E. M., Bruna, J. M., & de la Hoz, L. (1999). Changes in the components of dry-fermented sausages during ripening. *Critical Reviews in Food Science and Nutrition*, *39*, 329–367.

Pepe, G., Sommella, E., Ventre, G., Scala, M. C., Adesso, S., Ostacolo, C., Marzocco, S., Novellino, E., & Campiglia, P. (2016). Antioxidant peptides released from gastrointestinal digestion of "Stracchino" soft cheese: Characterization, in vitro intestinal protection and bioavailability. *Journal of Functional Foods*, *26*, 494–505. Available from https://doi.org/10.1016/j.jff.2016.08.021, http://www.elsevier.com/wps/find/journaldescription.cws_home/717426/description#description.

Saiga, A., Okumura, T., Makihara, T., Katsuta, S., Shimizu, T., Yamada, R., & Nishimura, T. (2003). Angiotensin I-converting enzyme inhibitory peptides in a hydrolyzed chicken breast muscle extract. *Journal of Agricultural and Food Chemistry*, *51*(6), 1741–1745. Available from https://doi.org/10.1021/jf020604h.

Sangsawad, P., Roytrakul, S., & Yongsawatdigul, J. (2017). Angiotensin converting enzyme (ACE) inhibitory peptides derived from the simulated in vitro gastrointestinal digestion of cooked chicken breast. *Journal of Functional Foods*, *29*, 77–83. Available from https://doi.org/10.1016/j.jff.2016.12.005, http://www.elsevier.com/wps/find/journaldescription.cws_home/717426/description#description.

Sayd, T., Chambon, C., & Santé-Lhoutellier, V. (2016). Quantification of peptides released during in vitro digestion of cooked meat. *Food Chemistry*, *197*, 1311–1323.

Sforza, S., Pigazzani, A., Motti, M., Porta, C., Virgili, R., Galaverna, G., Dossena, A., & Marchelli, R. (2001). Oligopeptides and free amino acids in Parma hams of known cathepsin B activity. *Food Chemistry*, *75*(3), 267–273. Available from https://doi.org/10.1016/S0308-8146(01)00224-2.

Simonetti, A., Gambacorta, E., & Perna, A. (2016). Antioxidative and antihypertensive activities of pig meat before and after cooking and in vitro gastrointestinal digestion: Comparison between Italian autochthonous pig Suino Nero Lucano and a modern crossbred pig. *Food Chemistry*, *212*, 590–595. Available from https://doi.org/10.1016/j.foodchem.2016.06.029, http://www.elsevier.com/locate/foodchem.

Sun, X., Acquah, C., Aluko, R. E., & Udenigwe, C. C. (2020). Considering food matrix and gastrointestinal effects in enhancing bioactive peptide absorption and bioavailability. *Journal of Functional Foods*, *64*, 103680. Available from https://doi.org/10.1016/j.jff.2019.103680.

Toldrá, F., Aristoy, M. C., & Flores, M. (2000). Contribution of muscle aminopeptidases to flavor development in dry-cured ham. *Food Research International*, *33*(3-4), 181–185. Available from https://doi.org/10.1016/S0963-9969(00)00032-6.

Toldrá, F., Gallego, M., Reig, M., Aristoy, M. C., & Mora, L. (2020). Bioactive peptides generated in the processing of dry-cured ham. *Food Chemistry* (321), 126689. Available from https://doi.org/10.1016/j.foodchem.2020.126689.

Wang, B., & Li, B. (2017). Effect of molecular weight on the transepithelial transport and peptidase degradation of casein-derived peptides by using Caco-2 cell model. *Food Chemistry*, *218*, 1–8.

Wang, J., Lu, S., Li, R., Wang, Y., & Huang, L. (2020). Identification and characterization of antioxidant peptides from Chinese dry-cured mutton ham. *Journal of the Science of Food and Agriculture*, *100*(3), 1246–1255. Available from https://doi.org/10.1002/jsfa.10136, http://onlinelibrary.wiley.com/journal/10.1002/(ISSN)1097-0010.

Warner, R. D., Wheeler, T. L., Ha, M., Li, X., Bekhit, A. E. D., Morton, J., Vaskoska, R., Dunshea, F. R., Liu, R., Purslow, P., & Zhang, W. (2022). Meat tenderness: Advances in biology, biochemistry, molecular mechanisms and new technologies. *Meat Science*, *185*. Available from https://doi.org/10.1016/j.meatsci.2021.108657, http://www.elsevier.com/locate/meatsci.

Yee, S. (1997). In vitro permeability across Caco-2 cells (colonic) can predict in vivo (small intestinal) absorption in man - Fact or myth. *Pharmaceutical Research*, *14*(6), 763–766. Available from https://doi.org/10.1023/A:1012102522787.

Yin, R., Xu, Y., Wang, X., Yang, L., & Zhao, D. (2022). Role of dipeptidyl peptidase 4 inhibitors in antidiabetic treatment. *Molecules (Basel, Switzerland)*, *27*(10), 3055. Available from https://doi.org/10.3390/molecules27103055.

CHAPTER

9 Bioactive peptides released from pseudo-cereal proteins during gastrointestinal digestion

María Cristina Añón, Susan Fiorella Garcia Filleria, Agustina Estefania Nardo, Alejandra Viviana Quiroga, Mariela Rodriguez, Ana Clara Sabbione, Adriana Scilingo, Santiago Emmanuel Suárez and Valeria Tironi

Laboratory of Food Proteins Research, Development and Innovation (LIDiPA), Center of Research and Development in Food Criotechnology, National Council of Scientific and Techical Research (CONICET), Scientific Research Commission of Buenos Aires Province (CICPBA), Faculty of Exact Sciences, National University of La Plata (UNLP), La Plata, Argentina

9.1 Pseudo-cereals: general aspects, proteins, and their susceptibility to proteolysis by enzymes of the gastrointestinal tract

Pseudo-cereals is the name given to a group of dicotyledonous plants since, although cereals are monocotyledons, they share some characteristics with them (Usman et al., 2022). In this way, these plants produced starch-rich grains with texture, palatability, and cooking methods similar to those of cereals (Morales et al., 2021). However, the different botanical origins are reflected in differences in the chemical, physical, and processing properties. Among these, the smaller seed kernel size, the specific starch structure (small starch granule size and architecture, differing content of amylose), the protein composition (gluten-free, differing protein fractions and amino acid composition, greater nutritional quality protein), the dietary fiber and secondary plant metabolites composition, could be mentioned.

The most studied and widely cultivated pseudo-cereals are quinoa, amaranth, and buckwheat. Amaranth (*Amaranthus* spp.), quinoa (*Chenopodium quinoa* Willd), and also the less known cañihua (*Chenopodium pallidicaule* Aellen) belong to the Amaranthaceae family; they are annual herbaceous flowering plants mainly cultivated in America (Ecuador, Chile, Argentina, Peru, Colombia, Mexico, and Bolivia). In other way, buckwheat species (common buckwheat *Fagopyrum esculentum* Moench, tartary buckwheat *Fagopyrum tataricum*, and others) belong to the Polygonaceae family and are mainly grown in China, India, and some European nations (Bender & Schönlechner, 2021; Thakur et al., 2021;

Protein Digestion-Derived Peptides. DOI: https://doi.org/10.1016/B978-0-443-19141-1.00009-1

Usman et al., 2022). A more detailed description of these pseudo-cereal grains could be found in (Bender & Schönlechner, 2021; Thakur et al., 2021) and in the references cited in these reviews.

Chia (*Salvia hispanica* L.) is also considered as a pseudo-cereal. It is an annual herbaceous plant belonging to the Lamiaceae family from America that is currently grown in Argentina, Mexico, Bolivia, Guatemala, Ecuador, and Australia, among others. Chia has differential characteristics concerning the other pseudo-cereals, such as its high content of oil (rich in omega-3 fatty acids) and the presence of the mucilage, a gelling, thickener, and emulsifier polysaccharide (Cotabarren et al., 2022; Julio et al., 2022).

Amaranth, quinoa, buckwheat, and chia, are all considered ancient grains. These crops could be regarded as superfoods due to their high nutritional and nutraceutical values (Repo-Carrasco-Valencia, 2022). Researchers from different countries have directed their research toward their potential use as sources of bioactive and health-promoting compounds as ingredients in functional foods. Among their bioactive compounds, an important amount of studies have focused on bioactive peptides, as showed several recent reviews (Balakrishnan & Schneider, 2022; Capraro et al., 2021; Martínez-Villaluenga et al., 2020; Morales et al., 2021; Orona-Tamayo & Paredes-López, 2017; Usman et al., 2022). Thus peptides in hydrolysates, gastrointestinal digests, and fermented products with diverse properties including antioxidant, antiinflammatory, antihypertensive, hypocholesterolemic, anticoagulant, and antidiabetic activities, among others, are increasing the relevance of pseudo-cereals proteins as food ingredients for the prevention and/or management of chronic diseases related to oxidative stress, cardiovascular pathologies, and/or diabetes. However, the number of studies in which a simulated gastrointestinal digestion (SGID) process is performed is scarce, as will be described in the following sections.

Pseudo-cereal protein content varies between 11% and 19% w/w (Alonso-Miravalles & O'Mahony, 2018; Chiang et al., 2022; López et al., 2018). Seed and grain proteins are usually classified in the literature according to different criteria. One of them classifies the proteins according to their physiological function in the plant; for example, as storage proteins, structural and metabolic proteins (such enzymes) and protective or defensive proteins (López et al., 2018). These categories are usually used in combination with the classification according to the solubility given by Osborne (albumins, globulins, prolamins, and glutelins). Pseudo-cereals, as other seeds, are rich in storage proteins primarily composed of globulins of the type 11S and 7S (amaranth, chia, quinoa) and 13S (quinoa and buckwheat) (Alonso-Miravalles & O'Mahony, 2018; Orona-Tamayo & Paredes-López, 2017; Orona-Tamayo et al., 2017; Scanlin & Lewis, 2017).

The relative content of different protein fractions (Table 9.1) may differ according to the origin/source and grain variety, but also according to the extraction method, the particle properties and the physical part of the seed used to obtain the pseudo-cereal protein isolate. Amaranth grains contain about 11%–19% w/w of protein, with 33%–48% globulins (mainly 11S and 7S) (Orona-Tamayo & Paredes-López,

Table 9.1 Protein fractions, digestibility, and sequenced storage proteins of pseudo-cereals.

Pseudo-cereal	% protein	In vitro protein digestibility[a]	% albumin	% globulins	% prolamins	% glutenin	Sequenced globulins—storage proteins[b]	References
Amaranth	11%–19%	76.0	6%–10%	33%–48%	7%–9.1%	20%–29%	11S ABM66807.1 CAA57633.1 ABM66807.1	Janssen et al. (2017), Martínez-Villaluenga et al. (2020), Orona-Tamayo and Paredes-López (2017)
Quinoa	11%–16%	81.5	13%	51%–60%	ND	1%–7%	11S AAS67036.1 AAS67037.1 ABI94735.1 ABI94736.1 XP_021768838.1 XP_021770184.1 XP_021774642.1 XP_021716228.1 XP_021720934.1 XP_021755284.1 XP_021733866.1 13S XP_021721911.1 XP_021752233.1 XP_021752668.1 12S XP_021728691.1 XP_021764762.1 7S XP_021757446.1 XP_021753809.1 XP_021757450.1	Dakhili et al. (2019), Janssen et al. (2017), Martínez-Villaluenga et al. (2020)

(Continued)

Table 9.1 Protein fractions, digestibility, and sequenced storage proteins of pseudo-cereals. *Continued*

Pseudo-cereal	% protein	In vitro protein digestibility[a]	% albumin	% globulins	% prolamins	% glutenin	Sequenced globulins—storage proteins[b]	References
Buckwheat	12.5%–14.8%	73.56	44%	20%–30%	1%–7%	12%–20%	13S A9NJG2.1 AAP15457.1 AAQ56206.1 BAO50856.1 BAO50857.1 BAO50858.1 BAO50859.1 BAO50860.1 BAO50861.1 BAO50862.1 BAO50864.1 BAO50865.1 BAO50866.1 BAO50867.1 BAO50868.1 BAO50869.1 BAO50870.1 BAO50871.1 BAO50872.1 BBK26531.1 O23878.1 O23880.1 Q9XFM4.1	Chiang et al. (2022), Janssen et al. (2017), Martínez-Villaluenga et al. (2020), Ninomiya et al. (2018)
Chia	15%–24%	49.4	3%–17%	7%–52%	12%–53%	14%–23%	11S XP_047952105.1 XP_047952381.1 XP_047953006.1 XP_047954700.1 XP_047955639.1 XP_047956179.1 XP_047960258.1 XP_047974399.1 XP_047976489.1	Grancieri et al. (2019), Orona-Tamayo et al. (2017)

[a] *Malik and Singh (2022).*
[b] *National Center for Biotechnology Information (NCBI). Available from: https://www.ncbi.nlm.nih.gov/.*

2017). Quinoa seeds contain about 11%–16% w/w of proteins, the main fraction is globulin (~50% of the total protein), being the 11S the majority. The albumin fraction represents 13% of the total protein content and the main protein in this fraction is the albumin 2S; and it has a low percentage of prolamins (1%–7.0% w/w) (Dakhili et al., 2019). Fewer studies have been carried out on chia proteins, and large differences have been found in the proportion of each fraction depending on the variety and stringency of the species studied (López et al., 2018). Globulins are the major components (52% w/w of total protein content) (Grancieri et al., 2019; Sandoval-Oliveros & Paredes-López, 2013). Buckwheat proteins are composed of albumin and globulin (50%–60% w/w), glutelin (12%–20% w/w) and prolamin (1%–7% w/w) (Chiang et al., 2022; Janssen et al., 2017). A more exhaustive description of pseudo-cereal proteins and their amino acid profile can be found in (López et al., 2018), in (Malik & Singh, 2022), and in the references cited in these reviews.

The digestibility of proteins depends on their structure and the matrix in which they are found. Studying the digestibility of a protein isolate is not the same as studying the digestibility of a protein in a food matrix. The evaluation of pseudo-cereal protein digestibility was extensively reviewed by (Malik & Singh, 2022) and is summarized in Table 9.1. The release of bioactive peptides during gastrointestinal digestion depends on how much the digestive enzymes can access the cleavage sites present in the protein sequences, so the profile of bioactive peptides obtained will also depend on these factors and also the presence of antinutritional factors and other secondary metabolites, such as saponins in quinoa and polyphenols (Hussain et al., 2021).

In contrast to legumes and cereals, the complete sequence of the main compounds of the different protein fractions has not yet been thoroughly determined (Table 9.1). In particular, this adds a difficulty to the sequencing of peptides purified from digests (it is necessary to use mass spectrometry methods that do not require a search in protein sequence databases) and limits in silico studies (or bioinformatics approach). On the other hand, although there are computational tools that allow simulating the cleavage site of the different enzymes that act in gastrointestinal digestion, these methods have the limitation that they do not take into account how accessible these sites are in the three-dimensional structure of the proteins and their use in pseudo-cereal proteins is even more limited by the lack of the determination of the amino acid sequence of the major proteins.

9.2 Bioactivities of peptides from simulated gastrointestinal digestion of purified proteins and protein isolates/concentrates

This section presents the published information regarding the bioactivities of peptides derived from pseudo-cereals obtained through studies carried out on simple samples, that is, protein isolates and/or purified protein fractions. The conditions used in the various studies are summarized in Table 9.2.

Table 9.2 Bioactivities of peptides obtained by SGID of total proteins or protein fractions from amaranth, quinoa, buckwheat, and chia.

Matrix	SGID protocol	Bioactivity	Research finding	References
Amaranth				
PC, alcalase-H	Pepsin 1/100 w/v, 37°C, 2 h	Antihypertensive	Inhibition of ACE	Tiengo et al. (2009)
A. cruentus Alegria	Pancreatin 1/50 w/v, 37°C, 4 h		$IC_{50} = 0.439$ and 0.123 mg/mL	
7S and 11S globulins	Pepsin 1:10 w/w, 37°C, 1 h		Inhibition of ACE	Quiroga et al. (2012)
A. hypochondriacus	Pancreatin 1:300, 37°C, 1 h		$IC_{50} = 0.17$ and 0.17 mg/mL	
PC *A. caudatus*	INFOGEST		Inhibition of ACE $IC_{50} = 0.088$ mg/mL	Vilcacundo et al. (2019)
			Identification of active sequences: FLISCLL, SVFDEELS, DFIILE	
PI *A. cruentus*	Mixture trypsin/ α-chymotrypsin/	Hypocholesterolemic	HMG-CoA reductase inhibitory capacity of GGV, IVG, VGVL	Soares et al. (2015)
	Peptidase (10 min, pH 8, 37°C)			
PI *A. manteggazianus*	Salivary fluid, pH 6.8, 37°C, 5 min		Displacement of cholesterol from micelles	Sisti et al. (2019)
	Pepsin, pH 2, 37°C, 1 h		$IC_{50} = 2.1$ mg/mL	
	Pancreatin, pH 7, 37°C, 1 h			

PI *A. hypochondriacus* (cv Mercado)	Pepsin 1/10, pH 2, 37°C, 1 h Pancreatin 1/10, pH 6, 37°C, 1 h	Anticoagulant	Inhibition of coagulation common pathway (fibrin clot formation) $IC_{50} = 0.23$ mg/mL	Sabbione et al. (2016)
			Separation of active fraction $IC_{50} = 0.07$ mg/mL	
			In silico identification of potentially active fractions: IQAEAGLTEV, TEVWDSNEQE, IDTANHANQLDK, LAGKPQQEHSGEHQ, DVYTPE, NDQGQSVFDEELS, AFEDGFE, DNLNDPK, LVDGNDPR, AESSQIDTGSK	
			Active fraction: Inhibition of tissue factor expression, $IC_{50} = 0.39$ mg/mL	Sabbione et al. (2018)
			Decreased TF membrane translocation (49%) in monocytes	
			Inhibition of the NF-κB pathway	
PI *A. manteggazianus*	Pepsin 1/10, pH 2, 37°C, 1 h	Antioxidant	Scavenging ROO· (ORAC), $IC_{50} = 0.024$ mg/mL	Orsini Delgado et al. (2011)
	Pancreatin 1/10, pH 6, 37°C, 1 h		Inhibition of OH· (HORAC), $IC_{50} = 0.9$ mg/mL	Delgado et al. (2015)
			Most active fractions: 6.1 and 0.2 kDa	
			Identification of active sequences (RP-HPLC, ESI-MS/MS, ORAC): AWEEREQGSR ($IC_{50} = 6.7$ μg/mL), YLAGKPQQEH ($IC_{50} = 16$ μg/mL), IYIEQGNGITGM ($IC_{50} = 17$ μg/mL), TEVWDSNE ($IC_{50} = 20$ μg/mL)	Orsini Delgado et al. (2016)
			LDL lipid oxidation: increment in the lag time and in the propagation	García Fillería and Tironi (2017)

(Continued)

Table 9.2 Bioactivities of peptides obtained by SGID of total proteins or protein fractions from amaranth, quinoa, buckwheat, and chia. *Continued*

Matrix	SGID protocol	Bioactivity	Research finding	References
			Rate of conjugated dienes, prevention of protein oxidation.	
			Digest, most active fractions: 0.59 and 1.4 kDa, most active sequences: HVIKPPSRA, GDRFQDQHQ, and KFNRPETT	
			Intracellular ROS inhibition (Caco-2-TC7): digest ($IC_{50} = 2.2$ mg/mL)	García Fillería and Tironi (2021)
			Most active fractions: 7.4 and 0.52 kDa, most active sequences: TEVWDSNEQ, IYIEQGNGITGM, GDRFQDQHQ, LAGKPQQEHSGEHQ, KFNRPETT	García Fillería and Tironi (2021)
			Diverse effects on SOD, GPx, and GSH	
			Modification and partial absorption by intestinal cells	
PI *A. manteggazianus*	INFOGEST		Scavenging ROO· (ORAC), $IC_{50} = 0.024$ mg/mL	Rodríguez et al. (2020)
			Inhibition of OH· (HORAC), $IC_{50} = 1.61$ mg/mL	
Alcalase/neutrase-H	Pepsin 1/100 w/v, 37°C, pH 2.5, 2 h		Scavenging of $ABTS^{+}$, 1.71 μmol TE/mg	Chirinos et al. (2020)
A. caudatus	Pancreatin 1/50 w/v, 37°C, pH 7, 4 h			
Albumins, globulins	Pepsin 1/100 w/w, pH 2, 37°C, 10 min	Antiinflammatory	Protection of Caco-2 cells against IL-1β-induced inflammation	Capraro et al. (2021)
A. retroflexus	Pancreatin 1/100 w/w, pH 8, 37°C, 10 min			

PI *A. manteggazianus*	Pepsin 1/10, pH 2, 37°C, 1 h	Antiproliferative	Antiproliferative effect on HT-29 cells (IC_{50} = 0.30 mg/mL)	Sabbione et al. (2019)
	Pancreatin 1/10, pH 6, 37°C, 1 h		Induction of apoptosis and necrosis (increasing caspase-3 activity)	
Extracted proteins	Pepsin 1:12.5, pH 2, 37°C, 15 min		Antiproliferative effect on human breast cancer cells MDA-MB-231	Taniya et al. (2020)
Heat denaturation	Pancreatin 1:62.5, pH 7, 37°C, 5 min		(IC_{50} = 0.048 mg/mL)	
			Induction of DNA fragmentation, membrane integrity loss, phosphatidylserine translocation, and increasing caspase-3 activity	
			Inhibition of cell migration	
			Scavenging of DPPH and reduction activity	
PI *A. caudatus*	INFOGEST	Antidiabetic	Inhibition of the DPP-IV (IC_{50} = 0.28 mg/mL) and a-amylase	Vilcacundo et al. (2019)
			Potentially active sequences: VEEGNM (DPP-IV), FLISCLL, SVFDEELS,	
			DFIILE, HVIKPPS (a-amylase), NRPET (both enzymes)	
Quinoa (*Chenopodium quinoa* Willd.)				
Neutrase-H of PC	Pepsin 1/100 w/v, 37°C, 2 h;	Antihypertensive	Inhibition of ACE	Chirinos et al. (2020)
	Pancreatin 1/50 w/v, 37°C, 4 h		IC_{50} = 0.07 mg/mL	
	Pepsin 5% w/w, pH 2, 37°C, 1 h		Inhibition of ACE	Guo et al. (2020)
	Pancreatin 5% w/w, pH 7.8, 37°C, 1 h		IC_{50} = 0.36 mg/mL	
			Identification of active sequences: NWFPLPR, FHPFPR, and NIFRPF	
			Inhibition of ACE	

(*Continued*)

Table 9.2 Bioactivities of peptides obtained by SGID of total proteins or protein fractions from amaranth, quinoa, buckwheat, and chia. *Continued*

Matrix	SGID protocol	Bioactivity	Research finding	References
RGQVIYVL (papain-H of bran albumin)	Pepsin 1/50, pH 2, 37°C, 1.5 h			(Zheng et al., 2019)
	Pancreatin 1/30, pH 7.4, 37°C, 3 h		$IC_{50} = 38\ \mu M$	
SAPPP (alcalase/trypsin-H of bran albumin)	Pepsin 1/50, pH 2, 37°C, 1.5 h		Inhibition of ACE	Wei et al. (2022)
	Pancreatin 1/30, pH 7.4, 37°C, 3 h		$IC_{50} = 915\ \mu M$	
PC	INFOGEST	Antioxidant	Scavenging ROO· (ORAC), 2.39 μmol TE/mg protein	Vilcacundo et al. (2018)
			Most active fraction <5 kDa	
Neutrase-H of PC	Pepsin 1/100 w/v, 37°C, pH 2.5, 2 h		Scavenging of $ABTS^{+}$, 1.54 μmol TE/mg	Chirinos et al. (2020)
	Pancreatin 1/50 w/v, 37°C, pH 7, 4 h			
Albumins and globulins	Pepsin 1/100 w/w, pH 2, 37°C, 10 min	Antiinflammatory	Protection of Caco-2 cells against inflammation induced by IL-1β	Capraro et al. (2021)
	Pancreatin 1/100 w/w, pH 8, 37°C, 10 min			
PC	INFOGEST	Antiproliferative	Colon cancer cell viability: Caco-2 ($IC_{50} = 0.191$ mg protein/mL), HT-29 ($IC_{50} = 0.212$ mg/mL), HCT-116 ($IC_{50} = 0.208$ mg/mL)	Vilcacundo et al. (2018)
			Identification of active sequences from 11S globulin: LWREGM, DKDYPK, DVYSPEAG, IFQEYI, RELGEWGI	
PC	Pepsin 5% w/w, pH 2, 1 h, 37°C		Antiproliferative activity on Caco-2 cells	Fan et al. (2022)
	Pancreatin 5% w/w, pH 7, 1 h, 37°C		Most active sequences: FHPFPR, NWFPLPR and HYNPYFPGGA	
			Decreasing HDAC1, NF-κB, IL-6, IL-8, Bcl-2, increasing caspase-3	

Buckwheat				
FI	Pepsin 1:20 w/w, pH 2, 3 h, 37°C	Anticoagulant	Inhibition of platelet aggregation	Yu et al. (2016)
	Trypsin/pancreatin 1:20 w/w, pH 7.5, 3 h, 37°C		IC_{50} = 0.328 mg protein/mL	
PI	Pepsin 4% w/w, pH 2, 37°C, 1 h	Antioxidant	Scavenging $ABTS^{+}\cdot$, Fe^{3+} reducing power, chelation Fe^{2+}, prevention lipid oxidation	Ma and Xiong (2009)
	Pancreatin 4% w/w, pH 7.5, 37°C, 2 h		Most active fraction 300–500 Da ($ABTS^{+}\cdot$, HOSC)	Ma et al. (2010)
			Identified sequences: WPL, VPW, VFPW, and PW	
Albumins and globulins *Fagopyrum esculentum*	Pepsin 1/100 w/w, pH 2, 37°C, 10 min	Antiinflammatory	Protection of Caco-2 cells against inflammation induced by IL-1β	Capraro et al. (2021)
	Pancreatin 1/100 w/w, pH 8, 37°C, 10 min			
Albumins	Pepsin 1/100 w/w, pH 2, 37°C, 2 h	Antidiabetic	Retention of α-amylase inhibition given by glycopeptides	Ninomiya et al. (2018)
	Trypsin 1/100 w/w, pH 8, 37°C, 2–6 h			

(Continued)

Table 9.2 Bioactivities of peptides obtained by SGID of total proteins or protein fractions from amaranth, quinoa, buckwheat, and chia. *Continued*

Matrix	SGID protocol	Bioactivity	Research finding	References
Chia (*Salvia hispanica*)				
Protein fractions	Pepsin 1/40, pH 2, 37°C, 3 h	Antihypertensive	Inhibition of ACE	Orona-Tamayo et al. (2015)
	Trypsin/pancreatin (1:1) 1: 40, pH 7.5, 37°C, 3 h		Most active fractions: globulins and albumins	
Expeller proteins	Pepsin 1/10, pH 2, 1 h, 37°C	Anticoagulant	Inhibition of intrinsic and extrinsic pathway of coagulation	Valicenti (2022)
	Pancreatin 1/10, pH 7.5, 1 h, 37°C		(increasing aPTT and PT times)	
			Most active fraction <3 kDa	
PI and protein fractions	Pepsin 1/20 w/w, pH 2, 2 h, 37°C	Antioxidant	Scavenging of $O_2^{-\cdot}$, NO, DPPH, and H_2O_2	Grancieri et al. (2019)
	Pancreatin 1/20 w/w, pH = 7.5, 2 h, 37°C		Glutelins most active against $O_2^{-}\cdot$	
			Albumin and total protein most actives against DPPH	

PC: protein concentrate; PI: protein isolate; H: hydrolysate. ACE: angiotensin-converting enzyme; HMG-CoA: 3-hidroxi-3-metilglutaril-coenzima A; TF: tissue factor; NF-κB: nuclear factor regulator of κB light chain expression; ORAC: oxygen radical absorption capacity; HORAC: hydroxyl radical averting capacity; TE: Trolox equivalent; LDL: low-density lipoproteins; SOD: superoxide dismutase; GPx: glutathione peroxidase; GSH: reduced glutathione; DPP-IV: dipeptidyl peptidase IV; IL: interleukin; HDCA1: histone deacetylase 1; HOSC: hydroxyl radical scavenging capacity; aPTT: activated partial thromboplastin time; PT: prothrombin time.

9.2.1 Antihypertensive activity

Cardiovascular diseases (CVD) are the leading cause of mortality and morbidity worldwide, accounting for nearly 30% of global deaths (Yu et al., 2016). Elevated blood pressure is the main metabolic risk factor that contributes to the development of CVD. Blood pressure (BP) is considered high, giving rise to arterial hypertension when its value reaches 140/90 mm Hg or higher. The regulation of BP is complex since it involves several intertwined metabolic pathways. The renin-angiotensin system plays a fundamental role in the regulation of BP. This system is mainly regulated by two proteases: renin and angiotensin I-converting enzyme (ACE). Both enzymes act in a chain to obtain the peptide angiotensin II which acts as a powerful vasoconstrictor, by binding to the angiotensin II receptor type 1 (AT1) of the vascular wall located in peripheral tissues around the body and brain, to cause contraction of blood vessels (Aluko, 2015). The most common type of hypertension manifests as an increase in BP due to an unknown cause and can be improved and controlled by modifying behavior factors, being defined as a controllable risk factor for CVD (Norris & J, 2013).

Pseudo-cereals are an important source of antihypertensive peptides (Morales et al., 2021; Nardo et al., 2020). In recent decades, various research groups have analyzed their bioaccessibility using SGID process. In many studies, gastrointestinal fluids have been simulated by the sequential action of pepsin at acidic pH (gastric phase), followed by the action of pancreatin at neutral pH (intestinal phase) (Chirinos et al., 2020; Guo et al., 2020; Quiroga et al., 2012; Tiengo et al., 2009; Wei & Lee, 2002; Zheng et al., 2019). However, each research group used different pHs, enzyme types and activities, enzyme/substrate ratios, and reaction times. Only in the study carried out by (Vilcacundo et al., 2019), the consensus protocol of COST INFOGEST (Minekus et al., 2014) was used. In all these cases, it was possible to release peptides from protein fractions, protein isolates or concentrates, and hydrolysates with the capacity to inhibit ACE.

In amaranth (Table 9.2), Tiengo et al. (2009) demonstrated that the digests of *A. cruentus* variety BR Alegria protein concentrates had a lower capacity to inhibit ACE than the digests of the concentrates previously hydrolyzed with alcalase (0.44 ± 0.02 mg protein/mL and 0.12 ± 0.01 mg protein/mL, respectively). The SGID process did not modify the ACE inhibitory capacity of the alcalase hydrolyzates (0.17 ± 0.01 mg protein/mL), which could imply a resistance of the peptides present in the hydrolyzates to the enzymes pepsin and pancreatin and/or the appearance during the SGID process of new peptides with similar antihypertensive activity. (Quiroga et al., 2012) using a different amaranth specie (*A. hypochondriacus*) and different SDIG conditions demonstrated that both globulins contained encrypted peptide sequences with a high capacity to inhibit ACE, similar for both globulins (IC_{50} 0.17 ± 0.03 and 0.17 ± 0.01 mg protein/mL). More recently, (Vilcacundo et al., 2019), using the consensus protocol, identified 13 peptides with different biological activities in digests of amaranth (*A. caudatus*) concentrates. The IC_{50} value obtained for ACE inhibition of the digest after the gastric stage was 39.0 ± 3.0 μg/mL, while at the end of the intestinal stage, it was 88.0 ± 14.0 μg/mL indicating an increment in this activity after intestinal digestion. It is noteworthy in this study the high inhibitory activity of ACE that presented the undigested concentrates (IC_{50} 79.0 ± 1.0 μg/mL). Among the peptides released during the SGID process and using information contained in databases, the authors identified 3 peptide sequences, FLISCLL, SVFDEELS, and DFIILE, capable of inhibiting ACE.

In the case of quinoa (Table 9.2), Chirinos et al. (2020) used the protocol described by (Tiengo et al., 2009) and as substrate protein concentrates previously hydrolyzed with neutrase. Like Tiengo et al., they observed that the ability to inhibit ACE of the hydrolysates was not modified after the SGID process (IC_{50} = 0.08 ± 0.01 and 0.07 ± 0.01 mg/mL, respectively) which does not mean that the antihypertensive peptides present are the same in both samples. On the other hand, (Guo et al., 2020) used the sequential action of pepsin-pancreatin in the SGID process and, compared with the control sample, the quinoa protein digests exhibited a significantly higher ACE inhibitory effect (undigested protein isolate, IC_{50} = 12.70 mg protein/mL, digested protein isolate IC_{50} = 0.36 mg protein/mL). The highest inhibitory activity was attributed to peptides with molecular weight (MM) <5 kDa. Through the use of HPLC–MS/MS and in silico analysis, 3 promising bioactive peptides could be identified, NWFPLPR, FHPFPR, and NIFRPF. They were synthesized, validating their ACE inhibitory capacity (IC_{50} = 16.77 μM, 34.92, and 32.40 μM, respectively).

Zheng et al. (2019) and Wei et al. (2022) identified two new antihypertensive peptides, RGQVIYVL and SAPPP, derived from quinoa bran albumin hydrolyzates obtained by the action of papain and alcalase-trypsin, respectively. The synthesized peptides were digested by pepsin and pancreatin (Table 9.2). Both peptides, RGQVIYVL and SAPPP, showed similar levels of ACE inhibitory activity before and after being subject to SGID process (IC_{50} = 38.16 and 915 μM, respectively). This suggests that they could effectively retain the activity after the gastrointestinal digestion.

Besides, Orona-Tamayo et al. (2015) also demonstrated that digests from the different protein fractions present in chia seeds were capable of inhibiting ACE, particularly those from albumins and globulins. Digests were obtained by extensive hydrolysis of proteins fractions, previously treated for 5 min at 80°C and pH 2, using the sequential action of pepsin and a mixture of trypsin and pancreatin (Table 9.2), reaction that released the greatest amount of peptides from the globulin fraction, followed by the albumin fraction and finally glutelins and prolamins.

9.2.2 Hypocholesterolemic activity

The combination of elevated plasma cholesterol levels with other risk factors, such as arterial hypertension, diabetes, smoking, consumption of unhealthy diets, and physical inactivity, significantly increase overall cardiovascular risk. The use of food components to treat hypercholesterolemia appears as an interesting alternative to the use of conventional drugs, such as statins, which present unwanted side effects. In this sense, in vitro and in vivo studies have shown that different components of pseudo-cereals, particularly phytosterols, and fiber, and to a lesser extent protein/peptides, have a significant ability to reduce cholesterol levels (Morales et al., 2021). Despite this, there are practically no studies that analyze the hypocholesterolemic effect of hydrolysates obtained by SGID process.

Soares et al. (2015) subjected a protein isolate from *A. cruentus* to SGID in order to analyze the digest obtained and its potential to inhibit HMG-CoA reductase (a key enzyme in the synthesis of cholesterol). Two different SGID protocols were used (Table 9.2): (1) pepsin and then trypsin-enriched pancreatin (Qiao et al., 2002) and (2) multienzyme mixture containing trypsin (5.16 mg), α-chymotrypsin (13.76 mg) and peptidase (1.28 mg), to which 250 mg of substrate were added (Hsu et al., 1977). The digestion process was stopped by the addition of 10% w/v trichloroacetic acid (TCA) or by cooling. The use of the multienzyme mixture followed by the precipitation of the digest with TCA was more effective in terms of recovery of peptides and a better separation of the remaining undigested protein. The peptide profile of the digests and their identification were determined by RP-HPLC and LC–ESI-MS/MS, respectively. It was possible to identify 3 tripeptides and 2 tetrapeptides, GGV, IVG, LVG, VGVI, and VGVL, which were synthesized and used to analyze their HMG-CoA reductase inhibitory capacity. Only three of the identified sequences, GGV, IVG, and VGVL, presented inhibitory activity. It should be noted that except for LVG, a sequence present in amaranth albumin, none of the others were encrypted in the amaranth protein sequences recorded at the time of carrying out this study. These sequences had not been previously reported as potential hypocholesterolemic peptides either.

Recently Sisti et al. (2019) demonstrated that digests of protein isolates from *A. manteggazianus* can displace cholesterol from model micelles prepared by sonication of mixtures of cholesterol, phosphatidylcholine, sodium taurocholate, and oleic acid, in the absence or presence of protein. For the SDGI, these authors followed the protocols of Gawlik-Dziki et al. (2009), Minekus et al. (2014), and Versantvoort et al. (2005) with some modifications (Table 9.2). Briefly, 300 mg of lyophilized protein isolate were suspended in 6 mL of salivary-like fluid (2.38 g/L Na_2HPO_4, 0.19 g/L KH_2PO_4, 0.3 g/L NaCl, pH 6.8, 60 mg of α-amylase), then 12 mL of gastric-like fluid (2.5 g/L pepsin in 0.03 M NaCl, pH 2) were added and later, 12 mL of duodenal-like fluid (3 g/L pancreatin in 0.1 M $NaHCO_3$ pH 7). The presence of undigested and digested proteins increased the size of the micelle formed in the absence of protein, which implies the interaction of proteins and peptides with some of the micellar components. The proteins/peptides could act as destabilizing agents of the micelle or displace any of the components that form it. These authors also observed that the proteins/peptides were capable of displacing cholesterol from the micelle, a displacement that increased as the amount of protein increased in a nonlinear manner. The IC_{50} for the micelles prepared with amaranth protein isolate was 0.2 ± 0.1 mg/mL while that corresponding to its digest, was 2.1 ± 0.03 mg/mL, an order of magnitude larger. This result would indicate that when protein size is reduced due to proteolysis that occurs during SGID, more protein is required to avoid cholesterol inclusion in the micelle.

9.2.3 Antithrombotic activity

Thrombosis, one of the triggering causes of CVD, produces clotting inside blood vessels without the occurrence of any vascular injury. This process could comprise an exacerbation in platelet activation, adhesion, and aggregation, or in the coagulation cascade process, where several serine proteases, such as thrombin enzyme, are involved (Loganathan et al., 2013).

Nowadays, the importance of using anticoagulant and antiplatelet therapies for thrombosis prevention has become increasingly recognized, leading to remarkable progress in the study of antithrombotic compounds from edible natural sources. These studies have been focused on traditional sources, such as milk (Masood & Khosravi-Darani, 2015), eggs (Cho et al., 2003, 2009), fish and mussels (Chen et al., 2019; Nasri et al., 2012; Qiao et al., 2018), pork meat (Shimizu et al., 2009), and cereals (Cavazos & Gonzalez de Mejia, 2013; Mason et al., 2020). Nevertheless, with the exception of amaranth, pseudo-cereals have not received much attention in the study of antithrombotic compounds since it is one of the less studied biological activities.

The antithrombotic activity of amaranth proteins and peptides has been described in numerous research papers (Ayala-Niño, Rodríguez-Serrano, González-Olivares et al., 2019; Ayala-Ninõ, Rodríguez-Serrano, Jiménez-Alvarado et al., 2019; Sabbione, Ibañez, et al., 2016; Sabbione, Nardo, et al., 2016; Sabbione et al., 2015, 2018). Among in vitro assays performed, most of them study the ability to inhibit thrombin, an enzyme that exerts a main role in the coagulation cascade. (Sabbione et al., 2015) found no inhibition of coagulation of amaranth (*A. hypochondriacus*) protein isolates, whereas Ayala-Niño, Rodríguez-Serrano, González-Olivares et al. (2019) and Ayala-Ninõ, Rodríguez-Serrano, Jiménez-Alvarado et al. (2019) evidenced a small percentage of inhibition when testing amaranth proteins. However, hydrolysis of these proteins via bacterial fermentation (Ayala-Ninõ, Rodríguez-Serrano, Jiménez-Alvarado et al., 2019) or by using commercial enzymes (Ayala-Niño, Rodríguez-Serrano, González-Olivares et al., 2019; Sabbione et al., 2015, 2016) showed significant inhibition of coagulation. Moreover, Sabbione et al. (2015) performed hemostasis clinical global tests, prothrombin time (PT, extrinsic pathway of coagulation), activated partial thromboplastin time (aPTT, intrinsic pathway), and thrombin time (TT, common pathway), to assess the potential antithrombotic activity of amaranth proteins and protein hydrolyzates (alcalase 20 min, pH 10, 37°C and trypsin 20 min, 37°C) in different stages of the coagulation cascade. The authors described no changes in PT but found increments in TT for the hydrolyzate and in the aPTT for the isolate and the hydrolyzate. Coagulation studies reported so far evidence that hydrolysis is necessary to increase antithrombotic activity of amaranth proteins. In this sense, Sabbione et al. (2016) subjected amaranth proteins to a simplified SGID with sequential steps of pepsin and pancreatin (Table 9.2). Hydrolyzates with a degree of hydrolysis of 51% exerted an in vitro inhibition of coagulation ($IC_{50} = 0.23 \pm 0.02$ mg protein/mL), using in vitro turbidimetric microplate's method. This assay studies the coagulation cascade common pathway in a simplified manner, using thrombin and fibrinogen reagents to evaluate fibrin clot formation. Moreover, the authors identified an active fraction after FPLC molecular exclusion separation with an IC_{50} value of 0.07 ± 0.01 mg protein/mL. After sequencing this fraction, 32 peptides from 11S globulin and 11 peptides from lectin were found. Considering the affinity and recognition mechanism of thrombin physiological substrates, bioinformatic tools were used to select those peptides that could be thrombin inhibitors: IQAEAGLTEV, TEVWDSNEQE, IDTANHANQLDK, LAGKPQQEHSGEHQ, DVYTPE, NDQGQSVFDEELS, AFEDGFE, DNLNDPK, LVDGNDPR, and AESSQIDTGSK (Sabbione et al., 2016). To provide a closer approach to the evaluation of this antithrombotic activity, Sabbione et al. (2018) analyzed the effect of the active fraction on tissue factor (TF) expression and activity. TF is a major molecule involved in the thrombosis process. In this study, the fraction caused inhibition of TF expression ($IC_{50} = 0.39$ mg protein/mL). An immunocytochemical fluorescence confocal microscopy analysis showed that treated monocyte cells decreased TF membrane translocation by 49% indicating a decrease of active TF to initiate the coagulation cascade. Furthermore, a cytokine array study suggested that the mechanism of action implied the inhibition of the NF-κB pathway.

Regarding quinoa proteins, only in silico approaches have been performed to study antithrombotic activity. (Guo et al., 2020) used a bioactive peptide (BIOPEP) database and computer-assisted simulation to provide a method for the theoretical estimation of potential bioactivities of quinoa protein sequences. The authors provided the frequency of occurrence of antithrombotic peptides in 11S and 13S globulins with given activity in quinoa globulins. In this sense, Shi et al. (2019) described the bioactive properties of quinoa protein hydrolysates, obtained using papain, pepsin, and pancreatin for 0, 30, 60, 90, and 120 min. After LC–MS/MS analysis, only one peptide released after 120-min pepsin hydrolysis (YIEQGNGISGLMIPG) was associated with antithrombotic activities with in silico tools. However, this in silico approach needs further in vitro and in vivo investigation.

To the present, there is only one report of antithrombotic activity in buckwheat crops. (Yu et al., 2016) studied in vitro inhibition of platelet aggregation of peptides released from oat (*Avena sativa* L.), highland barley (*Hordeum vulgare* Linn.), and buckwheat (*Fagopyrum esculentum* Moench) protein isolates after SGID (pepsin, trypsin and pancreatin; Table 9.2). Inhibition of platelet aggregation was evaluated according to Seo et al. (2012). Buckwheat hydrolyzates presented an IC_{50} value of 0.328 mg protein/mL, confirming the presence of antiplatelet peptides encrypted in buckwheat proteins.

Regarding chia antithrombotic activity, Grancieri et al. (2019) used the UniProt database to analyze the bioactive potential of chia proteins and their amino acid sequences with BIOPEP database. The analyses of amino acid sequences from 20 different chia seed proteins showed peptides with antithrombotic capacity. Moreover, Ozón et al. (2022) described in vitro antithrombotic activity of chia expeller hydrolysates obtained after alcalase and flavourzyme sequential digestion. The authors performed PT and aPTT hemostasis clinical global tests to assess the risk of bleeding and thrombosis. Chia hydrolyzates showed an increase in the aPTT and PT times in a dose-dependent manner, exhibiting a higher effect over the intrinsic pathway. This behavior is comparable to that observed in amaranth protein (Sabbione et al., 2015). Valicenti (2022) studied the effect of chia expeller proteins over coagulation after SGID (Table 9.2). Chia hydrolysates were fractionated by ultrafiltration, F >10 kDa, F 3–10 kDa, and F <3 kDa, and hemostasis clinical global tests were performed. All the samples showed a delay in aPTT and PT times, indicating an inhibition in coagulation, affecting intrinsic and extrinsic pathways. The fraction <3 kDa had the greatest effect in lengthening the time required for clot formation in aPTT and PT, indicating that the low-molecular-weight peptides presented greater antithrombotic capacity.

The results described are promising and indicate that ancient grain proteins could be used as a potential functional ingredient in diets with antithrombotic peptides released after ingestion. Whereas in amaranth seed proteins in vitro, in vivo, and in silico research studies have been performed, in quinoa, buckwheat, and chia proteins, the results described represent the first reports of potentially antithrombotic peptides, and more in vitro and in vivo studies should be done to confirm this hypothesis.

9.2.4 Antioxidant activity

Cells are continually exposed to exogenous and endogenous factors that lead to the generation of reactive oxygen species (ROS) and reactive nitrogen species (RNS), highly reactive molecules are responsible for oxidative damage to biological macromolecules. ROS have physiological functions related with cell signaling and are responsible for cell aging. Under physiological conditions, there is a balance between the generation and removal of ROS but when this balance is broken, caused by a deficiency of antioxidants or by an increase in the production of ROS, oxidative stress occurs. The incorporation of dietary antioxidants has emerged as a tool to prevent oxidative stress and its consequences on health (Lobo et al., 2010). Antioxidants can exert their cellular activity directly and/or indirectly. Direct antioxidant capacity is the ability to eliminate ROS and RNS through hydrogen or electron donation and to inhibit their formation by means of pro-oxidant metal chelation, while indirect antioxidant capacity is the ability to prevent oxidative stress by inducing the expression of phase II detoxifying and antioxidant enzymes (superoxide dismutase [SOD], and glutathione peroxidase [GPx], among others)(Wu et al., 2017). Among the chronic pathologies associated with oxidative stress, it can be mentioned cancer, CVD related to the oxidation of

LDL, cataracts, and neuronal degenerative diseases, such as Parkinson's and Alzheimer's (Wei et al., 2022).

Peptides released by SGID of proteins from pseudo-cereals are a good alternative as dietary antioxidants. Regarding protein isolates and purified protein fractions, simplified SGID protocols have been commonly used. In this sense, several studies have analyzed the release of antioxidant peptides from protein isolates, concentrates, or protein fractions isolated from amaranth, quinoa, buckwheat and chia (Capraro et al., 2021; Chirinos et al., 2020; Grancieri et al., 2019; Usman et al., 2022; Vilcacundo et al., 2018, 2019).

Orsini Delgado et al. (2011) submitted amaranth (*A. mantegazzianus*) protein isolate to SGID using pepsin and pancreatin (Table 9.2). Digestion produced an increase in antioxidant activity measured by in vitro methods, such as the oxygen radical absorption capacity (ORAC) and the hydroxyl radical averting capacity (HORAC); the IC_{50} values for the digest were 0.024 mg/mL and 0.90 mg/mL, respectively (Delgado et al., 2015). Fractions from the digested protein isolate separated according to their molecular mass (MM) by gel filtration FPLC (Superdex 30, optimum separation range <10 kDa) were also evaluated by ORAC and HORAC being the most active those corresponding to MM between 6.1 and 0.2 kDa. On the other hand, several sequences were identified after RP-HPLC and ESI-MS/MS, and synthetic peptides were evaluated. Four peptides presented interesting ORAC activity, such as AWEEREQGSR, YLAGKPQQEH, IYIEQGNGITGM, and TEVWDSNE, with IC_{50} values of 6.7, 16, 17, and 20 μg/mL, respectively (Orsini Delgado et al., 2016). Since there are several possible mechanisms and targets of action, other types of antioxidant activity assays are required to characterize these compounds. Ex vivo assays were performed to assess the effect of digested amaranth protein isolate during Cu^{+2}/H_2O_2 induced LDL oxidation. When the evolution of conjugated dienes (CD) was evaluated, the digest (0.06 mg/mL) produced an increment in the lag time (70%) and in the propagation rate (40%). For FPLC fractions between 0.59 and 1.4 kDa, no formation of CD was registered indicating that LDL lipid oxidation was completely inhibited. The most active synthetic peptides against LDL oxidation (concentration between 0.04 and 1 mg/mL) were HVIKPPSRA, GDRFQDQHQ, and KFNRPETT (García Fillería & Tironi, 2017). Also, in cellular assays, the effect of a pretreatment of Caco-2 TC7 cells with the digested protein isolate followed by a treatment with 500 μM H_2O_2 (1 h) was evaluated. When intracellular ROS inhibition was determined using 2′,7′-dichlorofluorescin diacetate (DCFH-DA), an IC_{50} value of 2.2 mg/mL was obtained for the complete digest while all FPLC fractions led to a decrease in the intracellular ROS being the most active those corresponding to MM between 0.52 and 7.4 kDa. Regarding the synthetic peptides, TEVWDSNEQ, IYIEQGNGITGM, GDRFQDQHQ, LAGKPQQEHSGEHQ, and KFNRPETT were the most effective. These ROS inhibition effects were accompanied by different behaviors when SOD and GPx activities and reduced glutathione (GSH) content were evaluated, suggesting that different peptides showed different actions. On the other hand, when simulated intestinal absorption across Caco-2 TC7 monolayer was evaluated for the digest (1.6 mg/mL), there was a passage of molecules with diverse hydrophilicity/hydrophobicity characteristics.

Regarding the FPLC fractions and the synthetic peptides, different behaviors were found: passage of intact peptides as well as modification by peptidases with or without passage across the cell monolayer. These results suggest potential activity at local intestinal and/or systemic sites (Fillería et al., 2021; García Fillería & Tironi, 2021).

Rodríguez et al. (2020) compared the release of peptides and the antioxidant activity of amaranth protein isolate digested using the previously described SGID protocol (Orsini Delgado et al., 2011) versus the COST INFOGEST consensus protocol (Minekus et al., 2014). Both digests showed the same behavior when ORAC assay was evaluated ($IC_{50} = 0.024$ mg/mL) while for HORAC a lower potency was determined after the consensus protocol ($IC_{50} = 1.61$ mg/mL). Only minor differences in the polypeptide/peptide composition (as determined by SDS-PAGE, tricine–SDS-PAGE, and gel filtration FPLC) and in the antioxidant activity of FPLC fractions (as determined by ORAC and HORAC) between the digests obtained applying the different SGID protocols were found. The authors concluded that for matrices whose main components are proteins (such as protein isolates) a simplified method of SGID works well and it could be chosen for its simplicity.

The increase in the antioxidant activity of other pseudo-cereals proteins submitted to SGID has also been reported. (Vilcacundo et al., 2018) evaluated the in vitro chemopreventive properties of peptides released from the quinoa (*Chenopodium quinoa* Willd) protein concentrate by the digestion protocol reported by (Minekus et al., 2014). Samples from different times of digestion were subjected to ultrafiltration and fractions <5 kDa and >5 kDa were obtained. After digestion, an increase in the activity measured by ORAC was found. The peptides released during the gastric phase presented an ORAC value of 1.03 μmol Trolox equivalents (TE)/mg protein. After the complete intestinal phase, 2.39 μmol TE/mg protein was found, and it was demonstrated that during de first 60 min of this phase the peptides released were very active as radical scavengers and resisted further hydrolysis by pancreatin maintaining their activity. Regarding the fractions obtained by ultrafiltration, the ORAC values of fractions containing short peptides (<5 kDa) were higher than those of fractions comprising large peptides (> 5 kDa). (Chirinos et al., 2020) demonstrated that the hydrolysis with neutrase (0.385 UA/g, 120 min) of quinoa protein concentrate, and sequential hydrolysis with alcalase (0.385 UA/g protein) and neutrase (0.385 UA/g) (240 min total) of amaranth (*A. caudatus*) protein concentrate, both at 50°C, releases peptides with $ABTS^{+}\cdot$ scavenging activity. This activity for both hydrolysates did not display changes after submitting to SGID with pepsin followed by pancreatin (Table 9.2) with $ABTS^{+}\cdot$ scavenging values of 1.54 ± 0.09 μmol TE/mg for digest quinoa protein and 1.71 ± 0.05 μmol TE/mg for digest amaranth protein, suggesting that they can maintain their antioxidant activity after digestion.

Peptides from buckwheat protein isolate released during a two-stage SGID (pepsin and pancreatin, Table 9.2) showed, after complete digestion, an increase in the ability to chelate Fe^{2+} (45%), scavenge 2,20-azinobis (3-ethylbenzothiszoline-6-sulfonic acid) ($ABTS^{+}\cdot$) (87%), and prevent lipid peroxidation (45%) determined by the thiobarbituric acid reactive substances (TBARS) assay (Ma & Xiong, 2009). Interesting results regarding pepsin digestion were found. For example, a decrease in the activity measured by Fe^{+2} chelation compared to the nonhydrolyzed protein was attributed to the alteration of the protein's native structure that would eliminate iron binding sites. On the other hand, the complete digest was subjected to Sephadex G-25 gel filtration and fractions around 456 and 362 Da showed the highest $ABTS^{+}\cdot$ scavenging activity, while the last one was the most effective in the hydroxyl radical scavenging capacity (HOSC) assay, 86% more efficient than the complete protein digest. By LC–MS/MS the authors identified WPL, VPW, VFPW, and PW as the prominent peptides in these fractions (Ma et al., 2010).

On the other hand, (Grancieri et al., 2019) evaluated the antioxidant activity of peptides from digested total protein and isolated protein fractions obtained from the mucilage and fat-free chia flour (*Salvia hispanica* L.). For SGID, a simplified hydrolysis with pepsin and pancreatin (Table 9.2) was carried out. The prolamin fraction reached the highest degree of hydrolysis, followed by glutelin and globulin. Digested albumin and total protein showed the lowest values. After SGID, all proteins produced numerous peptides that showed potential activities as hypotensive, hypoglycemic, antioxidant, and antiinflammatory. Regarding antioxidant properties, digested total protein, albumin, globulin, and glutelin at 1 mg/mL showed their more relevant scavenging capacity for superoxide (O_2^{-}), nitric oxide (NO), and DPPH (1,1-diphenyl-2-picrylhydrazyl), while a concentration of 0.1 mg/mL was the best in each sample for H_2O_2 scavenging. Digested glutelin showed the best results for superoxide scavenging capacity, and digested albumin and total protein for DPPH.

9.2.5 Antiinflammatory or immune-modulatory capacity

Inflammation is a normal immune defense to noxious stimuli, such as pathogenic microorganisms, irritants, and damaged cells. It is an innate immunity mechanism. An unregulated inflammatory response can lead to tissue damage and the development of chronic inflammatory diseases. Several food-derived compounds are able to modulate the immune response in humans (Zhu et al., 2020).

Recently, (Capraro et al., 2021) investigated the potential effects of quinoa (*Chenopodium quinoa* Willd.), amaranth (*A. retroflexus* L.), and buckwheat (*Fagopyrum esculentum* Moench) proteins and their digests on chronic inflammation and oxidative stress. SDGI was carried out using the sequential action of pepsin and pancreatin (Table 9.2). The immunomodulatory capacity of undigested proteins was determined by activating the NF-κB pathway by transiently transfecting intestinal epithelial Caco-2 cells with an NF-κB luciferase reporter; while in the case of the digests the expression of IL-8 by qRT-PCR was quantified. The undigested protein fractions of the analyzed pseudo-cereals did not show any pro-inflammatory activity; they only showed the ability to reduce the level of inflammation after stimulation of the intestinal epithelial Caco-2 cells with the pro-inflammatory cytokine IL-1β. The albumin and globulin with very low net charge at pH 8 showed the highest level of immunomodulatory activity. The digests from the different samples did not exert any inflammatory activity in Caco-2 cells and were only able to protect Caco-2 cells against inflammation induced by IL-1β. The highest immunomodulatory capacity corresponded to the digested fractions of quinoa and buckwheat albumin and globulin.

In the study of (Grancieri et al., 2019), the potential antiinflammatory activity of total protein and protein-isolated fractions digested from chia seeds was analyzed. Several of the released peptides from total and fractioned proteins interacted with antiinflammatory markers. In particular, TGPSPTAGPPAPGGGTH, a peptide from digested chia albumin showed the best interaction with lipoxygenase (LOX-1) and toll-like receptor (TLR4). Secondly, digested glutelin, globulin, and prolamin were the best chia protein fractions to inhibit 5-LOX, and digested albumin peptides showed the best activity to inhibit the inducible nitric oxide synthase enzyme (iNOS), for scavenging nitric oxide (NO), a key mediator of inflammation, and to reduce the activity of cyclooxygenase (COX-2). In the inflammatory process, the receptors present in immune cells, such as TLR and LOX can be activated, sending signals for activation of the nuclear factor kappa B (NF-κB). This promotes the activation of the iNOS and COX-2, which will produce NO and prostaglandins (PG), respectively, which will propagate the inflammatory process. This process can be aggravated by ROS, such as superoxide and hydrogen peroxide. The LOX receptor may also induce platelet and macrophage aggregation, which together with the excess of ROS, leads to the formation of atheroma plaques, which are directly associated with the complications of CVD. According to these facts and the assays related to specific inflammation markers and antioxidant capacity, the authors pointed out chia protein as a source of peptides with antiatherosclerotic effects.

9.2.6 Antiproliferative activity

The literature regarding the antiproliferative activity of pseudo-cereal peptides derived from SGID is scarce, and only four articles deal with this topic.

Considering amaranth, (Sabbione et al., 2019) studied in vitro the antiproliferative effect on human colon cancer cell line HT-29 of protein isolated before and after being subjected to a SGID. SGID was performed according to the simplified protocol previously described (Orsini Delgado et al., 2011). These authors found that the nonhydrolyzed and the digested amaranth protein isolate have an antiproliferative effect on human colon cancer cell line HT-29 ($IC_{50} = 1.35 \pm 0.12$ and 0.30 ± 0.07 mg protein/mL, respectively). Both samples induced apoptosis and necrosis due to an increase in caspase-3 activity. However, this effect was more important in the digest suggesting that SGID treatment releases potential bioactive peptides improving their effect compared to the protein isolate. According to these results, *A. mantegazzianus* proteins have encrypted peptides that are released after SGID and would exert a potential antiproliferative activity over the epithelial tumor cell line HT-29. (Taniya et al., 2020) studied the in vitro anticancer activity of peptides derived from heat-denatured (100°C, 10 min) amaranth proteins obtained by extraction of seed in buffer (50 mM phosphate buffer, pH 7 with 0.1 mM EDTA and 30 mM KCl). SGID of proteins was carried out using pepsin and pancreatin sequence (Table 9.2). The results indicated that the digested heat-denatured proteins were capable of inhibiting cell growth in a dose-dependent manner with an IC_{50} value of 48.3 ± 0.2 μg/mL. This sample also induced DNA fragmentation, membrane integrity loss, phosphatidylserine translocation, and an increase of the caspase-3 activity in the treated cells. Besides, the digest would have antimetastatic potential since it inhibited cell migration across an artificial wound created in the cell monolayer.

Regarding quinoa proteins, (Vilcacundo et al., 2018) studied the effect of peptides released from a quinoa protein concentrate after SGID over colon cancer cell viability, using three human colorectal cancer cell lines (Caco-2, HT-29, and HCT-116), with successful results. To obtain the peptides, these authors employed the consensus digestion protocol COST INFOGEST (Minekus et al., 2014). At first, they evaluated the effect on Caco-2 cell viability. Sample activity increased as the digestion progressed. While the nonhydrolyzed proteins showed no inhibitory effect at the highest concentration tested (4 mg protein/mL), the gastric digest (120 min) showed an IC_{50} value of 0.843 ± 0.001 mg protein/mL and after gastric and duodenal digestion, IC_{50} values were even lower (0.191 ± 0.003 mg protein/mL after 60 min and 0.180 ± 0.001 mg protein/mL after 120 min). Considering these results, more potent peptides inhibiting Caco-2 cell viability were released after SGID, although the authors do not discard the possible contribution of digestive enzymes and bile salts on the activity. HT-29 and HCT-116 cells were also sensitive to the effect of peptides from gastroduodenal digest (60 min), IC_{50} values were 0.212 ± 0.002 mg protein/mL and 0.208 ± 0.002 mg protein/mL, respectively. This digest was fractionated by RP-HPLC and 17 peptides from 11S globulin and other quinoa proteins were identified by LC–MS/MS. According the authors, the sequences LWREGM, DKDYPK, DVYSPEAG, IFQEYI, and RELGEWGI could act as anticancer compounds given their antioxidant effects and the demonstrated relationship between inhibition of oxidative stress and the start of carcinogenesis. However, further work is needed to investigate the possible mechanism of action of these chemopreventive peptides as well as their bioavailability and the possibility of reaching the target organ. (Fan et al., 2022) studied the anticolon cancer activity of peptides isolated from SGID of quinoa protein concentrate in Caco-2 cells. In order to simulate GID, a quinoa protein concentrate was dissolved in deionized water (1:100, g/mL) and the pH was adjusted to 2. Then the solution was hydrolyzed with trypsin (5% w/w) followed by a simulated intestinal fluid containing pancreatin (Table 9.2). The digest (proteolysis degree = 20.46%) was filtered through a 5 kDa ultrafiltration membrane and fractions MW <5 kDa and MW >5 kDa were obtained. The antiproliferative activity of all samples was tested, finding that MW <5 kDa possessed stronger antiproliferative activity. Thus they identify the peptides present in this fraction by ESI-Q-TOF–MS/MS. Thirteen quinoa peptides were predicted to have high bioactivity using the PeptideRanker software. The authors tested the inhibition of cancer cell proliferation and histone deacetylase 1 (HDAC1) inhibitory activity assay, since inhibitors of class I HDAC activity are potent inducers of growth arrest, differentiation, and apoptosis of colon cancer in vitro and in vivo. They found that FHPFPR, NWFPLPR, and HYNPYFPGGA were the most promising peptides. Besides, these peptides decrease the expression of HDAC1, NF-κB, IL-6, IL-8, Bcl-2 and increase the caspase-3 expression so they could modulate cancer-related gene expression.

9.2.7 Antidiabetic activity

Diabetes mellitus is a metabolic disorder characterized by an elevated blood glucose level (hyperglycemia) which represents another important cause of death worldwide. Type 2 diabetes (T2D) is the most frequent (90%) and its treatment includes the use of drugs that increase glucose uptake in tissues, decrease gluconeogenesis; stimulate insulin secretion; or inhibit α-glucosidase and α-amylase (hypoglycemic effect by reduction of the intestinal glucose absorption) and dipeptidyl peptidase IV (DPP −IV) (insulin-mimetic effect by enhancing the synthesis of insulin). Diverse bioactive peptides have exerted these kind of effects (Valenzuela Zamudio & Segura Campos, 2022). These authors reviewed the potential role of amaranth, quinoa, and chia in the management and prevention of T2D; the studies are scarce and only two investigations including SGID have been performed. (Vilcacundo et al., 2017) submitted quinoa protein to SGID (harmonized protocol, (Minekus et al., 2014)) and fractionated by ultrafiltration through a hydrophilic 5 kDa cut-off membrane. Fractions <5 kDa showed a high α-amylase inhibitory activity (IC_{50} = 0.32 mg/mL), and a less potent α-glucosidase inhibitory activity (IC_{50} = 1.64 mg/mL). Three novel peptides derived from 11S seed storage globulin B showed the potential to inhibit enzymes involved in the digestion of dietary carbohydrates. Especially, the peptide IQAEGGLT showed inhibitory activity of α-glucosidase (IC_{50} = 109 mM) associated with the presence of hydrophobic amino acids, and also of DPP-IV (IC_{50} = 268 mM) attributed to the presence of isoleucine as N-terminal amino acid. However, this peptide presented low-amylase inhibitory activity, which was attributed to the absence of aromatic amino acids. As was previously mentioned in section *2.1*, (Vilcacundo et al., 2019) studied peptides released after SGID (COST INFOGEST protocol, different digestion times) from *A. caudatus* proteins with different bioactivities, including the antidiabetic activity evaluating the DPP-IV and α-amylase inhibition. The digests were also ultrafiltrated through a hydrophilic 5 kDa cut-off membrane. The most active digest against DPP-IV was obtained after 60 min digestion (IC_{50} = 0.28 mg/mL, similar for both MM ranges). Regarding α-amylase, the <5 kDa fraction after digestion of 120 min was the most potent (IC_{50} = 0.84 mg/mL). Additionally, the 60-min digest was selected based on its multifunctionality, and fractionated by semipreparative RP-HPLC, and the obtained peptide fractions were analyzed by LC−MS/MS. DPP-IV inhibitory activity was attributed to the peptides NRPET and VEEGNM, while the α-amylase activity was attributed to the sequences FLISCLL, SVFDEELS,

DFIILE, NRPET, and HVIKPPS, according to molecular features discussed in the previous research (Vilcacundo et al., 2017). Authors considered that the sequence NRPET could potentially exhibit both, DPP-IV and α-amylase inhibitory activities.

Chiang et al. (2022) made an overview of the bioactive components present in buckwheat that could exert a blood glucose-lowering effect in the human body, including peptides among them. Again, studies performed on SGID products are very scarce. (Ninomiya et al., 2018) hydrolyzed purified albumin fractions containing α-amylase inhibitors by using a SGID process according to (Ma & Xiong, 2009) with some modifications (Table 9.2), demonstrating that although buckwheat α-amylase inhibitors were hydrolyzed to low-molecular-weight peptides by digestive enzymes, the 100% of the α-amylase inhibition was retained after digestion. This activity was also retained after heating. The authors concluded that glycopeptides were responsible for the inhibition through a competitive mechanism. They also demonstrated that this buckwheat albumin fraction suppressed blood glucose level elevation after starch loading in vivo.

9.3 Bioactivities of peptides from simulated gastrointestinal digestion of flour and food matrices

In this section, the bioactivities of peptides derived of SGID of more complex pseudo-cereals matrix, such as flours and derived products, are described. The conditions used in the various studies are summarized in Table 9.3.

9.3.1 Antihypertensive activity

In the case of pseudo-cereals, practically no studies of antihypertensive peptides from SGID of flour and derived food matrices have been carried out. The effect of the SGID process has been studied in 20:80 oil–water emulsions containing 1% or 2% of a 50:50 mixture of amaranth protein isolate and amaranth protein hydrolysate obtained by alcalase action (Suárez & Añón, 2019). These authors subjected the emulsions to the SGID process described by (Malaki Nik et al., 2011) in which the following stages were performed: emulsions 5 mL/gastric fluid with pepsin (12 mL) followed by the addition of 12 mL of intestinal fluid (Table 9.3). Emulsion stability during SGID and the ability to inhibit ACE of the digests were analyzed. The emulsions were susceptible to aggregation and coalescence phenomena during the SGID process, evidenced by the large increase in the oil droplet size. The destabilization of the emulsion was attributed to the chaotropic action of the bile salt on the interface and the proteolytic and lipolytic action of pancreatin and lipase, respectively. The ACE inhibitory effect of the amaranth hydrolysate employed to prepare the active emulsion, the 1% w/v emulsion, and the digested 1% w/v emulsion was assessed. The three samples analyzed presented ACE inhibitory activity with IC_{50} values of 0.14 ± 0.02, 0.29 ± 0.03,

Table 9.3 Bioactivities of peptides obtained by SGID of food matrices derived from pseudo-cereals.

Matrix	SGID protocol	Bioactivity	Research finding	References
Amaranth				
20:80 oil–water emulsion, 1% 50:50 PI/alcalase-H *A. hypochondriacus*	Pepsin (2.5 mg/mL, 12 mL)/emulsion 5 mL, pH 2, 37°C, 1 h 12 mL (3 mg/mL pancreatin, 1.5 mg/mL lipase), pH 7, 37°C, 1 h	Antihypertensive	Inhibition of ACE $IC_{50} = 0.13$ mg protein/mL	Suárez and Añón (2019)
F, cookies *A. hypochondriacus*	INFOGEST		Inhibition of ACE $IC_{50} = 0.08$ and 0.23 mg protein/mL	Sabbione et al. (2019)
F *A. manteggazianus*	Salivary fluid, pH 6.8, 37°C, 5 min Pepsin, pH 2, 37°C, 1 h Pancreatin, pH 7, 37°C, 1 h	Hypocholesterolemic	Displacement of cholesterol from micelles $IC_{50} = 0.1$ mg protein/mL	Sisti et al. (2019)
F, cookies *A. hypochondriacus*	INFOGEST	Anticoagulant	Inhibition of coagulation common pathway (fibrin clot formation), $IC_{50} = 1.20$, 0.22 mg protein/mL	Sabbione et al. (2019)
Ice cream	INFOGEST		Inhibition of coagulation common pathway (fibrin clot formation), $IC_{50} = 3.4$ mg protein/mL	Malgor et al. (2020)
F, extruded F	Pepsin 1:20 w/w, pH 2, 37°C, 3 h	Antioxidant	Scavenging ROO· (ORAC), 150–200 μmol TE/mg protein	Montoya-Rodrıguez et al. (2014)

A. hypochondriacus	Pancreatin 1/20 w/w, pH 7.5, 37°C, 3 h			
F *A. manteggazianus*	INFOGEST		Scavenging ROO· (ORAC), $IC_{50} = 0.034$ mg/mL	Rodríguez et al. (2020)
			Inhibition of OH· (HORAC), $IC_{50} = 1.33$ mg/mL	
Germinated F	Pepsin 1:20 w/w, pH 2, 37°C, 3 h		Scavenging ROO· (ORAC), 74 mmol TE/ 100 g F	Sandoval-Sicairos et al. (2021)
A. hypochondriacus	Pancreatin 1/20 w/w, pH 7.5, 37°C, 3 h		Most active fraction 3–10 kDa (33.6 mmol TE/ 100 g)	
F, beverage			Scavenging ROO· (ORAC), $IC_{50} = 0.05$ and 0.008 mg/mL	Rodríguez (2022)
A. manteggazianus			Intracellular ROS inhibition, reduced GSH content and SOD activity (Caco-2-TC7)	
F, extruded F	Pepsin 1:20 w/w, pH 2, 37°C, 3 h	Antiinflammatory	Inhibition LPS-induced inflammation in human THP-1 cells.	Montoya-Rodríguez et al. (2014)
A. hypochondriacus	Pancreatin 1/20 w/w, pH 7.5, 37°C, 3 h		macrophages-like and mouse RAW 264.7 macrophages	
			Prevention of activation of NF-kβ signaling	

(*Continued*)

Table 9.3 Bioactivities of peptides obtained by SGID of food matrices derived from pseudo-cereals. *Continued*

Matrix	SGID protocol	Bioactivity	Research finding	References
Buckwheat				
Heat-treated F	Pepsin 1/100, pH 2, 4 h, 37°C	Anticoagulant	Inhibition of ACE, $IC_{50} = 0.14$ mg protein/mL	Li et al. (2002)
Tartary	Chymotrypsin and trypsin 1:100, pH 6.8, 4 h, 37°C		Identification of active sequences: VKFY, AY, LF, YV, YQ, YQY, PSY, LGI, ITY, INSQ ($IC_{50} = 4-628$ μM)	
Albumins	Pepsin 1/20 w/w, pH 2, 3 h, 37°C 1:1 trypsin/pancreatin 1/20 w/w, pH 7.5, 3 h, 37°C	Antidiabetic	DPP-IV inhibitory activity, $IC_{50} = 1.98$ mg/mL	Wang et al. (2015)
Canihua (*C. pallidicaule*, C) and quinoa (*C. quinoa*, Q)				
Cookies, replacement of wheat F (15% QF, 25% CF)	INFOGEST 2.0	Antioxidant	Scavenging ROO· (ORAC): 195.26 μmol TE/g	Paucar-Menacho et al. (2022)

F: flour, PI: protein isolate, H: hydrolysate. ACE: angiotensin-converting enzyme; ORAC: oxygen radical absorption capacity; HORAC: hydroxyl radical averting capacity; TE: Trolox equivalent; ROS: reactive oxygen species; SOD: superoxide dismutase; GSH: reduced glutathione; LPS: lipopolysaccharide; NF-κB: nuclear factor regulator of κB light chain expression; DPP-IV: dipeptidyl peptidase IV.

and 0.13 ± 0.07 mg/mL, respectively. It should be considered that the estimation of IC_{50} was based on the total protein concentration regardless of the molecular mass of the (poly)-peptides present. It should be marked that the original emulsion contains 50% of protein isolate whose ACE inhibitory activity according to Vecchi & Añon (2009) is very low. According to the results, it is possible to obtain stable flocculated emulsions formulated with amaranth proteins + bioactive peptides, which can act as surfactants and/or are protected by the network of flocs present in the initial emulsion, and maintain their ACE inhibitory capacity after the SGID process.

Sabbione et al. (2019) simulated the GID of amaranth flour and cookies according to the consensus protocol (Minekus et al., 2014) with some modifications. The degree of hydrolysis reached during the GID process was 38.2 ± 0.8 and 41.0 ± 2.0%, respectively. Both the digests obtained from the flour and the cookies were capable of inhibiting the ACE but with different potency (IC_{50} = 0.08 ± 0.01 mg protein/mL and 0.23 ± 0.03 mg protein/mL, respectively). These results suggest that the enzymes have different access possibilities to the proteins depending on the complexity of the matrix that undergoes the GID process, probably releasing different peptides with different bioactivity in each case. The release of bioactive peptides from different foods is an interesting result since the organism itself would be in charge of converting, in this case, amaranth proteins into a bioactive after ingestion.

Li et al. (2002) analyzed the ACE inhibitory activity of digests obtained from heat-treated tartary buckwheat flour. In the SGID process used pepsin and then chymotrypsin and trypsin (Table 9.3). Their results show that the digests obtained at the end of the gastric stage had the same ACE inhibitory capacity as undigested flour, while for the intestinal digests were significantly higher (IC_{50} = 0.36, 0.35, and 0.14 mg protein/mL). The hypotensive action of flour digests was confirmed in vivo where the reduction of systolic blood pressure of spontaneously hypertensive rats was verified. From the digests, a fraction of di- and tripeptides with ACE inhibitory activity was isolated by chromatography, which was purified by successive chromatographic fractionations. From the most active fractions, it was possible to identify 9 di-tripeptides and 2 active tetrapeptides (VKFY, AY, LF, YV, YQ, YQY, PSY, LGI, ITY, INSQ) with IC_{50} in the range of 628 to 4 μM, some of which have been previously reported.

9.3.2 Hypocholesterolemic activity

Sisti et al. (2019) also analyzed the effect of the presence of *A. mantegazzianus* flour and its digest on the formation of model micelles and the displacement of cholesterol from them. Digests were prepared in the same way as previously described for protein isolates in section *2.2*. The presence of undigested or digested flour during the formation of the micelles led to a highly polydispersed particle size distribution, to which at least 3 populations contribute. The detected particle sizes were much larger than those obtained in the presence of protein isolate. According to (Sisti et al., 2019), the increase in size could be associated with an effect of the presence of fiber that could act as a bridge between the micelles, perhaps due to interaction with bile acids. Both undigested and digested flour showed the ability to displace cholesterol from the model micelles in a dose-dependent manner. The IC_{50} values were 0.1 ± 0.3 and 0.71 ± 0.07 mg/mL, respectively, for digested and undigested flour. Both the flour and the digested amaranth meal showed higher activity than the protein isolate samples, previously described, containing the same amount of protein. The dietary fiber present in amaranth flour would be responsible for this increase in activity since dietary fiber is characterized by binding to many components of the diet, cholesterol and bile acids among them, decreasing its absorption.

9.3.3 Antithrombotic activity

Antithrombotic peptides released from food matrices after SGID have only been reported in amaranth grains. To present, three food matrices have been studied. (Sabbione et al., 2019) elaborated gluten-free cookies with whole-meal amaranth flour. Both products, flour and cookies, exhibited antithrombotic activity after SGID (Minekus et al., 2014). The in vitro turbidimetric microplates method showed that digestion of these amaranth products was able to release peptides capable of exerting potential antithrombotic activity with IC_{50} values of 1.20 ± 0.18 for the flour and 0.22 ± 0.04 mg protein/mL for the cookie. Amaranth-digested flour exhibited less activity, indicating that the peptides released after flour digestion are different from those obtained in the cookies. Even though amaranth flour is the protein source, the processes applied, such as cooking or the presence of other ingredients in the food matrix could modify the structure of the proteins, producing higher amounts of active peptides. The third product studied was an amaranth lemon ice cream formulated with amaranth protein isolate containing 85% w/w protein (Malgor et al., 2020). Under acidic conditions (60 min, 20°C), using lemon juice, an endogenous aspartic protease modifies proteins, enhancing their foaming capacity (Sabbione et al., 2016). After subjecting the ice cream to SGID (Minekus et al., 2014), in vitro antithrombotic activity was evidenced using the turbidimetric microplates method ($IC_{50} = 3.4 \pm 0.1$ mg protein/mL). These studies show the versatility and bioactivity of amaranth seed components and demonstrate for the first time that food made with amaranth flour and protein isolates exert antithrombotic potential. In this sense, amaranth cookies and ice cream could be an alternative way of incorporating potentially health-beneficial products.

9.3.4 Antioxidant activity

As has been previously discussed, several researches have analyzed the release of antioxidant peptides from protein isolates or concentrates, or protein fractions isolated from amaranth, quinoa, or buckwheat. However, as far as can be found in the literature, few studies evaluate the antioxidant activity of peptides derived from the SGID of flour and food matrices derived from pseudo-cereals.

Regarding the application of SGID on amaranth flours, (Montoya-Rodríguez et al., 2014) demonstrated antioxidant and antiinflammatory activity of peptides obtained by an in-house protocol (pepsin and pancreatin, Table 9.3) applied to flour and extruded flour of *A. hypochondriacus*. These authors reported an increase in soluble protein with hydrolysis time; however, the extruded flour required longer hydrolysis times to reach maximum soluble protein values than the unprocessed flour. In addition, the latter presented a higher degree of proteolysis. The antioxidant activity measured by ORAC showed higher values for the unprocessed flour after digestion with pepsin; the antioxidant capacity varied throughout the hydrolysis with pancreatin, reaching values in the range of 150–200 μmol TE/mg of soluble protein for both flours. In a recent work, (Sandoval-Sicairos et al., 2021) applied the same SGID process on germinated amaranth flour (*A. hypochondriacus*). The digest obtained after 90 min of incubation with pancreatin presented an ORAC activity of 74 mmol TE/100 g germinated amaranth flour gastrointestinal digest; the fraction with MW between 3 and 10 kDa presented the highest activity (33.6 mmol TE/100 g). Interestingly, after a methanol wash, the antioxidant activity of the digest decreased to 13 mmol TE/100 g, suggesting the presence of methanol soluble compounds with antioxidant activity (i.e. polyphenols). (Rodríguez et al., 2020) evaluated the SGID of amaranth flour (*A. mantegazzianus*) by applying the COST INFOGEST consensus protocol compared to the in-house digestion protocol described in section *2.4*. The presence of other components and changes in the digestion method influenced the proteolysis, composition, and antioxidant activity (ORAC/HORAC) of the flour digest. The digest of flour obtained by the consensus method presented an ORAC activity of 1.2 ± 0.2 μmol TE/mg soluble protein or 101 ± 16 mmol TE/100 g digest, 100 times lower than those reported by (Montoya-Rodrıguez et al., 2014), but slightly higher than that obtained with germinated flour (Sandoval-Sicairos et al., 2021). In addition, the ORAC activity ($IC_{50} = 0.034 \pm 0.005$ mg/mL) was comparable with the results of (Orsini Delgado et al., 2011) using another digestion method on protein isolate of *A. mantegazzianus* (0.024 ± 0.002 mg/mL), as well as with the values obtained after SGID of this protein isolate with the consensus protocol (0.024 ± 0.006 mg/mL) (section *2.4*). Accordingly, neither the original matrix nor the digestion method affected the ORAC activity of *A. manteggazianus* peptides. In the case of the HORAC assay, amaranth flour showed a lower IC_{50} value (1.33 ± 0.08 mg/mL) than the protein isolate (2.3 ± 0.2 mg/mL), suggesting that the flour would contribute some active soluble components not present in the protein isolate, such as some low MM polypeptides, lost during the preparation of the protein isolate. In addition, (Rodríguez & Tironi, 2020) observed the loss of some phenolic compounds in the protein isolate compared to the flour.

Recently, (Rodríguez, 2022) prepared a beverage (B) from the solubilization of *A. manteggazianus* seed proteins (3.5% protein), and the polypeptide composition and antioxidant properties of B after SGID (consensus protocol) were studied in comparison with amaranth flour (BD and FD). Bioaccessible fractions (HD_b and BD_b) were obtained after treatment with cholestyramine resin to remove bile salts due to their cytotoxicity and oxidative effect. SGID produced an increase in ORAC activity for both samples; the IC_{50} values were 0.05 ± 0.01 and 0.008 ± 0.004 mg/mL for HD_b and BD_b fractions, respectively, indicating that the bioaccessible fraction derived from B had the highest ORAC potency. These fractions showed low cytotoxicity values (measured by LDH release) and produced high intracellular ROS inhibition (about 80%), increments in the reduced GSH content and the SOD activity, with no effect on the GPx enzyme in Caco-2-TC7 cells exposed to H_2O_2. These results present amaranth beverage as a potential functional food with antioxidant properties that need to be verified in vivo.

Paucar-Menacho et al. (2022) evaluated the effect of replacing wheat flour (WF) with germinated canihua (*Ch. pallidicaule Aellen*), kiwicha (*A. caudatus* L.) and quinoa (*C. quinoa* Willd.) (SCF, SKF, and SQF, respectively) in cookies. Optimal ternary flour blends were identified that maximize the bioactive compounds content and the antioxidant activity of the cookies while minimizing the phytic acid content. Thus the optimal cookie formulation (15% SQF, 25% SCF and 60% WF) was selected and submitted to SGID using the INFOGEST 2.0 method (Table 9.3) (Brodkorb et al., 2019). For this baked product, higher antioxidant activity was observed in gastric and intestinal digests compared to the control cookie (100% WF). The ORAC activities (expressed as μmol TE/g) were 25.51 and 136.25 before digestion; 41.69 and 119.47 after gastric digestion; 103.59 and 195.26 after intestinal digestion, for the control and for the optimal formulation cookie added with quinoa and canihua, respectively.

It should be noted that in the previously mentioned investigations, aqueous fractions where the peptides would be soluble were analyzed. However, there are several works in the literature in which the evaluation of the antioxidant activity of flours and derived matrices is frequently linked to polyphenolic compounds that are isolated from those matrices using organic solvents of different polarities (Fernández-López et al., 2020; Jan et al., 2015; Rodríguez & Tironi, 2020; Szawara-Nowak et al., 2016; Xu et al., 2019; Zielińska, 2019). It cannot be ruled out, however, that these extracts also contain some peptides, as well as, that aqueous extracts could contain some free or bound polyphenolic compounds.

9.3.5 Antiinflammatory and antidiabetic activities

Only one work has been reported in which the antiinflammatory activity of peptides from the SGID of unprocessed and extruded amaranth flour is reported (Montoya-Rodrıguez et al., 2014), which was also mentioned in the previous section since antioxidant activity was also evaluated. Their antiinflammatory effect was evaluated on LPS-induced human THP-1 macrophages-like and LPS-induced mouse RAW 264.7 macrophages. The results obtained indicated that amaranth digests inhibited LPS-induced inflammation in both types of macrophages by preventing activation of NF-kβ signaling. The values obtained for both cell lines indicated that unprocessed and extruded amaranth hydrolysates had a better effect on mouse macrophages than on human THP-1 macrophage-like cells. The extrusion of the flour improved antiinflammatory effect of amaranth hydrolyzates in both cells, which might be attributed to the production of bioactive peptides during processing.

Also one study on antidiabetic activity was found in the literature. After SGID of buckwheat (*F. esculentum*) flour (pepsin and trypsin/pancreatin, Table 9.3), digest presented DPP-IV inhibitory activity (IC_{50} = 1.98 mg/mL) (Wang et al., 2015).

9.4 Conclusion

Although an important number of studies demonstrating various in vitro bioactivities of peptides derived from pseudo-cereals proteins can be found in the literature, most of these studies do not include an SGID process which is essential to assess their bioaccessibility and bioactivity. This chapter presents results found in the literature related to proteins, flours, or food matrices from pseudo-cereals subjected to SGID, focusing on the most frequently studied bioactivities: antihypertensive, hypocholesterolemic, antithrombotic, antioxidant, antiinflammatory, antiproliferative, and antidiabetic.

Regarding total or fractioned protein fractions, available information—Table 9.2—suggests that simplified two steps SGID protocols focused on protein hydrolysis can be applied. However, studies using the consensus INFOGEST protocol or comparable studies are very scarce. Amaranth proteins have been the most studied; all aforementioned in vitro activities have been demonstrated for amaranth protein isolates in different degrees of depth.

Bioactivity studies of peptides released after SGID of meals or food matrices derived from pseudo-cereals are even scarcer, being again amaranth the most frequently studied as can be noted in Table 9.3. No studies have been carried out that analyze the antiproliferative activity of peptides from the digestion of formulated foods based on amaranth, buckwheat, chia, and quinoa. To our knowledge, no study to date evaluates the bioactivities of peptides derived from the SGID of matrices obtained from chia.

Although there are some similarities between the protein components of the different pseudo-cereals, the composition and polypeptide structures are different. This fact, added to the presence of various other components, changes the digestibility and digestion products resulting from each ingredient or food matrix studied. Although the bioactivity study of peptides released by SGID from pseudo-cereal proteins remains as an underexploited field compared to proteins from other sources, the existing results confirm their great potential as a natural antioxidant source, being able to positively impact the health of the population. However, further studies focused on demonstrating the bioavailability and mechanisms of action of these peptides should be needed to confirm their in vivo effects.

References

Alonso-Miravalles, L., & O'Mahony, J. (2018). Composition, protein profile and rheological properties of pseudocereal-based protein-rich ingredients. *Foods*, *7*(5), 73. Available from https://doi.org/10.3390/foods7050073.

Aluko, R. E. (2015). Antihypertensive peptides from food proteins. *Annual Review of Food Science and Technology*, *6*(1), 235–262. Available from https://doi.org/10.1146/annurev-food-022814-015520.

Ayala-Niño, A., Rodríguez-Serrano, G. M., González-Olivares, L. G., Contreras-López, E., Regal-López, P., & Cepeda-Saez, A. (2019). Sequence identification of bioactive peptides from amaranth seed proteins (Amaranthus hypochondriacus spp.). *Molecules*, *24* (17). Available from https://doi.org/10.3390/molecules24173033. Available from, https://www.mdpi.com/1420-3049/24/17/3033/pdf.

Ayala-Ninõ, A., Rodríguez-Serrano, G. M., Jiménez-Alvarado, R., Bautista-Avila, M., Sánchez-Franco, J. A., González-Olivares, L. G., & Cepeda-Saez, A. (2019). Bioactivity of peptides released during lactic fermentation of amaranth proteins with potential cardiovascular protective effect: An in vitro study. *Journal of Medicinal Food*, *22*(10), 976–981. Available from https://doi.org/10.1089/jmf.2019.0039, http://www.liebertonline.com/jmf.

Balakrishnan, G., & Schneider, R. G. (2022). The role of amaranth, quinoa, and millets for the development of healthy, sustainable food products—A concise review. *Foods*, *11* (16), 2442. Available from https://doi.org/10.3390/foods11162442.

Bender, D., & Schönlechner, R. (2021). Recent developments and knowledge in pseudocereals including technological aspects. *Acta Alimentaria*, *50*(4), 583–609. Available from https://doi.org/10.1556/066.2021.00136, https://akjournals.com/view/journals/066/066-overview.xml.

Brodkorb, A., Egger, L., Alminger, M., Alvito, P., Assunção, R., Ballance, S., Bohn, T., Bourlieu-Lacanal, C., Boutrou, R., Carrière, F., Clemente, A., Corredig, M., Dupont, D., Dufour, C., Edwards, C., Golding, M., Karakaya, S., Kirkhus, B., Le Feunteun, S., ... Recio, I. (2019). INFOGEST static in vitro simulation of gastrointestinal food digestion. *Nature Protocols*, *14*(4), 991–1014. Available from https://doi.org/10.1038/s41596-018-0119-1, http://www.natureprotocols.com/.

Capraro, J., Benedetti, S., Heinzl, G., Scarafoni, A., & Magni, C. (2021). Bioactivities of pseudocereal fractionated seed proteins and derived peptides relevant for maintaining human well-being. *International Journal of Molecular Sciences*, *22*(7), 3543. Available from https://doi.org/10.3390/ijms22073543.

Cavazos, A., & Gonzalez de Mejia, E. (2013). Identification of bioactive peptides from cereal storage proteins and their potential role in prevention of chronic diseases. *Comprehensive Reviews in Food Science and Food Safety*, *12*(4), 364–380. Available from https://doi.org/10.1111/1541-4337.12017.

Chen, H., Cheng, S., Fan, F., Tu, M., Xu, Z., & Du, M. (2019). Identification and molecular mechanism of antithrombotic peptides from oyster proteins released in simulated gastro-intestinal digestion. *Food and Function*, *10*(9), 5426–5435. Available from https://doi.org/10.1039/c9fo01433k, http://pubs.rsc.org/en/journals/journal/fo.

Chiang, J. H., Hua, X. Y., Yu, A. H. M., Peh, E. W. Y., See, E., & Jeyakumar Henry, C. (2022). A review on buckwheat and its hypoglycemic bioactive components in food systems. *Food Reviews International*. Available from https://doi.org/10.1080/87559129.2022.2103706, http://www.tandf.co.uk/journals/titles/87559129.asp.

Chirinos, R., Pedreschi, R., Velásquez-Sánchez, M., Aguilar-Galvez, A., & Campos, D. (2020). In vitro antioxidant and angiotensin I-converting enzyme inhibitory properties of enzymatically hydrolyzed quinoa (Chenopodium quinoa) and kiwicha (Amaranthus caudatus) proteins. *Cereal Chemistry*, *97*(5), 949–957. Available from https://doi.org/10.1002/cche.10317, http://onlinelibrary.wiley.com/journal/10.1002/(ISSN)1943-3638/issues.

Cho, H. J., Ham, H. S., Lee, D. S., & Park, H. J. (2003). Effects of proteins from hen egg yolk on human platelet aggregation and blood coagulation. *Biological and Pharmaceutical Bulletin*, *26*(10), 1388–1392. Available from https://doi.org/10.1248/bpb.26.1388, http://www.jstage.jst.go.jp/article/bpb/26/10/1388/_pdf.

Cho, H. J., Kittaka, R., Abdou, A. M., Kim, M., Kim, H. S., Lee, D. H., & Park, H. J. (2009). Inhibitory effects of oligopeptides from hen egg white on both human platelet aggregation and blood coagulation. *Archives of Pharmacal Research*, *32*(6), 945–953. Available from https://doi.org/10.1007/s12272-009-1618-y.

Cotabarren, J., Rosso, A., & Obregón, W. (2022). *Parisi, Chía proteins as a source of bioactive peptides to enhance human health benefits (chapter 12). Native crops in Latin America: Biochemical, processing, and nutraceutical aspects*. CRC Press.

Dakhili, S., Abdolalizadeh, L., Hosseini, S. M., Shojaee-Aliabadi, S., & Mirmoghtadaie, L. (2019). Quinoa protein: Composition, structure and functional properties. *Food Chemistry*, *299*125161. Available from https://doi.org/10.1016/j.foodchem.2019.125161.

Delgado, M. C. O., Galleano, M., Añón, M. C., & Tironi, V. A. (2015). Amaranth peptides from simulated gastrointestinal digestion: Antioxidant activity against reactive species. *Plant Foods for Human Nutrition*, *70*(1), 27–34. Available from https://doi.org/10.1007/s11130-014-0457-2, https://link.springer.com/journal/11130.

Fan, X., Guo, H., Teng, C., Zhang, B., Blecker, C., & Ren, G. (2022). Anti-colon cancer activity of novel peptides isolated from in vitro digestion of quinoa protein in Caco-2 cells. *Foods*, *11*(2), 194. Available from https://doi.org/10.3390/foods11020194.

Fernández-López, J., Viuda-Martos, M., Sayas-Barberá, M. E., Navarro-Rodríguez de Vera, C., Lucas-González, R., Roldán-Verdú, A., Botella-Martínez, C., & Pérez-Alvarez, J. A. (2020). Chia, quinoa, and their coproducts as potential antioxidants for the meat industry. *Plants*, *9*(10), 1359. Available from https://doi.org/10.3390/plants9101359.

Fillería, S. G., Nardo, A. E., Paulino, M., & Tironi, V. (2021). Peptides derived from the gastrointestinal digestion of amaranth 11S globulin: Structure and antioxidant functionality. *Food Chemistry: Molecular Sciences*, *3*. Available from https://doi.org/10.1016/j.fochms.2021.100053, http://www.journals.elsevier.com/food-chemistry-molecular-sciences.

García Fillería, S., & Tironi, V. (2021). Intracellular antioxidant activity and intestinal absorption of amaranth peptides released using simulated gastrointestinal digestion with Caco-2 TC7 cells. *Food Bioscience*, *41*. Available from https://doi.org/10.1016/j.fbio.2021.101086, http://www.journals.elsevier.com/food-bioscience/.

García Fillería, S. F., & Tironi, V. A. (2017). Prevention of in vitro oxidation of low density lipoproteins (LDL) by amaranth peptides released by gastrointestinal digestion. *Journal of Functional Foods*, *34*, 197–206. Available from https://doi.org/10.1016/j.jff.2017.04.032, http://www.elsevier.com/wps/find/journaldescription.cws_home/717426/description#description.

Gawlik-Dziki, U., Dziki, D., Baraniak, B., & Lin, R. (2009). The effect of simulated digestion in vitro on bioactivity of wheat bread with Tartary buckwheat flavones addition. *LWT - Food Science and Technology*, *42*(1), 137–143. Available from https://doi.org/10.1016/j.lwt.2008.06.009.

Grancieri, M., Martino, H. S. D., & Gonzalez de Mejia, E. (2019). Digested total protein and protein fractions from chia seed (Salvia hispanica L.) had high scavenging capacity and inhibited 5-LOX, COX-1-2, and iNOS enzymes. *Food Chemistry*, *289*, 204–214. Available from https://doi.org/10.1016/j.foodchem.2019.03.036, http://www.elsevier.com/locate/foodchem.

Guo, H., Hao, Y., Richel, A., Everaert, N., Chen, Y., Liu, M., Yang, X., & Ren, G. (2020). Antihypertensive effect of quinoa protein under simulated gastrointestinal digestion and peptide characterization. *Journal of the Science of Food and Agriculture*, *100*(15), 5569–5576. Available from https://doi.org/10.1002/jsfa.10609, http://onlinelibrary.wiley.com/journal/10.1002/(ISSN)1097-0010.

Hsu, H. W., Vavak, D. L., Satterlee, L. D., & Miller, G. A. (1977). A multienzyme technique for estimating protein digestibility. *Journal of Food Science*, *42*(5), 1269–1273. Available from https://doi.org/10.1111/j.1365-2621.1977.tb14476.x.

Hussain, M. I., Farooq, M., Syed, Q. A., Ishaq, A., Al-Ghamdi, A. A., & Hatamleh, A. A. (2021). Botany, nutritional value, phytochemical composition and biological activities of quinoa. *Plants*, *10*(11). Available from https://doi.org/10.3390/plants10112258, https://www.mdpi.com/2223-7747/10/11/2258/pdf.

Jan, U., Gani, A., Ahmad, M., Shah, U., Baba, W. N., Masoodi, F. A., Maqsood, S., Gani, A., Wani, I. A., & Wani, S. M. (2015). Characterization of cookies made from wheat flour blended with buckwheat flour and effect on antioxidant properties. *Journal of Food Science and Technology*, *52*(10), 6334–6344. Available from https://doi.org/10.1007/s13197-015-1773-8, http://www.springerlink.com/content/121580/.

Janssen, F., Pauly, A., Rombouts, I., Jansens, K. J. A., Deleu, L. J., & Delcour, J. A. (2017). Proteins of amaranth (Amaranthus spp.), buckwheat (Fagopyrum spp.), and quinoa (Chenopodium spp.): A food science and technology perspective. *Comprehensive Reviews in Food Science and Food Safety*, *16*(1), 39–58. Available from https://doi.org/10.1111/1541-4337.12240, http://onlinelibrary.wiley.com/journal/10.1111/(ISSN)1541-4337.

Julio, L. M., Ixtaina, V. Y., & Tomás, M. C. (2022). *Development of delivery systems of bioactive compounds using chia seed by-products. Native crops in Latin America: Biochemical, processing, and nutraceutical aspects* (pp. 265–286). CRC Press Available from. Available from http://www.tandfebooks.com/doi/book/9781000541502.

Li, Chun-Hui, Matsui, Toshiro, Matsumoto, Kiyoshi, Yamasaki, Rikio, & Kawasaki, Terukazu (2002). Latent production of angiotensin I-converting enzyme inhibitors from buckwheat protein. *Journal of Peptide Science*, *8*(6), 267–274. Available from https://doi.org/10.1002/psc.387.

Lobo, V., Patil, A., Phatak, A., & Chandra, N. (2010). Free radicals, antioxidants and functional foods: Impact on human health. *Pharmacognosy Reviews*, *4*(8), 118. Available from https://doi.org/10.4103/0973-7847.70902.

Loganathan, C., Sakkiah, S., Lee, K. W., Kabilan, S., & Meganathan, C. (2013). Pharmacophore design, virtual screening, molecular docking and optimization approaches to discover potent thrombin inhibitors. *Combinatorial Chemistry and High Throughput Screening*, *16*(9), 702–720. Available from https://doi.org/10.2174/13862073113169990007.

López, D. N., Galante, M., Robson, M., Boeris, V., & Spelzini, D. (2018). Amaranth, quinoa and chia protein isolates: Physicochemical and structural properties. *International Journal of Biological Macromolecules*, *109*, 152–159. Available from https://doi.org/10.1016/j.ijbiomac.2017.12.080, http://www.elsevier.com/locate/ijbiomac.

Ma, Y., & Xiong, Y. L. (2009). Antioxidant and bile acid binding activity of buckwheat protein in vitro digests. *Journal of Agricultural and Food Chemistry*, *57*(10), 4372–4380. Available from https://doi.org/10.1021/jf803670u, http://pubs.acs.org/doi/pdfplus/10.1021/jf803670u.

Ma, Y., Xiong, Y. L., Zhai, J., Zhu, H., & Dziubla, T. (2010). Fractionation and evaluation of radical scavenging peptides from in vitro digests of buckwheat protein. *Food Chemistry*, *118*(3), 582–588. Available from https://doi.org/10.1016/j.foodchem.2009.05.024.

Malaki Nik, A., Wright, A. J., & Corredig, M. (2011). Impact of interfacial composition on emulsion digestion and rate of lipid hydrolysis using different in vitro digestion models. *Colloids and Surfaces B: Biointerfaces*, *83*(2), 321–330. Available from https://doi.org/10.1016/j.colsurfb.2010.12.001.

Malgor, M., Sabbione, A. C., & Scilingo, A. (2020). Amaranth lemon sorbet, Elaboration of a potential functional food. *Plant Foods for Human Nutrition*, *75*(3), 404–412. Available from https://doi.org/10.1007/s11130-020-00818-y, https://link.springer.com/journal/11130.

Malik, A. M., & Singh, A. (2022). Pseudocereals proteins - A comprehensive review on its isolation, composition and quality evaluation techniques. *Food Chemistry Advances*, *1*100001. Available from https://doi.org/10.1016/j.focha.2021.100001.

Martínez-Villaluenga, C., Peñas, E., & Hernández-Ledesma, B. (2020). Pseudocereal grains: Nutritional value, health benefits and current applications for the development of gluten-free foods. *Food and Chemical Toxicology*, *137*111178. Available from https://doi.org/10.1016/j.fct.2020.111178.

Mason, E., L'Hocine, L., Achouri, A., Pitre, M., & Karboune, S. (2020). Health promoting bioactive properties of novel hairless canary seed flour after in vitro gastrointestinal digestion. *Foods*, *9*(7), 932. Available from https://doi.org/10.3390/foods9070932.

Masood, R., & Khosravi-Darani, K. (2015). Biopeptides in milk: Opiate and antithrombotic effects. *Mini-Reviews in Medicinal Chemistry*, *15*(10), 872–877. Available from https://doi.org/10.2174/1389557515666150519104219, http://www.benthamdirect.org/pages/all_b_bypublication.php.

Minekus, M., Alminger, M., Alvito, P., Ballance, S., Bohn, T., Bourlieu, C., Carrière, F., Boutrou, R., Corredig, M., Dupont, D., Dufour, C., Egger, L., Golding, M., Karakaya, S., Kirkhus, B., Le Feunteun, S., Lesmes, U., MacIerzanka, A., MacKie, A., . . . Brodkorb, A. (2014). A standardised static in vitro digestion method suitable for food-an international consensus. *Food and Function*, *5*(6), 1113–1124. Available from https://doi.org/10.1039/c3fo60702j, http://www.rsc.org/Publishing/Journals/FO/about.asp.

Montoya-Rodrıguez, A., Mejıa, V. P., Dia, C., Reyes-Moreno, J., & Milan-Carrillo. (2014). Extrusion improved the anti-inflammatory effect of amaranth (Amaranthus hypochondriacus) hydrolysates in LPS-induced human THP-1 macrophage-like and mouse RAW 264.7 macrophages by preventing activation of NF-kβ signaling. *Molecular Nutrition Food Research*, *58*, 1028–1041. Available from https://doi.org/10.1002/mnfr.201300764.

Montoya-Rodríguez, A., Milán-Carrillo, J., Dia, V. P., Reyes-Moreno, C., & González de Mejía, E. (2014). Pepsin-pancreatin protein hydrolysates from extruded amaranth inhibit markers of atherosclerosis in LPS-induced THP-1 macrophages-like human cells by reducing expression of proteins in LOX-1 signaling pathway. *Proteome Science*, *12*(1). Available from https://doi.org/10.1186/1477-5956-12-30, http://www.proteomesci.com/content/12/1/30.

Morales, D., Miguel, M., & Garcés-Rimón, M. (2021). Pseudocereals: A novel source of biologically active peptides. *Critical Reviews in Food Science and Nutrition*, *61*(9),

1537–1544. Available from https://doi.org/10.1080/10408398.2020.1761774, https://www.tandfonline.com/loi/bfsn20.

Nardo, A. E., Suárez, S., Quiroga, A. V., & Añón, M. C. (2020). Amaranth as a source of antihypertensive peptides. *Frontiers in Plant Science*, *11*. Available from https://doi.org/10.3389/fpls.2020.578631, https://www.frontiersin.org/journals/plant-science.

Nasri, R., Amor, I. B., Bougatef, A., Nedjar-Arroume, N., Dhulster, P., Gargouri, J., Châabouni, M. K., & Nasri, M. (2012). Anticoagulant activities of goby muscle protein hydrolysates. *Food Chemistry*, *133*(3), 835–841. Available from https://doi.org/10.1016/j.foodchem.2012.01.101.

Ninomiya, K., Ina, S., Hamada, A., Yamaguchi, Y., Akao, M., Shinmachi, F., Kumagai, H., & Kumagai, H. (2018). Suppressive effect of the α-amylase inhibitor albumin from buckwheat (Fagopyrum esculentum Moench) on postprandial hyperglycaemia. *Nutrients*, *10*(10), 1503. Available from https://doi.org/10.3390/nu10101503.

Norris, R., & J, R. (2013). *Antihypertensive peptides from food proteins*. InTech. Available from http://doi.org/10.5772/51710.

Orona-Tamayo, D., & Paredes-López, O. (2017). *Amaranth part 1-sustainable crop for the 21st century: Food properties and nutraceuticals for improving human health sustainable protein sources* (pp. 239–256). Mexico: Elsevier Inc. Available from http://www.sciencedirect.com/science/book/9780128027783.

Orona-Tamayo, D., Valverde, M. E., Nieto-Rendón, B., & Paredes-López, O. (2015). Inhibitory activity of chia (Salvia hispanica L.) protein fractions against angiotensin I-converting enzyme and antioxidant capacity. *LWT - Food Science and Technology*, *64*(1), 236–242. Available from https://doi.org/10.1016/j.lwt.2015.05.033.

Orona-Tamayo, D., Valverde, M. E., & Paredes-López, O. (2017). *Chia—The new golden seed for the 21st century* (pp. 265–281). Elsevier BV. Available from http://doi.org/10.1016/b978-0-12-802778-3.00017-2.

Orsini Delgado, M. C., Nardo, A., Pavlovic, M., Rogniaux, H., Añón, M. C., & Tironi, V. A. (2016). Identification and characterization of antioxidant peptides obtained by gastrointestinal digestion of amaranth proteins. *Food Chemistry*, *197*, 1160–1167. Available from https://doi.org/10.1016/j.foodchem.2015.11.092, http://www.elsevier.com/locate/foodchem.

Orsini Delgado, M. C., Tironi, V. A., & Añón, M. C. (2011). Antioxidant activity of amaranth protein or their hydrolysates under simulated gastrointestinal digestion. *LWT - Food Science and Technology*, *44*(8), 1752–1760. Available from https://doi.org/10.1016/j.lwt.2011.04.002.

Ozón, B., Cotabarren, J., Valicenti, T., Graciela Parisi, M., & David Obregón, W. (2022). Chia expeller: A promising source of antioxidant, antihypertensive and antithrombotic peptides produced by enzymatic hydrolysis with alcalase and flavourzyme. *Food Chemistry*, *380*132185. Available from https://doi.org/10.1016/j.foodchem.2022.132185.

Paucar-Menacho, L. M., Simpalo-López, W. D., Castillo-Martínez, W. E., Esquivel-Paredes, L. J., & Martínez-Villaluenga, C. (2022). Improving nutritional and health benefits of biscuits by optimizing formulations based on sprouted pseudocereal grains. *Foods*, *11*(11). Available from https://doi.org/10.3390/foods11111533, https://www.mdpi.com/2304-8158/11/11/1533/pdf?version = 1653381910.

Qiao, M., Tu, M., Wang, Z., Mao, F., Chen, H., Qin, L., & Du, M. (2018). Identification and antithrombotic activity of peptides from blue mussel (Mytilus edulis) protein.

International Journal of Molecular Sciences, *19*(1), 138. Available from https://doi.org/10.3390/ijms19010138.

Qiao, Y., Gumpertz, M., & Van Kempen, T. (2002). Stability of pepsin (EC 3.4.23.1) during in vitro protein digestibility assay. *Journal of Food Biochemistry*, *26*(4), 355–375. Available from https://doi.org/10.1111/j.1745-4514.2002.tb00759.x, http://www.blackwellpublishing.com/journal.asp?ref = 0145-8884&site = 1.

Quiroga, A. V., Aphalo, P., Ventureira, J. L., Martínez, E. N., & Añón, M. C. (2012). Physicochemical, functional and angiotensin converting enzyme inhibitory properties of Amaranth (Amaranthus hypochondriacus) 7S globulin. *Journal of the Science of Food and Agriculture*, *92*(2), 397–403. Available from https://doi.org/10.1002/jsfa.4590.

Repo-Carrasco-Valencia, R. (2022). *Native crops in Latin America: Biochemical, processing, and nutraceutical aspects*. CRC Press.

Rodríguez, M. (2022). Evaluation of amaranth flour as antioxidant ingredient for the preparation of functional food matrices [Evaluación de harina de amaranto como ingrediente antioxidante para la preparación de matrices alimentarias funcionales].

Rodríguez, M., García Fillería, S. F., & Tironi, V. A. (2020). Simulated gastrointestinal digestion of amaranth flour and protein isolate: Comparison of methodologies and release of antioxidant peptides. *Food Research International*, *138*. Available from https://doi.org/10.1016/j.foodres.2020.109735, http://www.elsevier.com/inca/publications/store/4/2/2/9/7/0.

Rodríguez, M., & Tironi, V. A. (2020). Polyphenols in amaranth (A. manteggazianus) flour and protein isolate: Interaction with other components and effect of the gastrointestinal digestion. *Food Research International*, *137*109524. Available from https://doi.org/10.1016/j.foodres.2020.109524.

Sabbione, A. C., Ibañez, S. M., Martínez, E. N., Añón, M. C., & Scilingo, A. A. (2016). Antithrombotic and antioxidant activity of amaranth hydrolysate obtained by activation of an endogenous protease. *Plant Foods for Human Nutrition*, *71*(2), 174–182. Available from https://doi.org/10.1007/s11130-016-0540-y, http://www.kluweronline.com/issn/0921-9668/.

Sabbione, A. C., Luna-Vital, D., Scilingo, A., Añón, M. C., & González de Mejía, E. (2018). Amaranth peptides decreased the activity and expression of cellular tissue factor on LPS activated THP-1 human monocytes. *Food and Function*, *9*(7), 3823–3834. Available from https://doi.org/10.1039/c8fo00323h, http://pubs.rsc.org/en/journals/journal/fo.

Sabbione, A. C., Nardo, A. E., Añón, M. C., & Scilingo, A. (2016). Amaranth peptides with antithrombotic activity released by simulated gastrointestinal digestion. *Journal of Functional Foods*, *20*, 204–214. Available from https://doi.org/10.1016/j.jff.2015.10.015, http://www.elsevier.com/wps/find/journaldescription.cws_home/717426/description#description.

Sabbione, A. C., Scilingo, A., & Añón, M. C. (2015). Potential antithrombotic activity detected in amaranth proteins and its hydrolysates. *LWT - Food Science and Technology*, *60*(1), 171–177. Available from https://doi.org/10.1016/j.lwt.2014.07.015, http://www.elsevier.com/inca/publications/store/6/2/2/9/1/0/index.htt.

Sabbione, A. C., Suárez, S., Añón, M. C., & Scilingo, A. (2019). Amaranth functional cookies exert potential antithrombotic and antihypertensive activities. *International Journal of Food Science and Technology*, *54*(5), 1506–1513. Available from http://onlinelibrary.wiley.com/journal/10.1111/(ISSN)1365-2621.

Sandoval-Oliveros, M. R., & Paredes-López, O. (2013). Isolation and characterization of proteins from chia seeds (Salvia hispanica L.). *Journal of Agricultural and Food Chemistry*, *61*(1), 193–201. Available from https://doi.org/10.1021/jf3034978.

Sandoval-Sicairos, E. S., Milán-Noris, A. K., Luna-Vital, D. A., Milán-Carrillo, J., & Montoya-Rodríguez, A. (2021). Anti-inflammatory and antioxidant effects of peptides released from germinated amaranth during in vitro simulated gastrointestinal digestion. *Food Chemistry*, *343*. Available from https://doi.org/10.1016/j.foodchem.2020.128394, http://www.elsevier.com/locate/foodchem.

Scanlin, L., & Lewis, K. A. (2017). *Quinoa as a sustainable protein source: Production, nutrition, and processing sustainable protein sources* (pp. 223–238). Elsevier Inc. Available from http://www.sciencedirect.com/science/book/9780128027783.

Seo, E. J., Ngoc, T. M., Lee, S. M., Kim, Y. S., & Jung, Y. S. (2012). Chrysophanol-8-O-glucoside, an anthraquinone derivative in rhubarb, has antiplatelet and anticoagulant activities. *Journal of Pharmacological Sciences*, *118*(2), 245–254. Available from https://doi.org/10.1254/jphs.11123FP, http://www.jstage.jst.go.jp/article/jphs/118/2/245/_pdf.

Shi, Z., Hao, Y., Teng, C., Yao, Y., & Ren, G. (2019). Functional properties and adipogenesis inhibitory activity of protein hydrolysates from quinoa (Chenopodium quinoa Willd.). *Food Science & Nutrition*, *7*(6), 2103–2112. Available from https://doi.org/10.1002/fsn3.1052.

Shimizu, M., Sawashita, N., Morimatsu, F., Ichikawa, J., Taguchi, Y., Ijiri, Y., & Yamamoto, J. (2009). Antithrombotic papain-hydrolyzed peptides isolated from pork meat. *Thrombosis Research*, *123*(5), 753–757. Available from https://doi.org/10.1016/j.thromres.2008.07.005.

Sisti, M. S., Scilingo, A., & Añón, M. C. (2019). Effect of the incorporation of amaranth (Amaranthus mantegazzianus) into fat- and cholesterol-rich diets for Wistar rats. *Journal of Food Science*, *84*(11), 3075–3082. Available from https://doi.org/10.1111/1750-3841.14810, http://onlinelibrary.wiley.com/journal/10.1111/(ISSN)1750-3841.

Soares, R. A. M., Mendonça, S., de Castro, L. Í. A., Menezes, A. C. C. C. C., & Arêas, J. A. G. (2015). Major peptides from amaranth (Amaranthus cruentus) protein inhibit HMG-CoA reductase activity. *International Journal of Molecular Sciences*, *16*(2), 4150–4160. Available from https://doi.org/10.3390/ijms16024150, http://www.mdpi.com/1422-0067/16/2/4150/pdf.

Suárez, S., & Añón, M. C. (2019). Amaranth proteins emulsions as delivery system of angiotensin-I converting enzyme inhibitory peptides. *Food Hydrocolloids*, *90*, 154–161. Available from https://doi.org/10.1016/j.foodhyd.2018.11.046.

Szawara-Nowak, D., Bączek, N., & Zieliński, H. (2016). Antioxidant capacity and bioaccessibility of buckwheat-enhanced wheat bread phenolics. *Journal of Food Science and Technology*, *53*(1), 621–630. Available from https://doi.org/10.1007/s13197-015-2074-y, http://www.springerlink.com/content/121580/.

Taniya, M. S., Reshma, M. V., Shanimol, P. S., Krishnan, Gayatri, & Priya, S. (2020). Bioactive peptides from amaranth seed protein hydrolysates induced apoptosis and antimigratory effects in breast cancer cells. *Food Bioscience*, *35*100588. Available from https://doi.org/10.1016/j.fbio.2020.100588.

Thakur, P., Kumar, K., & Dhaliwal, H. S. (2021). Nutritional facts, bio-active components and processing aspects of pseudocereals: A comprehensive review. *Food Bioscience*, *42*101170. Available from https://doi.org/10.1016/j.fbio.2021.101170.

Tiengo, A., Faria, M., & Netto, F. M. (2009). Characterization and ACE-inhibitory activity of amaranth proteins. *Journal of Food Science*, *74*(5), H121–H126. Available from https://doi.org/10.1111/j.1750-3841.2009.01145.x.

Usman, M., Patil, P. J., Mehmood, A., Rehman, A., Shah, H., Haider, J., Xu, K., Zhang, C., & Li, X. (2022). Comparative evaluation of pseudocereal peptides: A review of their nutritional contribution. *Trends in Food Science & Technology*, *122*, 287–313. Available from https://doi.org/10.1016/j.tifs.2022.02.009.

Valenzuela Zamudio, F., & Segura Campos, M. R. (2022). Amaranth, quinoa and chia bioactive peptides: A comprehensive review on three ancient grains and their potential role in management and prevention of Type 2 diabetes. *Critical Reviews in Food Science and Nutrition*, *62*(10), 2707–2721. Available from https://www.tandfonline.com/loi/bfsn20.

Valicenti, T. (2022). *Obtención y caracterización de hidrolizados con actividad antitrombótica a partir de expeller de chía* (degree dissertation).

Vecchi, B., & Añon, M. C. (2009). ACE Inhibitory Tetrapeptides from *Amaranthus hypochondriacus* 11S Globulin. *Phytochemistry*, *70*(7), 864–870. Available from https://doi.org/10.1016/j.phytochem.2009.04.006.

Versantvoort, C. H. M., Oomen, A. G., Van De Kamp, E., Rompelberg, C. J. M., & Sips, A. J. A. M. (2005). Applicability of an in vitro digestion model in assessing the bioaccessibility of mycotoxins from food. *Food and Chemical Toxicology*, *43*(1), 31–40. Available from https://doi.org/10.1016/j.fct.2004.08.007.

Vilcacundo, R., Martínez-Villaluenga, C., & Hernández-Ledesma, B. (2017). Release of dipeptidyl peptidase IV, α-amylase and α-glucosidase inhibitory peptides from quinoa (Chenopodium quinoa Willd.) during in vitro simulated gastrointestinal digestion. *Journal of Functional Foods*, *35*, 531–539. Available from https://doi.org/10.1016/j.jff.2017.06.024, http://www.elsevier.com/wps/find/journaldescription.cws_home/717426/description#description.

Vilcacundo, R., Martínez-Villaluenga, C., Miralles, B., & Hernández-Ledesma, B. (2019). Release of multifunctional peptides from kiwicha (Amaranthus caudatus) protein under in vitro gastrointestinal digestion. *Journal of the Science of Food and Agriculture*, *99*(3), 1225–1232. Available from https://doi.org/10.1002/jsfa.9294, http://onlinelibrary.wiley.com/journal/10.1002/(ISSN)1097-0010.

Vilcacundo, R., Miralles, B., Carrillo, W., & Hernández-Ledesma, B. (2018). In vitro chemopreventive properties of peptides released from quinoa (Chenopodium quinoa Willd.) protein under simulated gastrointestinal digestion. *Food Research International*, *105*, 403–411. Available from https://doi.org/10.1016/j.foodres.2017.11.036, http://www.elsevier.com/inca/publications/store/4/2/2/9/7/0.

Wang, F., Yu, G., Zhang, Y., Zhang, B., & Fan, J. (2015). Dipeptidyl peptidase IV inhibitory peptides derived from oat (Avena sativa L.), buckwheat (Fagopyrum esculentum), and highland barley (Hordeum vulgare trifurcatum (L.) Trofim) proteins. *Journal of Agricultural and Food Chemistry*, *63*(43), 9543–9549. Available from https://doi.org/10.1021/acs.jafc.5b04016, http://pubs.acs.org/journal/jafcau.

Wei, Y., Liu, Y., Li, Y., Wang, X., Zheng, Y., Xu, J., Sang, S., & Liu, Y. (2022). A Novel antihypertensive pentapeptide identified in quinoa bran globulin hydrolysates: Purification, in silico characterization, molecular docking with ACE and stability against different food-processing conditions. *Nutrients*, *14*(12), 2420. Available from https://doi.org/10.3390/nu14122420.

Wei, Y. H., & Lee, H. C. (2002). Oxidative stress, mitochondrial DNA mutation, and impairment of antioxidant enzymes in aging. *Experimental Biology and Medicine*, *227*(9), 671–682. Available from https://doi.org/10.1177/153537020222700901, http://www.uk.sagepub.com/journals/Journal202180.

Wu, J., Huo, J., Huang, M., Zhao, M., Luo, X., & Sun, B. (2017). Structural characterization of a tetrapeptide from sesame flavor-type baijiu and its preventive effects against AAPH-induced oxidative stress in HepG2 cells. *Journal of Agricultural and Food Chemistry*, *65*(48), 10495–10504. Available from https://doi.org/10.1021/acs.jafc.7b04815, http://pubs.acs.org/journal/jafcau.

Xu, X., Luo, Z., Yang, Q., Xiao, Z., & Lu, X. (2019). Effect of quinoa flour on baking performance, antioxidant properties and digestibility of wheat bread. *Food Chemistry*, *294*, 87–95. Available from https://doi.org/10.1016/j.foodchem.2019.05.037, http://www.elsevier.com/locate/foodchem.

Yu, G., Wang, F., Zhang, B., & Fan, J. (2016). In vitro inhibition of platelet aggregation by peptides derived from oat (Avena sativa L.), highland barley (Hordeum vulgare Linn. var. nudum Hook. f.), and buckwheat (Fagopyrum esculentum Moench) proteins. *Food Chemistry*, *194*, 577–586. Available from https://doi.org/10.1016/j.foodchem.2015.08.058, http://www.elsevier.com/locate/foodchem.

Zheng, Y., Wang, X., Zhuang, Y., Li, Y., Tian, H., Shi, P., & Li, G. (2019). Isolation of novel ACE-inhibitory and antioxidant peptides from quinoa bran albumin assisted with an in silico approach: Characterization, in vivo antihypertension, and molecular docking. *Molecules*, *24*(24), 4562. Available from https://doi.org/10.3390/molecules24244562.

Zhu, W., Ren, L., Zhang, L., Qiao, Q., Farooq, M. Z., & Xu, Q. (2020). The potential of food protein-derived bioactive peptides against chronic intestinal inflammation. *Mediators of Inflammation*, *2020*. Available from https://doi.org/10.1155/2020/6817156, http://www.hindawi.com/journals/mi/.

Zielińska, D. (2019). *The bioaccessible reducing capacity of buckwheat-enhanced wheat breads estimated by electrochemical method*. IntechOpen. Available from http://doi.org/10.5772/intechopen.84716.

CHAPTER

10 Bioactive peptides released from legumes during gastrointestinal digestion

Cristina Martínez-Villaluenga[1], Elena Peñas[1], Martin Mondor[2,3], Ruixian Han[4] and Alan Javier Hernandez-Alvarez[4]

[1]*Institute of Food Science, Technology and Nutrition, Spanish National Research Council, Madrid, Spain*

[2]*Saint-Hyacinthe Research and Development Centre, Agriculture and Agri-Food Canada, Saint-Hyacinthe, QC, Canada*

[3]*Institute of Nutrition and Functional Foods (INAF), Université Laval, Quebec, QC, Canada*

[4]*School of Food Science & Nutrition, University of Leeds, Leeds, United Kingdom*

10.1 Introduction

Legumes are plants from the Fabaceae (or Leguminosae) botanical family that comprises nonoil seed (pulses) and oil-seed crops in both fresh and dried forms (Didinger & Thompson, 2021; Hughes et al., 2022). Pulses include lentil (*Lens culinaris* L.), bean (*Phaseolus* spp.), pea (*Pisum sativum* L.), chickpea (*Cicer arietinum* L.), fava bean (*Vicia faba* L.), lupin (*Lupinus* spp.), Bambara groundnuts (*Vigna subterranean*), cowpea (*Vigna unguiculata* L.), pigeon pea (*Cajanus cajan* L.), vetch (*Vicia sativa*), and other minor pulse crops. Oilseeds comprise soybean (*Glycine max* L.) and peanut (*Arachis hypogaea* L.) among others.

Annually, 89 million tons of legumes are produced, with Asia being the largest producer. Grain legumes occupy an important place in the diet worldwide, especially in low-income countries and in those regions in which animal protein intake is limited (nonavailability or by cultural/religious habits) due to their valuable nutritional properties. They are an economic source of fiber, proteins, minerals (iron, zinc, calcium), vitamins (thiamine, riboflavin, biotin, folic acid, and niacin) and are low in saturated fats (Hughes et al., 2022). Moreover, legumes are rich in low-digestible starch and are considered low-glycemic grains. Regular legume consumption has been shown to increase the diet's nutritional density (Mitchell et al., 2021) and has been associated with the prevention of cardiovascular diseases type 2 diabetes, and obesity (de Almeida et al., 2022; Jenkins et al., 2012; Jeong et al., 2019). These studies have increased the interest in including legumes in the diet and producing ingredients derived from them.

From an environmental perspective, legumes improve soil fertility by fixing atmospheric nitrogen due to the presence of nitrogen-fixing microorganisms in

Protein Digestion-Derived Peptides. DOI: https://doi.org/10.1016/B978-0-443-19141-1.00010-8

their nodules and can be cultivated by intercropping with little water use (Venkidasamy et al., 2019). Moreover, they align with the current trend related to reducing carbon emissions (Zhang & Chai, 2022). Due to their nutritional and environmental benefits, legumes are part of nutritious and sustainable dietary patterns (Hughes et al., 2022; Margier et al., 2018).

Legumes are excellent sustainable protein sources (17%–40%) and represent valuable alternatives to animal-based proteins. Legume proteins are mainly storage proteins categorized according to their solubility into globulins (soluble in low-salt solutions), albumins (water soluble), prolamins (ethanol/water soluble), and glutelins (soluble in alkaline solutions) (Juárez-Chairez et al., 2022; Sánchez-Chino et al., 2015). Legume proteins are rich in the essential amino acid lysine, but they have a low content of tryptophan and sulfur amino acids and exhibit lower digestibility and bioavailability than animal proteins due to their secondary structure. Food processing is a valuable approach for modifying legume proteins, consequently improving their digestibility. Different processing techniques have been successfully applied to improve protein digestibility and utilization including thermal treatments (cooking, pressure cooking, baking, autoclaving, microwave irradiation), fermentation, germination, extrusion, high-hydrostatic pressure, etc. (Drulyte & Orlien, 2019).

Several recent studies have shown that during gastrointestinal digestion, legume proteins can release peptides that might exert their biological functions through interaction with different receptors in the organism. These amino acid sequences, known as bioactive peptides, are inactive fragments within the precursor protein and exert their functional activities in vivo after their release through enzyme application (enzymatic hydrolysis) or use of microorganisms (fermentation), as well as during their gastrointestinal digestion in vivo. Compared to proteins of animal origin, the bioactivity of peptides derived from legume protein digestion is less understood (Asledottir et al., 2023). To exert their activity, resistant digestion peptides should be absorbed in intestine and transported via the bloodstream to reach their target receptors. These peptides can exert local effects by interacting with digestive receptors or systemic effects whether they resist the digestion process and are absorbed in the intestine, metabolized, and transported to target tissues/organs in an active form (Fernández-Tomé & Hernández-Ledesma, 2020).

Gastrointestinal digestion can be in vitro simulated, trying to mimic physiological conditions (agitation, temperature, pH, enzyme activity, and composition of gastric and intestinal fluids) and the sequence of events in the gastrointestinal system (oral, gastric, and intestinal). In fact, in vitro gastrointestinal digestion models are increasingly used to simulate the digestive process naturally occurring in the human body to elucidate how legume proteins are in vivo degraded under physiological conditions. Due to the diversity of factors affecting in vitro digestion, standardized simulated static (Brodkorb et al., 2019; Minekus et al., 2014) or semidynamic gastrointestinal digestion models (Mulet-Cabero et al., 2020) have been developed. Since these models are less invasive and less expensive than in vivo studies, they are the preferred methods to study the influence of digestive processes on protein digestibility (Makran et al., 2022). Several recent studies have demonstrated that peptides released during simulated gastrointestinal digestion from legume proteins can exert an array of bioactivities, including antihypertensive, antiinflammatory, antioxidant, antidiabetic, immune-stimulating, and anticarcinogenic (Asledottir et al., 2023; Correa et al., 2022; Fleury et al., 2022; González-Montoya et al., 2018; Martineau-Côté, Achouri, Wanasundara, et al., 2022).

Although the structure-activity relationship has not been yet fully clarified, many legume-bioactive peptides have common structural features, including short length (2–20 amino acids), presence of hydrophobic amino acid residues as well as of lysine, arginine, and proline (Carbonaro & Nucara, 2022; López-Barrios et al., 2014).

Scientific knowledge regarding the bioactive properties of legume peptides released by in vivo gastrointestinal digestion is nearly nonexistent. It should be taken into account that is very difficult to elucidate the biological activity of peptides released during digestion from legumes in the context of a diet since, on the one hand, legumes contain various bioactive compounds that may be responsible for the observed biological activity and, on the other hand, legumes are consumed with other foods that may also release bioactive peptides during digestion. Moreover, the pharmacokinetics of the peptides released during gastrointestinal digestion is not well known, so it is almost impossible to know if they are absorbed and transported to the target organ (Malaguti et al., 2014).

The conventional methods used for isolating bioactive peptides and evaluating their activities are time-consuming and expensive. Therefore during the last years bioinformatics approaches have emerged to speed up the analysis of bioactive peptides. Bioinformatics combines biological mass data using biology, computer science, mathematics, and statistical techniques to predict the physicochemical, bioactive, and sensory properties of bioactive peptides (Agyei et al., 2018).

This chapter provides an overview of the more recently available information regarding the production of bioactive peptides from legume proteins during gastrointestinal digestion. Special emphasis will be paid to the in vitro and in vivo evidence of the bioactivity of legume-derived peptides released during gastrointestinal processes as well as on the role of food processing to increase the digestibility of legumes. The advances and prospects of bioinformatics for exploring the structure and functions of legume-bioactive peptides formed during digestive processes will be also reviewed.

10.2 Legume proteins

Legumes are recognized sources of sustainable plant-based protein since their protein content ranges from 18% to 40% on a dry matter basis (Schutyser et al.,

2015; Vaz Patto et al., 2015; Venkidasamy et al., 2019). In recent years legume proteins have attracted much interest in the food industry due to their high nutritional value, affordability, and bioavailability.

According to their composition and function, legume proteins can be classified into simple (yield amino acids and sometimes small amounts of carbohydrates after their hydrolysis) or conjugated (attached to another chemical group). According to their function, legume proteins can be categorized into storage, structural, and biologically active proteins. Storage proteins are the most abundant in legumes, which are accumulated inside protein bodies during seed development and hydrolyzed during seed germination, providing ammonia, carbon, and amino acids to the emerging plant (Carbonaro & Nucara, 2022). The storage proteins can be further classified, according to their solubility: globulins (soluble in weak salt solutions), albumins (water soluble), prolamins (soluble in 60% ethanol), and glutelins (soluble in dilute alkali solutions). Globulins are the most abundant storage proteins in legumes that account for 60%–80% of the total proteins, while albumins account for 10%–30% of the total proteins (Eze et al., 2022). The relative amounts can differ notably between legume species (Table 10.1), and also due to the variety and growing conditions, being the albumin/globulin ratio as high as 0.5–0.6 (Sadeghi et al., 2023; Vogelsang-O'dwyer et al., 2022). In contrast, glutelins and prolamins are minor proteins in legumes.

Globulins are rich in aspartic and glutamic acids (Venkidasamy et al., 2019). They are classified based on their sedimentation coefficient as 7S and 11S, with sedimentation coefficients ranging from 7–9S and 10.5–13S, respectively. These proteins are categorized as vicilin-like (7S) and legumin-like proteins (Shrestha

Table 10.1 Storage protein fractions (%) in legumes.

Storage protein	Albumin	Globulin	Glutelin	Prolamin	References
Lentils	11–61	27–70	3–47	2–3	Boye et al. (2010), Neves and Lourenço (1995), Sulieman et al. (2008), Sadeghi et al. (2023)
Pea	10–31	55–80	3–4	4–5	Sadeghi et al. (2023), Lam et al. (2018), Lu et al. (2020)
Mung bean	25	60–62	13	0.9	Shrestha et al. (2023a), Yi-Shen et al. (2018)
Chickpea	8–12	53–60	19–25	3–7	Grasso et al. (2022)
Fava bean	1–3	70–80	12–18	2–4	Martineau-Côté, Achouri, Karboune, et al. (2022)
Winged bean	26	32	13	4	Makeri et al. (2017)
Soybean	33	35	12	4	Makeri et al. (2017)
Albumin	7–11	24–32	5–16		Gulewicz et al. (2008)

et al., 2023b). In soybean, globulins 7S and 11S are known as β-conglycinin and glycinin (Din et al., 2021), while in pulses they are named vicilin and legumin, respectively (Carbonaro & Nucara, 2022). Globulin 11S is a hexamer of a molecular weight of about 320−400 kDa, with 6 polypeptide subunits linked by noncovalent bonds. Each subunit contains an acidic (~40 kDa) and a basic (~20 kDa) polypeptide chain covalently linked (Shrestha et al., 2023b). Meanwhile, globulin 7S is commonly a trimer of about 150−190 kDa, containing various subunits linked by hydrophobic interactions and hydrogen bonds (Carbonaro & Nucara, 2022; Shrestha et al., 2023a; Vogelsang-O'dwyer et al., 2022). The 7S/11S ratio varies between legume species, variety, and cultivation conditions, from 0.4 to 2.78 in soybean, lentil, chickpea and pea proteins (Barac et al., 2010; De Santis et al., 2021; Nishinari et al., 2014; Sadeghi et al., 2023; Scippa et al., 2010; Vreeke et al., 2023). Generally, legumin-like proteins are richer in sulfur-containing amino acids, while vicilin-like proteins exhibit better techno-functional properties.

Regarding albumins, they mainly comprise metabolic proteins ranging from 5 to 80 kDa and include enzymatic and nonenzymatic proteins (Boye et al., 2010; Shevkani et al., 2019), containing high amounts of cysteine and methionine. Prolamins have high amount of proline and glutamine, while glutelins are characterized by higher concentration of methionine and cysteine than the other storage proteins (Boye et al., 2010).

Structural proteins are ribosomal, chromosomal, and membrane proteins (Duranti, 2006; Onyango, 2022), while biologically active proteins include lectins, lipoxygenase, enzymes, and enzyme inhibitors. Enzyme inhibitors are considered antinutritional factors because they negatively affect the digestive and absorption processes. The most important protein antinutritional compounds in legumes are trypsin and chymotrypsin inhibitors, lectins, and ribosome inactivating proteins (Eze et al., 2022). Legume protease inhibitors block digestive enzymes by competitive binding. There are two families of protease inhibitors: the Kunitz family and the Bowman−Birk family, which can be present in different isoforms depending on legume species and variety (Muzquiz et al., 2012). Their content varies from 6 to 84 trypsin inhibitor units/mg, depending on the legume type (Pedrosa et al., 2021). Protease inhibitors are resistant to acidic conditions and the action of proteolytic enzymes present in the gastrointestinal tract (Clemente et al., 2008). However, they are thermal-labile compounds not causing important digestive problems since usually they are cooked prior to consumption. Emerging scientific evidence suggests that protease inhibitors may also exert health-promoting properties including protective/suppressive effect in colorectal and other type of cancers development (Oliás et al., 2019; Panzeri et al., 2020; Pedrosa et al., 2021). On the other hand, lectins have been shown to bind to epithelial intestinal cells, negatively affecting digestion process (Gharibzahedi et al., 2022). In contrast, recent studies have shown that legume lectins can also exert anticancer and immunomodulatory activities (Adamcová et al., 2021; Katoch & Tripathi, 2021).

10.3 Legume protein digestion

10.3.1 In vitro digestion models

Understanding how legume proteins are digested is crucial to gain insights on their nutritional value. In addition, identification of protein breakdown products such as peptides and amino acids physiologically relevant may provide evidence for potential health effects (Santos-Hernández et al., 2020).

Legume digestibility was recently evaluated in the context of adult human nutrition using the harmonized in vitro INFOGEST protocol (Santos-Hernández et al., 2020). In this study, protein digestion of soybean, garden pea, grass pea, and lentil protein isolates was monitored as well as the identification of digestion products was carried out. Legume globulins were resistant to gastric digestion and subsequently hydrolyzed into amino acids and peptides in the small intestine. Large-size peptides (6–20 kDa) were detected at the end of gastric digestion whereas intestinal digests were characterized by 21%–24% of free amino acids (particularly leucine, lysine, phenylalanine, and arginine) and 66%–76% of small protein fragments up to 6 kDa, respectively. This distribution of the nitrogen fraction into free amino acids and peptides was similar to that found for milk proteins, except for soybean, where the percentage of insoluble nitrogen was higher (12%).

The study of Gilani and Sepehr demonstrated a significant impact of age in legume protein digestibility (Gilani & Sepehr, 2003). Specifically, in this study digestibility of fava bean protein was reduced from 82% to 77% in young rats as compared to older rats.

The immature newborn digestive system differs from the adults one in terms of the composition of digestive fluids, pH reached in different digestive compartments, and enzyme activities (Hamosh, 1996). Therefore the development of legume-based infant formulas requires an in vitro digestion system that mimics the newborn physiology to evaluate protein digestibility. This newborn digestion model was used in the investigations of Roux et al. who studied the impact of 50% milk replacement by plant-based proteins (pea, fava bean, rice, and potato) in the protein digestibility of infant formulas (Roux et al., 2020). Interestingly, the amino acid bioaccessibility and degree of protein hydrolysis in the infant formula enriched with legume proteins were close to the milk reference. However, milk protein replacement by rice and potato proteins reduced the protein digestibility in infant formulas, suggesting that legume proteins were good alternatives for formulation.

The digestibility of pea proteins was also studied by using in vitro digestion systems simulating elderly gastrointestinal conditions (Melchior et al., 2023). Aged gut conditions deeply affected pea proteolysis leading to a general decrease (10% reduction) of protein digestibility in comparison to the adult model. More recently Hernández-Olivas et al. studied the influence of oral, gastric and/or intestinal alterations appearing with ageing on the luminal digestion of different cereal/pseudocereal grains and legumes (Hernández-Olivas et al., 2021). In healthy gastrointestinal conditions, oat, spelt, and quinoa proteins were more digestible than chickpea, lentils, soya bean, and white bean; however, when intestinal conditions are suboptimal (as occurred in the elderly), digestion of cereal and pseudocereals was more compromised than in legumes. In addition, a preferential release of essential amino acids as compared to that of nonessential amino acids was observed in the elderly as compared to the adult digestion model.

10.3.2 In vivo digestion models

Currently the literature is primarily focusing on in vitro protein digestibility, the impact of food structure, and absorption of amino acid after GI digestion, while few studies have measured the release of legume peptides in vivo and clinical trials due to the complexities and a wide diversity of constraints, including costs, ethical issues, reproducibility, and extended duration of trials.

Recently a minimally invasive dual-isotope technique has been developed to determine the true ileal indispensable amino acid (IAA) digestibility of chickpea, yellow pea, and mung bean (hulled and dehulled) protein (Kashyap et al., 2019). The detection of labeled amino acids in plasma from intrinsically labeled (^{2}H) food protein is compared to simultaneously ingested labeled (^{13}C) spirulina protein of known digestibility. Results from this study revealed that the true ileal IAA digestibilities of legumes were similar (63.2% for mung bean) or even higher (74.6% and 71.6% for chickpea and yellow pea, respectively) than spirulina protein (64.0%). Moreover, dehulling improved the true ileal IAA digestibility of mung bean (70.9%).

Some papers used in vivo mastication (buccal digestion) together with in vitro digestion to investigate the digestibility of legume proteins. For example, Alpos et al. (2021) used this protocol and found up to twofold increase (rate constant increased from 0.046 to 0.093/min) in the digestion rate of cooked black bean after pulse electric field treatment (2.3 kV/cm, 134 kJ/kg, 20 μs, and 50 Hz), compared with the untreated one. They also highlighted that slow chewing (duration $>$ 29 s) lead to up to 51% more protein being hydrolyzed during GI digestion, compared with fast chewers (duration $<$25 s). With the similar methods, Sun et al. (2023) investigated peptides released from red kidney bean (fermentation/no-fermentation) during the GI digestion. They found an increase of 1.8-fold (from 7.1 to 12.7 mg/mL) in peptides level and 2.5-fold (from 435.1 to 1168.9 mg/100 g) in free amino acids level as a result of 29-h fermentation, because of a reduction in cotyledon cell well integrity.

Only a few studies used rat models to investigate the digestibility/nutritional response of consumed legumes. For example, Rubio and Clemente (2009) claimed only free Val (91 and 132 μM, compared with 56 μM) and Cys (56 and 56 μM, compared with 16 μM) in arterial plasma and free Cys (55 and 58 μM, compared with 29 μM) in portal plasma were found at higher levels in rats fed chickpea and lupin seeds (150 g protein/kg) compared to lactalbumin, which indicates legume proteins had slower digestion rate than animal proteins.

Similarly, all chickpea fractions with/without heat treatment (albumin, glutelin, globulin, and vicilin) (78.2%–94.7%) showed lower apparent digestibility than casein (96.1%–98.2%) (Tavano et al., 2008). In addition, Pirman et al. (2006) investigated the protein metabolism in growing rats fed cooked bean (229 g protein/kg) and cooked lentils (190 g protein/kg) for 20 days, with casein diet (193 g protein/kg) used as a control. They claimed that ingestion of cooked legumes (beans and lentils) increased absolute protein synthesis in small intestine (+246 mg/day and +211 mg/day) and large intestine (+148 and +118 mg/day). However, they had a negative effect on gastrocnemius muscles (−12 and −8 mg/day) and soleus muscles (−1.5 and −1.1 mg/day), compared to those of control groups (casein diet). Furthermore, protein content in fava bean extract was found to be positively correlated with superoxide dismutase activity (0.679), reduced glutathione level (0.805) and glutathione reductase activity (0.616) in colon tissues using rat models (El-Sayed et al., 2022).

A recent paper reported an increase of 49% and 83% in total amino acid in human serum after ingestion of bread, for which chickpea powder replaced 30% (20.2 g protein/100 g) and 60% (25.5 g protein/100 g) of white wheat bread flour, compared with the control (17.0 g protein/100 g) (Perez-Moral et al., 2023). Their findings indicate that legume proteins were digested to small fragments in human GI and degraded into free amino acid residues, which were found in human serum.

However, these papers were not specifically related to bioactive peptides. Therefore well-designed research for exploring and validating the release of legume peptides and their biological activity in vivo is needed, as these would provide more convincing data on the health benefits of legumes protein consumption.

10.3.3 Factors affecting legume protein digestibility

Intrinsic factors including legumes antinutritional content and protein structure influence protein digestibility (Kashyap et al., 2019). Legume proteins are less digestible than animal proteins due to the high concentration of antinutritional compounds. Protein complexes may be affected by the cross-linking effects of phytic acid, tannins, and polyphenols. This decrease in protein solubility makes the protein complexes less vulnerable to digestive tract proteolytic action.

The structural features of legume proteins play a role in their stability to digestion, as well as on the release of bioactive peptides. The tight globular shape of globulins, which is stabilized by disulfide bonds, creates an intrinsic structural barrier that prevents enzyme access (Drulyte & Orlien, 2019). In this regard, it was shown that pea seeds with a higher proportion of globulins and a lower proportion of albumins might help to boost protein digestibility (Drulyte & Orlien, 2019). Legume proteins mainly possess high content of β-structures (parallel and antiparallel β-sheet and β-turn) and low proportion of α-helices (Shevkani et al.,

2019), traits are also present in cereal grain proteins. For instance, soybean and lentil contain at least 30% of β-sheet in their secondary structure, while common bean and chickpea exhibited higher values (37%–44%). In contrast, α-helix constitute 18%–20% of the secondary structure of pulses and 12% in soybean (Carbonaro et al., 2012). Bai et al. (2016) observed a negative correlation between protein digestibility and the percentage of the β-sheet in protein secondary structures. Similarly, Carbonaro et al. (2012) demonstrated an inverse linear correlation between digestibility and β-sheet content of legume proteins attributed to the hydrophobic character of this secondary structure, which involves branched (Val, Thr, and Ile) and aromatic (Tyr, Trp, and Phe) amino acids. The abundance of hydrophobic amino acids between monomers in oligomeric legume proteins has been also suggested to decrease the susceptibility of legume proteins to digestive proteolysis. The high rigidity of the β-sandwich structure (two opposing antiparallel β-sheets) of lectins has been also hypothesized as the most important reason for the resistance of legume lectins to gastrointestinal digestion (Carbonaro et al., 2015). The abundance of cysteine residues and the formation of a high number of disulfide bonds in legume proteins such as Bowman–Birk inhibitor and glycinin are also related with the high resistance to gastrointestinal digestion to these proteins (Faris et al., 2008; Santos-Hernández et al., 2020). Moreover, the presence of glycan moieties in some legume proteins such as β-conglicinin of soybean also limits the accessibility of digestive enzymes to cleavage sites, thus reducing their digestibility (Amigo-Benavent et al., 2011; De Angelis et al., 2017).

Processing methods have been reported to affect the digestibility of legume seed proteins by reducing antinutritional factors (tannins, phytic acid, and trypsin inhibitors) as it has been extensively reviewed (Drulyte & Orlien, 2019). Processing methods can also disrupt the compact cell and protein structure and the entire matrix, thus improving or reducing their protein digestibility and the bioavailability of peptides and amino acids. In contrast, in recent years, novel pulse-based ingredients consisting exclusively of individual cotyledon cells have been developed to preserve the natural encapsulation of macronutrients and to ameliorate amylolysis during digestion and reduce the glycemic index of food products. The effect of several food processing techniques on legume digestibility has been discussed in detail in Section 12.4.

10.4 The effect of food processing on legumes protein digestibility

10.4.1 Mechanical disintegration

Three types of methods are applied for the mechanical disintegration of legumes: dehulling, splitting, and grinding. In dehulling, phenolic compounds, dietary fiber, and some minerals are partially removed improving protein digestibility (Kashyap et al., 2019). Dehulling and splitting do not induce the microstructural organization of legume seeds; however, milling causes the breakdown of parenchyma cells that favors the diffusion of intracellular material to digestive fluids and subsequently protein digestibility during digestion (Pallares et al., 2021). Recently legume powders produced by unconventional treatments are receiving much attention. These unconventional treatments allow the isolation of individual cells from pulse cotyledons through thermal processing of pulse seeds (60–100°C for 40–120 min), wet sieving, and drying (Pallares et al., 2021). The preservation of cellular integrity was associated to lower in vitro and in vivo starch digestibility compared to conventional legume flour. Lately, a few studies have assessed the in vitro protein digestion of isolated cotyledon cells from cooked legumes (Duijsens et al., 2022; Guevara-Zambrano et al., 2023; Gwala et al., 2020; Pallares Pallares et al., 2021). Results from these studies demonstrated that the protein digestion kinetic is higher in mechanically broken cotyledon cells than intact cotyledon cells, which are surrounded by the cell wall decreasing protein hydrolysis during digestion. Additionally, it has been reported that the reduction in protein hydrolysis observed for intact cotyledon cells occurred to a much lesser extent than starch hydrolysis and does not influence amino acid release in the small intestine. Indeed, Gwala et al. reported a high susceptibility of proteins to gastric digestion in Bambara groundnut cotyledon cells being protein hydrolysis even much more effective during the intestinal phase (Gwala et al., 2020). Recent studies support that digestive properties of cellular legume-based ingredients greatly depend on manufacturing conditions such as cooking time (Duijsens et al., 2022). Although cooking time, isolation technique and drying method increase susceptibility of amylolysis in cellular lentil powders, their effect on proteolysis seem to be limited.

10.4.2 Thermal treatment

Thermal treatment applied to legumes during cooking leads to protein denaturation and unfolding, exposing hydrophobic amino acid residues buried in the core of the native protein (Lefèvre et al., 2022). Protein aggregation may also occur during heating as new interactions are formed between unfolded proteins reducing the protease access to specific amino acids that are trapped inside the aggregate structures. Moreover, conformational changes in the secondary structure of legume proteins may occur upon thermal treatment. In soybean, heat processing produced an increase of α-helix and an overall decrease of β-sheets which is associated to increased protein digestibility (Neji et al., 2022). In contrast, Carbonaro et al. (2005) demonstrated that in vivo digestibility of isolated fava bean globulins (78.8%) was lower after heating at 120°C for 20 min than native globulins (95.1%) due to thermal-induced changes in protein secondary structure. Structural changes depend on the temperature and the moisture used during thermal treatment (Chávez-Murillo et al., 2018). As an example, fava bean protein digestibility increased from 76.2% to 82.4% when heated at 65°C (annealing) and 60% moisture but diminished to 73.3% at 120°C and 30% heat moisture (Chávez-Murillo et al., 2018). Furthermore, the α-helix: β-sheet increased following mild thermal conditions (65°C and 60% moisture) and decreased at 120°C and 30% heat moisture.

On the other hand, heat-sensitive antinutritional factors are known to be reduced during thermal processing; for instance, trypsin inhibitory activity is reduced up to 70% in pea heated at 95°C for 15 minutes (Hall & Moraru, 2022). Therefore heat-induced protein denaturation and antinutrient deactivation may enhance digestibility as a result of an increased susceptibility to digestive enzyme (Tang et al., 2022).

Plant-based infant formula production involves spray-drying, pasteurization, and sterilization that induces not only protein denaturation and aggregation but also glycation (Corzo-Martínez et al., 2010). Glycation blocks the lysine residues in polypeptide chains that are the specific cleavage sites of trypsin, and this may negatively impact protein digestion (Corzo-Martínez et al., 2010). Recently Tang et al. (2023) explored the effect of glycation on protein structures of soy and pea infant formulas and its impact for infant digestibility. In the presence of free glucose, dry heating (60 °C) negatively affected gastric digestion due to protein glycation and aggregation in legume infant formulas. However, the impact of glycation and aggregation on protein digestibility was smaller in the intestinal phase because of the broad specificity of pancreatic enzymes.

10.4.3 High-pressure processing

High-pressure processing (HPP) is a nonthermal process usually applied alternatively to pasteurization because unlike thermal processing, HPP preserves the natural sensory and nutritional attributes of the food matrix (Rastogi et al., 2007). Pressure levels used in typical HPP treatments (200–600 MPa) cause a rearrangement of protein molecular structures toward reduced volume states that are stabilized by hydrogen bonding and breakdown of electrostatic and hydrophobic interactions (Mozhaev et al., 1996). As such, HPP-induced structural and functional changes of food polymers have an impact in the protein susceptibility to digestive enzymes (Deshpande & Damodaran, 1989), and consequently in the digestibility of pressure-treated products. In this line, Hall and Moraru (2022) indicated that HPP (600 MPa/5°C/5 min) induced polypeptide chains dissociation, denaturation, and rearrangement of pea proteins structure which resulted in aggregation formed by weak noncovalent forces. Therefore aggregates generated during HHP treatment are easy to breakdown during digestion becoming proteins more susceptible to proteolysis. In other words, HHP enhanced gastric proteolysis but did not impact in the overall in vitro digestibility of pea protein concentrate.

10.4.4 Extrusion

Extrusion is used to texturize legume proteins from protein concentrates (Saldanha do Carmo et al., 2021) In this process, proteins are mixed with water, heated and sheared so as to form thin fibers. Extrusion process affects the food particle size as well as the secondary and tertiary structures of legume proteins (Chen et al., 2011). Moreover, new interactions of proteins with other food matrix components take place, for instance, protein-polysaccharide interactions (Nasrollahzadeh et al., 2023). Structural assembly within protein changes as consequence of the triple action of shear force, pressure, and high temperature causing protein denaturation and aggregation. Many studies provide evidence on the positive effect of extrusion on protein digestibility (Omosebi et al., 2018; Palanisamy et al., 2019). For instance, extrusion of lupin, soybean, and mung bean increased protein digestibility due to the inactivation of antinutrients and to the thermal denaturation of proteins. Mung bean extrudates showed improved digestibility attributed to increased α-helix and reduced β-sheet secondary structures. Nevertheless, the effect of extrusion on legume protein digestibility is highly dependent on processing factors including feed moisture and barrel temperature (Yu et al., 2013). An extrusion process with low feed moisture may produce an extensive protein denaturation and increased degree of aggregation and cross-linking. Increases in feed moisture during wet extrusion (from 55% to 68%) of lupin protein concentrate have been recommended to reduce protein aggregation and enhance protein digestibility and total amino acid content (Palanisamy et al., 2019; Webb et al., 2023). Barrel temperature increase may induce Maillard reaction and thermal cross-linking, which in turn decreases protein digestibility as described previously during the extrusion of mung bean protein isolates (Brishti et al., 2020).

10.4.5 Fermentation

Fermentation is commonly associated with the improvement of protein digestibility in legumes (Coda et al., 2015; Xiao et al., 2015). Lactic acid bacteria metabolism stimulates endogenous proteases that are involved in the hydrolysis of legume storage proteins to release amino acids and short peptides (Coda et al., 2015; Zhang et al., 2022). More recently several investigations demonstrated that fermentation affects the secondary structure of legume proteins (Dai et al., 2022; Yasar et al., 2020). Particularly, Liu et al. (2023) showed that lactobacillus fermentation of chickpea flour increased protein solubility and enhanced digestibility facilitating the degradation of spherical protein structures into looser states characterized by aromatic acid residues more exposed and a higher formation of intermolecular disulfide bonds. These changes were responsible for the effective hydrolysis of fermented chickpea proteins during digestion. The study of Sun et al. was the first to explain the mechanism by which solid-state fermentation enhances protein digestibility in intact cotyledon cells from kidney bean (Sun et al., 2023). Solid-state fermentation caused a disruption of cellular integrity, favored the diffusion of protein matrix, and promoted the enzymatic hydrolysis of major globulins. Subsequently, oral, gastric, and intestinal digestion of fermented kidney bean cotyledon cells showed a higher cell disintegration, proteolysis, and release of peptides and amino acids as compared to unfermented counterparts.

10.4.6 Germination

Germination is an ancient and sustainable biochemical process responsible for the chemical and structural changes that occur in grains generating new functional and nutritional components (Mir et al., 2021). Moreover, germination of legumes offers a great opportunity to improve protein digestibility as the process involves an enzymatic activation that results in seed structure disintegration, protein hydrolysis (Chen et al., 2019; Sjöö & Nilsson, 2017), and antinutrient degradation (Nosworthy et al., 2018). Improved digestibility of proteins in sprouted legumes may concomitantly facilitate the release of bioactive peptides, providing additional health benefits (Ohanenye et al., 2020). A recent study showed that seed permeability had an influence on the speed of the germination process and consequently on the degree of antinutrient and seed microstructure degradation and endogenous seed proteolysis, and ultimately in protein digestibility and the release of bioactive peptides. In particular, dehulled red lentil was characterized by a rapid imbibition rate, radicle emergence, and sprout development, which was associated with a greater cellular structure degradation, a lower antinutrient content, and a higher protein digestibility, and amino acid bioaccessibility as compared to black lentil, zero-tannin gray lentil and low tannin/zero vicine and convicine fava bean (Bautista-Expósito et al., 2021). In contrast, low tannin/zero vicine and convicine fava bean had a lower protein digestibility than lentil sprouts due to a slow germination, a higher antinutrient content, and higher cellular integrity. Peptidome analysis of intestinal digests from lentil and fava bean sprouts allowed the identification of digestion-resistant peptides with reported bioactivity (Bautista-Expósito et al., 2021). Interestingly, legume sprouts with higher protein digestibility showed a smaller number of total and bioactive peptides with a lower mean MW. In addition, lentil and fava bean vicilins were more resistant to digestion than legumins.

10.4.7 Protein extraction methods

Legume seeds often are used as raw materials for the production of protein-rich ingredients including protein flours (up to 65% protein), concentrates (65%–90% protein), and isolates (>90% protein) (Vogelsang-O'dwyer et al., 2022). Protein-rich legume flours are usually produced by dry fractionation method consisting of air classification and fractionation (Pelgrom et al., 2015), while protein concentrates and isolates are commonly obtained using alkaline extraction-isoelectric precipitation or ultrafiltration, and salt-induced extraction dilutive precipitation. All these protein extraction methods influence protein digestibility (Neji et al., 2022). In general, protein extraction methods result in the rearrangement of protein structure, exposure of cleavage sites, increased susceptibility to digestive enzymes and therefore improve protein digestibility (Neji et al., 2022). For instance, protein digestibility increased from 77% to 85% and from 72% to 76% for lentil and soybean after the production of protein isolates from legume flours

(Aryee & Boye, 2016; Nguyen et al., 2015). Nevertheless, protein digestibility may vary depending on the type of method applied for protein extraction. Falade and Akeem (2020) observed a lower digestibility for protein isolates from African mesquite beans obtained by isoelectric precipitation compared to micellization precipitation (Falade & Akeem, 2020). This was attributed to alkaline pH reached in isoelectric precipitation resulting in cross-linking of amino acids, formation of lysinoalanine and dehydroalanine via serine and cysteine degradation.

Furthermore, protein extraction has been found to reduce the content of antinutritional factors that might subsequently lead to improved digestibility (Neji et al., 2022). Alkaline extraction, isoelectric precipitation, and ultrafiltration seemed to lower trypsin inhibitors in lentil protein concentrates and isolates (Barbana & Boye, 2013; Joehnke et al., 2021). Moreover, the size of the protein bodies seems to be associated with their digestibility. Protein extracted from lentils using isoelectric precipitation had a lower amount of large molar mass proteins ($>$699 kDa), compared with the dry-milled flour, which is often less digestible (Joehnke et al., 2021).

10.5 Bioactivity of legume peptides/protein hydrolysates over digestion

Gastrointestinal digestion plays an important role in releasing peptide fragments from protein in food mixtures. Briefly, proteases in the gastric phase, that is, pepsin, and intestinal phase, that is, trypsin and chymotrypsin, or pancreatin, in a sequential order hydrolyze protein into peptides or free amino acids. Various studies use a combination of these proteases to simulate the protein digestion process in the human body to investigate the release and stability of peptides. However, peptide profiles in digestates will be largely affected by protease activity, processing conditions and residence time at each digestion stage. Also, when simulating physiological digestion, chemical and mechanical processing should also be accurately simulated for the digestion of flours and protein ingredients in order to mimic mastication and peristaltic movements which have an impact on digestion rate and absorption. In vivo models are always a gold standard for better understanding close physiological digestion conditions, but these face a wide variety of challenges including complicated, expensive, and time-consuming procedures. Therefore a rapid, straightforward, and cost-effective protocol is needed, which could also provide an opportunity to investigate the change of peptide profiles during oral, gastric, and intestinal phases, respectively (Zhou et al., 2023). Recently a standard static digestive protocol "INFOGEST" was published, which standardized salt concentrations and enzyme activities involved in this type of research (Brodkorb et al., 2019). Researchers have employed the INFOGEST protocol to hydrolyze legume proteins in order to obtain more convincing and reproducible findings (Nitride et al., 2022). In this section, the biological activities (including antioxidant, antihypertensive, antidiabetic, antiinflammatory, antiallergic, and cholesterol-lowering) of legume proteins after subjected to gastrointestinal digestion will be reviewed. Most relevant studies only evaluated the biological activities at the protein digestate level, as a mixture, but here insight on the bioactive peptides that have been individually identified after digestion will be provided.

Antioxidant activities of different digested legume proteins have been reported in several studies. Cowpea protein hydrolysates showed oxygen radical absorbance capability (ORAC) (~480 μM Trolox eq, TE/g protein), ferric reducing antioxidant power (FRAP) (~16 mM TE/g protein), as well as superoxide radical scavenging activity (~16%), while the antioxidant capability (radical scavenging and reducing power) is negatively affected by thermal (90°C, 30 min) and ultrasonic (25% amplitude, 10 min) pretreatment (Quansah et al., 2013). Researchers also found that chickpea protein hydrolysates presented copper chelating activity (~28%) (Torres-Fuentes et al., 2011) and iron chelating activity (~10%) (Torres-Fuentes et al., 2012) after pepsin and pancreatin treatment. Stanisavljević et al. (2013) claimed protein hydrolysates from three pea and five grass pea cultivars presented DPPH scavenging activity (7.0%–9.2%) and H_2O_2 scavenging activity (15.3%–20.6%), while the digested legume flours presented stronger antioxidant activity, which is largely contributed by the presence polyphenol compounds (0.3–1.2 mg of gallic acid/g sample). The same finding was also observed for common bean digestion enhanced Folin-Ciocalteu reducing capability (FC) (up to ~100 GAE/g), DPPH radical scavenging activity (up to ~90 μM TE/g) and FRAP (up to ~20 μM TE/g) in most common bean protein hydrolysates regardless of the seed preparation conditions (two cultivars, cooked or germinated for 36/72 h) (Rossi et al., 2022).

At cellular level, germinated albumins and globulins fractions of chickpea protein presented 21.2% and 35.0% cellular antioxidant activity, and after adding 2 mg Na_2SeO_3 /100 g seeds, these percentages increased to 35.7% and 56.2%, respectively (Serrano-Sandoval et al., 2019). Cellular antioxidant activity was also found in digested common bean, and two fractions (phaseolin and lectin). After pepsin and pancreatin hydrolysis, cellular antioxidant capabilities were found in protein isolate hydrolysates (29%–73%), phaseolin hydrolysates (80%–85%) and lectin hydrolysates (27%–46%). These hydrolysates also presented copper chelating activity (up to 30%, up to 35% and up to 50%, respectively) and iron chelating activity (5%–23%, up to 81% and 7%, respectively) (Carrasco-Castilla et al., 2012b). The authors also claimed that digested common bean and its phaseolin fraction presented antioxidant activity via β-carotene blenching (73.0% and 69.5%), reducing power (0.3 and 0.1), ABTS scavenging capability (1.8 and 5.5 mM TEAC/mg protein), copper chelating (50.5% and 37.5%) and iron chelating (22.9% and 81.2%) activity (Carrasco-Castilla et al., 2012a).

Legume peptides possessing antihypotensive activities were also reported in the literature. Boschin et al. (2014) used pepsin to release ang iotensin-converting enzyme (ACE) inhibitory peptides from chickpea (IC_{50}: 673 μg/mL), common bean

(IC_{50}: 633 μg/mL), lentil (IC_{50}: 606 μg/mL), lupin (IC_{50}: 226–268 μg/mL), pea (IC_{50}: 595 μg/mL), and soybean (IC_{50}: 224 μg/mL), which proves the possibility of ACE inhibitory peptides released from digestion. After pepsin-pancreatin hydrolysis, the ACE inhibitory activity of heat-treated (121.1°C, 15–50 min) common beans, pinto beans, and green lentils was promoted 1.3–2.5, 2.2–4.7, and 10.9–23.5 times, respectively. However, the effect of heat treatment was not discussed in their paper (Akıllıoğlu & Karakaya, 2009). Rui et al. (2012) also claimed that in vitro gastrointestinal digestion could release new ACE inhibitory peptides from navy beans, black beans, and small red beans, pretreated with alcalase/papain cohydrolysis. Similar findings were also shown by (Pertiwi et al., 2020)'s work, ACE inhibition of koro kratok (*Phaseolus lunatus L.*) increased from 26.4% to 49.3% after pepsin-pancreatin hydrolysis. Similarly, soy protein gained ACE inhibition (IC_{50} 0.3 mg/mL) after 180 min of digestion (Lo & Li-Chan, 2005).

Apart from antioxidant and antihypertensive properties, some studies also claimed antidiabetic activities for digested legume protein hydrolysates. For instance, IC_{50} value of Dipeptidyl Peptidase-IV (DPP-IV) inhibitory activity of germinated soybean was 1.5 mg/mL after pepsin and pancreatin hydrolysis, with similar effects found in those treated with α-amylase (1.7 mg/mL) and α-glucosidase (2.9 and 3.7 mg/mL) (González-Montoya et al., 2018). In addition, DPP-IV (IC_{50}: 0.2–2.2 mg/mL) inhibition and α-glucosidase inhibition (32%–66% at 10 mg protein/mL) were found in chickpea (five varieties) after pepsin-pancreatin digestion (Martínez & Mejia, 2021). Antiinflammatory activities were also reported for legume digestates. In (Milán-Noris et al., 2018)'s study, average IC_{50} value of nitric oxide (NO) inhibition in lipopolysaccharide (LPS)-stimulated RAW 264.7 macrophages of digested desi and kabuli chickpea was 2.0 mg/mL (germination pretreatment) and 3.6 mg/mL (cooking pretreatment). They also identified 24 peptides (including VLLEEQEQKPK, DVGQKTKEVGQDIQAK, LHQNIGSSSSPDIYNPQAGRIK and LHQNIGSSSSPDIYNPQAGR) from the most promising peptide fraction, which had an IC_{50} value of 93 μg/mL. In BALB/c mouse macrophages, the fraction with M_w 5–10 kDa in digested *Mucuna pruriens L.* was able to modulate inflammation via reducing NO (50.2%), H_2O_2 (61.0%), Interleukin-6 (IL-6) (60.3%) and tumor necrosis factor-alpha (TNF-α) (69.5%) level (Martínez Leo et al., 2018).

Even the antiallergic activity of protein hydrolysates obtained using gastrointestinal enzymes has been reported for legume proteins. The highest inhibition of β-hexosaminidase (80.3%) was found in pancreatin treated mung bean hydrolysates. However, this paper employed pepsin, trypsin and pancreatin separately, rather than a combination or sequential hydrolysis. Therefore when a combination of enzymes acts on mung bean proteins, antiallergenic activity of the resulted peptides might be different from what was reported (Budseekoad et al., 2019). On the other hand, Marques et al. (2018) claimed that cowpea bean peptides with M_w < 3 kDa have significant modulation effect on the expression of cholesterol-related genes (ATP-binding cassette transporters A1, ABCA1) (up to ~ 85%) and Niemann-pick C1-like 1 (NPC1L1) (up to ~ 90%) at 2.5 and 5.0 mg/mL concentration in Caco-2 cells, and

MELNAVSVVHS from cowpea bean protein have significant modulation effect on the sterol regulatory element binding protein 2 (SREBP2) (up to ~ 70%), HMG CoA reductase (HMGCR) (up to ~ 60%) and (LDL receptors) LDLR (up to ~ 75%) mRNA levels at 50 μM and 100 μM.

During GI digestion, a large number of peptides were released from parental proteins. Peptides tend to have various biological activities. Therefore it is highly possible that legume digests would present more than one biological activities. Table 10.2 summarized the multiple biological activities of legume protein hydrolysates after gastrointestinal digestion with/without pretreatments. As shown in Table 10.2, antioxidant and ACE inhibitory activities are the two most frequent biological activities studied in legume digestates. After GI digestion, an increase in DPP-IV and ACE inhibitory activity was found in pea (IC_{50} from 2.3 mg/mL to 1.5 mg/mL, IC_{50} from ND to 29.5 μg/mL) and fava bean (IC_{50} from ND to 0.5 mg/mL, IC_{50} from ND to 52.0 μg/mL) (Dugardin et al., 2020).

It is noticed that the digestion does not always promote biological activities, especially in prehydrolyzed proteins. Savinase-treated lentil showed a decreased ACE inhibition, but increased ORAC activity (Garcia-Mora et al., 2015; García-Mora et al., 2017). Similarly, alcalase and themolysin-treated Bambara protein hydrolysates showed decrease in ACE inhibition (−24.5% and −4.0%), DPP-IV inhibition (−35.3% and −16.4%) as well as Ferrous chelating activity (−0.25 and −0.40 mg TE/mg) after digestion (Mune et al., 2018).

So far, several legume peptides released via gastrointestinal digestion have been sequenced and reported. Eight peptides were identified in digested fermented bean (*Phaseolus vulgaris L.*) (INEGSLLLPH, FVVAEQAGNEEGFE, SGGGGGGVAGAATASR, GSGGGGGGGFGGPRR, INEGSLLLPH, GGYQGG GYGGNSGGGYGNRG, GGSGGGGGSSSGRRP, and GDTVTVEFDTFLSR), among which INEGSLLLPH is labeled as an ACE inhibitor and FVVAEQAGEEFGE and GDTVTVEFDTFLSR as alpha-amylase inhibitors (Jakubczyk et al., 2017). Olagunju et al. (2022) identified and synthesized 10 pigeon pea peptide sequences (GSH, VVSVP, SGEKI, KAVSE, TKDIG, ISTVN, TSRLG, IGVKSI, GRIST, AGVTVS, and AGSVLS). Among them, AGVTVS presented the most potential biological activity, including ACE inhibition (~100%), renin (~82%), hydroxyl radical scavenging activity (66.81%), superoxide anion radical scavenging activity (~100%), and FRAP (0.008 mM Fe^{2+}/mg).

10.6 The role of bioinformatic analysis in legume-based bioactive peptide release from gastrointestinal digestion

As shown in Fig. 10.1, bioinformatic tools, such as BIOPEP (https://biochemia.uwm.edu.pl/en/biopep-uwm-2/) and Peptide Cutter (https://web.expasy.org/peptide_cutter/) provide great opportunities for predicting the peptides release from

Table 10.2 Summary of multiple biological activities of legume peptides/protein hydrolysates released from gastrointestinal digestion.

Protein sources	Pretreatment	The major biological activities detected after GI digestion	References
Common bean (Black, Pinto, Red, Navy, and Great Northern) (*Phaseolus vulgaris*)	Raw or precooked (boiling for 15 min)	ORAC • The difference in antioxidant activity between precooked and raw beans was only observed in black bean (611 mM TE/g protein and 340 mM TE/g protein) and pinto bean (468 mM TE/g protein and 298 mM TE/g) • Other beans: 300–500 mM TE/g protein ACE inhibitory activity • No significant difference was observed between precooked and raw beans • IC_{50} of bean hydrolysates ranges from 0.2 to 0.5 mg protein/mL DPP-IV inhibitory activity • Precooked beans showed a higher DPP-IV inhibitory activity • Lowest IC_{50} value = 0.093 and 0.095 mg protein/mL (Navy bean, precooked and raw, most promising) • IC_{50} value of other beans: ~1 to ~0.1 mg protein/mL α-Amylase inhibitory activity • Precooked beans showed reduced α-amylase inhibitory activity, in comparison to raw beans, with the exception of pinto bean, which has no significant difference	Mojica, and Chen, and de Mejía (2015)

		• Raw Red (36% inhibition relative to acarbose/mg protein), Navy (31% inhibition relative to acarbose/mg protein), and Great Northern beans (27% inhibition relative to acarbose/mg protein) • Raw beans: ~5% to ~23% inhibition relative to acarbose/mg protein • Precooked beans: ~15% to 36% inhibition relative to acarbose/mg protein α-Glucosidase inhibitory activity • Precooked black beans showed lower α-glucosidase inhibitory activity, in comparison to raw beans • No significant difference was observed among other beans • Raw beans: 60%–70% inhibition relative to acarbose/mg protein • Precooked beans: 40%–60% inhibition relative to acarbose/mg protein	
Lentil seeds (*Lens culinaris* var. Castellana)	Savinase 16L (pH 8, 40°C for 2 h)	ORAC • Increased 10–14 μM TE/μM peptides ACE inhibitory activity • Decreased 11–21 μM IC_{50} value of peptides	García-Mora et al. (2017)

(Continued)

Table 10.2 Summary of multiple biological activities of legume peptides/protein hydrolysates released from gastrointestinal digestion. *Continued*

Protein sources	Pretreatment	The major biological activities detected after GI digestion	References
Adzuki bean (*V. angularis*)	Soaking (for 12 h) and boiling (for 20 min), then protein sequential fractionation (albumins, globulins, prolamins, and glutelins)	ACE inhibitory activity • Crude extract (IC_{50}: 21.2 mg/mL); albumins (IC_{50}: 6.2 mg/mL); globulins (IC_{50}: 1.0 mg/mL); prolamins (IC_{50}: 0.17 mg/mL); and glutelins (IC_{50}: 3.2 mg/mL) ABTS radical scavenging activity • Crude extract (IC_{50}: 0.7 mg/mL); albumins (IC_{50}: 0.9 mg/mL); globulins (IC_{50}: 0.5 mg/mL); prolamins (IC_{50}: 0.1 mg/mL); and glutelins (IC_{50}: 1.1 mg/mL) DPPH radical scavenging activity • Crude extract (IC_{50}: 0.4 mg/mL); albumins (IC_{50}: 0.8 mg/mL); globulins (IC_{50}: 0.1 mg/mL); prolamins (IC_{50}: 0.02 mg/mL); and glutelins (IC_{50}: 1.7 mg/mL) Fe^{2+}-chelating activity • Crude extract (IC_{50}: 16.9 mg/mL); albumins (IC_{50}: 72.4 mg/mL); globulins (IC_{50}: 5.4 mg/mL); prolamins (IC_{50}: 2.5 mg/mL); and glutelins (IC_{50}: 34.0 mg/mL) Cu^{2+}-chelating activity • Crude extract (IC_{50}: 9.1 mg/mL); albumins (IC_{50}: 36.7 mg/mL); globulins (IC_{50}: 5.2 mg/mL); prolamins (IC_{50}: 0.9 mg/mL); and glutelins (IC_{50}: 16.9 mg/mL)	Durak et al. (2013)

Common bean (*Phaseolus vulgaris* L. var Eureka)	Soaking (for 12 h) and fermentation (*L. plantarum 299v* at 22, 30, and 37°C) and time (3 h, 3 days, and 7 days),	Fermentation conditions selected for producing peptides with the most promising biological activities • Fermentation at 22°C for 3 h, 3.5–7 kDa • Fermentation at 30°C for 3 days, 3.5–7 kDa ACE inhibitory activity • IC_{50} of peptide fraction I–IV (22°C for 3 h, 3.5–7 kDa): 0.4–10.7 μg/mL • IC_{50} of Peptide fraction 3 and 4 (30°C for 3 days, 3.5–7 kDa): 0.4 and 0.5 μg/mL. ACE inhibitory activity was not detected in other two fractions α-Amylase inhibitory activity • IC_{50} of peptide fraction I–IV (22°C for 3 h, 3.5–7 kDa): 0.04–0.65 μg/mL • IC_{50} of peptide fraction 1–4 (30°C for 3 days, 3.5–7 kDa): 0.5–2.1 μg/mL Lipase inhibitory activity • IC_{50} of peptide fraction I–IV (22°C for 3 h, 3.5–7 kDa): 1.0–10.2 μg/mL • IC_{50} of peptide fraction 1–4 (30°C for 3 days, 3.5–7 kDa): 0.1 and 15.2 μg/mL	Jakubczyk et al. (2017)
Dry common bean (*Phaseolus vulgris*)	Subtilisin A (pH 8.0, 25°C for 24 h)	Iron chelating activity • Common bean flour (BF): +0.08 mg EDTA/g; common bean hydrocolloid extract (BHE): +2.74 mg EDTA/g; common bean protein water extract (BP1): +13.93 EDTA/g, and common	Aydemir et al. (2022)

(*Continued*)

Table 10.2 Summary of multiple biological activities of legume peptides/protein hydrolysates released from gastrointestinal digestion. *Continued*

Protein sources	Pretreatment	The major biological activities detected after GI digestion	References
		bean protein PBS extract (BP2): +26.83 EDTA/g. Free radical scavenging activity • BF: +19.4 μM TE/g; BHE: +177.2 μM TE/g; BP1: +1761.9 μM TE/g; and BP2: + 1122.8 μM TE/g ACE inhibitory activity • BF: −26.2%; BHE: −26.6%; BP1: −36.6%; and BP2: −0.2% α-Amylase inhibitory activity • BF: +37.3%; BHE: +25.1%; BP1: −15.1% and BP2: −29.8% α-Glucosidase inhibitory activity • BF: +83.5%; BHE: +63.7%; BP1: +80.2% and BP2: +74.5% Antimicrobial activity • BP1: −1.1 mm (*S. aureus*) and +1.64 mm (*E. coli*), and BP2: −1.83 mm (*S. aureus*) and +1.95 mm (*E. coli*)	
Lentil (*Lens culinaris* var. Meritt), black turtle bean (*Phaseolus vulgaris* var. Turtle) and black soybean (*Glycine max*)	Soaking (lentil, 1 h; black turtle, 1 or 2 h and black soybean, 4 h) Boiling or steaming under regular, 5 psi (108°C) and 15 psi (121°C) for 5, 10, 15, 20, 30 and 40 min	DPPH radical scavenging • A reduction in DPPH radical scavenging were found in all digested legumes ACE inhibitory activity • Lentil (16.1%–31.0%) (strongest ACE inhibition: 15 psi steaming for 40-min treatment)	Zhang and Chang (2019)

		• Black soybean (27.8%–47.0%) (strongest ACE inhibition: 0 psi boiling for 90-min treatment) • Black turtle bean (7.78%–44.1%) (strongest ACE inhibition: 5 psi boiling for 10-min treatment)	
Green gram (*Vigna radiata*), horse gram (*Macrotyloma uniflorum*), lentil (*Lens culinaris*), chickpea (*Cicer arietinum*), cowpea (*Vigna unguiculata*), black pea (*Pisum sativum*) and white pea (*Pisum sativum*)	Raw, soaking (for 16 h); germination (24, 48, and 72 h), and soaking followed by pressure cooking (15 lb psi at 121°C) for 10 min	ACE inhibitory activity • IC_{50} values decreased 0.1–0.3 mg/mL (lentil, horse gram and white pea) and 0.5–0.7 mg/mL (green gram, cowpea, chickpea and black bean) after digestion • Processing increased ACE inhibition in green gram (soaking), cowpea (soaking, germination 24 h, germination 48 h, germination 72 h and soaking followed by cooking), Black bean (soaking and germination 24 h) and white pea (soaking, germination 24 h, germination 48 h, and soaking followed by cooking) ABTS radical scavenging activity • Two- to fourfold increase after digestion • Processing increased ABTS radical scavenging activity in chickpea (soaking and germination 48 h) and black pea (soaking) Total antioxidant capability • Two- to threefold increased after digestion • Processing increased total antioxidant capability in green gram (germination	Jamdar et al. (2017)

(Continued)

Table 10.2 Summary of multiple biological activities of legume peptides/protein hydrolysates released from gastrointestinal digestion. *Continued*

Protein sources	Pretreatment	The major biological activities detected after GI digestion	References
		48 h and germination 72 h), lentil (germination 24 h, germination 48 h, and germination 72 h), cowpea (germination 24 h, germination 48 h, and germination 72 h), and white pea (germination 24 h and germination 72 h) Ferric ion reducing antioxidant power • Increased (chickpea, black pea and white pea) after digestion • No processing increased ferric ion reducing antioxidant power Metal ion chelating activity • All increased after digestion, especially black pea 2.5- to 3-fold, pea twofold • Processing increased metal ion chelating activity in lentil (soaking) and white pea (soaking) ROS scavenging activity • Decreased after digestion was found in black pea, white pea, or chickpea • Processing increased ROS scavenging activity in green gram (soaking, germination 24 and 72 h) and horse gram (soaking, germination 24 h and soaking followed by cooking).	

Pigeon pea	n/a	DPPH radical scavenging activity • 41.0% at 1 mg/mL Hydroxyl radical scavenging activity • 16.2% at 1 mg/mL Superoxide anion radical scavenging activity • 33.5% at 1 mg/mL Metal ion chelating activity • 11.2% at 1 mg/mL FRAP • 0.04 mM Fe^{2+}/mg at 1 mg/mL ACE inhibitory activity • 61.8% at 1 mg/mL Renin inhibitory activity • 10.6% at 1 mg/mL	Olagunju et al. (2022)
Lentil seeds (*Lens culinaris* var. Castellana)	High pressure (300 MPa), Savinase (pH 8, 15 min) and ultrafiltration (M_w < 3 kDa)	ACE inhibitory activity • IC_{50} value increased from 0.2 to 0.3 mg peptide/mL ORAC • Twofold increase after digestion	Garcia-Mora et al. (2015)
Bambara bean (*Vigna subterranea*)	Alcalase (pH 7, 55°C); thermolysin (pH 7, 55°C); and trypsin (pH 8, 70°C) for 24 h	ACE inhibitory activity • Alcalase (−24.5%); thermolysin (−4.0%); and trypsin (+3.2%) at 100 μg/mL DPP-IV inhibitory activity • Alcalase (−35.3%); thermolysin (−16.4%); and trypsin (+21.3%) at 1 mg/mL	Mune Mune et al. (2018)

(*Continued*)

Table 10.2 Summary of multiple biological activities of legume peptides/protein hydrolysates released from gastrointestinal digestion. *Continued*

Protein sources	Pretreatment	The major biological activities detected after GI digestion	References
		DPPH radical scavenging activity • Alcalase (+0.6 μg TE/mg); thermolysin (+3.1 μg TE/mg); and trypsin (−0.9 μg TE/mg) Ferrous chelating activity • Alcalase (−0.3 mg TE/mg); thermolysin (−0.4 mg TE/mg); and trypsin (−0.9 mg TE/mg)	
Adzuki bean (*Vigna angularis*)	Soaking (for 6 h) and then boiling (60 min); autoclaving (121°C for 10 min); autoclaving (121°C for 60 min)	ABTS radical scavenging activity • 98.9 (boiling); 115.4 (10-min autoclaving) and 123.7 (20-min autoclaving) mg Trolox per 100 g raw seed FRAP • 5.5 (boiling); 6.8 (10-min autoclaving) and 7.7 (60-min autoclaving) mg Trolox per 100 g raw seed Ferrous chelating activity • 0.5 (boiling); 0.5 (10-min autoclaving) and 0.5 (60-min autoclaving) mg EDTA per 100 g raw seed ORAC • 24.1 (boiling); 26.4 (10-min autoclaving); and 19.5 (60-minute autoclaving) mg Trolox per 100 g raw seed	Sangsukiam and Duangmal (2022)

		ACE inhibitory activity • 31.6 (boiling), 46.6 (10-min autoclaving), and 60.6 (60-min autoclaving) % per 100 g raw seed DPP-IV inhibitory activity • 14.8 (boiling), 25.8 (10-min autoclaving), and 27.1 (60-min autoclaving) % per 100 g raw seed	
Pea (*Pisum sativum*), hydrolyzed pea, fava bean (*Vicia faba*)	n/a	DPP-IV inhibitory activity • IC_{50}: 1.0–1.5 (pea); 0.7 (hydrolyzed pea) and 0.5 (fava bean) mg/mL Opioid receptor binding • ED_{50}: 2.4–14.9 (pea); 10.3 (hydrolyzed pea) and 1.5 (fava bean) mg/mL ACE inhibitory activity • IC_{50}: 29.5–221.8 (pea); 113.6 (hydrolyzed pea) and 52.0 (fava bean) μg/mL H_2O_2 production inhibiting activity • IC_{50}: 0.6 – ND (pea); 1.2 (hydrolyzed pea) and 0.7 (fava bean) mg/mL HO production inhibiting activity • IC_{50}: 1.0–2.1 (pea); 2.5 (hydrolyzed pea) and 1.5 (fava bean) mg/mL	Dugardin et al. (2020)

n/a: not provided in the publication.

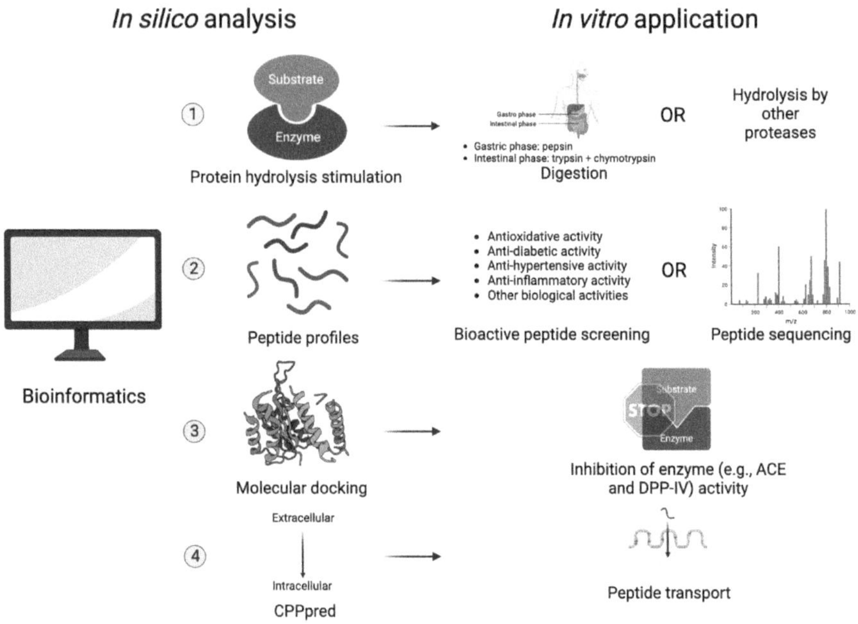

FIGURE 10.1 An overview of bioinformatic analysis for legume peptide research.

The different types of in silico analysis are indicated with numbers: (1) in silico hydrolysis: mimicking GI digestive processing (pepsin + trypsin + chymotrypsin) and other protease hydrolysis; (2) predicting peptide profiles: screening bioactive peptides and aligning the amino acid sequences of peptides based on mass analysis data; (3) molecular docking: predicting the binding sites of peptides in targeted enzymes (e.g., ACE and DPP-IV) to explore the role of peptides in modulating enzyme activities; and (4) cell penetrating capability prediction: analyzing the transport of peptides (e.g., in Caco-2 cell monolayer).

parental proteins. Such bioinformatic functionalities could be used to mimic the GI digestion, where pepsin, trypsin, and chymotrypsin have been selected as digestive proteases. (Martineau-Côté, Achouri, Wanasundara, et al., 2022) used the BIOPEP tool to screen peptide profiles from fava bean, soybean and pea after conducting in silico gastrointestinal digestion. Consequently, bioactive peptides with a range of biological activities including DPP-IV inhibition, ACE inhibition, DPP-III inhibition, α-glucosidase inhibition, antioxidant, stimulation and renin inhibition, were predicted to be released. Among them, 11 peptides (NYDEGSEPR, PVNRPGEPQ, LDNINALEPDH, TETWNPNHPE, TETWNPNHPEL, EEEDEDEPR, KEEEDEDEPR, VIPTEPPH, VIPTEPPHA, VVIPTEPPHA, and VVIPTEPPH) have been identified from fava bean digestate (in vitro). In addition, Jakubczyk et al. (2019) claimed that the peptides exerting hypotension (ACE inhibition, promote vasoactive substance release) and antidiabetic (DPP-IV, increasing glucose uptake) properties were predicted in fermented fava bean seeds digestates. Despite the advantages, rarely studies published have used in silico digestion to investigate the peptide profiles released from legumes. The complexity of protein subunits in real protein sources, together with the easy-to-follow standard digestive protocols, which provides more convincing in vitro data, would be the reason that explains researchers' preference of in vitro digestion over in silico prediction. However, in silico digestion is a good strategy for evaluation of peptide stability and monitoring of change in biological activity. For instance, García-Mora et al. (2017) used Peptide Cutter to predict the stability of 8 peptides identified from Savinase-treated lentil protein. All these peptides were further hydrolyzed into small fractions, in which some small fragments were recorded as bioactive peptides (e.g., QS, released from DLAIPVNRPGQLQSF, is recorded to be a DPP-IV inhibitor).

In several papers, the database of NCBI (https://www.ncbi.nlm.nih.gov/) and UniportKB (https://www.uniprot.org/) was used to match peptide sequences in digestates after peptidome and mass analysis. Nitride et al. (2022) matched ALEPDHR, TETWNPNHPEL, and DGH as the sequences of peptides in *ex vivo* gastro-duodenal digestion of chickpea flour. In addition, Mojica and de Mejía (2015) identified 137 potential peptide sequences in 15 common bean cultivars (*Phaseolus vulgaris* L.) from Mexico and Brazil. With the aid of in silico tools and databases (BLAST tool and BIOPEP), six tetrapeptides (SGAM, DSSG, LLAH, YVAT, EPTE, and KPKL) were determined as the most significant bioactive peptides. Following similar steps, Martineau-Côté, Achouri, Wanasundara, et al. (2022) identified 11 peptides in fava bean digestates. Subsequently, these peptide sequences were used to screen for bioactive fragments. For instance, LDNINALEPDH is one of the peptides identified. In this sequence, fragment ALEP is an ACE inhibitor and fragments EP, AL, DN, IN, and NA are DPP-IV inhibitors.

Molecular docking is a novel in silico tool for simulating the interaction between peptides and targeted protein(s) to predict the behavior of peptides in regulating enzyme activity (e.g., ACE and DPP-IV). This tool together with in silico prediction of cell-penetrating capability (http://bioware.ucd.ie/~compass/biowareweb/Server_pages/cpppred.php) is expected to provide more information on bioactivity of legume peptides (which have not been recorded) and peptide absorption. However, there is a lack of published studies using bioinformatic tools on legume peptides released during digestion.

10.7 Conclusion

Legumes are an important and high-quality protein source for human consumption. This chapter started with an overview of legume proteins and followed by elucidating an overall positive effect of processing on protein digestibility. After that, a wide range of biological activities of legume protein hydrolysates and peptide fractions has been summarized, while seldom papers validated the bioactive properties of purified legume peptides. Also, it is important to highlight that in some cases, polyphenols (found in legumes digestates) contributed to stronger biological activities (e.g., antioxidant) than peptides.

With advanced in silico techniques, the theoretical predictions of peptides released from legume proteins during GI digestion (e.g., digestibility, peptide profiles, potential bioactive peptides, and peptide absorption) can be made rapidly. It should be noted that the validation of these bioinformatic analyses, as well as findings from clinical trials for physiological release of peptides are largely missing.

10.8 Challenges and future perspectives

Currently a plenty of research revealed legumes are marvelous sources of high-quality proteins, with satisfactory digestibility values and peptide fragments with potential biological activities. With a remarkable increasing in investigating legume peptides released during GI digestion, INFOGEST protocol is recommended as a static digestion model for future studies to produce more comparable and convincing data.

Most papers only explored the biological activities of protein hydrolysates and/or peptide fractions rather than purified peptides, which may lead to an overestimation in biological capabilities of peptides. Other compounds (e.g., polyphenols) would also present similar biological activity therefore exerting a synergistic effect thus clarification of the contribution of each peptide(s) on overall biological capabilities is highly desired.

Epidemiological studies indicated a correlation between legume consumption and a reducing risk in developing lifestyle-related diseases. While strong in vivo evidence is still needed to validate the health benefits of legume digestates, especially bioactivity and bioavailability of peptides after digestion and during transportation to targeted organs. In addition, legumes always need to be processed and/or cooked before human consumption. Targeted studies on the selection of specific processing techniques for promoting the bioactive peptides release during GI digestion are needed.

Furthermore, advanced in silico methodologies create a great opportunity for peptides research, which could be used to rapidly provide theoretical data for legume peptides release during GI digestion. Meanwhile, in parallel in vitro/in vivo validation is always recommended to match the predicted data produced by in silico models.

References

Adamcová, A., Laursen, K. H., & Ballin, N. Z. (2021). Lectin activity in commonly consumed plant-based foods: Calling for method harmonization and risk assessment. *Foods*, *10*(11). Available from https://doi.org/10.3390/foods10112796, https://www.mdpi.com/2304-8158/10/11/2796/pdf.

Agyei, D., Tsopmo, A., & Udenigwe, C. C. (2018). Bioinformatics and peptidomics approaches to the discovery and analysis of food-derived bioactive peptides. *Analytical and Bioanalytical Chemistry*, *410*(15), 3463–3472. Available from https://doi.org/10.1007/s00216-018-0974-1, link.springer.de/link/service/journals/00216/index.htm.

Akıllıoğlu, H. G., & Karakaya, S. (2009). Effects of heat treatment and in vitro digestion on the angiotensin converting enzyme inhibitory activity of some legume species. *European Food Research and Technology*, *229*(6), 915–921. Available from https://doi.org/10.1007/s00217-009-1133-x.

Alpos, M., Leong, S. Y., Liesaputra, V., Martin, C. E., & Oey, I. (2021). Understanding in vivo mastication behaviour and in vitro starch and protein digestibility of pulsed electric field-treated black beans after cooking. *Foods*, *10*(11). Available from https://doi.org/10.3390/foods10112540, https://www.mdpi.com/2304-8158/10/11/2540/pdf.

Amigo-Benavent, M., Clemente, A., Ferranti, P., Caira, S., & Del Castillo, M. D. (2011). Digestibility and immunoreactivity of soybean β-conglycinin and its deglycosylated form. *Food Chemistry*, *129*(4), 1598–1605. Available from https://doi.org/10.1016/j.foodchem.2011.06.015.

Aryee, A. N. A., & Boye, J. I. (2016). Improving the digestibility of lentil flours and protein isolate and characterization of their enzymatically prepared hydrolysates. *International Journal of Food Properties*, *19*(12), 2649–2665. Available from https://doi.org/10.1080/10942912.2015.1123269.

Asledottir, T., Vegarud, G. E., Picariello, G., Mamone, G., Lea, T. E., Røseth, A., Ferranti, P., & Devold, T. G. (2023). Bioactive peptides identified in pea and faba bean after in vitro digestion with human gastrointestinal enzymes. *Journal of Functional Foods*, *102*. Available from https://doi.org/10.1016/j.jff.2023.105445.

Aydemir, L. Y., Diblan, S., Aktas, H., & Cakitli, G. (2022). Changes in bioactive properties of dry bean extracts during enzymatic hydrolysis and in vitro digestion steps. *Journal of Food Measurement and Characterization*, *16*(5), 3682–3698. Available from https://doi.org/10.1007/s11694-022-01484-1.

Bai, M., Qin, G., Sun, Z., & Long, G. (2016). Relationship between molecular structure characteristics of feed proteins and protein In vitro digestibility and solubility. *Asian-Australas Journal of Animal Science*, *29*(8), 1159–1165. Available from https://doi.org/10.5713/ajas.15.0701.

Barac, M., Cabrilo, S., Pesic, M., Stanojevic, S., Zilic, S., Macej, O., & Ristic, N. (2010). Profile and functional properties of seed proteins from six pea (Pisum sativum) genotypes. *International Journal of Molecular Sciences*, *11*(12), 4973–4990. Available from https://doi.org/10.3390/ijms11124973.

Barbana, C., & Boye, J. I. (2013). In vitro protein digestibility and physico-chemical properties of flours and protein concentrates from two varieties of lentil (Lens culinaris). *Food and Function*, *4*(2), 310–321. Available from https://doi.org/10.1039/c2fo30204g.

Bautista-Expósito, S., Vandenberg, A., Peñas, E., Frias, J., & Martínez-Villaluenga, C. (2021). Lentil and fava bean with contrasting germination kinetics: A focus on digestion of proteins and bioactivity of resistant peptides. *Frontiers in Plant Science*, *12*. Available from https://doi.org/10.3389/fpls.2021.754287.

Boschin, G., Scigliuolo, G. M., Resta, D., & Arnoldi, A. (2014). ACE-inhibitory activity of enzymatic protein hydrolysates from lupin and other legumes. *Food Chemistry*, *145*, 34–40. Available from https://doi.org/10.1016/j.foodchem.2013.07.076.

Boye, J., Zare, F., & Pletch, A. (2010). Pulse proteins: Processing, characterization, functional properties and applications in food and feed. *Food Research International*, *43*(2), 414–431. Available from https://doi.org/10.1016/j.foodres.2009.09.003.

Brishti, F. H., Chay, S. Y., Muhammad, K., Ismail-Fitry, M. R., Zarei, M., Karthikeyan, S., & Saari, N. (2020). Effects of drying techniques on the physicochemical, functional, thermal, structural and rheological properties of mung bean (Vigna radiata) protein isolate powder. *Food Research International, 138*. Available from https://doi.org/10.1016/j.foodres.2020.109783.

Brodkorb, A., Egger, L., Alminger, M., Alvito, P., Assunção, R., Ballance, S., Bohn, T., Bourlieu-Lacanal, C., Boutrou, R., Carrière, F., Clemente, A., Corredig, M., Dupont, D., Dufour, C., Edwards, C., Golding, M., Karakaya, S., Kirkhus, B., Le Feunteun, S., ... Recio, I. (2019). INFOGEST static in vitro simulation of gastrointestinal food digestion. *Nature Protocols, 14*(4), 991−1014. Available from https://doi.org/10.1038/s41596-018-0119-1.

Budseekoad, S., Takahashi Yupanqui, C., Alashi, A. M., Aluko, R. E., & Youravong, W. (2019). Anti-allergic activity of mung bean (Vigna radiata (L.) Wilczek) protein hydrolysates produced by enzymatic hydrolysis using non-gastrointestinal and gastrointestinal enzymes. *Journal of Food Biochemistry, 43*(1). Available from https://doi.org/10.1111/jfbc.12674.

Carbonaro, M., Grant, G., & Cappelloni, M. (2005). Heat-induced denaturation impairs digestibility of legume (Phaseolus vulgaris L andVicia faba L) 7S and 11S globulins in the small intestine of rat. *Journal of the Science of Food and Agriculture, 85*(1), 65−72. Available from https://doi.org/10.1002/jsfa.1940.

Carbonaro, M., Maselli, P., & Nucara, A. (2012). Relationship between digestibility and secondary structure of raw and thermally treated legume proteins: A Fourier transform infrared (FT-IR) spectroscopic study. *Amino Acids, 43*(2), 911−921. Available from https://doi.org/10.1007/s00726-011-1151-4.

Carbonaro, M., Maselli, P., & Nucara, A. (2015). Structural aspects of legume proteins and nutraceutical properties. *Food Research International, 76*(1), 19−30. Available from https://doi.org/10.1016/j.foodres.2014.11.007.

Carbonaro, M., & Nucara, A. (2022). Legume proteins and peptides as compounds in nutraceuticals: A structural basis for dietary health effects. *Nutrients, 14*(6). Available from https://doi.org/10.3390/nu14061188.

Carrasco-Castilla, J., Hernández-Álvarez, A. J., Jiménez-Martínez, C., Jacinto-Hernández, C., Alaiz, M., Girón-Calle, J., Vioque, J., & Dávila-Ortiz, G. (2012a). Antioxidant and metal chelating activities of Phaseolus vulgaris L. Var. Jamapa protein isolates, phaseolin and lectin hydrolysates. *Food Chemistry, 131*(4), 1157−1164. Available from https://doi.org/10.1016/j.foodchem.2011.09.084.

Carrasco-Castilla, J., Hernández-Álvarez, A. J., Jiménez-Martínez, C., Jacinto-Hernández, C., Alaiz, M., Girón-Calle, J., Vioque, J., & Dávila-Ortiz, G. (2012b). Antioxidant and metal chelating activities of peptide fractions from phaseolin and bean protein hydrolysates. *Food Chemistry, 135*(3), 1789−1795. Available from https://doi.org/10.1016/j.foodchem.2012.06.016.

Chávez-Murillo, C. E., Veyna-Torres, J. I., Cavazos-Tamez, L. M., de la Rosa-Millán, J., & Serna-Saldívar, S. O. (2018). Physicochemical characteristics, ATR-FTIR molecular interactions and in vitro starch and protein digestion of thermally-treated whole pulse flours. *Food Research International, 105*, 371−383. Available from https://doi.org/10.1016/j.foodres.2017.11.029.

Chen, F. L., Wei, Y. M., & Zhang, B. (2011). Chemical cross-linking and molecular aggregation of soybean protein during extrusion cooking at low and high moisture content. *LWT − Food Science and Technology, 44*(4), 957−962. Available from https://doi.org/10.1016/j.lwt.2010.12.008.

Chen, L., Wu, J., Li, Z., Liu, Q., Zhao, X., & Yang, H. (2019). Metabolomic analysis of energy regulated germination and sprouting of organic mung bean (Vigna radiata) using NMR spectroscopy. *Food Chemistry*, *286*, 87–97. Available from https://doi.org/10.1016/j.foodchem.2019.01.183.

Clemente, A., Jimenez, E., Marin-Manzano, M. C., & Rubio, L. A. (2008). Active Bowman-Birk inhibitors survive gastrointestinal digestion at the terminal ileum of pigs fed chickpea-based diets. *Journal of the Science of Food and Agriculture*, *88*(3), 513–521. Available from https://doi.org/10.1002/jsfa.3115.

Coda, R., Melama, L., Rizzello, C. G., Curiel, J. A., Sibakov, J., Holopainen, U., . . . Sozer, N. (2015). Effect of air classification and fermentation by Lactobacillus plantarum VTT E-133328 on faba bean (Vicia faba L.) flour nutritional properties. *International Journal of Food Microbiology*, *193*, 34–42. Available from https://doi.org/10.1016/j.ijfoodmicro.2014.10.012.

Correa, J. L., Zapata, J. E., & Hernández-Ledesma, B. (2022). Release of bioactive peptides from Erythrina edulis (Chachafruto) proteins under simulated gastrointestinal digestion. *Nutrients*, *14*(24). Available from https://doi.org/10.3390/nu14245256.

Corzo-Martínez, M., Soria, A. C., Belloque, J., Villamiel, M., & Moreno, F. J. (2010). Effect of glycation on the gastrointestinal digestibility and immunoreactivity of bovine β-lactoglobulin. *International Dairy Journal*, *20*(11), 742–752. Available from https://doi.org/10.1016/j.idairyj.2010.04.002.

Dai, C., Hou, Y., Xu, H., Huang, L., Dabbour, M., Mintah, B. K.... (2022). MaEffect of solid-state fermentation by three different Bacillus species on composition and protein structure of soybean mealJournal of the. *Science of Food and Agriculture*, *102*(2), 557–566. Available from https://doi.org/10.1002/jsfa.11384.

de Almeida, A. P., Juvanhol, L. L., Bersch-Ferreira, Â. C., Torreglosa, C. R., Marcadenti, A., Weber, B., Bressan, J., & Hermsdorff, H. H. M. (2022). Adiposity and insulin resistance mediate the inverse association between legume intake and blood pressure: A cross-sectional analysis in secondary cardiovascular prevention. *British Journal of Nutrition*, *128*(12), 2353–2362. Available from https://doi.org/10.1017/S0007114521005018.

De Angelis, E., Pilolli, R., Bavaro, S. L., & Monaci, L. (2017). Insight into the gastro-duodenal digestion resistance of soybean proteins and potential implications for residual immunogenicity. *Food and Function*, *8*(4), 1599–1610. Available from https://doi.org/10.1039/c6fo01788f.

De Santis, M. A., Rinaldi, M., Menga, V., Codianni, P., Giuzio, L., Fares, C., & Flagella, Z. (2021). Influence of organic and conventional farming on grain yield and protein composition of chickpea genotypes. *Agronomy*, *11*(2). Available from https://doi.org/10.3390/agronomy11020191.

Deshpande, S. S., & Damodaran, S. (1989). Structure-digestibility relationship of legume 7S proteins. *Journal of Food Science*, *54*(1), 108–113. Available from https://doi.org/10.1111/j.1365-2621.1989.tb08579.x.

Didinger, C., & Thompson, H. J. (2021). Defining nutritional and functional niches of legumes: A call for clarity to distinguish a future role for pulses in the dietary guidelines for Americans. *Nutrients*, *13*(4). Available from https://doi.org/10.3390/nu13041100.

Din, J. U., Sarwar, A., Li, Y., Aziz, T., Hussain, F., Shah, S. M. M., Yi, G., & Liu, X. (2021). Separation of storage proteins (7S and 11S) from soybean seed, meals and protein isolate using an optimized method via comparison of yield and purity. *Protein Journal*, *40*(3), 396–405. Available from https://doi.org/10.1007/s10930-021-09990-9.

Drulyte, D., & Orlien, V. (2019). The effect of processing on digestion of legume proteins. *Foods*, *8*(6). Available from https://doi.org/10.3390/foods8060224.

Dugardin, C., Cudennec, B., Tourret, M., Caron, J., Guérin-Deremaux, L., Behra-Miellet, J., Lefranc-Millot, C., & Ravallec, R. (2020). Explorative screening of bioactivities generated by plant-based proteins after in vitro static gastrointestinal digestion. *Nutrients*, *12*(12). Available from https://doi.org/10.3390/nu12123746.

Duijsens, D., Pälchen, K., De Coster, A., Verkempinck, S. H. E., Hendrickx, M. E., & Grauwet, T. (2022). Effect of manufacturing conditions on in vitro starch and protein digestibility of (cellular) lentil-based ingredients. *Food Research International*, *158*. Available from https://doi.org/10.1016/j.foodres.2022.111546.

Durak, A., Baraniak, B., Jakubczyk, A., & Świeca, M. (2013). Biologically active peptides obtained by enzymatic hydrolysis of Adzuki bean seeds. *Food Chemistry*, *141*(3), 2177–2183. Available from https://doi.org/10.1016/j.foodchem.2013.05.012.

Duranti, M. (2006). Grain legume proteins and nutraceutical properties. *Fitoterapia*, *77*(2), 67–82. Available from https://doi.org/10.1016/j.fitote.2005.11.008.

El-Sayed, S. M., Nossier, M. I., & Nossier, A. I. (2022). Faba beans with enhanced antioxidant activity ameliorate acetic acid-induced colitis in experimental rats. *Food and Function*, *13*(22), 11865–11878. Available from https://doi.org/10.1039/d2fo02782h.

Eze, C. R., Kwofie, E. M., Adewale, P., Lam, E., & Ngadi, M. (2022). Advances in legume protein extraction technologies: A review. *Innovative Food Science and Emerging Technologies*, *82*. Available from https://doi.org/10.1016/j.ifset.2022.103199.

Falade, K. O., & Akeem, S. A. (2020). Physicochemical properties, protein digestibility and thermal stability of processed African mesquite bean (Prosopis africana) flours and protein isolates. *Journal of Food Measurement and Characterization*, *14*(3), 1481–1496. Available from https://doi.org/10.1007/s11694-020-00398-0.

Faris, R. J., Wang, H., & Wang, T. (2008). Improving digestibility of soy flour by reducing disulfide bonds with thioredoxin. *Journal of Agricultural and Food Chemistry*, *56*(16), 7146–7150. Available from https://doi.org/10.1021/jf801136n.

Fernández-Tomé, S., & Hernández-Ledesma, B. (2020). Gastrointestinal Digestion of food proteins under the effects of released bioactive peptides on digestive health. *Molecular Nutrition & Food Research*, *64*(21). Available from https://doi.org/10.1002/mnfr.202000401.

Fleury, L., Deracinois, B., Dugardin, C., Nongonierma, A. B., FitzGerald, R. J., Flahaut, C., Cudennec, B., & Ravallec, R. (2022). In vivo and in vitro comparison of the DPP-IV inhibitory potential of food proteins from different origins after gastrointestinal digestion. *International Journal of Molecular Sciences*, *23*(15). Available from https://doi.org/10.3390/ijms23158365.

García-Mora, P., Martín-Martínez, M., Angeles Bonache, M., González-Múniz, R., Peñas, E., Frias, J., & Martinez-Villaluenga, C. (2017). Identification, functional gastrointestinal stability and molecular docking studies of lentil peptides with dual antioxidant and angiotensin I converting enzyme inhibitory activities. *Food Chemistry*, *221*, 464–472. Available from https://doi.org/10.1016/j.foodchem.2016.10.087.

Garcia-Mora, P., Peñas, E., Frias, J., Gomez, R., & Martinez-Villaluenga, C. (2015). High-pressure improves enzymatic proteolysis and the release of peptides with angiotensin I converting enzyme inhibitory and antioxidant activities from lentil proteins. *Food Chemistry*, *171*, 224–232. Available from https://doi.org/10.1016/j.foodchem.2014.08.116.

Gharibzahedi, S. M. T., Smith, B., & Altintas, Z. (2022). Bioactive and health-promoting properties of enzymatic hydrolysates of legume proteins: A review. *Critical Reviews in*

Food Science and Nutrition. Available from https://doi.org/10.1080/10408398.2022.2124399.

Gilani, G. S., & Sepehr, E. (2003). Protein digestibility and quality in products containing antinutritional factors are adversely affected by old age in rats. *Journal of Nutrition*, *133* (1), 220–225. Available from https://doi.org/10.1093/jn/133.1.220, http://jn.nutrition.org.

González-Montoya, M., Hernández-Ledesma, B., Mora-Escobedo, R., & Martínez-Villaluenga, C. (2018). Bioactive peptides from germinated soybean with anti-diabetic potential by inhibition of dipeptidyl peptidase-IV, α-amylase, and α-glucosidase enzymes. *International Journal of Molecular Sciences*, *19*(10). Available from https://doi.org/10.3390/ijms19102883.

Grasso, N., Lynch, N. L., Arendt, E. K., & O'Mahony, J. A. (2022). Chickpea protein ingredients: A review of composition, functionality, and applications. *Comprehensive Reviews in Food Science and Food Safety*, *21*(1), 435–452. Available from https://doi.org/10.1111/1541-4337.12878.

Guevara-Zambrano, J. M., Verkempinck, S. H. E., Muriithi, B., Duijsens, D., Hendrickx, M. E., Loey, A. M. V., & Grauwet, T. (2023). Protein accessibility level affects macronutrient digestion kinetics of plant-based shakes. *Food Hydrocolloids*, *137*. Available from https://doi.org/10.1016/j.foodhyd.2022.108428.

Gulewicz, P., Martínez-Villaluenga, C., Frias, J., Ciesiołka, D., Gulewicz, K., & Vidal-Valverde, C. (2008). Effect of germination on the protein fraction composition of different lupin seeds. *Food Chemistry*, *107*(2), 830–844. Available from https://doi.org/10.1016/j.foodchem.2007.08.087.

Gwala, S., Pallares Pallares, A., Pälchen, K., Hendrickx, M., & Grauwet, T. (2020). In vitro starch and protein digestion kinetics of cooked Bambara groundnuts depend on processing intensity and hardness sorting. *Food Research International*, *137*. Available from https://doi.org/10.1016/j.foodres.2020.109512.

Hall, A. E., & Moraru, C. I. (2022). Comparative effects of high pressure processing and heat treatment on in vitro digestibility of pea protein and starch. *Npj Science of Food*, *6*(1). Available from https://doi.org/10.1038/s41538-021-00116-0.

Hamosh, M. (1996). Digestion in the newborn. *Clinics in Perinatology*, *23*(2), 191–209. Available from https://doi.org/10.1016/s0095-5108(18)30238-0.

Hernández-Olivas, E., Muñoz-Pina, S., Andrés, A., & Heredia, A. (2021). Age-related gastrointestinal alterations of legumes and cereal grains digestibility. *Food Bioscience*, *41*. Available from https://doi.org/10.1016/j.fbio.2021.101027.

Hughes, J., Pearson, E., & Grafenauer, S. (2022). Legumes—a comprehensive exploration of global food-based dietary guidelines and consumption. *Nutrients*, *14*(15). Available from https://doi.org/10.3390/nu14153080.

Jakubczyk, A., Karaś, M., Złotek, U., Szymanowska, U., Baraniak, B., & Bochnak, J. (2019). Peptides obtained from fermented faba bean seeds (Vicia faba) as potential inhibitors of an enzyme involved in the pathogenesis of metabolic syndrome. *LWT*, *105*, 306–313. Available from https://doi.org/10.1016/j.lwt.2019.02.009.

Jakubczyk, A., Karaś, M., Złotek, U., & Szymanowska, U. (2017). Identification of potential inhibitory peptides of enzymes involved in the metabolic syndrome obtained by simulated gastrointestinal digestion of fermented bean (Phaseolus vulgaris L.) seeds. *Food Research International*, *100*, 489–496. Available from https://doi.org/10.1016/j.foodres.2017.07.046.

Jamdar, S. N., Deshpande, R., & Marathe, S. A. (2017). Effect of processing conditions and in vitro protein digestion on bioactive potentials of commonly consumed legumes. *Food Bioscience*, *20*, 1–11. Available from https://doi.org/10.1016/j.fbio.2017.07.007.

Jenkins, D. J. A., Kendall, C. W. C., Augustin, L. S. A., Mitchell, S., Sahye-Pudaruth, S., Blanco Mejia, S., Chiavaroli, L., Mirrahimi, A., Ireland, C., Bashyam, B., Vidgen, E., De Souza, R. J., Sievenpiper, J. L., Coveney, J., Leiter, L. A., & Josse, R. G. (2012). Effect of legumes as part of a low glycemic index diet on glycemic control and cardiovascular risk factors in type 2 diabetes mellitus: A randomized controlled trial. *Archives of Internal Medicine*, *172*(21), 1653–1660. Available from https://doi.org/10.1001/2013.jamainternmed.70Canada, http://archinte.jamanetwork.com/data/Journals/INTEMED/26027/iiq120070_1653_1660.pdf.

Jeong, D., Han, J. A., Liu, Q., & Chung, H. J. (2019). Effect of processing, storage, and modification on in vitro starch digestion characteristics of food legumes: A review. *Food Hydrocolloids*, *90*, 367–376. Available from https://doi.org/10.1016/j.foodhyd.2018.12.039.

Joehnke, M. S., Jeske, S., Ispiryan, L., Zannini, E., Arendt, E. K., Bez, J., Sørensen, J. C., & Petersen, I. L. (2021). Nutritional and anti-nutritional properties of lentil (Lens culinaris) protein isolates prepared by pilot-scale processing. *Food Chemistry: X*, *9*. Available from https://doi.org/10.1016/j.fochx.2020.100112, https://www.journals.elsevier.com/food-chemistry-x.

Juárez-Chairez, M. F., Cid-Gallegos, M. S., Meza-Márquez, O. G., & Jiménez-Martínez, C. (2022). Biological functions of peptides from legumes in gastrointestinal health. A review legume peptides with gastrointestinal protection. *Journal of Food Biochemistry*, *46*(10). Available from https://doi.org/10.1111/jfbc.14308, http://onlinelibrary.wiley.com/journal/10.1111/(ISSN)1745-4514.

Kashyap, S., Varkey, A., Shivakumar, N., Devi, S., Reddy B H, R., Thomas, T., Preston, T., Sreeman, S., & Kurpad, A. V. (2019). True ileal digestibility of legumes determined by dual-isotope tracer method in Indian adults. *The American Journal of Clinical Nutrition*, *110*(4), 873–882. Available from https://doi.org/10.1093/ajcn/nqz159.

Katoch, R., & Tripathi, A. (2021). Research advances and prospects of legume lectins. *Journal of Biosciences*, *46*(4). Available from https://doi.org/10.1007/s12038-021-00225-8.

Lam, A. C. Y., Can Karaca, A., Tyler, R. T., & Nickerson, M. T. (2018). Pea protein isolates: Structure, extraction, and functionality. *Food Reviews International*, *34*(2), 126–147. Available from https://doi.org/10.1080/87559129.2016.1242135, http://www.tandf.co.uk/journals/titles/87559129.asp.

Lefèvre, C., Bohuon, P., Lullien-Pellerin, V., & Mestres, C. (2022). Modeling the thermal denaturation of the protein–water system in pulses (lentils, beans, and chickpeas). *Journal of Agricultural and Food Chemistry*, *70*(32), 9980–9989. Available from https://doi.org/10.1021/acs.jafc.2c03553.

Liu, Y., Zhu, S., Li, Y., Sun, F., Huang, D., & Chen, X. (2023). Alternations in the multi-level structures of chickpea protein during fermentation and their relationship with digestibility. *Food Research International*, *165*. Available from https://doi.org/10.1016/j.foodres.2022.112453.

Lo, W. M. Y., & Li-Chan, E. C. Y. (2005). Angiotensin I converting enzyme inhibitory peptides from in vitro pepsin – pancreatin digestion of soy protein. *Journal of Agricultural and Food Chemistry*, *53*(9), 3369–3376. Available from https://doi.org/10.1021/jf048174d.

López-Barrios, L., Gutiérrez-Uribe, J. A., & Serna-Saldívar, S. O. (2014). Bioactive peptides and hydrolysates from pulses and their potential use as functional ingredients. *Journal of Food Science*, *79*(3), R273–R283. Available from https://doi.org/10.1111/1750-3841.12365, http://www3.interscience.wiley.com/journal/118509799/issueyear?year = 2008.

Lu, Z. X., He, J. F., Zhang, Y. C., & Bing, D. J. (2020). Composition, physicochemical properties of pea protein and its application in functional foods. *Critical Reviews in Food Science and Nutrition*, *60*(15), 2593–2605. Available from https://doi.org/10.1080/10408398.2019.1651248, https://www.tandfonline.com/loi/bfsn20.

Makeri, M. U., Mohamed, S. A., Karim, R., Ramakrishnan, Y., & Muhammad, K. (2017). Fractionation, physicochemical, and structural characterization of winged bean seed protein fractions with reference to soybean. *International Journal of Food Properties*, *20*, 2220–2236. Available from https://doi.org/10.1080/10942912.2017.1369101, http://www.tandf.co.uk/journals/titles/10942912.asp.

Makran, M., Miedes, D., Cilla, A., Barberá, R., Garcia-Llatas, G., & Alegría, A. (2022). Understanding the influence of simulated elderly gastrointestinal conditions on nutrient digestibility and functional properties. *Trends in Food Science & Technology*, *129*, 283–295. Available from https://doi.org/10.1016/j.tifs.2022.10.005.

Malaguti, M., Dinelli, G., Leoncini, E., Bregola, V., Bosi, S., Cicero, A. F. G., & Hrelia, S. (2014). Bioactive peptides in cereals and legumes: Agronomical, biochemical and clinical aspects. *International Journal of Molecular Sciences*, *15*(11), 21120–21135. Available from https://doi.org/10.3390/ijms151121120, http://www.mdpi.com/1422-0067/15/11/21120/pdf.

Margier, M., Georgé, S., Hafnaoui, N., Remond, D., Nowicki, M., Du Chaffaut, L., Amiot, M.-J., & Reboul, E. (2018). Nutritional composition and bioactive content of legumes: Characterization of pulses frequently consumed in France and effect of the cooking method. *Nutrients*, *10*(11). Available from https://doi.org/10.3390/nu10111668.

Marques, M. R., Maureira, A. D. C., Fontanari, G. G., Pimenta, D. C., Soares-Freitas, R. M., Hirata, M. H., Hirata, R. D. C., & Arêas, J. A. G. (2018). Transport of cowpea bean derived peptides and their modulator effects on mRNA expression of cholesterol-related genes in Caco-2 and HepG2 cells. *Food Research International*, *107*, 165–171. Available from https://doi.org/10.1016/j.foodres.2018.01.031.

Martineau-Côté, D., Achouri, A., Karboune, S., & L'Hocine, L. (2022). Faba bean: An untapped source of quality plant proteins and bioactives. *Nutrients*, *14*(8). Available from https://doi.org/10.3390/nu14081541.

Martineau-Côté, D., Achouri, A., Wanasundara, J., Karboune, S., & L'Hocine, L. (2022). Health beneficial bioactivities of faba bean gastrointestinal (in vitro) digestate in comparison to soybean and pea. *International Journal of Molecular Sciences*, *23*(16). Available from https://doi.org/10.3390/ijms23169210.

Martínez, K. A. A., & Mejia, E. G. (2021). Comparison of five chickpea varieties, optimization of hydrolysates production and evaluation of biomarkers for type 2 diabetes. *Food Research International*, *147*.

Martínez Leo, E. E., Arana Argáez, V. E., Acevedo Fernández, J. J., Puc, R. M., & Segura Campos, M. R. (2018). Effect of enzymatic digestion of protein derivatives obtained from Mucuna pruriens L. On production of proinflammatory mediators by BALB/c mouse macrophages. *Applied Biochemistry and Biotechnology*, *186*(3), 597–612. Available from https://doi.org/10.1007/s12010-018-2740-4, http://www.springer.com/humana + press/journal/12010.

Melchior, S., Moretton, M., Alongi, M., Calligaris, S., Cristina Nicoli, M., & Anese, M. (2023). Comparison of protein in vitro digestibility under adult and elderly conditions: The case study of wheat, pea, rice, and whey proteins. *Food Research International*, *163*. Available from https://doi.org/10.1016/j.foodres.2022.112147.

Milán-Noris, A. K., Gutiérrez-Uribe, J. A., Santacruz, A., Serna-Saldívar, S. O., & Martínez-Villaluenga, C. (2018). Peptides and isoflavones in gastrointestinal digests contribute to the anti-inflammatory potential of cooked or germinated desi and kabuli chickpea (Cicer arietinum L.). *Food Chemistry*, *268*, 66–76. Available from https://doi.org/10.1016/j.foodchem.2018.06.068.

Minekus, M., Alminger, M., Alvito, P., Ballance, S., Bohn, T., Bourlieu, C., Carrière, F., Boutrou, R., Corredig, M., Dupont, D., Dufour, C., Egger, L., Golding, M., Karakaya, S., Kirkhus, B., Le Feunteun, S., Lesmes, U., Maclerzanka, A., MacKie, A., ... Brodkorb, A. (2014). A standardised static in vitro digestion method suitable for food-an international consensus. *Food and Function*, *5*(6), 1113–1124. Available from https://doi.org/10.1039/c3fo60702j, http://www.rsc.org/Publishing/Journals/FO/about.asp.

Mir, S. A., Farooq, S., Shah, M. A., Sofi, S. A., Dar, B. N., Hamdani, A. M., & Khaneghah, A. M. (2021). An overview of sprouts nutritional properties, pathogens and decontamination technologies. *LWT*, *141*. Available from https://doi.org/10.1016/j.lwt.2021.110900.

Mitchell, D. C., Marinangeli, C. P. F., Pigat, S., Bompola, F., Campbell, J., Pan, Y., Curran, J. M., Cai, D. J., Jaconis, S. Y., & Rumney, J. (2021). Pulse intake improves nutrient density among us adult consumers. *Nutrients*, *13*(8). Available from https://doi.org/10.3390/nu13082668, https://www.mdpi.com/2072-6643/13/8/2668/pdf.

Mojica, L., Chen, K., & de Mejía, E. G. (2015). Impact of commercial precooking of common bean (Phaseolus vulgaris) on the generation of peptides, after pepsin-pancreatin hydrolysis, capable to inhibit dipeptidyl peptidase-IV. *Journal of Food Science*, *80*(1), H188–H198. Available from https://doi.org/10.1111/1750-3841.12726, http://www3.interscience.wiley.com/journal/118509799/issueyear?year = 2008.

Mojica, L., & de Mejía, E. G. (2015). Characterization and comparison of protein and peptide profiles and their biological activities of improved common bean cultivars (Phaseolus vulgaris L.) from Mexico and Brazil. *Plant Foods for Human Nutrition*, *70*(2), 105–112. Available from https://doi.org/10.1007/s11130-015-0477-6, http://www.kluweronline.com/issn/0921-9668/.

Mozhaev, V. V., Heremans, K., Frank, J., Masson, P., & Balny, C. (1996). High pressure effects on protein structure and function. *Proteins: Structure, Function and Genetics*, *24*(1), 81–91, 10.1002/(SICI)1097-0134(199601)24:1 < 81::AID-PROT6 > 3.0.CO;2-R.

Mulet-Cabero, A. I., Egger, L., Portmann, R., Ménard, O., Marze, S., Minekus, M., Le Feunteun, S., Sarkar, A., Grundy, M. M. L., Carrière, F., Golding, M., Dupont, D., Recio, I., Brodkorb, A., & Mackie, A. (2020). A standardised semi-dynamic: In vitro digestion method suitable for food-an international consensus. *Food and Function*, *11*(2), 1702–1720. Available from https://doi.org/10.1039/c9fo01293a.

Mune Mune, M. A., Minka, S. R., & Henle, T. (2018). Investigation on antioxidant, angiotensin converting enzyme and dipeptidyl peptidase IV inhibitory activity of Bambara bean protein hydrolysates. *Food Chemistry*, *250*, 162–169. Available from https://doi.org/10.1016/j.foodchem.2018.01.001.

Muzquiz, M., Varela, A., Burbano, C., Cuadrado, C., Guillamón, E., & Pedrosa, M. M. (2012). Bioactive compounds in legumes: Pronutritive and antinutritive actions.

Implications for nutrition and health. *Phytochemistry Reviews*, *11*(2-3), 227–244. Available from https://doi.org/10.1007/s11101-012-9233-9.

Nasrollahzadeh, F., Roman, L., Skov, K., Jakobsen, L. M. A., Trinh, B. M., Tsochatzis, E. D., Mekonnen, T., Corredig, M., Dutcher, J. R., & Martinez, M. M. (2023). A comparative investigation of seed storage protein fractions: The synergistic impact of molecular properties and composition on anisotropic structuring. *Food Hydrocolloids*, *137*. Available from https://doi.org/10.1016/j.foodhyd.2022.108400.

Neji, C., Semwal, J., Kamani, M. H., Máthé, E., & Sipos, P. (2022). Legume protein extracts: The relevance of physical processing in the context of structural, techno-functional and nutritional aspects of food development. *Processes*, *10*(12). Available from https://doi.org/10.3390/pr10122586.

Neves, V. A., & Lourenço, E. J. (1995). Isolation and in vitro hydrolysis of globulin G1 from lentils (Lens culinaris, Medik). *Journal of Food Biochemistry*, *19*(2), 109–120. Available from https://doi.org/10.1111/j.1745-4514.1995.tb00525.x.

Nguyen, T. T. P., Bhandari, B., Cichero, J., & Prakash, S. (2015). Gastrointestinal digestion of dairy and soy proteins in infant formulas: An in vitro study. *Food Research International*, *76*, 348–358. Available from https://doi.org/10.1016/j.foodres.2015.07.030.

Nishinari, K., Fang, Y., Guo, S., & Phillips, G. O. (2014). Soy proteins: A review on composition, aggregation and emulsification. *Food Hydrocolloids*, *39*, 301–318. Available from https://doi.org/10.1016/j.foodhyd.2014.01.013.

Nitride, C., Vegarud, G. E., Comi, I., Devold, T. G., Røseth, A., Marti, A., Iametti, S., Mamone, G., Picariello, G., Alfieri, F., Adalgisa Nicolai, M., Mills, C., & Ferranti, P. (2022). Effect of sprouting on the proteome of chickpea flour and on its digestibility by ex vivo gastro-duodenal digestion complemented with jejunal brush border membrane enzymes. *Food Research International*, *154*. Available from https://doi.org/10.1016/j.foodres.2022.111012.

Nosworthy, M. G., Medina, G., Franczyk, A. J., Neufeld, J., Appah, P., Utioh, A., Frohlich, P., & House, J. D. (2018). Effect of processing on the in vitro and in vivo protein quality of red and green lentils (Lens culinaris). *Food Chemistry*, *240*, 588–593. Available from https://doi.org/10.1016/j.foodchem.2017.07.129.

Ohanenye, I. C., Tsopmo, A., Ejike, C. E. C. C., & Udenigwe, C. C. (2020). Germination as a bioprocess for enhancing the quality and nutritional prospects of legume proteins. *Trends in Food Science and Technology*, *101*, 213–222. Available from https://doi.org/10.1016/j.tifs.2020.05.003.

Olagunju, A. I., Alashi, A. M., Omoba, O. S., Enujiugha, V. N., & Aluko, R. E. (2022). Pigeon pea penta-and hexapeptides with antioxidant properties also inhibit renin and angiotensin-I-converting enzyme activities. *Journal of Food Biochemistry*.

Oliás, R., Becerra-Rodríguez, C., Soliz-Rueda, J. R., Moreno, F. J., Delgado-Andrade, C., & Clemente, A. (2019). Glycation affects differently the main soybean Bowman-Birk isoinhibitors, IBB1 and IBBD2, altering their antiproliferative properties against HT29 colon cancer cells. *Food and Function*, *10*(9), 6193–6202. Available from https://doi.org/10.1039/c9fo01421g.

Omosebi, M. O., Osundahunsi, O. F., & Fagbemi, T. N. (2018). Effect of extrusion on protein quality, antinutritional factors, and digestibility of complementary diet from quality protein maize and soybean protein concentrate. *Journal of Food Biochemistry*, *42*(4). Available from https://doi.org/10.1111/jfbc.12508.

Onyango, E. (2022). *Legume protein: Properties and extraction for food applications.* IntechOpen. Available from https://doi.org/10.5772/intechopen.100393.

Palanisamy, M., Franke, K., Berger, R. G., Heinz, V., & Töpfl, S. (2019). High moisture extrusion of lupin protein: Influence of extrusion parameters on extruder responses and product properties. *Journal of the Science of Food and Agriculture, 99*(5), 2175–2185. Available from https://doi.org/10.1002/jsfa.9410.

Pallares Pallares, A., Gwala, S., Pälchen, K., Duijsens, D., Hendrickx, M., & Grauwet, T. (2021). Pulse seeds as promising and sustainable source of ingredients with naturally bioencapsulated nutrients: Literature review and outlook. *Comprehensive Reviews in Food Science and Food Safety, 20*(2), 1524–1553. Available from https://doi.org/10.1111/1541-4337.12692.

Panzeri, D., Guzzetti, L., Sacco, G., Tedeschi, G., Nonnis, S., Airoldi, C., Labra, M., Fusi, P., Forcella, M., & Regonesi, M. E. (2020). Effectiveness of: Vigna unguiculata seed extracts in preventing colorectal cancer. *Food and Function, 11*(7), 5853–5865. Available from https://doi.org/10.1039/d0fo00913j.

Pedrosa, M. M., Guillamón, E., & Arribas, C. (2021). Autoclaved and extruded legumes as a source of bioactive phytochemicals: A review. *Foods, 10*(2). Available from https://doi.org/10.3390/foods10020379.

Pelgrom, P. J. M., Boom, R. M., & Schutyser, M. A. I. (2015). Method development to increase protein enrichment during dry fractionation of starch-rich legumes. *Food and Bioprocess Technology, 8*, 1495–1502.

Perez-Moral, N., Saha, S., Pinto, A. M., Bajka, B. H., & Edwards, C. H. (2023). In vitro protein bioaccessibility and human serum amino acid responses to white bread enriched with intact plant cells. *Food Chemistry, 404*. Available from https://doi.org/10.1016/j.foodchem.2022.134538.

Pertiwi, M. G. P., Marsono, Y., & Indrati, R. (2020). In vitro gastrointestinal simulation of tempe prepared from koro kratok (Phaseolus lunatus L.) as an angiotensin-converting enzyme inhibitor. *Journal of Food Science and Technology, 57*(5), 1847–1855. Available from https://doi.org/10.1007/s13197-019-04219-1.

Pirman, T., Combe, E., Ribeyre, M. C., Prugnaud, J., Stekar, J., & Mirand, P. P. (2006). Differential effects of cooked beans and cooked lentils on protein metabolism in intestine and muscle in growing rats. *Annals of Nutrition and Metabolism, 50*(3), 197–205. Available from https://doi.org/10.1159/000090741.

Quansah, J. K., Udenigwe, C. C., Saalia, F. K., & Yada, R. Y. (2013). The effect of thermal and ultrasonic treatment on amino acid composition, radical scavenging and reducing potential of hydrolysates obtained from simulated gastrointestinal digestion of cowpea proteins. *Plant Foods for Human Nutrition, 68*(1), 31–38. Available from https://doi.org/10.1007/s11130-013-0334-4.

Rastogi, N. K., Raghavarao, K. S. M. S., Balasubramaniam, V. M., Niranjan, K., & Knorr, D. (2007). Opportunities and challenges in high pressure processing of foods. *Critical Reviews in Food Science and Nutrition, 47*(1), 69–112. Available from https://doi.org/10.1080/10408390600626420.

Rossi, G. B., Seraglio, S. K. T., Honaiser, T. C., Toaldo, I. M., Costa, A. Cd. O., Faria, J. Cd, & Arisi, A. C. M. (2022). Protein profile and antioxidant capacity of processed seeds from two common bean (Phaseolus vulgaris L.) cultivars. *International Journal of Food Science and Technology, 57*(7), 3934–3943. Available from https://doi.org/10.1111/ijfs.15537.

Roux, L. L., Chacon, R., Dupont, D., Jeantet, R., Deglaire, A., & Nau, F. (2020). In vitro static digestion reveals how plant proteins modulate model infant formula digestibility. *Food Research International, 130.* Available from https://doi.org/10.1016/j.foodres.2019.108917.

Rubio, L. A., & Clemente, A. (2009). In vivo (rat) and in vitro (caco-2 cells) absorption of amino acids from legume protein isolates as compared to lactalbumin or casein. *Archives of Animal Nutrition, 63*(5), 413–426. Available from https://doi.org/10.1080/17450390903169237.

Rui, X., Boye, J. I., Simpson, B. K., & Prasher, S. O. (2012). Angiotensin I-converting enzyme inhibitory properties of Phaseolus vulgaris bean hydrolysates: Effects of different thermal and enzymatic digestion treatments. *Food Research International, 49*(2), 739–746. Available from https://doi.org/10.1016/j.foodres.2012.09.025.

Sadeghi, R., Colle, M., & Smith, B. (2023). Protein composition of pulses and their protein isolates from different sources and in different isolation pH values using a reverse phase high performance liquid chromatography method. *Food Chemistry, 409.* Available from https://doi.org/10.1016/j.foodchem.2022.135278.

Saldanha do Carmo, C., Knutsen, S. H., Malizia, G., Dessev, T., Geny, A., Zobel, H., Myhrer, K. S., Varela, P., & Sahlstrøm, S. (2021). Meat analogues from a faba bean concentrate can be generated by high moisture extrusion. *Future Foods, 3.* Available from https://doi.org/10.1016/j.fufo.2021.100014.

Sánchez-Chino, X., Jiménez-Martínez, C., Dávila-Ortiz, G., Álvarez-González, I., & Madrigal-Bujaidar, E. (2015). Nutrient and nonnutrient components of legumes, and its chemopreventive activity: A review. *Nutrition and Cancer, 67*(3), 401–410. Available from https://doi.org/10.1080/01635581.2015.1004729.

Sangsukiam, T., & Duangmal, K. (2022). Changes in bioactive compounds and health-promoting activities in adzuki bean: Effect of cooking conditions and in vitro simulated gastrointestinal digestion. *Food Research International, 157.* Available from https://doi.org/10.1016/j.foodres.2022.111371.

Santos-Hernández, M., Alfieri, F., Gallo, V., Miralles, B., Masi, P., Romano, A., Ferranti, P., & Recio, I. (2020). Compared digestibility of plant protein isolates by using the INFOGEST digestion protocol. *Food Research International, 137.* Available from https://doi.org/10.1016/j.foodres.2020.109708.

Schutyser, M. A. I., Pelgrom, P. J. M., van der Goot, A. J., & Boom, R. M. (2015). Dry fractionation for sustainable production of functional legume protein concentrates. *Trends in Food Science & Technology., 45*(2), 327–335. Available from https://doi.org/10.1016/j.tifs.2015.04.013.

Scippa, G. S., Rocco, M., Ialicicco, M., Trupiano, D., Viscosi, V., Di Michele, M., Arena, S., Chiatante, D., & Scaloni, A. (2010). The proteome of lentil (Lens culinaris Medik.) seeds: Discriminating between landraces. *Electrophoresis, 31*(3), 497–506. Available from https://doi.org/10.1002/elps.200900459, Italy.

Serrano-Sandoval, S. N., Guardado-Félix, D., & Gutiérrez-Uribe, J. A. (2019). Changes in digestibility of proteins from chickpeas (Cicer arietinum L.) germinated in presence of selenium and antioxidant capacity of hydrolysates. *Food Chemistry, 285*, 290–295. Available from https://doi.org/10.1016/j.foodchem.2019.01.137.

Shevkani, K., Singh, N., Chen, Y., Kaur, A., & Yu, L. (2019). Pulse proteins: Secondary structure, functionality and applications. *Journal of Food Science and Technology.* Available from https://doi.org/10.1007/s13197-019-03723-8.

Shrestha, S., van 't Hag, L., Haritos, V. S., & Dhital, S. (2023a). Lentil and mungbean protein isolates: Processing, functional properties, and potential food applications. *Food Hydrocolloids*, *135*. Available from https://doi.org/10.1016/j.foodhyd.2022.108142.

Shrestha, S., van 't Hag, L., Haritos, V., & Dhital, S. (2023b). Comparative study on molecular and higher-order structures of legume seed protein isolates: Lentil, mungbean and yellow pea. *Food Chemistry*, *411*. Available from https://doi.org/10.1016/j.foodchem.2023.135464.

Sjöö, M., & Nilsson, L. (2017). *Starch in food: Structure, function and applications: Second edition* (pp. 1–893). Elsevier Inc. Available from http://www.sciencedirect.com/science/book/9780081008683.

Stanisavljević, N., Jovanović, Z., Čupić, T., Lukić, J., Miljuš Dukić, J., Radović, S., & Mikić, A. (2013). Extractability of antioxidants from legume seed flour after cooking and in vitro gastrointestinal digestion in comparison with methanolic extraction of the unprocessed flour. *International Journal of Food Science and Technology*, *48*(10), 2096–2104. Available from https://doi.org/10.1111/ijfs.12191.

Sulieman, M. A., Hassan, A. B., Osman, G. A., El Tyeb, M. M., El Khalil, E. A., El Tinay, A. H., & Babiker, E. E. (2008). Changes in total protein digestibility, fractions content and structure during cooking of lentil cultivars. *Pakistan Journal of Nutrition*, *7*(6), 801–805. Available from https://doi.org/10.3923/pjn.2008.801.805Sudan.

Sun, W., He, J., Wang, H., Zhang, Q., Li, W., & Rui, X. (2023). Solid-state fermentation alters the fate of red kidney bean protein during buccal and gastrointestinal digestion: Relationship with cotyledon cell wall integrity. *Food Chemistry*, *410*. Available from https://doi.org/10.1016/j.foodchem.2022.135370.

Tang, J., Wichers, H. J., & Hettinga, K. A. (2022). Heat-induced unfolding facilitates plant protein digestibility during in vitro static infant digestion. *Food Chemistry*, *375*. Available from https://doi.org/10.1016/j.foodchem.2021.131878.

Tang, J., Wichers, H. J., & Hettinga, K. A. (2023). Glycation of soy and pea proteins influences infant gastric digestibility more than intestinal digestibility. *Food Hydrocolloids*, *136*. Available from https://doi.org/10.1016/j.foodhyd.2022.108251.

Tavano, O. L., Da Silva, S. I., Démonte, A., & Neves, V. A. (2008). Nutritional responses of rats to diets based on chickpea (Cicer arietinum L.) seed meal or its protein fractions. *Journal of Agricultural and Food Chemistry*, *56*(22), 11006–11010. Available from https://doi.org/10.1021/jf8010799.

Torres-Fuentes, C., Alaiz, M., & Vioque, J. (2011). Affinity purification and characterisation of chelating peptides from chickpea protein hydrolysates. *Food Chemistry*, *129*(2), 485–490. Available from https://doi.org/10.1016/j.foodchem.2011.04.103.

Torres-Fuentes, C., Alaiz, M., & Vioque, J. (2012). Iron-chelating activity of chickpea protein hydrolysate peptides. *Food Chemistry*, *134*(3), 1585–1588. Available from https://doi.org/10.1016/j.foodchem.2012.03.112.

Vaz Patto, M. C., Amarowicz, R., Aryee, A. N. A., Boye, J. I., Chung, H. J., Martín-Cabrejas, M. A., & Domoney, C. (2015). Achievements and challenges in improving the nutritional quality of food legumes. *Critical Reviews in Plant Sciences*, *34*, 105–143. Available from https://doi.org/10.1080/07352689.2014.897907.

Venkidasamy, B., Selvaraj, D., Nile, A. S., Ramalingam, S., Kai, G., & Nile, S. H. (2019). Indian pulses: A review on nutritional, functional and biochemical properties with future perspectives. *Trends in Food Science and Technology*, *88*, 228–242. Available from https://doi.org/10.1016/j.tifs.2019.03.012.

Vogelsang-O'dwyer, M., Sahin, A. W., Arendt, E. K., & Zannini, E. (2022). Enzymatic hydrolysis of pulse proteins as a tool to improve techno-functional properties. *Foods*, *11*(9). Available from https://doi.org/10.3390/foods11091307.

Vreeke, G. J. C., Meijers, M. G. J., Vincken, J. P., & Wierenga, P. A. (2023). Towards absolute quantification of protein genetic variants in Pisum sativum extracts. *Analytical Biochemistry*, *665*. Available from https://doi.org/10.1016/j.ab.2023.115048.

Webb, D., Li, Y., & Alavi, S. (2023). Chemical and physicochemical features of common plant proteins and their extrudates for use in plant-based meat. *Trends in Food Science & Technology*, *131*, 129–138. Available from https://doi.org/10.1016/j.tifs.2022.11.006.

Xiao, Y., Xing, G., Rui, X., Li, W., Chen, X., Jiang, M., & Dong, M. (2015). Effect of solid-state fermentation with Cordyceps militaris SN-18 on physicochemical and functional properties of chickpea (Cicer arietinum L.) flour. *LWT–Food Science and Technology*, *63*(2), 1317–1324. Available from https://doi.org/10.1016/j.lwt.2015.04.046.

Yasar, S., Tosun, R., & Sonmez, Z. (2020). Fungal fermentation inducing improved nutritional qualities associated with altered secondary protein structure of soybean meal determined by FTIR spectroscopy. *Measurement*, *161*, 107895. Available from https://doi.org/10.1016/j.measurement.2020.107895.

Yi-Shen, Z., Shuai, S., & Fitzgerald, R. (2018). Mung bean proteins and peptides: Nutritional, functional and bioactive properties. *Food and Nutrition Research*, *62*. Available from https://doi.org/10.29219/fnr.v62.1290.

Yu, L., Ramaswamy, H. S., & Boye, J. (2013). Protein rich extruded products prepared from soy protein isolate-corn flour blends. *LWT – Food Science and Technology*, *50* (1), 279–289. Available from https://doi.org/10.1016/j.lwt.2012.05.012.

Zhang, J., & Chai, L. (2022). Trade-off between human health and environmental health in global diets. *Resources, Conservation and Recycling*, *182*. Available from https://doi.org/10.1016/j.resconrec.2022.106336.

Zhang, Y., & Chang, S. K. C. (2019). Comparative studies on ACE inhibition, degree of hydrolysis, antioxidant property and phenolic acid composition of hydrolysates derived from simulated in vitro gastrointestinal proteolysis of three thermally treated legumes. *Food Chemistry*, *281*, 154–162. Available from https://doi.org/10.1016/j.foodchem.2018.12.090.

Zhang, J., Liu, M., Zhao, Y., Zhu, Y., Bai, J., Fan, S., Zhu, L., Song, C., & Xiao, X. (2022). Recent developments in fermented cereals on nutritional constituents and potential health benefits. *Foods*, *11*(15), 2243. Available from https://doi.org/10.3390/foods11152243.

Zhou, H., Tan, Y., & McClements, D. J. (2023). Applications of the INFOGEST in vitro digestion model to foods: A review. *Annual Review of Food Science and Technology*, *14*, 135–156. Available from https://doi.org/10.1146/annurev-food-060721-012235.

CHAPTER

Bioactive peptides released from Cucurbitaceae seed proteins during gastrointestinal digestion

11

Elsa Marisa Ferreira Vieira

REQUIMTE/LAQV/ISEP, Polytechnic of Porto – School of Engineering, Porto, Portugal

Abbreviations

GI	gastrointestinal
MW	molecular weight
ACE	angiotensin-converting enzyme
DH	degree of hydrolysis
RIP	ribosome-inactivating protein
E/S	enzyme/substrate
BIOPEP-UWM	food-derived peptides with in silico-predicted biological activity
UF	ultrafiltration
IMAC	immobilized metal affinity chromatography
GF	gel filtration chromatography
SDS-PAGE	sodium dodecyl sulfate-polyacrylamide gel electrophoresis
QTOF-MS with ESI	quadrupole time-of-flight mass spectrometer with an electrospray ionization

11.1 Introduction

The Cucurbitaceae family comprises about 130 genera and ~1000 species, including important crops, such as cucumber (*Cucumis sativus* L.), melon (*Cucumis melo* L.), wax gourd (*Benincasa hispida*), apple gourd (*Benincasa fistulosa*), bitter gourd (*Momordica charantia*), bottle gourd (*Lagenaria siceraria*), chayote (*Sechium edule*), watermelon (*Citrullus lanatus*), pumpkins, and squashes (*Cucurbita* spp.) (Guo et al., 2020; Mukherjee et al., 2022; Ozuna & León-Galván, 2017). Based on their morphological, cytological, and floral characteristics, species are classified into two major subfamilies: Cucurbitoideae and Zanonioideae (Mukherjee et al., 2022). Most of them are climbers and have characteristic tendrils and pepo, being recognized for their notable diversity in fruit

Protein Digestion-Derived Peptides. DOI: https://doi.org/10.1016/B978-0-443-19141-1.00011-X

FIGURE 11.1 Representation of some.

Cucurbitaceae species, identified by their scientific and (common) name [1]*Cucumis sativus* (cucumber), [2]*Cucumis melo* (muskmelon), [3]*Benincasa hispida* (wax gourd), [4]*Solena amplexicaulis* (creeping cucumber), [5]*Lagenaria siceraria* (bottle gourd), [6]Citrullus lanatus (watermelon), [7]*Citrullus colocynthis* (bitter apple), [8]*Coccinia grandis* (ivy gourd), [9]*Cucurbita maxima* (pumpkin), [10]*Cucurbita moschata* (crookneck pumpkin), [11]*Cucurbita pepo* (field pumpkin), [12]*Cucurbita foetidissima* (missouri gourd), [13]*Cucurbita ficifolia* (black seed pumpkin), [14]*Trichosanthes anguina* (snake gourd), [15]*Trichosanthes cucumerina* (snake gourd), [16]*Trichosanthes dioica* (pointed gourd), [17]*Trichosanthes tricuspidata* (bitter snake gourd), [18]*Gymnopetalum chinensis* (jin gua), [19]*Sechium* edule (chayote, mirliton squash), [20]*Cyclanthera pedata* (klaya), [21]*Marah macrocarpa* (wild cucumber), [22]*Luffa cylindrica* (sponge gourd), [23]*Luffa acutangula* (ridge gourd), [24]*Schizopepon bryoniifolius* (cracking melon), [25]*Ecballium elaterium* (squirting cucumber), [26]*Telfairia pedate* (fluted gourd), [27]*Momordica charantia* (bitter gourd), [28]*Momordica cochinchinensis* (gac), [29]*Momordica dioica* (spine gourd), [30]*Siraitia grosvenorii* (luohan guo), [31]*Thladiantha globicarpa* (Himalayan tuber-gourd), [32]*Gynostemma pentaphyllum* (jiaogulan), [33]*Hemsleya chinensis* (common name is unknown), [34]*Neoalsomitra integrifoliola* (common name is unknown), [35]*Actinostemma tenerum* (tender ray-flower gourd), and [36]*Bolbostemma biglandulosum* (common name is unknown).

Adapted from Guo, J., Xu, W., Hu, Y., Huang, J., Zhao, Y., Zhang, L., Huang, C.-H., Ma, H. (2020). Phylotranscriptomics in Cucurbitaceae reveal multiple whole-genome duplications and key morphological and molecular innovations. Molecular Plant, 13*(8), 1117–1133. doi:10.1016/j.molp.2020.050.011.*

size, shape, flavor, and color (Guo et al., 2020). Fig. 11.1 illustrates some representative species of Cucurbitacea family.

According to Food and Agriculture Organization of the United Nations (FAO) (FAO, 2019), the world production of Cucurbitaceae in 2020 was led by China, India, Turkey, Ukraine, and Russia Federation, and included 107.6 Mt (megatons) of watermelon, 92.8 Mt of cucumber, 28.5 Mt of other melons, and 28.0 Mt of pumpkins and squashes (Table 11.1). Although more than 100 countries produce these

Table 11.1 Worldwide and top 10 producers (Megatos) of Cucurbitaceae in 2020.

Cucumber	Mt	% (% for Top 10)	Pumpkin and squashes	Mt	% (% for Top 10)	Watermelon	Mt	% (% for Top 10)	Cantaloupes and other melons	Mt	% (% for Top 10)	Melon seeds	Mt	%
Worldwide	92.8		Worldwide	28.0		Worldwide	106.7		Worldwide	28.5		Worldwide	0.93	
Top 10	82.7	89.2	Top 10	19.6	69.8	Top 10	80.2	75.2	Top 10	22.7	79.8	Top 10	0.93	99.2
China	72.8	78.5 (88.1)	China	7.5	26.7 (38.3)	China	60.2	56.5 (75.1)	China	13.9	48.7 (61.0)	Nigeria	0.56	59.5
Turkey	1.9	2.1 (2.3)	India	5.1	18.3 (26.1)	Turkey	3.5	3.3. (4.4)	Turkey	1.7	6.1 (7.6)	Sudan	0.07	7.6
Russian Fed.	1.7	1.8 (2.0)	Ukraine	1.3	4.5 (6.5)	India	2.8	2.6 (3.5)	India	1.3	4.7 (5.8)	Congo	0.06	6.8
Iran	1.2	1.3 (1.5)	Russian Fed.	1.1	4.1 (5.8)	Iran	2.7	2.6 (3.4)	Iran	1.3	4.5 (5.6)	Cameroon	0.05	5.2
Mexico	1.2	1.3 (1.4)	United States	1.1	3.8 (5.4)	Algeria	2.3	2.1 (2.9)	Kazakhstan	1.2	4.1 (5.1)	Mali	0.04	4.8
Ukraine	1.0	1.1 (1.2)	Spain	0.8	2.8 (4.0)	Brazil	2.2	2.0 (2.7)	Afghanistan	0.8	2.8 (3.5)	African Rep.	0.04	4.1
Uzbekistan	0.8	0.9 (1.0)	Mexico	0.8	2.7 (3.9)	United States	1.7	1.6. (2.2)	United States	0.7	2.4 (3.1)	China	0.03	3.2
Spain	0.8	0.9 (1.0)	Turkey	0.7	2.5 (3.6)	Senegal	1.7	1.6 (2.1)	Guatemala	0.7	2.3 (2.9)	Iran	0.03	3.1
United States	0.6	0.7 (0.8)	Bangladesh	0.7	2.4 (3.4)	Russian Fed.	1.6	1.5 (2.0)	Brazil	0.6	2.2 (2.7)	Chad	0.02	2.5
Egypt	0.6	0.7 (0.7)	Italy	0.6	2.1 (3.1)	Egypt	1.5	1.4 (1.9)	Mexico	0.6	2.2 (2.7)	South Sudan	0.02	2.4

Data are from FAO (http://www.fao.org/faostat/en/#data/QC (FAO, 2019) accessed in January 10, 2023).

crops, the top 10 producing countries contribute greater than 70% of the global yield. China is the main producer of each of these crops, contributing for 78.5% production of cucumber (representing 88.1% production by the top 10), 56.5% of watermelon (75.1% of top 10), 48.7% of melons (61.0% of top 10), and 26.7% of pumpkin and squash (38.3% of top 10). Melon seed (egusi) is another Cucurbitaceae crop recorded by FAO, with a worldwide production (led by Nigeria) of 0.93 Mt (Fao, 2019/2019).

Although leaves and roots are often used in some culinary practices, fruits, and seeds are the preferable consumed parts of Cucurbitaceae species (Grumet et al., 2021; Mukherjee et al., 2022). Cucurbitacea seeds present high contents of proteins, fat, vitamins, and minerals and are consumed raw, roasted, cooked, or used for oil extraction and medicinal purposes (Dhillon et al., 2016; Hussain et al., 2022; Salehi et al., 2019, 2021). They present important bioactive compounds, such as carotenoids, tocopherols, phenols, terpenoids, triterpenoids (e.g., cucurbitacins), saponins, sterols, fatty acids, and functional polysaccharides and proteins/peptides, which explain the pharmacological effects evidenced by Cucurbitaceae seeds, including antidiabetic, analgesic, nephroprotective, and anticancer properties (Mukherjee et al., 2022; Salehi et al., 2021). This fact has prompted a huge interest in its applications in folk medicine. For instance, in the Iberian Peninsula, Argentina, and India, raw *Cucurbita maxima* seeds are ingested for the treatment of intestinal worms and constipation. In Mauritius and Southern Italy, dried *Cucurbita maxima* seeds are consumed to treat renal failure and prostatitis. Accordingly, in Ethiopia and Palestine, *Cucurbita pepo* seeds are respectively used for gonorrhea and breast cancer treatment (Salehi et al., 2019). Another example is the use of *Citrullus lanatus* (watermelon) seed extracts to cure cancer, cardiovascular diseases, hypertension, edema, and urinary tract problems, in various parts of the world (Yadav et al., 2011). Hence, Cucurbitaceae seeds proteins have gained special interest in the development of nutraceuticals and functional foods due to their great potential as a source of bioactive peptides (Salehi et al., 2019; Wong et al., 2021; Ozuna & León-Galván, 2017) provided a comprehensive review of the potential of Cucurbitaceae seed protein isolates/hydrolysates as a source of bioactive peptides. These authors compiled information about the protein isolation processes and its subsequent enzymatic hydrolysis and summarized the nutritional composition and bioavailability of Cucurbitaceae seed hydrolysates, but relatively little information was elucidated regarding the bioavailability of Cucurbitaceae-derived peptides and their effectiveness on human health. Taking this into account, this chapter aims to delineate the recent developments in Cucurbiateceae seed proteins hydrolysates and derived bioactive peptides and discuss the bioaccessibility of Cucurbitaceae seed-derived bioactive peptides, using the Web of Science, Scopus, and Google search engines.

11.2 Cucurbitaceae seed proteins as a source of bioactive peptides

Cucurbitaceae seeds and their defatted cakes are rich in proteins with a well-balanced essential amino acid composition. This aspect has encouraged the production of protein hydrolysates and bioactive peptides as alternatives to animal proteins, which are usually associated with a high carbon footprint. Although intact Cucurbitaceae seed proteins can present biological activity, its derived peptides have been proven to have a higher potential for in vivo bioactivities, due to their ability to be absorbed more easily through the small intestine epithelium (Singh et al., 2022). Bioactive peptides derived from Cucurbitaceae seeds originate from the breakdown of proteins by endogenous and exogenous peptidases. They have been shown to present a positive biological activity in the human metabolism system, such as antidiabetic, antihypertensive, anticancer, antiinflammatory, antimicrobial, and antioxidant activities (Meena et al., 2020; Ozuna & León-Galván, 2017; Priyanto et al., 2021; Singh et al., 2023). Several bioactive peptides have been released from Cucurbitaceae seed protein hydrolysates, as summarized in and discussed in the next section.

The release of bioactive peptides from Cucurbitaceae seed proteins involves various processing steps, which are divided into four main stages: protein extraction, proteolysis, fractionation, and peptides characterization (Fig. 11.2).

Seedsare separated from Cucurbitaceae fruit, dehulled, grounded to increase the seed's surface area, and defatted with organic solvents to help the hydrolysis efficacy. The final product is referred to as seed meal or seed flour and is used for the protein extraction. The protein recovery from seed material is a challenging process due to the presence of large vacuoles, the rigidity and thickness of cell walls, the heterogeneity of proteins, as well as the presence of interferent compounds, namely lipids, polysaccharides, and phenolic compounds.(Olivares-Galván et al., 2020). Several chemical (e.g., salt, alcohol, or alkali treatments, followed by isoelectric precipitation), physical [e.g., ultrasonic (Wen et al., 2019), pulsed electric field (Naik et al., 2022)], and enzymatic methods, alone or in combination, have been applied to produce Cucurbitaceae seeds protein hydrolysates and derived bioactive peptides. In a general way, the chemical extraction involves extracting proteins in a solvent, and its precipitation using the isoelectric pH method. The Osborne method (Osborne, 1924/1924) has been widely employed for the extraction and fractionation of Cucurbitaceae proteins based on their solubility in different media: water (albumins), salt solution (globulins), alkaline solution (glutelins), and alcoholic solution (prolamins). This methodology has been applied for the extraction of proteins from pumpkin (Dash & Ghosh, 2017b), bottle gourd (Dash et al., 2018), melon (Siddeeg et al., 2015), and watermelon (Dash & Ghosh, 2017a) seeds. For pumpkin seeds, the globulin fraction held the highest protein content (46%), followed by glutelins (39%), albumins (23%), and prolamins (12%) (Dash et al., 2018). For melon seeds, the glutelin fraction was the richest one (81%), while the prolamin content was 6% (Siddeeg et al., 2015). The globulin fractions from bottle gourd, watermelon, and pumpkin seeds presented the highest amount of hydrophobic amino acids and antioxidant capacity; in the case of melon seed proteins, the glutelin fraction presented the highest presence of polyphenols and yielded the highest antioxidant activity (Dash et al., 2020).

Without any sequential hydrolysis, some Cucurbitaceae seed protein concentrates/isolates, consisting of about 80%–90% of protein, were reported to present

Product	Step	Operation	Method
Cucurbitacea seeds	***Seeds material***	*Debulling*	
		Disintegration	*sieving*
		Defatting	*n-hexane*
		Cell membrane disruption	Solvent extraction Ultrasonic extraction Pulsed electric field extraction High-pressure extraction
		Fractionation	Alkali/alcohol/water/salt extraction followed by isoelectric precipitation
		Concentration	Centrifugation
Cucurbitacea seed protein powder			
	Enzymatic hydrolysis	In silico digestion	BIOPEP-UWM™
		OR	
		Enzymatic treatment	Endoproteaes Exoproteases Endo + exoproteases
		Optimization	E/S ratio, pH, temperature, time
		Heating	95-100 °C, 5-15 min
	Protein hydrolysate recovery	Precipitation	Ammonium sulphate, alcohol, acid
		Enrichment	Solid-phase extraction (SPE), Ultrafiltration (Ul
		Dehydration	Spray drying, lyophilization, evaporation
		Storage (power)	Temperatures between -18 °C and 4 °C
Cucurbitacea seed protein hydrolysates			
	Characterization	MW distribution, amino acids composition, hydrophobicity,	
	Bioaccessibility	Simulated GI (Static, dynamic, semi-dynamic models)	
	Bioactive properties	Antioxidant, antidiabetic, anti-hypertensive, anti-inflammatory, anticancer, antifungal; Zinc-chelating, binding affinity to SARS-Co' viral protein, Influenza A inactivation	
	Fractionation	SPE, SEC, UF, HPLC, Semi-preparative HPLC, Ion-exchange HPL IMAC, GF	
Cucurbitacea seed-derived peptides	**Characterization, Purification and Identification**	SDS-PAGE, LC/MS/MS, MALDI-TOF MS, QTOF-MS with ESI Bioinformatic and computational tools (BIOPEP, PEPTIDERANKER, BITTERDB databases)	
	Bioaccessibility	Simulated GI (Static, dynamic, semi-dynamic models)	
	Bioavailability	Cell model studies In vivo and clinical studies	
	Bioactive properties	Antioxidant, antidiabetic, anti-hypertensive, anti-inflammatory, anticancer, antifungal; Zinc-chelating, binding affinity to SARS-Co' viral protein, Influenza A inactivation	

(*Continued*)

some bioactive and functional properties (Ozuna & León-Galván, 2017). Following the extraction process, the protein isolates (or the defatted seed meal) are submitted to enzymatic hydrolysis. Different proteases have been applied to produce bioactive hydrolysates (and derived bioactive peptides) from *Cucurbitaceae* seed proteins, including endoproteases (e.g., alcalase, pepsin, trypsin, protamex, neutrase, α-chymotrypsin, bromelain, papain), mixtures of endo- and exoproteases (e.g., flavourzyme) or sequential treatments (e.g., alcalase + flavourzyme; trypsin + pepsin; pepsin + trypsin). The combined treatment with endo and exoproteases usually produce shorter peptide fragments with higher bioactive potential (Ozuna & León-Galván, 2017). As hydrolysates are a complex mixture of free amino acids and peptides with diverse chain-length peptides and sequences, important parameters have been optimized to maximize the hydrolysis efficiency and control the production of the targeted bioactive peptides. The main parameters affecting the hydrolysis yield included the enzyme-to-substrate (E/S) ratio, pH, time, and temperature (Akbarian et al., 2022; Olivares-Galván et al., 2020; Singh et al., 2023). Generally, the hydrolytic efficiency is measured in terms of the degree of hydrolysis (DH), defined as the proportion of cleaved peptide bonds in a protein hydrolysate. DH strongly affects the molecular weight (MW) distribution of peptides and their bio-functional properties (Ozuna & León-Galván, 2017). The hydrolytic process is usually stopped by heating the reaction mixture up to 95°C–100°C for 5–15 min. The following step is the separation of the mixture of peptides present in the protein hydrolysate, referred to as fractionation, which is crucial to select the promising bioactive peptides for further purification and characterization. Several fraction methods have been applied to Cucurbitaceae seed protein hydrolysates and derived peptides, including solid-phase extraction, size exclusion chromatography, ultrafiltration (UF), high-performance liquid chromatography, hydrophilic interaction liquid chromatography, and next-generation chromatography (Ngamsuk et al., 2020). UF is a poor selective membrane separation technique that uses membrane fibers with various pore sizes and molecular weight cut-off (MWCO), usually from 3 to 10 kDa, to separate peptides according to their molecular size (Olivares-Galván et al., 2020).

FIGURE 11.2 General overview of the procedures involved in the production and characterization of Cucurbitaceae seed hydrolysates and derived bioactive peptides.

Legend: enzyme/substrate (E/S); gastrointestinal digestion (GI); molecular weight (MW); food-derived peptides with in silico-predicted biological activity (BIOPEP-UWM); solid-phase extraction (SPE); size exclusion chromatography (SEC); ultrafiltration (UF); high-performance liquid chromatography (HPLC); immobilized metal affinity chromatography (IMAC); gel filtration chromatography (GF); sodium dodecyl sulfate-polyacrylamide gel electrophoresis (SDS-PAGE); liquid chromatography–tandem mass spectrometry (LC/MS/MS); matrix-assisted laser desorption ionization-time-of-flight mass spectrometry (MALDI-TOF MS); quadrupole time-of-flight mass spectrometer with an electro-spray ionization (QTOF-MS with ESI).

The obtained supernatant, rich in bioactive peptides, is then lyophilized and characterized in terms of MW distribution, amino acid composition, bioactive properties, and bioaccessibility. Proteomic technologies plugged with mass spectrometric tools, such as matrix-assisted laser desorption/ionization time-of-flight mass spectrometry, capillary electrophoresis time-of-flight mass spectrometry, ultra-high-performance liquid chromatography quadrupole time-of-flight mass spectrometry, liquid chromatography–mass spectrometry, and gas chromatography–mass spectrometry have emerged to identify and characterize the Cucurbitacea seed-derived bioactive peptides (Singh et al., 2023).

The bioaccessibility of a protein hydrolysate or derived bioactive peptides evaluates its resistance to gastrointestinal (GI) enzymes and brush border membrane peptidases, being an indicator of absorption over the intestinal barrier and systemic circulation stability (Alblooshi et al., 2023; Samtiya et al., 2021). Simulated in vitro GI digestion models are used to mimic the human GI environment and investigate the complex influences of diverse factors on peptide availability. They are classified as static and dynamic in vitro models. The first ones are used to study the bioaccessibility of a target peptide using a sequential exposure to simulate digestion in different compartments: mouth, stomach, and intestine. The dynamic in vitro models reproduce the gradual transit of ingested hydrolysate/ bioactive peptide through the GI tract using a multicompartment computer-controlled system, providing more reliable and reproducible data (Lee et al., 2018). Moreover, the dynamic conditions give information about the effect of microbial metabolization and their modulating effect on target peptides, over longer periods of time under representative conditions (Duque-Soto et al., 2023). TIM (TNO GI model), DIDGI (dynamic GI digester), and ESIN (engineered stomach and small intestine) are some examples of dynamic in vitro models (Lee et al., 2018). Although the simulated digestion systems explain diverse aspects related to the metabolism and absorption of bioactive peptides, new methods, such as two-dimensional cell cultures and three-dimensional structure living cells, are recommended as being more physiologically relevant (Aguilar-Toala et al., 2022). Also, innovative bioinformatic tools have been used to predict the potential of a Cucurbitaceae seed protein as a precursor of bioactive peptides. These tools also evaluate the safety, allergenicity, and sensory qualities of bioactive peptides, as well as their bioavailability and effectiveness in the human system (Chai et al., 2021; Singh et al., 2022). Some of the most widely used bioinformatic tools and databases for discovering and developing bioactive peptides were recently reviewed (Singh et al., 2022; Singh et al., 2023). Some examples of these bioinformatic tools are the BIOPEP database, the PEPTIDERANKER database for the prediction of potential bioactivity, and the BITTERDB database for the prediction of tasting potential. In silico methodologies are economical, less laboratory intensive, and can predict the structural and functional activities of the bioactive peptides (Singh et al., 2023). For this reason, these tools have been used to better elucidate the mechanism of Cucurbitaceae bioactive peptides (Wen et al., 2021).

11.2.1 Pumpkin seed hydrolysates and derived bioactive peptides

The pumpkin family includes 12 species, embracing 4 domestic and economically important species: *Cucurbita moschata*, *C. maxima*, *C. pepo*, and *C. mixta* (Hussain et al., 2022). As mentioned before, the worldwide production of pumpkin is led by China (Table 11.1), being the *Cucurbita moschata* seeds one of the major oil seeds in the world (Vinayashree & Vasu, 2021). Pumpkin seeds are conventionally used to treat digestive disorders and prevent kidney stones (Mukherjee et al., 2022). They are an excellent source of proteins (21%–44% dw, dry weight), mostly represented by the storage proteins 2S globulin, 2S albumin, and 11S globulin, with high quality due to the presence of all essential amino acids (Bučko et al., 2016; Chelliah 2018; Nourmohammadi et al., 2017; Rezig et al., 2013; Rico et al., 2020). Almost of pumpkin seeds (around 90%) is for edible oil extraction; the largest by-product from this industrial process is the pumpkin seed meal, which contains 60%–70% dw of protein (Hussain et al., 2022; Liang et al., 2021). Pumpkin seeds are also a good source of dietary zinc (mean contents of 91.2 μg/g dw) (Ovca et al., 2011), making them a good source of zinc-chelating peptides (Lu et al., 2021; Peng et al., 2022). Several pumpkin seed hydrolysates and derived bioactive peptides have been prepared from pumpkin and pumpkin seed meal, displaying antioxidant (Nourmohammadi et al., 2017; Popović et al., 2017; Vaštag et al., 2011), antidiabetic (Park et al., 2015), antifungal (Abdel-Rahim et al., 2018; Vassiliou et al., 1998), antimicrobial (Barbieri et al., 2006; Wang & Ng, 2003), antiinflammatory (Perez Gutierrez, 2016), and hepatoprotective activities (Ozuna & León-Galván, 2017). However, only four studies reported the effects of GI digestion on the bio-functional properties of pumpkin-derived peptides, as summarized in Table 11.2. Liang et al. (2021) identified a novel tripeptide IAF from pumpkin seed proteins by in silico GI approach. Two sequences of pumpkin seed proteins, 11S globulin subunit beta, and 2S albumin, were selected to determine the angiotensin I-converting enzyme (ACE) inhibitory peptides occurrence by the BIOPEP-UWM database. A total of 111 peptides were released from the 2 protein sequences after in silico GI digestion with alcalase, including 20 tripeptides and 27 dipeptides. From the virtual screening, IAF was selected as the promising bioactive peptide, exhibiting a considerable ACE inhibitory ability (IC_{50} of 19.87 ± 0.50 μM). Wong et al. (2021) also performed in silico GI model (BIOPEP-UWM database) to assess the bioaccessibility of *Cucurbita maxima-derived* peptides with antiviral properties. Results showed that the pumpkin-derived peptide PW could bind more strongly to SARS-CoV-2 spike protein, presenting a higher antiviral activity than some previously reported anti-SARS-CoV-2 (Table 11.2). Literature also reports the bioaccessibility of zinc-chelating peptides released from pumpkin seed proteins. For instance, Lu et al. (2021) prepared pumpkin seed-derived peptides by treatment with bromelain, papain, flavourzyme, alcalase, and pepsin. It was found that the papain hydrolysate had the smallest particle size, highest hydrophobicity, the greatest zinc-chelating capability and retained more than 50% of zinc after

Table 11.2 Effect of gastrointestinal (GI) digestion on the bio-functional properties of Cucurbitaceae seed hydrolysates and/or derived peptides.

Seed source	Protein digestion	Protein hydrolysate or peptide sequence (MW, Da)	Bioactivity of protein hydrolysate/peptide	GI digestion model	Bioactivity post-GI digestion	Cell Transport model	Bioactivity posttransport	Bioactivity in vivo	References
Pumpkin (*Cucurbita moschata*)	In silico digestion (BIOPEP-UWM)	IAF	ACE-I activity	In silico digestion (SwissADME)	ACE-I activity IC_{50} (IAF) - 19.87 ± 0.50 μM	NE	NE	NE	(Liang et al., 2021)
Pumpkin (*Cucurbita maxima*)	In silico digestion (BIOPEP-UWM)	PW	Binding affinity to SARS-CoV-2 viral protein	In silico digestion (SwissADME)	Binding affinity to SARS-CoV-2 viral protein	NE	NE	NE	(Wong et al., 2021)
Pumpkin (*Cucurbita pepo L.*)	Comparative treatment using E/S of 2% (w/w) for 3 h: Bromelain (pH 6.5, 37°C) Alcalase (pH 8.0, 50°C) Flavourzyme (pH 7.0, 55°C) Papain (pH 7.0, 55°C)Trypsin (pH 7.5, 37°C)	Protein hydrolysates (<180 and >5000 Da)	Zinc-chelating ability 64.7 ± 9.1%; 62.5 ± 5.6%; 57.9 ± 3.8%; 76.3 ± 8.7%; 39.3 ± 3.0% for bromelain, alcalase, flavourzyme, papain and trypsin hydrolysates, respectively	Static model: pepsin (E/S ratio 1:100 w/w, pH 2.0, 37°C for 30 min) followed by pancreatin (E/S ratio 1:25 w/w, pH 7.5, 37°C for 180 min)	Zinc-chelating stability in the stomach: 90.06% for papain hydrolysate 83.23% for trypsin hydrolysateZinc-chelating stability in the intestine:52.99% for papain hydrolysate33.23% for trypsin hydrolysate	NE	NE	NE	(Lu et al., 2021)
Pumpkin (*Cucurbita pepo L.*)	Papain (E/S of 2% (w/w), pH 7.0, 50°C, for 3 h)	RPKHPLK-Zn(871 Da) RPKHPLSHDL-Zn(1198 Da)	Zinc-chelating ability:6.48 ± 0.69 mg/mg for RPKHPLK-Zn;92.87 ± 0.41 mg/mg for RPKHPLSHDL-Zn	Static model: pepsin (E/S ratio 1:100 w/w, pH 2.0, 37°C for 30 min) followed by pancreatin (E/S ratio 1:25 w/w, pH 7.5, 37°C for 180 min)	Zinc-chelating stability in the stomach:95.42% for RPKHPLK-Zn97.26% for RPKHPLSHDL-ZnZinc-chelating stability in the intestine:63.90% for RPKHPLK-Zn78.89% for RPKHPLSHDL-Zn	Caco-2 cell monolayer model	Zinc-chelating ability:33.70% for RPKHPLK-Zn and 40.37% for RPKHPLSHDL-Zn	NE	(Peng et al., 2022)

(Continued)

Table 11.2 Effect of gastrointestinal (GI) digestion on the bio-functional properties of Cucurbitaceae seed hydrolysates and/or derived peptides. *Continued*

Seed source	Protein digestion	Protein hydrolysate or peptide sequence (MW, Da)	Bioactivity of protein hydrolysate/ peptide	GI digestion model	Bioactivity post-GI digestion	Cell Transport model	Bioactivity posttransport	Bioactivity in vivo	References
Bitter melon *(Momordica charantia)*	Ultrasound pretreatment followed by comparative treatment using E/S ratio of 1:50 (w/w) for 12 h: Trypsin (pH 8, 37°C) Chymotrypsin (pH 8, 37°C) Pepsin (pH 1.5, 37°C)Alcalase (pH 8, 50°C) Thermolysin (pH 8, 60°C)	Thermolysin peptides: VSGAGRY(VY-7071 Da) VDSDVVKG(VG-8080 Da)	ACE-I activityIC$_{50}$ (VY-7) - 8.64 μMIC$_{50}$ (VG-8) - 13.3 μM	Static model: pepsin (E/S ratio 1:50 w/w, pH 2.0, 37°C for 90 min) followed by Corolase PP (E/S ratio 1:25 w/w, pH 7–8, 37°C for 240 min)	ACE-I activity:50% to 49% for VY-750% to 28% for VG-8	NE	NE	SHR model: The ingestion of VY-7 at 2 mg/kg bw reduces ($P < .01$) the SBP from 220 to 180 mmHg	(Priyanto et al., 2015)
Bitter melon *(Momordica charantia)*	Extraction with phosphate-buffered saline (pH 7.2), at 4°C overnight stirring	mcIRBP-19 peptide (∼7000 Da) RVRVWVTERGIVARPPTIG	Hypoglycemic activity via affecting insulin receptor signaling transduction pathway	Static model: pepsin (34.2 mM NaCl, 80 mM HCl, pH 3.0, 37°C for 30 min) followed by pancreatin (100 mM KH2PO4, 30 mM NaOH, pH 7.0, 37°C for 60 min)	Hypoglycemic activity:IVARPPTIG accounted for the majority (20.78%) of digested peptides	NE	NE	Type 1 diabetic mice model: IVARPPTIG peptide targeted insulin receptors and improved glucose tolerance	(Lo et al., 2017)
Bitter melon *(Momordica charantia)*	Three-step procedure purification (ammonium sulfate precipitation, affinity chromatography and size exclusion chromatography)	Mc7S globulin(57 kDa)	ACE-I activity	In silico model (BIOPEP-UWM)	ACE-I activity:Higher binding affinity for ACE than lisinopril	NE	NE	NE	(Kesari et al., 2020)

(Continued)

Table 11.2 Effect of gastrointestinal (GI) digestion on the bio-functional properties of Cucurbitaceae seed hydrolysates and/or derived peptides. *Continued*

Seed source	Protein digestion	Protein hydrolysate or peptide sequence (MW, Da)	Bioactivity of protein hydrolysate/peptide	GI digestion model	Bioactivity post-GI digestion	Cell Transport model	Bioactivity posttransport	Bioactivity in vivo	References
Watermelon (*Citrullus lanatus*)	Ultrasound pretreatment (28, 33, 40 kHz frequencies at 30°C–50°C) followed by alcalase hydrolysis (pH 8.0, 50°C, E/S of 1% (v/v) for 10, 30, 50, 70, and 90 min) and by Ultrafiltration (1 or 5 kDa-MWCO)	Protein hydrolysates: WSPHs-I (<1000 Da), WSPHs-II (1000–5000 Da) WSPHs-III (>5000 Da)	Antioxidant activity: DPPH$^•$ scavenging rate and Fe^{2+} chelating capacity of protein hydrolysates were significantly ($P < .05$) improved by ultrasound and ultrafiltration treatments RAW 264.7 cells were protected from H_2O_2-induced oxidative stress damage via activating the Nrf2/HO-1 pathway	Static model: pepsin (E/S ratio 1:30 g/mL, pH 2.0, 37°C for 120 min) followed by pancreatin (E/S ratio 1:25 w/w, pH 7.5, 37°C for 60 min)	Antioxidant stability in the stomach: DPPH$^•$ scavenging rate of WSPHs-I increased ($P < .05$) to 58.89%; Fe^{2+} chelating capacity did not changedAntioxidant stability in the intestine:DPPH$^•$ scavenging rate of WSPHs-I decreased to 42.12%; Fe^{2+} chelating capacity increased ($P < .05$) from 29.12% to 51.25%	NE	NE	NE	(Wen et al., 2019)

Not evaluated (NE); molecular weight (MW); gastrointestinal (GI); angiotensin I converting enzyme inhibition (ACE-I); systolic blood pressure (SBP); spontaneously hypertensive rat (SHR) model; body weight (bw); half maximal ACE inhibitory concentration (IC_{50}), enzyme/substrate (E/S); molecular weight cut-off (MWCO); 2,2-diphenyl-1-picryl-hydrazyl-hydrate scavenging activity (DPPH$^•$); nuclear factor erythroid 2-related factor transcription factor/hemoxygenase 1 (Nrf2/HO-1).
Alanine (A); cysteine (C); aspartic acid (D); glutamic acid (E); phenylalanine (F); glycine (G); histidine (H); isoleucine (I); lysine (K); leucine (L); methionine (M); asparagine (N); proline (P); glutamine (Q); arginine (R); serine (S); threonine (T); valine (V); tryptophan (W); tyrosine (Y); zinc (Zn).
BIOPEP-UWM (http://www.uwm.edu.pl/biochemia/index.php/en/biopep); SwissADME (http://www.swissadme.ch).

in vitro GI digestion. In subsequent work, the zinc-chelating peptides were evaluated for their in vitro bioaccessibility and Caco-2 cell monolayer permeability. Results showed that RPKHPLK-Zn and RPKHPLSHDL-Zn peptides presented higher GI stability than inorganic zinc salts in the simulated GI tract model, suggesting their efficacy in zinc absorption (Peng et al., 2022).

11.2.2 Bitter melon seeds hydrolysates and derived bioactive peptides

Bitter melon (*Momordica charantia* L.), also known as bitter gourd, is widely cultivated in Asia (especially in Taiwan), South America, and East Africa (Bora et al., 2023). Clinical studies have pointed to the pharmacological potential of bitter melon for the treatment of several chronic diseases. For instance, it was shown that the daily intake of bitter melon fruit for 4 weeks promoted hypoglycemic effects in patients with type 2 diabetes (Lo et al., 2013), being the insulin receptor binding peptide-19 (mcIRBP-19) responsible for the glucose uptake (Lo et al., 2014).

Bitter melon seeds are rich in oil ($\sim$27%–34% dw) and proteins ($\sim$28%–31% dw) and are preferably consumed cooked due to their hard outer shells (Naik et al., 2022). Seeds have shown great potential to produce protein hydrolysates and derived bioactive peptides with important bioactivities (Horax et al., 2011; Xu et al., 2022) namely anti-HIV (Lee-Huang et al., 1990), anticancer (Dia & Krishnan, 2016; Fang, Zhang, Ng, et al., 2012; Fang, Zhang, Wong, et al., 2012), antitumor (Li et al., 2009), antidiabetic (Liang et al., 2021; Murugesan et al., 2022; Priyanto et al., 2015), antioxidant (Kesari et al., 2020), and antihypertensive (Kesari et al., 2020; Lo et al., 2013, 2014) properties. The most studied proteins isolated from bitter melon seeds belong to the ribosome-inactivating proteins family (RIP) and include lectin, momordins, momorcharins, and momordica anti-HIV protein (Bora et al., 2023; Naik et al., 2022).

Priyanto et al. (2015) prepared different bitter melon seed hydrolysates by enzymatic treatment with trypsin, α-chymotrypsin, pepsin, alcalase, and thermolysin and screened the ACE inhibition activity at the same concentration of 1 mg/mL. Two peptides were isolated, VY-7 and VG-8, derived respectively from alpha-momorcharin (α-MMC) and momordica anti-HIV protein 30 (MAP30), and presenting IC_{50} values of 8.64 ± 0.60 and 13.30 ± 0.62 μM (Table 11.2). The in vitro GI stability studies showed that the ACE-I activity of VY-7 peptide slightly decreased from 50% to 49%, suggesting its resistance to GI peptidases. For the VG-8 peptide, the ACE-I activity decreased from 50% to 28%, suggesting an apparent degradation. The in vivo antihypertensive activity of VY-7 was further assessed by measurement the systolic blood pressure (SBP) of spontaneously hypertensive rats (SHRs), within after 24 h oral administration. Data showed that VY-7 peptide significantly ($P < .01$) reduced the average SBP of SHRs from 220 to 180 mmHg after 8 h of oral administration, demonstrating an antihypertensive effect. Jones et al. (2018) isolated a bioactive peptide (BG-4) from bitter melon seeds with antioxidant and antiinflammatory potential. The isolated peptide reduced the production of pro-inflammatory cytokines IL-6 and TNF-α in a dose-dependent manner. However, the BG-4 stability to GI digestion was not evaluated. More recently Kesari et al. (2020) isolated and characterized a 52 kDa glycoprotein (*Mc*7S) from bitter melon seeds. *Mc*7S displayed an increase in DPPH$^{\bullet}$ (2,2-diphenyl-1-picryl-hydrazyl-hydrate) scavenging activity from 0.0 to 1.5 mg/mL, which was comparable to ascorbic acid scavenging activity (0.5–1.0 mg/mL). Additionally, in silico digestion indicated that *Mc*7S released the peptide VFK (valin-phenylalanine-lysine) with ACE inhibitory activity.

Bitter melon peptides have also been identified with anticancer attributes. Fang, Zhang, Ng, et al. (2012) investigated the antitumor activity of lectin, a type II-RIP with a MW of 130 kDa. Lectin displayed potent cytotoxicity toward CNE-1 (well differentiated) and CNE-2 (poorly differentiated) human nasopharyngeal carcinoma cell lines, with respective IC_{50} values of 6.9 ± 0.2 and 7.4 ± 0.4 μmol/L. Also, Dia and Krishnan (2016) isolated a trypsin inhibitor (BG-4) from bitter melon seed and demonstrated its potential therapeutic effect against two human colon cancer cell lines, HCT-116 and HT-29. Data showed that BG-4 peptide could significantly inhibit the proliferation of both colon cancer cells through the downregulation of antiapoptotic proteins Bcl-2 and XIAP, and upregulation of proapoptotic proteins. Similarly, another study examined the anticancer activity of MAP30 protein, revealing that MAP30 positively modulated glycolysis and lipid metabolism, leading to a significant reduction in ovarian tumor dissemination (David W Chan et al., 2020).

The hypoglycemic and antidiabetic properties of bioactive peptides derived from bitter melon seeds have also been reported in the literature. In this topic, Lo et al. (2013, 2014) identified a hypoglycemic polypeptide, referred as mcIRPB (MW of 7 kDa) from an aqueous extract that was identified as a novel insulin receptor binding protein (IRBP). Simulated GI digestion of mcIRPB found that IVARPPTIG peptide accounted 20.78% of fragments and was resistant to GI enzymes, improving the glucose tolerance in type 1 diabetic mice. Moreover, a pilot study showed that daily ingestion of IVARPPTIG peptide for 30 days improved glucose tolerance and decreased the HbA1c levels via IR signaling pathway (Lo et al., 2017) More recently Murugesan et al. (2022) isolated an 11 kDa napin-like protein from bitter melon seed, referred as Mc-3.2-1, that significantly lower the blood glucose level in alloxan-diabetic rats by 55%.

11.2.3 Melon seed hydrolysates and derived bioactive peptides

Melon (*Cucumis melo* L.) is one of the most consumed fruit crops worldwide (Ribeiro et al., 2007), with an annual production of 30 million tons, as reported in Table 11.1. Its high antioxidant potential is mostly attributed to the phytochemicals (mainly phenolic compounds), amino acids, and organic acids composition (Rico et al., 2020). The protein fraction of melon seeds (17%–35% dw(Rico et al., 2020)) is represented by glutelin (81.80%), albumin (73.09%), and globulin

(50.95%) (Siddeeg et al., 2015), and has been explored for bioactive peptides production (Mukherjee et al., 2022; Ribeiro et al., 2007; Siddeeg et al., 2015). Ribeiro et al. (2007) isolated vicilin-like peptides with antifungal properties from melon seeds. Proteins were extracted with phosphate buffer, pH 5.4 and subjected to precipitation with 90% relative ammonium sulfate saturation. According to (Siddeeg et al., 2015) the glutelin fraction obtained by trypsin and pepsin digestion showed the highest content of the total phenolic (43.53 μg Gallic Acid Equivalents (GAE)/ mL) and the strongest antioxidant activity (assessed by hydroxyl radical scavenging, DPPH$^•$ scavenging activity, metal chelating activity, and reducing power assays). For both melon-derived bioactive peptides, no data is available concerning their resistance to GI digestion, as well as pharmacokinetics and toxicity responses in in vivo models.

11.2.4 Watermelon seeds hydrolysates and derived bioactive peptides

Watermelon (*Citrullus lanatus*) is the largest fruit crop in the world (Fao, 2019/2019), with extensive production in China (Table 11.1). Watermelon fruit contains vitamins A, B, C and E, minerals (K, Mg, Ca, and Fe), specific amino acids (citrulline and arginine), lycopene, and phenolic compounds, which exhibit a markedly antioxidant activity. Watermelon fruit has been used in traditional medicine for the treatment of urinary tract infections and renal stones (Rico et al., 2020; Zamuz et al., 2021). The roasted seeds are served as snacks or used to prepare soups, while the seed extracts are often used in folk medicine for acute eczema and urinary tract infections (Yadav et al., 2011). Watermelon seeds are excellent source of protein (18%–25% dw), with a well-balanced amino acid profile (Rico et al., 2020; Wen et al., 2019) and a good source of vitamin B-complex (B1, B2, B3, B6, B12) and minerals (Mg, Ca, K, Fe, P, Zn) (Yadav et al., 2011; Zia et al., 2021). Globulin represents the major protein fraction (27.5% dw) of watermelon seeds, followed by albumin (4% dw), prolamin (1.8% dw), and glutelin (0.7% dw) (Teugwa et al., 2013). All these protein fractions possess high potential for the preparation of bioactive peptides (Wen et al., 2021).

Dash and Ghosh (2017a) isolated globulin, glutelin, albumin, and prolamin proteins from watermelon seeds with respective concentrations of 46%, 39%, 23%, and 22%, rich in hydrophobic and acidic amino acids and showing antimicrobial and antioxidant activities. The same authors evaluated the in vivo antioxidant potential of the globulin fraction, at a dose of 1 gm/kb BW for 15 days, in ethanol-induced oxidative stressed rats. Data showed that the globulin fraction maintained normal cellular function and prevented oxidative stress by decreasing the MDA (product of lipid peroxidation) level and by restoring the antioxidant potential of catalase (Dash et al., 2020). The antioxidant potential of watermelon seed hydrolysates was also reported by Wen et al. (2019). These authors applied UAE (28, 33, 40 kHz frequencies at 30°C–50°C) and UF (1 or 5 kDa-MWCO) techniques to prepare a watermelon seed protein hydrolysate. The hydrolysate fraction with a MW lower than <1 kDa exhibited a cytoprotective effect by protecting RAW 264.7 cells from damage from H_2O_2-induced oxidative damage; good stability against oxidation at temperature under 100°C, and a strong antioxidant activity after simulated GI digestion. Moreover, the DPPH$^\bullet$ scavenging rate of hydrolysate increased to 58.89% after 2 h of pepsin digestion and decreased to 42.12% after 1 h of trypsin digestion; the Fe^{2+} chelating capacity was similar after pepsin digestion and decrease ($P < .05$) from 29.12% to 51.25% after trypsin digestion. Then, the same authors purified the antioxidant peptide from watermelon seed protein hydrolysate, referred to as RDPEER, reporting HepG2 cells protection from H_2O_2-induced oxidative damage (Wen et al., 2020). No information was however given regarding the stability of RDPEER peptide to GI digestion and its effective in vivo biological activity (Table 11.2).

11.2.5 Gac seeds hydrolysates and derived bioactive peptides

Gac (*Momordica cochinchinensis* Spreng.) is widely produced in Southeast Asian countries, including Thailand, Laos, Myanmar, Cambodia, Vietnam, Malaysia, India, and Taiwan (Do et al., 2019). Seeds are traditionally used in folk medicine to treat certain diseases, namely diabetes, eye disorders, fluxes, liver spleen disorders, and hemorrhoids (Ngamsuk et al., 2020). Literature reports that gac seed protein extracts (Nguyen et al., 2023), hydrolysates and derived peptides exert anticancer activity (Table 11.1). Some examples include 35 kDa (Tien et al., 2005) and 30 kDa proteins (Bolognesi et al., 1990), cochinin B (Chuethong et al., 2007), and MCoCC-1 and MCoCC-2 peptides (Chan et al., 2009). Ngamsuk et al. (2020) also screened the ACE-I activity of gac seed proteins hydrolysate. The gac seed proteins hydrolysate (<3 kDa), obtained by a combined treatment of pepsin, trypsin, and chymotrypsin, exhibited an IC_{50} of 70.0 ± 4 μg/mL, and the isolated peptide ALVY showed an IC_{50} of 7.03 ± 0.09 μM. Nevertheless, the GI stability of protein hydrolysate (<3 kDa) and the isolated ALVY peptide was not evaluated.

11.2.6 Bitter cucumber seeds hydrolysates and derived bioactive peptides

Citrullus colocynthis (L.) Schrad., commonly known as colocynth, bitter apple, bitter cucumber, desert gourd, or vine of Sodom, is commonly produced in North Africa coast. The protein fraction of bitter cucumber seed is around 8.25% dw and is composed of 60.2% nonessential amino acids and 39.8% essential amino acids (Alblooshi et al., 2023). Alblooshi et al. (2023) produced bitter cucumber seed protein hydrolysates using alcalase, bromelain, and chymotrypsin for 3, 6, and 9 h. Chymotrypsin digestion for 9 h promoted the highest DH (87.82%) and the strongest DPPH$^\bullet$ (65.7 mM TE/μg protein) and ABTS$^\bullet$ (525.2 mM TE/μg protein) values, as well as cholesteryl esterase inhibition (IC_{50} of 13.68 μg/mL) and pancreatic lipase inhibition (IC_{50} of 14.12 μg/mL) activities. Authors also observed that bromelain and alcalase treatments for 6 h were more effective in producing bitter cucumber seed protein hydrolysates with the highest α-glucosidase and α-amylase inhibitory activities (respective IC_{50} values of 13.27 μg/mL and 17 μg/mL). Bitter cucumber seed hydrolysates were not characterized regarding the resistance to GI digestion, as well as pharmacokinetics and toxicity along with studies on the in vivo models.

11.2.7 Bottle gourd seeds hydrolysates and derived bioactive peptides

Lagenaria siceraria, commonly known as bottle gourd or calabash gourd, is an important vegetable crop in India, with traditional use in the treatment of several diseases related to the skin and heart (Roopan et al., 2016). According to Akinsola et al. (2022) bottle gourd seeds are rich in oil (46.2% dw), protein (22.4% dw), phosphorous (554 mg/100 g dw), and several antinutrients, namely saponin (10.2 mg/100 g dw), cyanide (0.794 mg/100 g dw), tannin (1.31 mg/100 g dw), oxalate (10.3 mg/100 g dw) and phytate (19.3 mg/100 g dw). Bottle gourd seed protein hydrolysates have been claimed with various pharmacological activities, like antioxidant (Choudhary et al., 2013), antidiabetic (Bhattacharya & Das, 2012), antihyperglycemic (Teugwa et al., 2013), antiinflammatory (Gill et al., 2012), antimicrobial (Sood et al., 2012), and cardioprotective (Roopan et al., 2016). Dash et al. (2018) applied the Osborne method to isolate albumins, globulins, prolamins, and glutelins from bottle gourd seeds proteins. Protein fractions presented MW in the range of 10–260 kDa and the globulin fraction displayed the highest antioxidant activity, screened by $DPPH^{•}$, $ABTS^{•}$, nitric oxide radical scavenging, hydrogen peroxide scavenging, lipid peroxidation, reducing power and phosphomolybdate assays. Although the stability of the globulin hydrolysate to GI digestion was not evaluated, further work using in vivo models showed its effectiveness in reducing the risk of oxidative stressed, suggesting its potential use for the development of functional foods (Dash et al., 2020).

11.2.8 Wax gourd seeds hydrolysates and derived bioactive peptides

Wax gourd (*Benincasa hispida (Thunb) Cogn.*) is widely produced in Thailand and India, and routinely used in folk medicine for the management of peptic ulcer, hemorrhages, epilepsy, and other nervous disorders (Atiwetin et al., 2006; Gupta et al., 2021; Islam et al., 2021). Sharma et al. (2014) purified a 6.0-kDa peptide from wax gourd seeds showing potent inhibitory effects against various human bacterial and fungal pathogens, and antioxidant activity (70.8% $DPPH^{•}$ scavenging activity and 69.5% lipid peroxide inhibition) at a concentration level of 40 μg/mL. No data was reported about the stability of wax gourd seed hydrolysate to GI digestion, as well as effective biological activity based on the in vivo models.

11.3 Bioaccessibility of Cucurbitaceae seed hydrolysates and derived bioactive peptides

Bioaccessibility of a bioactive peptide refers to the capacity to cross the intestinal mucosa and reach the target organs in an active form to exert health-promoting effects (Zhao et al., 2022). This phenomenon depends on peptides' resistance to GI enzymes and brush border membrane peptidases during the absorption over the intestinal barrier, and its stability in the systemic circulation (Singh et al., 2022). Once solubilized within the GI fluids, the bioactive peptides are transported through the GI tract and absorbed by the epithelial cells (Zhao et al., 2022). Dipeptides and some tripeptides are usually transported primarily via a proton-coupled peptide transporter 1 (PepT1), oligopeptides across Caco-2 cell monolayers via a tight junction (TJs)-mediated paracellular pathway, while polypeptides with more than 10 amino acids and lipophilic peptides are generally transported via transcytosis (Wang et al., 2019, 2020).

The bioaccessibility of protein hydrolysates and derived bioactive peptides from Cucurbitaceae seed proteins has been preferably assessed by in vitro static models (Table 11.1). Pepsin, trypsin, pancreatic, chymotrypsin, and corolase PP have been commonly employed to simulate the human digestion process and evaluate the bioactive peptide resistance to GI degradation. The most evaluated properties of Cucurbitaceae-derived peptides after simulated GI digestion have been antioxidant (Wen et al., 2019), ACE-I (Liang et al., 2021; Priyanto et al., 2021), and zinc-chelating capacity (Lu et al., 2021; Peng et al., 2022). The extensive hydrolysis of bioactive peptides during GI digestion and absorption processes is the major concern. Thus additional in vitro and in vivo studies are required to establish the effective therapeutic role of bioactive peptides on human organisms (Singh et al., 2022).

Antioxidant peptides typically present MW between 0.5 and 1.5 kDa and a large amount of aromatic amino acids (e.g., histidine, tyrosine, and tryptophan) and hydrophobic amino acids (e.g., leucine, alanine, and phenylalanine) (Olivares-Galván et al., 2020). Wen et al. (2019) found that treating watermelon (*Citrullus lanatus*) hydrolysate (<1 kDa) with pepsin for 2 h resulted in a DPPH$^•$ scavenging rate increase ($P < .05$) to 58.89%, presumably due to the exposure of more hydrophobic amino acid residues. However, after trypsin digestion for 1 h, the DPPH$^•$ scavenging rate of hydrolysate decreased to 42.12%, probably due to the significant increase of hydrophilic components, that weakly react with fat-soluble DPPH radicals. The GI hydrolysate was not characterized in terms of its composition, structure, and hydrophobicity. Also, neither in vitro nor in vivo studies were carried out to ascertain the effective antioxidant capacity of watermelon seed hydrolysates.

ACE-inhibiting peptides are short-chain peptides (2–12 amino acids) that contain acidic (e.g., Asp and Glu) and hydrophobic amino acids (Görgüç et al., 2020). Peptides having hydrophobic amino acid residues valine (V) in the sequence have exhibited potent ACE-I action (Singh et al., 2022). This is the case of VY-7 and VG-8, two peptides found in bitter melon (*Momordica charantia*) seeds, after ultrasound pretreatment and thermolysin digestion. The simulated GI digestion with pepsin and corolase PP showed that the ACE-I activity increased 50%–49% for VY-7 and 50%–28% for VG-8. The in vivo models with SHR showed that the ingestion of VY-7 at 2 mg/kg bw was efficient in reducing ($P < .01$) the SBP from 220 to 180 mmHg (Priyanto et al., 2015).

Zinc-chelating peptides are ligands of peptides and zinc, usually incorporated into food systems to increase zinc ingestion. These peptides are absorbed directly into the small intestine through the peptide transport mechanism (Udechukwu et al., 2016) or via endocytosis (Regazzo et al., 2010). The presence of cysteine and histidine residues and acidic amino acids in the peptide sequence is related to improving the zinc-chelating rate (Peng et al., 2022). Also, the resistance of zinc-chelating peptides to gastric degradation is crucial to ensure that zinc ions can be released from the zinc-chelating peptides in the duodenum (Udechukwu et al., 2016). Pumpkin (*Cucurbita pepo* L.) protein hydrolysates (Lu et al., 2021) and derived peptides (Peng et al., 2022) from papain digestion have been reported to possess zinc-chelating ability. Two peptides, RPKHPLK and RPKHPLSHDL, were evaluated for their zinc-chelating capability, in vitro bioaccessibility, and absorption. Data from simulated GI digestion assay showed that RPKHPLSHDL-Zn had higher stability and chelation rate than RPKHPLK-Zn. Upon Caco-2 cell transport, it was found that RPKHPLSHDL-Zn and RPKHPLK-Zn peptides showed the highest increase rate in zinc transport, 40.37%, and 33.70%, respectively, suggesting that the zinc chelate of pumpkin seed polypeptide could be effectively absorbed and utilized by the gut (Peng et al., 2022).

11.4 Conclusions and future perspectives

Although Cucurbitaceae seed protein hydrolysates have been evaluated for a wide range of biological activities, only a few derived bioactive peptides were isolated and screened for in vitro antioxidant, antimicrobial, antiviral, antidiabetic, antihypertensive, anticancer, and zinc-chelating properties. The bioaccessibility of Cucurbitacea seed-derived peptides has been preferably assessed by in vitro static models, using pepsin, trypsin, pancreatin, chymotrypsin, and corolase PP enzymes; the resistant digested peptides have shown antioxidant, ACE inhibition, and zinc-chelating capacity (Fig. 11.3). As simulated GI digestion models may not fully reflect the true conditions that Cucurbitaceae seed-derived peptides face when consumed orally, animal studies and randomized human intervention trials should be conducted to establish their health benefits and allow their use for preventive or therapeutic treatments.

To date, only two Cucurbitaceae seed peptides, both obtained from Bitter melon (*Momordica charantia*) seeds, were evaluated regarding their in vivo efficacy. In both cases, animal models were used to assess the effective hypoglycemic and ACE-I activities of the isolated bioactive peptides, and no information was given regarding their potential toxicity, allergenicity, and bitterness. Taking this into account, in vivo and human clinical trials should be prioritized to evaluate the commercial application of Cucurbitaceae seed-derived bioactive peptides as nutraceuticals.

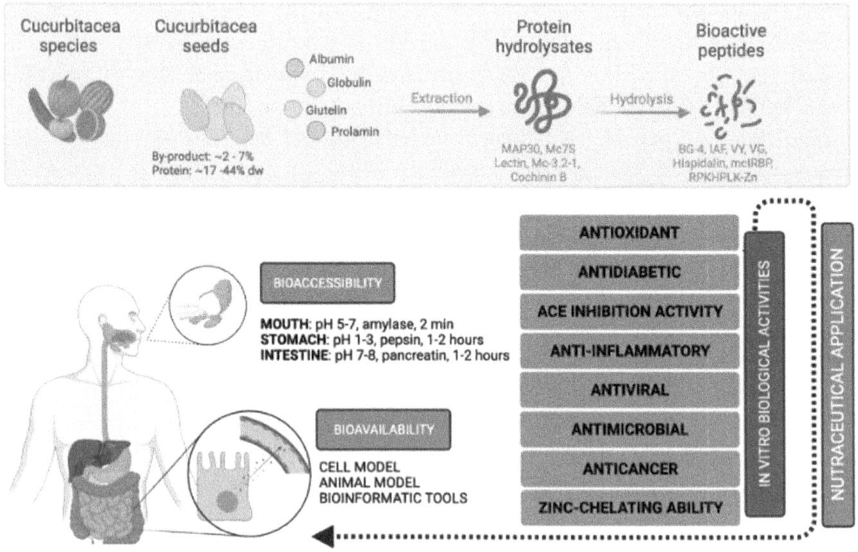

FIGURE 11.3

A schematic diagram showing the release of biologically active peptides from Cucurbitaceae seed proteins.

Funding

This work was financially supported through the projects UIDB/50006/2020 and UIDP/50006/2020 and LA/P/0008/2020, funded by FCT/MCTES through national funds.

Acknowledgments

This work received support from PT national funds (FCT/MCTES, Fundação para a Ciência e Tecnologia and Ministério da Ciência, Tecnologia e Ensino Superior) through the projects UIDB/50006/2020, UIDP/50006/2020, and LA/P/0008/2020. Elsa F. Vieira thanks FCT (Fundação para a Ciência e Tecnologia) for funding through the Scientific Employment Stimulus-Individual Call (Ref.CEECIND/03988/2018/CP1545/CT0006, https://doi.org/10.54499/CEECIND/03988/2018/CP1545/CT0006). The author also thanks the project SYSTEMIC "an integrated approach to the challenge of sustainable food systems: adaptive and mitigatory strategies to address climate change and malnutrition." The Knowledge hub on Nutrition and Food Security has received funding from national research funding parties in Belgium (FWO), France (INRA), Germany (BLE), Italy (MIPAAF), Latvia (IZM), Norway (RCN), Portugal (FCT), and Spain (AEI) in a joint action of JPI HDHL, JPI-OCEANS and FACCE-JPI launched in 2019 under the ERA-NET ERA-HDHL (no 696295).

Conflicts of interest

The authors declare that they have no known competing financial interests or personal relationships that could have appeared to influence the work reported in this chapter.

References

Abdel-Rahim, A. A., Osman, S. A., & Atif, A. (2018). Antimicrobial activity and some physiochemical properties of pumpkin (Cucurbita pepo L.) seed oil. *Gezira Journal of Engineering and Applied Science*, *10*(1).

Aguilar-Toala, J. E., Quintanar-Guerrero, D., Liceaga, A. M., & Zambrano- Zaragoza, M. L. (2022). Encapsulation of bioactive peptides: A strategy to improve the stability, protect the nutraceutical bioactivity and support their food applications. *RSC Advances*, *12*(11), 6449–6458. Available from https://doi.org/10.1039/d1ra08590e, http://pubs.rsc.org/en/journals/journal/ra.

Akbarian, M., Khani, A., Eghbalpour, S., & Uversky, V. N. (2022). Bioactive peptides: Synthesis, sources, applications, and proposed mechanisms of action. *International Journal of Molecular Sciences*, *23*(3). Available from https://doi.org/10.3390/ijms23031445.

Akinsola, A. F., Osasona, I., Akintayo, E. T., Siyanbola, T. O., & Omosebi, S. O. (2022). Nutritional evaluation of calabash gourd (L agenaria Siceraria) seeds and oil. *Journal of Culinary Science & Technology.*, *21*, 1–22. Available from https://doi.org/10.1080/15428052.2021.2016527.

Alblooshi, M., Devarajan, A. R., Singh, B. P., Ramakrishnan, P., Mostafa, H., Kamal, H., Mudgil, P., & Maqsood, S. (2023). Multifunctional bioactive properties of hydrolysates from colocynth (Citrullus colocynthis) seeds derived proteins: Characterization and biological properties. *Plant Physiology and Biochemistry*, *194*, 326–334. Available from https://doi.org/10.1016/j.plaphy.2022.110.026, http://www.journals.elsevier.com/plant-physiology-and-biochemistry/.

Atiwetin, P., Harada, S., & Kamei, K. (2006). Serine proteinase inhibitor from wax gourd (Benincasa hispida [Thunb] Cogn.) seeds. *Bioscience, Biotechnology, and Biochemistry*, *70*(3), 743–745. Available from https://doi.org/10.1271/bbb.70.743Japan, http://www.jstage.jst.go.jp/article/bbb/70/3/743/_pdf.

Barbieri, L., Polito, L., Bolognesi, A., Ciani, M., Pelosi, E., Farini, V., Jha, A., Sharma, N., Vivanco, J., & Chambery, A. (2006). Ribosome-inactivating proteins in edible plants and purification and characterization of a new ribosome-inactivating protein from Cucurbita moschata. *Biochimica et Biophysica Acta (BBA) - General Subjects*, *1760*(5), 783–792. Available from https://doi.org/10.1016/j.bbagen.2006.010.002.

Bhattacharya, S., & Das, B. (2012). Anti-diabetic activity of Lagenaria siceraria pulp and seed extract in normal and alloxan-induced diabetic rats. *International Journal of Pharmaceutical Sciences and Research*, *3*(9).

Bolognesi, A., Barbieri, L., Abbondanza, A., Falasca, A. I., Carnicelli, D., Battelli, M. G., & Stirpe, F. (1990). Purification and properties of new ribosome-inactivating proteins with RNA N-glycosidase activity. *Biochimica et Biophysica Acta (BBA) - Gene Structure and Expression*, *1087*(3), 293–302. Available from https://doi.org/10.1016/0167-4781(90)90002-j.

Bora, A. F. M., Kouame, K. J. E. P., Li, X., Liu, L., & Pan, Y. (2023). New insights into the bioactive polysaccharides, proteins, and triterpenoids isolated from bitter melon (Momordica charantia) and their relevance for nutraceutical and food application: A review. *International Journal of Biological Macromolecules*, *231*. Available from https://doi.org/10.1016/j.ijbiomac.2023.123173.

Bučko, S., Katona, J., Popović, L., Petrović, L., & Milinković, J. (2016). Influence of enzymatic hydrolysis on solubility, interfacial and emulsifying properties of pumpkin (Cucurbita pepo) seed protein isolate. *Food Hydrocolloids*, *60*, 271–278. Available from https://doi.org/10.1016/j.foodhyd.2016.040.005.

Chai, T. T., Koh, J. A., Wong, C. C. C., Sabri, M. Z., & Wong, F. C. (2021). Computational screening for the anticancer potential of seed-derived antioxidant peptides: A cheminformatic approach. *Molecules (Basel, Switzerland)*, *26*(23). Available from https://doi.org/10.3390/molecules26237396, https://www.mdpi.com/1420-3049/26/23/7396/pdf.

Chan, D. W., Yung, M. M. H., Chan, Y.-S., Xuan, Y., Yang, H., Xu, D., Zhan, J.-B., Chan, K. K. L., Ng, T.-B., & Ngan, H. Y. S. (2020). MAP30 protein from Momordica charantia is therapeutic and has synergic activity with cisplatin against ovarian cancer in vivo by altering metabolism and inducing ferroptosis. *Pharmacological Research*, *161*. Available from https://doi.org/10.1016/j.phrs.2020.105157.

Chan, L. Y., Wang, C. K. L., Major, J. M., Greenwood, K. P., Lewis, R. J., Craik, D. J., & Daly, N. L. (2009). Isolation and characterization of peptides from Momordica cochinchinensis seeds. *Journal of Natural Products*, *72*(8), 1453–1458. Available from https://doi.org/10.1021/np900174nAustralia, http://pubs.acs.org/doi/pdfplus/10.1021/np900174n.

Chelliah, R. (2018). Antihypertensive effect of peptides from sesame, almond, and pumpkin seeds: In-silico and in-vivo evaluation. *The Journal of Agricultural, Life and Environmental Sciences*, *30*(1). Available from 10.12972/jales.20180002.

Choudhary, R. R., Verma, H., & Sharma, N. K. (2013). Antioxidant and dehalogenase activity of Lagenaria sciceraria seed protein fraction. *Journal of Plant Science and Research*, *29*(2).

Chuethong, J., Oda, K., Sakurai, H., Saiki, I., & Leelamanit, W. (2007). Cochinin B, a novel ribosome-inactivating protein from the seeds of Momordica cochinchinensis. *Biological and Pharmaceutical Bulletin*, *30*(3), 428–432. Available from https://doi.org/10.1248/bpb.300.428.

Dash, P., & Ghosh, G. (2017a). Fractionation, amino acid profiles, antimicrobial and free radical scavenging activities of Citrullus lanatus seed protein. *Natural Product Research*, *31*(24), 2945–2947. Available from https://doi.org/10.1080/14786419.2017.1305385.

Dash, P., & Ghosh, G. (2017b). Proteolytic and antioxidant activity of protein fractions of seeds of Cucurbita moschata. *Food Bioscience*, *18*, 1–8. Available from https://doi.org/10.1016/j.fbio.2016.120.004.

Dash, P., Panda, P., & Ghosh, G. (2018). Free radical scavenging activities and nutritional value of Lagenaria siceraria: A nutriment creeper. *Iranian Journal of Science and Technology, Transactions A: Science*, *42*(4), 1743–1752. Available from https://doi.org/10.1007/s40995-017-0349-z.

Dash, P., Rath, G., & Ghosh, G. (2020). In vivo antioxidant potential of protein hydrolysates of some Cucurbitaceae seed. *Journal of Drug Delivery and Therapeutics*, *10*(3), 128–132. Available from https://doi.org/10.22270/jddt.v10i3.4091.

Dhillon, N. P. S., Sanguansil, S., Singh, S. P., Masud, M. A. T., Kumar, P., Bharathi, L. K., Yetişir, H., Huang, R., Canh, D. X., & McCreight, J. D. (2016). *Gourds: Bitter, bottle, wax, snake, sponge and ridge* (pp. 155–172). Springer Science and Business Media LLC. Available from 10.1007/7397_2016_24.

Dia, V. P., & Krishnan, H. B. (2016). BG-4, a novel anticancer peptide from bitter gourd (Momordica charantia), promotes apoptosis in human colon cancer cells. *Scientific Reports*, *6*. Available from https://doi.org/10.1038/srep33532.

Do, T. V. T., Fan, L., Suhartini, W., & Girmatsion, M. (2019). Gac (Momordica cochinchinensis Spreng) fruit: A functional food and medicinal resource. *Journal of Functional Foods*, *62*. Available from https://doi.org/10.1016/j.jff.2019.103512.

Duque-Soto, C., Quintriqueo-Cid, A., Rueda-Robles, A., Robert, P., Borrás-Linares, I., & Lozano-Sánchez, J. (2023). Evaluation of different advanced approaches to simulation of dynamic in vitro digestion of polyphenols from different food matrices—A systematic review. *Antioxidants*, *12*(1). Available from https://doi.org/10.3390/antiox12010101.

Fang, E. F., Zhang, C. Z. Y., Ng, T. B., Wong, J. H., Pan, W. L., Ye, X. J., Chan, Y. S., & Fong, W. P. (2012). Momordica Charantia lectin, a type II ribosome inactivating protein, exhibits antitumor activity toward human nasopharyngeal carcinoma cells in vitro and in vivo. *Cancer Prevention Research*, *5*(1), 109–121. Available from https://doi.org/10.1158/1940-6207.CAPR-11-0203HongKong, http://cancerpreventionresearch.aacrjournals.org/content/5/1/109.full.pdf + html.

Fang, E. F., Zhang, C. Z. Y., Wong, J. H., Shen, J. Y., Li, C. H., & Ng, T. B. (2012). The MAP30 protein from bitter gourd (Momordica charantia) seeds promotes apoptosis in liver cancer cells in vitro and in vivo. *Cancer Letters*, *324*(1), 66–74. Available from https://doi.org/10.1016/j.canlet.2012.050.005.

FAO. (2019). *FAOSTAT database*.

Gill, N. S., Singh, S., Arora, R., & Bali, M. (2012). Evaluation of ethanolic seed extract of Lagenaria siceraria for their therapeutic potential. *Journal of Medical Sciences*, *12*(3), 78–84. Available from https://doi.org/10.3923/jms.2012.78.84.

Grumet, R., McCreight, J. D., McGregor, C., Weng, Y., Mazourek, M., Reitsma, K., Labate, J., Davis, A., & Fei, Z. (2021). Genetic resources and vulnerabilities of major cucurbit crops. *Genes*, *12*(8). Available from https://doi.org/10.3390/genes12081222.

Guo, J., Xu, W., Hu, Y., Huang, J., Zhao, Y., Zhang, L., Huang, C.-H., & Ma, H. (2020). Phylotranscriptomics in Cucurbitaceae reveal multiple whole-genome duplications and key morphological and molecular innovations. *Molecular Plant*, *13*(8), 1117–1133. Available from https://doi.org/10.1016/j.molp.2020.050.011.

Gupta, P., Chikkala, S., & Kundu, P. (2021). Ash gourd and its applications in the food, pharmacological and biomedical industries. *International Journal of Vegetable Science*, *27*(1), 44–53. Available from https://doi.org/10.1080/19315260.2019.1699222.

Görgüç, A., Gençdağ, E., & Yılmaz, F. M. (2020). Bioactive peptides derived from plant origin by-products: Biological activities and techno-functional utilizations in food developments – A review. *Food Research International*, *136*. Available from https://doi.org/10.1016/j.foodres.2020.109504.

Horax, R., Hettiarachchy, N., Kannan, A., & Chen, P. (2011). Protein extraction optimisation, characterisation, and functionalities of protein isolate from bitter melon (Momordica charantia) seed. *Food Chemistry*, *124*(2), 545–550. Available from https://doi.org/10.1016/j.foodchem.2010.060.068.

Hussain, A., Kausar, T., Sehar, S., Sarwar, A., Ashraf, A. H., Jamil, M. A., Noreen, S., Rafique, A., Iftikhar, K., Quddoos, M. Y., Aslam, J., & Majeed, M. A. (2022). A comprehensive review of functional ingredients, especially bioactive compounds present in pumpkin peel, flesh and seeds, and their health benefits. *Food Chemistry Advances*, *1*. Available from https://doi.org/10.1016/j.focha.2022.100067.

Islam, M. T., Quispe, C., El-Kersh, D. M., Shill, M. C., Bhardwaj, K., Bhardwaj, P., Sharifi-Rad, J., Martorell, M., Hossain, R., Al-Harrasi, A., Al-Rawahi, A., Butnariu, M., Rotariu, L. S., Suleria, H. A. R., Taheri, Y., Docea, A. O., Calina, D., & Cho, W. C. (2021). A literature-based update on Benincasa hispida (Thunb.) Cogn.: Traditional uses, nutraceutical, and phytopharmacological profiles. *Oxidative Medicine and Cellular Longevity*, *2021*. Available from https://doi.org/10.1155/2021/6349041, http://www.hindawi.com/journals/oximed/.

Jones, L. D., Pangloli, P., Krishnan, H. B., & Dia, V. P. (2018). BG-4, a novel bioactive peptide from momordica charantia, inhibits lipopolysaccharide-induced inflammation in THP-1 human macrophages. *Phytomedicine: International Journal of Phytotherapy and Phytopharmacology*, *42*, 226–232. Available from https://doi.org/10.1016/j.phymed.2018.030.047.

Kesari, P., Pratap, S., Dhankhar, P., Dalal, V., Mishra, M., Kumar Singh, P., Chauhan, H., & Kumar, P. (2020). Structural characterization and in-silico analysis of Momordica charantia 7S globulin for stability and ACE inhibition. *Scientific Reports*, *10*(1). Available from https://doi.org/10.1038/s41598-020-58138-9.

Lee, E. H., Cha, K. H., Vuong, T. T., Kim, S. M., & Pan, C. H. (2018). Comparison of static and dynamic in vitro digestion models to estimate the bioaccessibility of lutein in lutein-rich foods. *Applied Biological Chemistry*, *61*(4), 441–447. Available from https://doi.org/10.1007/s13765-018-0378-0, http://link.springer.com/journal/volumesAndIssues/13765.

Lee-Huang, S., Huang, P. L., Nara, P. L., Chen, H. C., Kung, Hf, Huang, P., Huang, H. I., & Huang, P. L. (1990). MAP 30: A new inhibitor of HIV-1 infection and replication. *FEBS Letters*, *272*(1–2), 12–18. Available from https://doi.org/10.1016/0014-5793(90)80438-O.

Li, M., Chen, Y., Liu, Z., Shen, F., Bian, X., & Meng, Y. (2009). Anti-tumor activity and immunological modification of ribosome-inactivating protein (RIP) from Momordica charantia by covalent attachment of polyethylene glycol. *Acta Biochimica et Biophysica Sinica*, *41*(9), 792–799. Available from https://doi.org/10.1093/abbs/gmp068.

Liang, F., Shi, Y., Shi, J., Zhang, T., & Zhang, R. (2021). A novel angiotensin-I-converting enzyme (ACE) inhibitory peptide IAF (Ile-Ala-Phe) from pumpkin seed proteins: in silico screening, inhibitory activity, and molecular mechanisms. *European Food Research and Technology*, *247*(9), 2227–2237. Available from https://doi.org/10.1007/s00217-021-03783-1.

Lo, H. Y., Ho, T. Y., Li, C. C., Chen, J. C., Liu, J. J., & Hsiang, C. Y. (2014). A novel insulin receptor-binding protein from Momordica charantia enhances glucose uptake and glucose clearance in vitro and in vivo through triggering insulin receptor signaling pathway. *Journal of Agricultural and Food Chemistry*, *62*(36), 8952–8961. Available from https://doi.org/10.1021/jf5002099, http://pubs.acs.org/journal/jafcau.

Lo, H. Y., Ho, T. Y., Lin, C., Li, C. C., & Hsiang, C. Y. (2013). Momordica charantia and its novel polypeptide regulate glucose homeostasis in mice via binding to insulin receptor. *Journal of Agricultural and Food Chemistry*, *61*(10), 2461–2468.

Lo, H. Y., Li, C. C., Chen, F. Y., Chen, J. C., Hsiang, C. Y., & Ho, T. Y. (2017). Gastro-resistant insulin receptor-binding peptide from Momordica charantia improved the glucose tolerance in streptozotocin-induced diabetic mice via insulin receptor signaling pathway. *Journal of Agricultural and Food Chemistry*, *65*(42), 9266–9274. Available from https://doi.org/10.1021/acs.jafc.7b03583, http://pubs.acs.org/journal/jafcau.

Lu, D., Peng, M., Yu, M., Jiang, B., Wu, H., & Chen, J. (2021). Effect of enzymatic hydrolysis on the zinc binding capacity and in vitro gastrointestinal stability of peptides derived from pumpkin (*Cucurbita pepo* L.) seeds. *Frontiers in Nutrition*, *8*, 647782. Available from 10.3389/fnut.2021.647782.

Meena, S., Kanthaliya, B., Joshi, A., Khan, F., & Arora, J. (2020). Biologia futura: Medicinal plants-derived bioactive peptides in functional perspective—A review. *Biologia Futura*, *71*(3), 195–208. Available from https://doi.org/10.1007/s42977-020-00042-4.

Mukherjee, P. K., Singha, S., Kar, A., Chanda, J., Banerjee, S., Dasgupta, B., Haldar, P. K., & Sharma, N. (2022). Therapeutic importance of Cucurbitaceae: A medicinally important family. *Journal of Ethnopharmacology*, *282*. Available from https://doi.org/10.1016/j.jep.2021.114599.

Murugesan, A., Yadav, S. K. R., & Dixit, A. (2022). Anti-hyperglycemic activity of HPLC-fractionated Momordica charantia seed extract enriched in a novel napin-like protein in experimental diabetic rats and its validation with recombinant napin-like protein. *Current Research in Biotechnology*, *4*, 179–189. Available from https://doi.org/10.1016/j.crbiot.2022.030.001, https://www.journals.elsevier.com/current-research-in-biotechnology/.

Naik, M., Natarajan, V., Modupalli, N., Thangaraj, S., & Rawson, A. (2022). Pulsed ultrasound assisted extraction of protein from defatted Bitter melon seeds (Momardica charantia L.) meal: Kinetics and quality measurements. *LWT*, *155*. Available from https://doi.org/10.1016/j.lwt.2021.112997.

Ngamsuk, S., Huang, T. C., & Hsu, J. L. (2020). ACE inhibitory activity and molecular docking of gac seed protein hydrolysate purified by HILIC and RP-HPLC. *Molecules (Basel, Switzerland)*, *25*(20). Available from https://doi.org/10.3390/molecules25204635, https://www.mdpi.com/1420-3049/25/20/4635.

Nguyen, D., Holien, J., Dekiwadia, C., Thrimawithana, T., Piva, T., & Huynh, T. (2023). Momordica cochinchinensis (Gấc) seed extracts induce apoptosis and necrosis in melanoma cells. *Pharmaceuticals*, *16*(1). Available from https://doi.org/10.3390/ph16010100.

Nourmohammadi, E., SadeghiMahoonak, A., Alami, M., & Ghorbani, M. (2017). Amino acid composition and antioxidative properties of hydrolysed pumpkin (Cucurbita pepo L.) oil cake protein. *International Journal of Food Properties*, *20*(12), 3244–3255. Available from https://doi.org/10.1080/10942912.2017.1283516.

Olivares-Galván, S., Marina, M. L., & García, M. C. (2020). Extraction and characterization of antioxidant peptides from fruit residues. *Foods*, *9*(8). Available from https://doi.org/10.3390/foods9081018.

Osborne, T. B. (1924). *The vegetable proteins*. Longmans, Green and Company.

Ovca, A., Van Elteren, J. T., Falnoga, I., & Šelih, V. S. (2011). Speciation of zinc in pumpkin seeds (Cucurbita pepo) and degradation of its species in the human digestive tract. *Food Chemistry*, *128*(4), 839–846. Available from https://doi.org/10.1016/j.foodchem.2011.030.102.

Ozuna, C., & León-Galván, M. F. (2017). Cucurbitaceae seed protein hydrolysates as a potential source of bioactive peptides with functional properties. *BioMed Research International*, *2017*, 1–16. Available from https://doi.org/10.1155/2017/2121878.

Park, K. S., Kim, Y. T., Kim, H. S., Cha, J. S., & Park, K. H. (2015). Selection of the antibacterial agents for control against Pseudomonas syringae pv. syringae causing leaf spot disease on green pumpkin (Cucurbita moschata). *The Korean Journal of Pesticide Science*, *19*(2), 119–124. Available from https://doi.org/10.7585/kjps.2015.19.20.119.

Peng, M., Lu, D., Yu, M., Jiang, B., & Chen, J. (2022). Identification of zinc-chelating pumpkin seed (Cucurbita pepo L.) peptides and in vitro transport of peptide–zinc chelates. *Journal of Food Science*, *87*(5), 2048–2057. Available from https://doi.org/10.1111/1750-3841.16132.

Perez Gutierrez, R. M. (2016). Review of Cucurbita pepo (pumpkin) its phytochemistry and pharmacology. *Medicinal Chemistry (Shariqah (United Arab Emirates))*, *6*(1), 12–21.

Popović, L., Stolić, Ž., Čakarević, J., Torbica, A., Tomić, J., & Šijački, M. (2017). Biologically active digests from pumpkin oil cake protein: Effect of cross-linking by transglutaminase. *Journal of the American Oil Chemists' Society*, *94*(10), 1245–1251. Available from https://doi.org/10.1007/s11746-017-3041-8.

Priyanto, A. D., Putranto, A. W., & Hsu, J. L. (2021). The potential of various seeds as angiotensin-I converting enzyme inhibitory peptides derived from protein hydrolysate: A short review. *IOP Conference Series: Earth and Environmental Science*, *733*(1). Available from https://doi.org/10.1088/1755-1315/733/1/012130.

Priyanto, A. D., Doerksen, R. J., Chang, C. I., Sung, W. C., Widjanarko, S. B., Kusnadi, J., Lin, Y. C., Wang, T. C., & Hsu, J. L. (2015). Screening, discovery, and characterization of angiotensin-I converting enzyme inhibitory peptides derived from proteolytic hydrolysate of bitter melon seed proteins. *Journal of Proteomics*, *128*, 424–435. Available from https://doi.org/10.1016/j.jprot.2015.080.018, http://www.elsevier.com.

Regazzo, D., Mollé, D., Gabai, G., Tomé, D., Dupont, D., Leonil, J., & Boutrou, R. (2010). The (193-209) 17-residues peptide of bovine β-casein is transported through Caco-2 monolayer. *Molecular Nutrition & Food Research*, *54*(10), 1428–1435. Available from https://doi.org/10.1002/mnfr.200900443.

Rezig, L., Chibani, F., Chouaibi, M., Dalgalarrondo, M., Hessini, K., Guéguen, J., & Hamdi, S. (2013). Pumpkin (Cucurbita maxima) seed proteins: Sequential extraction processing and fraction characterization. *Journal of Agricultural and Food Chemistry*, *61*(32), 7715–7721. Available from https://doi.org/10.1021/jf402323u.

Ribeiro, S. F. F., Agizzio, A. P., Machado, O. L. T., Neves-Ferreira, A. G. C., Oliveira, M. A., Fernandes, K. V. S., Carvalho, A. O., Perales, J., & Gomes, V. M. (2007). A new peptide of melon seeds which shows sequence homology with vicilin: Partial characterization and antifungal activity. *Scientia Horticulturae*, *111*(4), 399–405. Available from https://doi.org/10.1016/j.scienta.2006.110.004.

Rico, X., Gullón, B., Alonso, J. L., & Yáñez, R. (2020). Recovery of high value-added compounds from pineapple, melon, watermelon and pumpkin processing by-products: An overview. *Food Research International*, *132*. Available from https://doi.org/10.1016/j.foodres.2020.109086.

Roopan, S. M., Devi Rajeswari, V., Kalpana, V. N., & Elango, G. (2016). Biotechnology and pharmacological evaluation of Indian vegetable crop Lagenaria siceraria: an overview. *Applied Microbiology and Biotechnology*, *100*(3), 1153–1162. Available from https://doi.org/10.1007/s00253-015-7190-0, link.springer.de/link/service/journals/00253/index.htm.

Salehi, B., Capanoglu, E., Adrar, N., Catalkaya, G., Shaheen, S., Jaffer, M., Giri, L., Suyal, R., Jugran, A. K., Calina, D., Docea, A. O., Kamiloglu, S., Kregiel, D., Antolak, H., Pawlikowska, E., Sen, S., Acharya, K., Selamoglu, Z., Sharifi-Rad, J., … Capasso, R.

(2019). Cucurbits plants: A key emphasis to its pharmacological potential. *Molecules (Basel, Switzerland)*, *24*(10). Available from https://doi.org/10.3390/molecules24101854.

Salehi, B., Quispe, C., Sharifi-Rad, J., Giri, L., Suyal, R., Jugran, A. K., Zucca, P., Rescigno, A., Peddio, S., Bobiş, O., Moise, A. R., Leyva-Gómez, G., Del Prado-Audelo, M. L., Cortes, H., Iriti, M., Martorell, M., Cruz-Martins, N., Kumar, M., & Zam, W. (2021). Antioxidant potential of family Cucurbitaceae with special emphasis on Cucurbita genus: A key to alleviate oxidative stress-mediated disorders. *Phytotherapy Research*, *35*(7), 3533–3557. Available from https://doi.org/10.1002/ptr.7045.

Samtiya, M., Acharya, S., Pandey, K. K., Aluko, R. E., Udenigwe, C. C., & Dhewa, T. (2021). Production, purification, and potential health applications of edible seeds' bioactive peptides: A concise review. *Foods*, *10*(11). Available from https://doi.org/10.3390/foods10112696.

Sharma, S., Verma, H. N., & Sharma, N. K. (2014). Cationic bioactive peptide from the seeds of Benincasa hispida. *International Journal of Peptides*, *2014*. Available from https://doi.org/10.1155/2014/156060, http://www.hindawi.com/journals/ijpep/.

Siddeeg, A., Xu, Y., Jiang, Q., & Xia, W. (2015). In vitro antioxidant activity of protein fractions extracted from seinat (Cucumis melo var. tibish) seeds. *CYTA - Journal of Food*, *13*(3), 472–481. Available from https://doi.org/10.1080/19476337.2014.1003199, http://www.tandfonline.com/toc/tcyt20/current.

Singh, B. P., Bangar, S. P., Alblooshi, M., Ajayi, F. F., Mudgil, P., & Maqsood, S. (2022). Plant-derived proteins as a sustainable source of bioactive peptides: Recent research updates on emerging production methods, bioactivities, and potential application. *Critical Reviews in Food Science and Nutrition*, 1–22. Available from https://doi.org/10.1080/10408398.2022.2067120.

Singh, P. P., Gupta, V., & Prakash, B. (2023). Recent advancement in functional properties and toxicity assessment of plant-derived bioactive peptides using bioinformatic approaches. *Critical Reviews in Food Science and Nutrition*, *63*(20), 4503–4521. Available from https://doi.org/10.1080/10408398.2021.2002807, https://www.tandfonline.com/loi/bfsn20.

Sood, A., Kaur, P., & Gupta, R. (2012). Phytochemical screening and antimicrobial assay of various seeds extract of Cucurbitaceae family. *International Journal of Applied Biology and Pharmaceutical Technology*, *3*.

Teugwa, C. M., Boudjeko, T., Tchinda, B. T., Mejiato, P. C., & Zofou, D. (2013). Anti-hyperglycaemic globulins from selected Cucurbitaceae seeds used as antidiabetic medicinal plants in Africa. *BMC Complementary and Alternative Medicine*, *13*. Available from https://doi.org/10.1186/1472-6882-13-63Cameroon, http://www.biomedcentral.com/1472-6882/13/63.

Tien, P. G., Kayama, F., Konishi, F., Tamemoto, H., Kasono, K., Hung, N. T. K., Kuroki, M., Ishikawa, S. E., Van, C. N., & Kawakami, M. (2005). Inhibition of tumor growth and angiogenesis by water extract of Gac fruit (Momordica cochinchinensis Spreng). *International Journal of Oncology*, *26*(4), 881–889. Available from https://doi.org/10.3892/ijo.26.40.881.

Udechukwu, M. C., Collins, S. A., & Udenigwe, C. C. (2016). Prospects of enhancing dietary zinc bioavailability with food-derived zinc-chelating peptides. *Food and Function*, *7*(10), 4137–4144. Available from https://doi.org/10.1039/c6fo00706f, http://www.rsc.org/Publishing/Journals/FO/about.asp.

Vassiliou, A. G., Neumann, G. M., Polya, G. M., & Condron, R. (1998). Purification and mass spectrometry-assisted sequencing of basic antifungal proteins from seeds of pumpkin (Cucurbita maxima). *Plant Science, 134*(2), 141–162.

Vaštag, Ž., Popović, L., Popović, S., Krimer, V., & Peričin, D. (2011). Production of enzymatic hydrolysates with antioxidant and angiotensin-I converting enzyme inhibitory activity from pumpkin oil cake protein isolate. *Food Chemistry, 124*(4), 1316–1321. Available from https://doi.org/10.1016/j.foodchem.2010.070.062.

Vinayashree, S., & Vasu, P. (2021). Biochemical, nutritional and functional properties of protein isolate and fractions from pumpkin (Cucurbita moschata var. Kashi Harit) seeds. *Food Chemistry, 340*. Available from https://doi.org/10.1016/j.foodchem.2020.128177.

Wang, B., Xie, N., & Li, B. (2019). Influence of peptide characteristics on their stability, intestinal transport, and in vitro bioavailability: A review. *Journal of Food Biochemistry, 43*(1). Available from https://doi.org/10.1111/jfbc.12571.

Wang, H. X., & Ng, T. B. (2003). Isolation of cucurmoschin, a novel antifungal peptide abundant in arginine, glutamate and glycine residues from black pumpkin seeds. *Peptides, 24*(7), 969–972. Available from https://doi.org/10.1016/s0196-9781(03)00191-8.

Wang, L., Ding, L., Du, Z., & Liu, J. (2020). Effects of hydrophobicity and molecular weight on the transport permeability of oligopeptides across Caco-2 cell monolayers. *Journal of Food Biochemistry, 44*(5). Available from https://doi.org/10.1111/jfbc.13188.

Wen, C., Zhang, J., Feng, Y., Duan, Y., Ma, H., & Zhang, H. (2020). Purification and identification of novel antioxidant peptides from watermelon seed protein hydrolysates and their cytoprotective effects on H2O2-induced oxidative stress. *Food Chemistry, 327*. Available from https://doi.org/10.1016/j.foodchem.2020.127059.

Wen, C., Zhang, J., Zhang, H., Duan, Y., & Ma, H. (2019). Effects of divergent ultrasound pretreatment on the structure of watermelon seed protein and the antioxidant activity of its hydrolysates. *Food Chemistry, 299*. Available from https://doi.org/10.1016/j.foodchem.2019.125165.

Wen, C., Zhang, J., Zhang, H., Duan, Y., & Ma, H. (2021). Study on the structure–activity relationship of watermelon seed antioxidant peptides by using molecular simulations. *Food Chemistry, 364*. Available from https://doi.org/10.1016/j.foodchem.2021.130432.

Wong, F. C., Ong, J. H., & Chai, T. T. (2021). SARS-CoV-2 spike protein-, main protease- and papain-like-protease-targeting peptides from seed proteins following gastrointestinal digestion: An in silico study. *Phytomedicine Plus, 1*(1). Available from https://doi.org/10.1016/j.phyplu.2020.100016, https://www.journals.elsevier.com/phytomedicine-plus.

Xu, B., Li, Z., Zeng, T., Zhan, J., Wang, S., Ho, C.-T., & Li, S. (2022). Bioactives of Momordica charantia as potential anti-diabetic/hypoglycemic agents. *Molecules (Basel, Switzerland), 27*(7). Available from https://doi.org/10.3390/molecules27072175.

Yadav, S., Tomar, A. K., Jithesh, O., Khan, M. A., Yadav, R. N., Srinivasan, A., Singh, T. P., & Yadav, S. (2011). Purification and partial characterization of low molecular weight vicilin-like glycoprotein from the seeds of Citrullus lanatus. *The Protein Journal, 30*(8), 575–580. Available from https://doi.org/10.1007/s10930-011-9362-9.

Zamuz, S., Munekata, P. E. S., Gullón, B., Rocchetti, G., Montesano, D., & Lorenzo, J. M. (2021). Citrullus lanatus as source of bioactive components: An up-to-date review.

Trends in Food Science and Technology, *111*, 208–222. Available from https://doi.org/10.1016/j.tifs.2021.030.002, http://www.elsevier.com/wps/find/journaldescription.cws_home/601278/description#description.

Zhao, C., Liu, D., Feng, L., Cui, J., Du, H., Wang, Y., Xiao, H., & Zheng, J. (2022). Research advances of in vivo biological fate of food bioactives delivered by colloidal systems. *Critical Reviews in Food Science and Nutrition*, 1–19. Available from https://doi.org/10.1080/10408398.2022.2154741.

Zia, S., Khan, M. R., Shabbir, M. A., & Aadil, R. M. (2021). An update on functional, nutraceutical and industrial applications of watermelon by-products: A comprehensive review. *Trends in Food Science and Technology*, *114*, 275–291. Available from https://doi.org/10.1016/j.tifs.2021.050.039, http://www.elsevier.com/wps/find/journaldescription.cws_home/601278/description#description.

CHAPTER

Bioactive peptides released from microalgae during gastrointestinal digestion 12

Ahmed A. Zaky[1], Samuel Paterson[2], Pilar Gómez-Cortés[2] and Blanca Hernández-Ledesma[2]

[1]*Department of Food Technology, Food Industries and Nutrition Research Institute, National Research Centre, Cairo, Egypt*

[2]*Department of Bioactivity and Food Analysis, Institute of Food Science Research (CIAL, CSIC-UAM, CEI UAM + CSIC), Madrid, Spain*

12.1 Introduction

Microalgae encompass a diverse range of predominantly autotrophic unicellular prokaryotic and eukaryotic organisms. Although there is a vast number of microalgal species and new species continue to be identified, only limited species have undergone extensive research. This select group comprises *Dunaliella* sp., *Chlorella* sp., *Isochrysis* sp., *Nannochloropsis* sp., *Haematococcus* sp., and *Arthrospira* sp. (Eilam et al., 2023). A small subset of these 200,000 microalgal species have a long consumption history, dating back to ancient times, and are recognized as safe food sources, known as GRAS (generally recognized as safe). These include *Arthrospira*, *Chlorella*, *Dunaliella*, *Haematococcus*, *Schizochytrium*, *Porphyridium cruentum*, and *Crypthecodinium cohnii*. The Food and Drug Administration (FDA, United States) has classified certain microalgae-derived products, such as oils from *Schizochytrium* and *Ulkenia*, whole microalgal protein powder, and a lipid ingredient sourced from *Chlorella* as GRAS. Additionally, various microalgal products have obtained approval in the European Union as novel food ingredients, such as *Tetraselmis chuii*, which was approved in 2014 (Eilam et al., 2023).

Microalgae are acknowledged for their remarkable diversity and serve as a valuable reservoir of bioactive molecules that play essential physiological roles in their ecosystem. They possess the ability to adjust their metabolism in response to environmental conditions. Consequently, microalgae hold significant potential as a natural source of bioactive compounds, including polysaccharides, lipids, essential fatty acids, proteins, peptides, carotenoids, vitamins, pigments, and phenolic compounds. These compounds exhibit promising health-related applications, attributed to their antioxidant, antimicrobial, antiinflammatory, anticancer, antidiabetic, antihypertensive, antihyperlipidemic, and antiobesity effects (Ampofo & Abbey, 2022; Barkia et al., 2019). Microalgae, owing to their ability to produce a wide array of primary and secondary

Protein Digestion-Derived Peptides. DOI: https://doi.org/10.1016/B978-0-443-19141-1.00012-1

metabolites, have gained significant attention as potential therapeutic agents for various health disorders. They have applications in the cosmetic, food, feed, and pharmaceutical industries (Januário et al., 1930; Zhou et al., 2022). Bioactive compounds derived from microalgae have demonstrated antiinflammatory properties, making them valuable in the treatment of different inflammation-associated diseases. These compounds function by inhibiting the production of proinflammatory cytokines and reducing the expression of inflammatory genes (Choo et al., 2020). A significant example is *Arthrospira platensis* (*Spirulina*), a highly nutritious blue–green microalgae that is widely utilized as a nutraceutical food supplement worldwide. In addition to its nutritional value, *Spirulina* exhibits therapeutic properties, including antiinflammatory activity (Tzachor et al., 2021). It contains a pigment-binding protein called phycocyanin, which hinders the formation of proinflammatory cytokines, such as tumor necrosis factor-alpha (TNF-α), diminishes the production of prostaglandin E2 (PGE2), and inhibits the expression of cyclooxygenase-2 (COX-2) (Finamore et al., 2017).

Microalgae are renowned for their ability to generate diverse bioactive peptides that possess a broad range of physiological and pharmacological activities. These bioactive peptides can be liberated from microalgae during enzymatic hydrolysis and/or gastrointestinal digestion. The objective of this chapter is to present a comprehensive overview of the bioactive peptides/proteins released from microalgae under conditions that simulate human gastrointestinal digestion.

12.2 Microalgae-based proteins and peptides

Microalgae can produce between 2.5 and 7.5 tons of protein per hectare annually. The protein content of microalgal biomass varies significantly, ranging from 30% to 80% by mass, depending on the species of microalgae and their growth conditions. For instance, *C. vulgaris*, *A. platensis*, *T. chuii*, *N. oceanica*, *D. salina*, and *Galdieria sulphuraria* contain 51%–58%, 60%–71%, 31%–46%, 35%–44%, 50%–80%, and 60%, respectively (Muys et al., 2019; Wild et al., 2018). Certain microalgae, such as *Arthrospira* and *Chlorella*, are considered excellent protein sources due to their favorable amino acid composition, and they are marketed as such (Muys et al., 2019). Essential amino acids, which the human body cannot synthesize and must obtain from external sources, are produced in significant quantities by various microalgal species. Microalgae-based proteins are particularly rich in essential amino acids, such as leucine (L), arginine (R), and lysine (K), constituting approximately 7% of total proteins. In addition, other essential amino acids, such as isoleucine (I), phenylalanine (F), threonine (T), and valine (V), typically make up approximately 4% of total proteins and are also present in substantial amounts (Kumar et al., 2022).

Microalgae-based proteins, including phycocyanin, as well as specific peptides, such as LNGDVW, found in *C. ellipsoidea*, play a significant role in providing potent health benefits. These proteins and peptides contribute to various physiological functions within cells and tissues. Additionally, essential amino acids, such as arginine (R), histidine (H), isoleucine (I), leucine (L), lysine (K), and methionine (M), are crucial for the proper functioning of cells and tissues. They are either essential nutrients that the human body cannot produce on its own or play essential roles in carrying out vital functions (Xiao & Guo, 2022). The presence of these compounds in microalgae-based proteins is important in supporting overall health and well-being.

Over the last few years, the interest in bioactive peptides derived from microalgae has grown significantly due to their potential health benefits. Currently approximately half of the global protein and peptide market relies on terrestrial plants as sources. However, it is projected that by 2054, microalgae-based proteins may replace these traditional sources (Khanra et al., 2018). Microalgal proteins and peptides are being explored as part of the ongoing trend to discover sustainable novel food ingredients and alternative proteins that offer health advantages. Suttisuwan et al. (2019) conducted a study demonstrating the ability of a peptide fraction derived from microalgae to reduce the expression of proinflammatory cytokine genes. This finding highlights the potential of microalgal peptides as a valuable resource for developing new functional food ingredients and medicinal compounds targeting inflammation. Current research opens up possibilities for leveraging microalgae-derived peptides to create innovative products with antiinflammatory properties and significant health benefits.

Chlorella-11 peptide derived from the green algae *C. pyrenoidosa* has been shown to exhibit various biological effects. This specific peptide sequence (VECYGPNRPNF) has demonstrated ability to reduce the production of the proinflammatory cytokine monocyte chemoattractant protein-1 (MCP-1) in RAW264.7 macrophages stimulated by bacterial lipopolysaccharide. Additionally, it inhibited the production of intercellular cell adhesion molecule-1, endothelial cell selection, and vascular cell adhesion molecule-1 in endothelial cells (Shih et al., 2013). The bioactive peptide also displayed a dose- and time-dependent reduction in nitric oxide (NO) production and the expression and activity of nuclear factor-kappa B (NF-κB) (Cherng et al., 2010; Sheih et al., 2010). Over the past decade, in vitro and in vivo studies have revealed the health benefits and biological activities of bioactive peptides derived from *A. platensis* (Zhang & Zhang, 2013). These peptides have demonstrated antioxidant, anticancer, antidiabetic, and antiobesity activities. A peptide isolated from *C. ellipsoidea* (LNGDVW) exhibited scavenging activity against peroxyl, 1,1-diphenyl-2-picrylhydrazyl (DPPH), and hydroxyl radicals, as reported by (Ko et al., 2012). Furthermore, two peptides derived from *N. oculata* (GMNNLTP and LEQ) displayed antihypertensive properties by inhibiting angiotensin I-converting enzyme (ACE) activity (Wu et al., 2008). These examples underscore the immense potential of microalgae-based peptides in promoting human health and their ability to exhibit diverse beneficial effects.

Phycocyanin, which can be extracted from *Spirulina* and other microalgal species, is a biliprotein known for its antioxidant and antiinflammatory properties. In a mouse model, phycocyanin was found to have inhibitory effects on immunoglobulin E activity and to reduce vascular permeability, thereby reducing inflammation (NEMOTO-KAWAMURA et al., 2004).

12.3 Release of bioactive peptides from microalgae

The production of microalgae-derived peptides involves the use of various commercial proteases, such as papain, trypsin, pepsin, and chymotrypsin. These proteases are employed to hydrolyze microalgal proteins and release the desired peptides. For example, chymotrypsin is used to release antioxidant peptides from marine *C. ellipsoidea* (Ko et al., 2012), while pepsin is used to liberate new antioxidant peptides, such as VECYGPNRPQF, from microalgae protein waste (Sheih et al., 2009). Food-grade enzymes, such as alcalase and flavourzyme, have been specifically chosen to hydrolyze the proteins of the red microalga *P. purpureum* and diatom *P. tricornutum*, which possess antidiabetic and antioxidant properties (Stack et al., 2018).

In addition to enzymatic hydrolysis, microalgae-derived peptides can also be produced through in situ microbial fermentation of the parent proteins. Bacterial proteinases, including subtilisin-A, have been utilized to release peptides from *A. maxima* biomass cultivated on sugarcane vinasse, both in laboratory and pilot-scale settings. This demonstrates the diverse methods and enzymes employed to generate microalgae-derived peptides with various applications (Montalvo et al., 2019).

12.3.1 Simulated gastrointestinal digestion

Simulated gastrointestinal digestion is a process that mimics the conditions of the human gastrointestinal tract. It involves the use of various enzymes and simulated salivary, gastric and intestinal fluids to mimic the digestive process. The purpose of simulated gastrointestinal digestion is to evaluate the release of bioactive compounds from food sources and to assess their potential health benefits. During simulated gastrointestinal digestion, bioactive peptides are released from microalgae through the action of various enzymes present in the digestive system. Proteases, such as pepsin in the stomach and pancreatic enzymes (trypsin, chymotrypsin, and elastase) in the small intestine, play a crucial role in breaking down proteins into smaller peptide fragments. These enzymes cleave specific peptide bonds within proteins, resulting in the release of bioactive peptides.

Several factors can influence the release of bioactive peptides from microalgae during simulated gastrointestinal digestion. These include:

1. Microalgae composition: The protein content and profile of microalgae can vary depending on the species, growth conditions, and processing methods. These factors can impact the type and abundance of bioactive peptides present in microalgae and subsequently influence their release during digestion.
2. Enzymatic activity: The activity and specificity of digestive enzymes during simulated gastrointestinal digestion can affect the extent and pattern of peptide release. Different enzymes may cleave proteins at specific sites, leading to the production of distinct bioactive peptide fragments.

3. Digestion conditions: Factors, such as pH, temperature, and digestion time, can also influence peptide release. Simulating the conditions of the stomach and small intestine accurately is essential for mimicking the digestive process and obtaining reliable results.
4. Processing techniques: Various processing techniques, such as drying, grinding, and extraction methods, can impact the release of bioactive peptides. For instance, cell disruption methods, such as sonication or high-pressure homogenization, may enhance the accessibility of proteins to digestive enzymes, increasing peptide release (Eilam et al., 2023).

Several studies have investigated the bioactive peptides released from microalgae under simulated gastrointestinal digestion. Each microalgal species has a unique protein composition, leading to the release of distinct bioactive peptides with specific functionalities (Table 12.1).

For example, a study by Li et al. (2021) investigated the bioactive peptides released from the green microalga *C. pyrenoidosa*. They found that simulated gastrointestinal digestion resulted in the release of peptides with ACE and dipeptidyl peptidase-IV (DPP-IV) inhibitory activities. Another investigation by Liu et al. (2022) studied the bioactive peptides liberated from the microalga *A. platensis*. This study discovered that simulated gastrointestinal digestion led to the release of peptides with ACE and DPP-IV inhibitory activity and antioxidant action. According to Moaveni et al. (2022), *S. limacinum* has the potential to be used as a high-protein bioactive food ingredient. The protein digestibility was tested under simulated conditions for the stomach phase (using pepsin) and the small intestine phase (using chymotrypsin and trypsin). The antioxidant activities of the hydrolyzate fractions through radical (DPPH), 2,2'-azino-bis(3-ethylbenzothiazoline-6-sulfonic acid, ABTS) scavenging and reducing power activity were assayed. Additionally, the protective impact of the purified peptide against oxidation of low-density lipoprotein (LDL) was identified. The protein hydrolyzates demonstrated increased reducing power and radical scavenging ability as well as considerable inhibition of LDL oxidation compared to the control. These outcomes are compatible with the work by Xia et al. (2019) that hydrolyzed *D. salina* protein under simulated gastrointestinal digestive conditions (pepsin followed by trypsin) and produced antioxidant peptides. They found that the total digest had high DPPH scavenging activity in the simulated in vitro digestion assay, rising to 86.5%, in comparison with the nondigested samples (46.8%). As reported by Sheih et al. (2009), pepsin hydrolysis of *C. vulgaris* protein yielded peptides with excellent antioxidant activity. Additionally, the purified peptide (VECYGPNRPQF) showed resistance to the action of gastrointestinal enzymes, and no in vitro cytotoxicity against human lung fibroblast cell lines (WI-38). The four identified peptide sequences were evaluated for potential cleavage by gastrointestinal tract enzymes from *N. oculata* as reported by Hayes et al. (2022). Simulated cleavage of the peptides with pepsin ($pH > 1.3$ and pH 2.0) (EC 3.4.23.1) and trypsin (EC 3.4.21.4) was carried out. The resulting peptide fragments were subsequently examined for

Table 12.1 Protein hydrolyzates and bioactive peptides obtained from microalgae biomass under simulated gastrointestinal conditions.

Microalga species	Digestion conditions			Bioactivity	Bioactive peptide	References
	Oral	Gastric	Intestinal			
Arthrospira sp.	–	Pepsin (E/S = 1/100–2 hours)	Trypsin + chymotrypsin (E/S = 1/100–2.5 hours)	Inhibition of Ang II-induced production of NO and ROS in endothelial EA. hy926 cells Downregulation of expression of iNOS and ET-1 Blockage of the activation of p38 MAPK	TMEPGKP	Heo et al. (2017)
	–	Pepsin–2 hours	Lipase + trypsin + chymotrypsin + α-amylase (2 hours)	*Arthrorpira*-enriched foods may be considered a good source of the potentially bioactive microalgae peptides	–	Donadio et al. (2021)
Arthrospira maxima	–	Pepsin (E/S = 1/100)–4 hours	Trypsin + chymotrypsin (E/S = 1/100–4 hours)	Inhibition of production and expression of IL-6 and MCP-1 in histamine-stimulated EA-hy926 endothelial cells Reduction of in vitro monocyte adhesion onto cells through inhibition of adhesion molecules P-selectin and E-selectin production Reduction of ROS production Downregulation of Egr-1 expression via histamine receptor and PKCd-dependent MAPKs activation pathway	LDAVNR; MMLDF	Vo and Kim (2013)

	–	Pepsin (E/S = 1/100)—4 hours	Trypsin + chymotrypsin (E/S = 1/100—4 hours)	Inhibition of release and production of histamine by antigen-stimulated RBL-2H3 mast cells Suppression of production and expression of IL-8 in histamine-stimulated EA.hy926 endothelial cells Reduction of ROS production in mast and endothelial cells	LDAVNR; MMLDF	Vo et al. (2013)
Arthrospira platensis	–	Pepsin—6 hours	Pancreatin (18 hours)	Improving the digestibility of microalgal biomass	–	Niccolai et al. (2019)
	–	In silico pepsin digestion	In silico trypsin digestion	Antioxidant activity ACE inhibitory activity DPP-IV inhibitory activity	–	Liu et al. (2022)
Chlorella vulgaris	–	In silico pepsin digestion	In silico trypsin + chymotrypsin digestion	ACE inhibition and antihypertensive activity in SHR of identified peptides	TTW; VHW	Xie et al. (2018)
	–	In silico pepsin digestion	In silico trypsin + chymotrypsin digestion	DPP-IV inhibitory activity of identified peptides in mice serum	AAR; VPA; VPW; IPL; IPR; PPL	Zhu et al. (2017)
	–	Pepsin (E/S = 1/100)—3 hours	Porcine pancreatin (E/S = 1/100—4 hours)	Radical (superoxide, peroxyl, DPPH, and ABTS) scavenging capacity No cytotoxicity observed in human lung fibroblasts cell lines (WI-38) in vitro	VECYGPNRPQF	Sheih et al. (2009)
	–	Pepsin—4 hours	Pancreatin (16 hours)	Cell-disrupted improved protein digestibility	–	Wild et al. (2018)

(Continued)

Table 12.1 Protein hydrolyzates and bioactive peptides obtained from microalgae biomass under simulated gastrointestinal conditions. *Continued*

Microalga species	Digestion conditions			Bioactivity	Bioactive peptide	References
	Oral	Gastric	Intestinal			
Chlorella pyrenoidosa	–	Pepsin (E/S = 1/50)—16 hours	Trypsin (E/S = 1/50—16 hours)	ACE inhibitory activity DPP-IV inhibitory activity	QIYTMGK; FLKPLGSGK; LLTKS; ILGCR; QHAGTKAK	Li et al. (2021)
Dunaliella salina	–	Pepsin (E/S = 3000 U/g protein) – 4 hours	Trypsin (4 hours)	Radical (DPPH) scavenging activity	ILTKAAIEGK; IIYFQGK, NDPSTVK; TVRPPQR	Xia et al. (2019)
Isochrysis zhanjiangensis	–	Pepsin (E/S = 1/100)—2.5 hours	Trypsin + chymotrypsin (E/S = 1/100—2.5 hours)	Radical (DPPH, hydroxyl, superoxide, and alkyl) scavenging capacity of the digest Attenuation of ethanol-induced cytotoxicity in HepG2 cells by inhibition of ROS production, upregulation of SOD and GSH, and downregulation of GGT protein expression	NDAEYGICGF	Chen et al. (2019)
	–	Pepsin (E/S = 1/100)—2.5 hours	Trypsin + chymotrypsin (E/S = 1/100—2.5 hours)	ACE inhibitory activity Inhibition of Ang II-induced vascular factor secretion and expression by blocking inflammation and apoptosis through NF-κB, Nrf2, MAPKs, and the Akt signal pathways in HUVEC	FEIHCC	Chen et al. (2020)
	–	Pepsin (E/S = 1/100)—2.5 hours	Trypsin + chymotrypsin (E/S = 1/	Inhibition of the upregulation of oxidized LOX-1 induced by ox-LDL Inhibition of ROS, inflammatory factors (IL-6, IL-1β, and TNF-α) and	EMFGTSSET	Pei et al. (2022)

			100—2.5 hours)	the expression of cell adhesion VCAP-1 and ICAM-1 Upregulation of Nrf2, HO-1, p-Akt, and Bcl-2 levels, and downregulation of pp65, p-IκB-α, p-p38, p-ERK, p-JNK, Bax, and c-c-9/-3		
Nannochloropsis oculata	–	In silico pepsin digestion	In silico trypsin digestion	ACE-I inhibition of identified peptides Inhibition of antithrombotic and calcium/CAMKII	NKFPYTTQ	Hayes et al. (2022)
	–	Lipase + pepsin—2 hours	Pancreatin + bile (2 hours)	pH-shift processing with protein solubilization at pH 7 or pH 10 increased the accessibility of lipids Protein accessibility rose from 32% to 47% when the product was prefrozen prior to digestion With adequate accessibility (>50%) of amino acids and TFAs, pH-shift processed microalga has possible as a functional food or feed ingredient	–	Cavonius et al. (2016)
Nostoc sphaeroides	–	Pepsin—6 hours	Pancreatin (18 hours)	Improving the digestibility of microalgal biomass	–	Niccolai et al. (2019)
Pavlova lutheri	–	Pepsin (E/S = 1/100)—2 hours	Trypsin + chymotrypsin (E/S = 1/100—2.5 hours)	Reduction of mRNA and protein expression levels of MMP-9 Inhibition of PMA-induced translocation of NF-κB by suppressing IκB-α phosphorylation Suppression of phosphorylation of JNK and MAPKs	LLAPPER	Ko et al. (2018)

(Continued)

Table 12.1 Protein hydrolyzates and bioactive peptides obtained from microalgae biomass under simulated gastrointestinal conditions. *Continued*

Microalga species	Digestion conditions			Bioactivity	Bioactive peptide	References
	Oral	Gastric	Intestinal			
Schizochytrium sp.	–	Pepsin (E/S = 3%, w/v)–1 hours	Trypsin (E/S = 3%, w/v–5 hours)	Increase of resistance to colitis in a DSS-colitis-induced mouse model Reduction of body weight loss, clinical colitis scores, colonic histological damage, and colonic inflammation Attenuation of the induction of proinflammatory cytokines and increased the induction of antiinflammatory cytokines Promotion of the proliferation of colonic crypt stem cells and progenitor cells required for crypt repair	–	Wang et al. (2018)
Schizochytrium limacinum	–	Pepsin (E/S = 1/100)–1 hours	Trypsin + chymotrypsin (E/S = 1/100–4 hours)	Radical (DPPH, ABTS) scavenging and reducing power activity Protective effects against LDL oxidation Highest effects in 5–10 kDa fraction	–	Moaveni et al. (2022)

Calmodulin-dependent kinase II (CAMKII); superoxide dismutase (SOD); glutathione (GSH); gamma-glutamyltransferase (GGT); angiotensin II; nuclear factor κB (NF-κB), nuclear erythroid 2-related factor 2 (Nrf2), mitogen-activated protein kinases (MAPKs); serine/threonine kinase (Akt); human vein endothelial cells (HUVECs); oxidized low-density lipoprotein receptor-1 (LOX-1), interleukin-6 (IL-6), interleukin-1β (IL-1β), and tumor necrosis factor-α (TNF-α); vascular cell adhesion protein 1 (VCAP-1) and intercellular cell adhesion molecule-1 (ICAM-1); nuclear red blood cell 2 related factor 2 (Nrf2), heme oxygenase-1 (HO-1), phospho-serine/threonine kinase (p-Akt), B-cell lymphoma 2 (bcl-2); cleaved caspase-9/-3 (c-c-9/-3), c-Jun N-terminal (JNK); matrix metalloproteinase-9 (MMP-9); phorbol 12-myristate 13-acetate (PMA); dextran sulfate sodium (DSS); inducible nitric oxide synthase (iNOS) and endothelin-1 (ET-1)trifluoracetic acid (TFA). nitric oxide (NO), reactive oxygen species (ROS), enzyme/substrate (E/S), somatic hybrid human cells (EA.hy926), monocyte chemoattractant protein-1 (MCP-1), early growth response factor 1 (Egr-1), protein kinase C (PkC), basophils from rat peripheral blood (RBL-2H3), dipeptidyl peptidase IV (DPP-IV), angiotensin-converting enzyme (ACE), low-density lipoprotein (LDL), 2,2'-azino-bis(3-ethylbenzothiazoline-6-sulfonic acid) (ABTS), 2, 2-diphenyl-1-picrylhydrazyl (DDPH), spontaneously hypertensive rat (SHR), B-cell lymphoma 2 associated X protein, (BAX), nuclear factor of kappa light polypeptide gene enhancer in B-cells inhibitor-alpha (IkB-α).

possible bioactivities, including ACE-I inhibition, antithrombotic, and calcium/calmodulin-dependent kinase II (CAMKII) inhibition. This implies that the discovered peptides may operate favorably to lower the possibility of thrombus formation after digestion and may inhibit CAMKII to improve cardiovascular health (Pellicena & Schulman, 2014). Similarly, numerous peptides previously identified as ACE-I inhibitors or having other bioactivities, such as the inhibition of DPP-IV and antioxidant action, were discovered (Chen et al., 2019, 2020; Xie et al., 2018; Zhu et al., 2017). In a recent study, Pei et al. (2022) assessed the protective impacts of polypeptides from the microalgae *I. zhanjiangensis* (EMFGTSSET, ETT) on endothelial injury in angiotensin II (Ang II)-induced human umbilical vein endothelial cells (HUVECs). The findings of this study demonstrated that ETT suppressed intracellular reactive oxygen species (ROS) fluorescence intensity and decreased the expression of associated cytokines, including interleukin (IL)-1β, IL-8, TNF-α, endothelial NO synthase (eNOS), COX-2, endothenin-1 (ET-1), angiotensin 1 (AT-1) receptor, and adhesion factor. It reduces endothelial damage by inhibiting apoptosis and inflammation via mitogen-activated protein kinases (MAPKs), NF-κB, and serine/threonine kinase (Akt) signal pathways. ETT can therefore become the primary ingredient of potential functional foods for the prevention of cardiovascular diseases (especially in controlling hypertension) and further increase the usefulness of *I. zhanjiangensis*. In another study by Heo et al. (2017), an ACE inhibitory peptide from a gastrointestinal digest of *Spirulina* sp. protein was isolated, and the peptide was examined for its possible inhibitory impact on endothelial dysfunction in a human umbilical vein cell line. An ACE inhibitory peptide of 7 amino acids (TMEPGKP) was obtained after purification. They reported that the ACE inhibitory peptide from *Spirulina* sp. protein hydrolyzate possessed a proline (P) residue within its C-terminal tripeptide sequence, which may be responsible for the peptide's ACE inhibitory effect. Overall, the interaction between the three subsites S1 (antepenultimate), S'1 (penultimate), and S'2 (ultimate), and the ACE active site, was discovered to be responsible for the effectiveness of the primary substrates of peptide composition, such as W, Y, F, P, and a hydrophobic amino acid at the C-terminus. In addition, they mentioned that the isolated peptide reduced the amount of NO and intracellular ROS produced by angiotensin (Ang) II. Treatment with such isolated peptide allowed for the recovery of alterations in cell form associated with intracellular ROS generation. These authors also stated that Ang II induces NO production via the expression of inducible NOS (iNOS) in vascular smooth muscle cells, with no impact on iNOS expression. It has been noted that increased NO production through the expression of iNOS initiates apoptotic and necrotic cell death in the adjacent tissue, as in atherosclerosis. Similar to Ang II, ET-1 is a recognized vasoconstriction agent that can cause endothelial dysfunction. In the same work, it was shown that iNOS and ET-1 expression in Ang II-stimulated EA.hy926 cells was decreased by the purified ACE inhibitory peptide (Heo et al., 2017). They also discovered that the purified peptide lowered the Ang II-induced phosphorylation of p38 MAPK.

Atherogenesis of atherosclerosis is thought to be triggered by histamine, a strong inflammatory mediator. In this regard, two bioactive peptides, P1 (LDAVNR; 686 Da) and P2 (MMLDF; 655 Da), isolated from the gastric enzymatic hydrolyzate of *A. maxima*, were tested for their ability to prevent early atherosclerotic reactions in EA. hy926 endothelial cells caused by histamine, according to (Vo & Kim, 2013). Both P1 and P2 showed inhibitory impacts on the production and expression of IL-6 and MCP-1. Moreover, P1 and P2 reduced the in vitro cell adhesion of monocytes onto endothelial cells by inhibiting the synthesis of adhesion molecules, such as P-selectin and E-selectin. In addition, it was noted that the presence of P1 or P2 decreased the generation of intracellular ROS. Notably, the inhibitory effects of P1 and P2 were discovered as a result of downregulating Egr-1 expression through PKCd-dependent MAPK activation and the histamine receptor. These findings imply that *S. maxima* peptides P1 and P2 are potent inhibitors of histamine-induced endothelial cell activation, which may help prevent the development of early atherosclerosis. In a different study by Vo et al. (2013), the same peptides (P1 and P2) from *S. maxima* were used effectively to preventing mast cell histamine release and synthesis, as well as endothelial cell IL-8 production. The inhibitory effects of these peptides were mediated by preventing cells from producing intracellular ROS.

The matrix metallopeptidase 9 (MMP-9) is regarded as a key target for the prevention of cancer metastasis, since it is essential for the stimulation of tumor migration and invasion (Lee et al., 2015). MMP-9 activity and expression are frequently studied in model systems using HT1080 fibrosarcoma cells. In this respect, Ko et al. (2018) isolated a number of peptides from the gastrointestinal digest of *Pavlova lutheri* and evaluated them to determine if they have any MMP-9 inhibitory impact on HT1080 fibrosarcoma cells that were generated by phorbol 12-myristate 13-acetate (PMA). They discovered that *P. lutheri* gastrointestinal digest possessed MMP-9 inhibitory action in PMA-induced HT1080 fibrosarcoma cells, and they purified and sequenced peptides responsible for the observed effects. LLAPPER was found to be the MMP-9 inhibitory peptide, and its molecular weight was 796.4 Da. In PMA-stimulated HT1080 fibrosarcoma cells, this peptide reduced the mRNA and protein expression levels of MMP-9. They also demonstrated that this peptide blocks PMA-induced translocation of NF-κB by repressing IκB-α phosphorylation and that it decreases the phosphorylation of the c-Jun N-terminal and p38 MAPKs. These findings suggest that this peptide may prove to be a useful natural preventive agent for MMP-related diseases.

Docosahexaenoic acid (DHA), an omega-3 poliinsaturated fatty acid well known for its bioactive potential, is now produced using the bioengineered marine microalgae *Schizochytrium* sp. Unfortunately, the remaining protein-rich material is not used effectively after lipid extraction. In this context, (Wang et al., 2018) reported that *Schizochytrium* sp. hydrolyzate (MESH), which shows a distinctive peptide profile as identified by ultra-performance liquid chromatography coupled with Q-TOF mass spectrometry (UPLC/Q-TOF-Ms), ameliorated bowel inflammation in mice. The mice treated with MESH were significantly less susceptible to develop colitis than the control mice, using a mouse model of experimental colitis caused by dextran sulfate sodium, as shown by significantly less body weight loss, clinical colitis scores, colonic histological damage, and colonic inflammation. Mechanistically, MESH reduced proinflammatory cytokine induction and boosted antiinflammatory cytokine induction. The colonic crypt stem cells and progenitor cells necessary for crypt healing were also multiplied by MESH. Overall, these findings highlight the potential of MESH as an antiinflammatory agent for colitis.

Moreover, proteins that are soluble and insoluble in biomass are separated during the pH-shift process, and the solubilized proteins are then precipitated and recovered. Using the INFOGEST digestion protocol with additional gastric lipase, *N. oculata* was subjected to the pH-shift process, followed by the digestion of various intermediates and product fractions of the process. The degrees of protein hydrolysis and fatty acid liberation were evaluated after digestion and compared to the quantities of peptide bonds and total fatty acids contained in the raw materials as indicators of protein and lipid accessibility (Cavonius et al., 2016). Their results demonstrated that the mammalian digestive enzymes used in the digestion model could not access the proteins or lipids of intact *Nannochloropsis* cells. Lipid accessibility was increased by cell disruption and, to a lesser extent, by additional pH-shift processing, such as protein solubilization at pH 7 or pH 10. Although prefreezing the product before digestion enhanced the protein accessibility from 32% to 47%, differences among the pH-shifted materials were not significant. These findings show the importance of microalgae processing to obtain substrates with advantageous nutritional properties and with potential application in the food and feed industry.

Spirulina-enriched foods, being a good source of bioactive peptides, were assessed by Donadio et al. (2021). Their investigation was focused on pasta and four different sorbets, and it combined a simulated digestion process with quantitative mass spectrometry analysis of the resulting mixtures. According to the outcomes, inclusion of microalgae into food matrixes has a specific impact on the production of various peptides, specifically altering the amounts of peptides that make up the β-subunit of C phycocyanin.

12.4 Conclusions

Microalgal proteins can be regarded as an alternative protein source with potential nutritional and biological characteristics after being broken down in the gastrointestinal tract, liberating essential amino acids and bioactive peptides (Fig. 12.1).

Simulated gastrointestinal digestion has become one of the most widely investigated strategies to evaluate bioactive peptides from microalgae proteins, as it is fast, affordable, safe, and has no ethical limitations. To encourage the use of these proteins as novel antioxidant, ACE inhibitor, antidiabetic, and immunostimulatory ingredients in functional foods and nutraceuticals, additional research should be undertaken with a specific focus on identifying the peptides responsible for the positive impacts, as well as on confirmation of their bioavailability and bioactivity in animal models and human intervention trials.

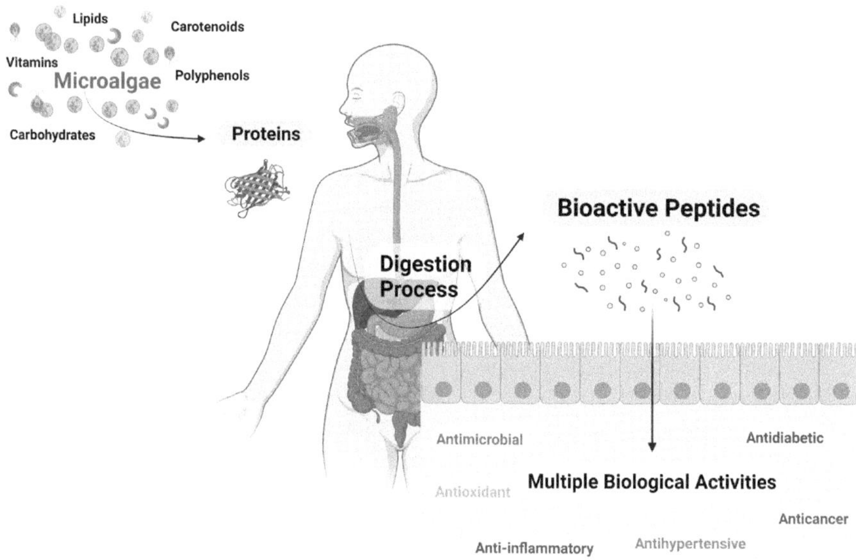

FIGURE 12.1

Biological activities of peptides released from microalgae during their transit through the gastrointestinal tract. Activities of microalgae-derived peptides released during their transit through the gastrointestinal tract.

References

Ampofo, J., & Abbey, L. (2022). Microalgae: Bioactive composition, health benefits, safety and prospects as potential high-value ingredients for the functional food industry. *Foods*, *11*(12). Available from https://doi.org/10.3390/foods11121744.

Barkia, I., Saari, N., & Manning, S. R. (2019). Microalgae for high-value products towards human health and nutrition. *Marine Drugs*, *17*(5). Available from https://doi.org/10.3390/md17050304, https://www.mdpi.com/1660-3397/17/5/304/pdf.

Cavonius, L. R., Albers, E., & Undeland, I. (2016). In vitro bioaccessibility of proteins and lipids of pH-shift processed Nannochloropsis oculata microalga. *Food & Function*, *7*(4), 2016–2024. Available from https://doi.org/10.1039/C5FO01144B.

Chen, J., Tan, L., Li, C., Zhou, C., Hong, P., Sun, S., & Qian, Z. J. (2020). Mechanism analysis of a novel angiotensin-I-converting enzyme inhibitory peptide from Isochrysis zhanjiangensis microalgae for suppressing vascular injury in human umbilical vein endothelial cells. *Journal of Agricultural and Food Chemistry*, *68*(15), 4411–4423. Available from https://doi.org/10.1021/acs.jafc.0c00925, http://pubs.acs.org/journal/jafcau.

Chen, M. F., Zhang, Y. Y., Di He, M., Li, C. Y., Zhou, C. X., Hong, P. Z., & Qian, Z. J. (2019). Antioxidant peptide purified from enzymatic hydrolysates of isochrysis zhanjiangensis and its protective effect against ethanol induced oxidative stress of HepG2 cells. *Biotechnology and Bioprocess Engineering*, *24*(2), 308–317. Available from https://doi.org/10.1007/s12257-018-0391-5, http://www.springerlink.com/content/1226-8372.

Cherng, J. Y., Liu, C. C., Shen, C. R., Lin, H. H., & Shih, M. F. (2010). Beneficial effects of chlorella-11 peptide on blocking LPS-induced macrophage activation and alleviating thermal injury-induced inflammation in rats. *International Journal of Immunopathology and Pharmacology*, *23*(3), 811–820. Available from https://doi.org/10.1177/039463201002300316.

Choo, W. T., Teoh, M. L., Phang, S. M., Convey, P., Yap, W. H., Goh, B. H., & Beardall, J. (2020). Microalgae as potential anti-inflammatory natural product against human inflammatory skin diseases. *Frontiers in Pharmacology*, *11*. Available from https://doi.org/10.3389/fphar.2020.01086, http://www.frontiersin.org/Pharmacology.

Donadio, G., Santoro, V., Dal Piaz, F., & De Tommasi, N. (2021). Food matrices affect the peptides produced during the digestion of arthrospira platensis-based functional aliments. *Nutrients*, *13*(11), 3919. Available from https://doi.org/10.3390/nu13113919.

Eilam, Y., Khattib, H., Pintel, N., & Avni, D. (2023). Microalgae—sustainable source for alternative proteins and functional ingredients promoting gut and liver health. *Global Challenges*.

Finamore, A., Palmery, M., Bensehaila, S., & Peluso, I. (2017). Antioxidant, immunomodulating, and microbial-modulating activities of the sustainable and ecofriendly spirulina. *Oxidative Medicine and Cellular Longevity*, *2017*, 1–14. Available from https://doi.org/10.1155/2017/3247528.

Hayes, M., Mora, L., & Lucakova, S. (2022). Identification of bioactive peptides from Nannochloropsis oculata using a combination of enzymatic treatment, in silico analysis and chemical synthesis. *Biomolecules*, *12*(12). Available from https://doi.org/10.3390/biom12121806.

Heo, S. Y., Ko, S. C., Kim, C. S., Oh, G. W., Ryu, B., Qian, Z. J., Kim, G., Park, W. S., Choi, I. W., Phan, T. T. V., Heo, S. J., Kang, D. H., Yi, M., & Jung, W. K. (2017). A heptameric peptide purified from Spirulina sp. gastrointestinal hydrolysate inhibits angiotensin I-converting enzyme- and angiotensin II-induced vascular dysfunction in human endothelial cells. *International Journal of Molecular Medicine*, *39*(5), 1072–1082. Available from https://doi.org/10.3892/ijmm.2017.2941, https://www.spandidos-publications.com/ijmm/39/5/1072/download.

Januário, A. P., Félix, R., Félix, C., Reboleira, J., Valentão, P., & Lemos, M. F. (1930). Red seaweed-derived compounds as a potential new approach for acne vulgaris care. *Pharmaceutics*, *13*(11).

Khanra, S., Mondal, M., Halder, G., Tiwari, O. N., Gayen, K., & Bhowmick, T. K. (2018). Downstream processing of microalgae for pigments, protein and carbohydrate in industrial application: A review. *Food and Bioproducts Processing*, *110*, 60–84. Available from https://doi.org/10.1016/j.fbp.2018.02.002, http://www.elsevier.com/wps/find/journaldescription.cws_home/713883/description#description.

Ko, S. C., Heo, S. Y., Choi, S. W., Qian, Z. J., Heo, S. J., Kang, D. H., Kim, N., & Jung, W. K. (2018). A heptameric peptide isolated from the marine microalga Pavlova lutheri suppresses PMA-induced secretion of matrix metalloproteinase-9 through the inactivation of the JNK, p38, and NF-κB pathways in human fibrosarcoma cells. *Journal of Applied Phycology*, *30*(4), 2367–2378. Available from https://doi.org/10.1007/s10811-018-1443-0, http://www.wkap.nl/journalhome.htm/0921-8971.

Ko, S. C., Kim, D., & Jeon, Y. J. (2012). Protective effect of a novel antioxidative peptide purified from a marine Chlorella ellipsoidea protein against free radical-induced oxidative stress. *Food and Chemical Toxicology*, *50*(7), 2294–2302. Available from https://doi.org/10.1016/j.fct.2012.04.022.

Kumar, R., Hegde, A. S., Sharma, K., Parmar, P., & Srivatsan, V. (2022). Microalgae as a sustainable source of edible proteins and bioactive peptides – Current trends and future prospects. *Food Research International*, *157*. Available from https://doi.org/10.1016/j.foodres.2022.111338.

Lee, W. T., Lee, T. H., Cheng, C. H., Chen, K. C., Chen, Y. C., & Lin, C. W. (2015). Antroquinonol from Antrodia Camphorata suppresses breast tumor migration/invasion through inhibiting ERK-AP-1- and AKT-NF-κB-dependent MMP-9 and epithelial-mesenchymal transition expressions. *Food and Chemical Toxicology*, *78*, 33–41. Available from https://doi.org/10.1016/j.fct.2015.01.012, http://www.elsevier.com/locate/foodchemtox.

Li, Y., Aiello, G., Fassi, E. M. A., Boschin, G., Bartolomei, M., Bollati, C., Roda, G., Arnoldi, A., Grazioso, G., & Lammi, C. (2021). Investigation of chlorella pyrenoidosa protein as a source of novel angiotensin I-converting enzyme (ACE) and Dipeptidyl Peptidase-IV (DPP-IV) inhibitory peptides. *Nutrients*, *13*(5). Available from https://doi.org/10.3390/nu13051624, https://www.mdpi.com/2072-6643/13/5/1624/pdf.

Liu, J., Bai, X., & Fu, P. (2022). In silico and in vitro assessment of bioactive peptides from Arthrospira platensis phycobiliproteins for DPP-IV inhibitory activity, ACE inhibitory activity, and antioxidant activity. *Journal of Applied Phycology*, *34*(3), 1497–1511. Available from https://doi.org/10.1007/s10811-022-02732-z.

Moaveni, S., Salami, M., Khodadadi, M., McDougall, M., & Emam-Djomeh, Z. (2022). Investigation of S. limacinum microalgae digestibility and production of antioxidant bioactive peptides. *LWT*, *154*.

Montalvo, G. E. B., Thomaz-Soccol, V., Vandenberghe, L. P. S., Carvalho, J. C., Faulds, C. B., Bertrand, E., Prado, M. R. M., Bonatto, S. J. R., & Soccol, C. R. (2019). Arthrospira maxima OF15 biomass cultivation at laboratory and pilot scale from sugarcane vinasse for potential biological new peptides production. *Bioresource Technology*, *273*, 103–113. Available from https://doi.org/10.1016/j.biortech.2018.10.081, http://www.elsevier.com/locate/biortech.

Muys, M., Sui, Y., Schwaiger, B., Lesueur, C., Vandenheuvel, D., Vermeir, P., & Vlaeminck, S. E. (2019). High variability in nutritional value and safety of commercially available Chlorella and Spirulina biomass indicates the need for smart production strategies. *Bioresource Technology*, *275*, 247–257. Available from https://doi.org/10.1016/j.biortech.2018.12.059, http://www.elsevier.com/locate/biortech.

Nemoto-Kawamura, C., Hirahashi, T., Nagai, T., Yamada, H., Katoh, T., & Hayashi, O. (2004). Phycocyanin enhances secretary IgA antibody response and suppresses allergic IgE antibody response in mice immunized with antigen-entrapped biodegradable microparticles. *Journal of Nutritional Science and Vitaminology*, *50*(2), 129–136. Available from https://doi.org/10.3177/jnsv.50.129.

Niccolai, A., Chini Zittelli, G., Rodolfi, L., Biondi, N., & Tredici, M. R. (2019). Microalgae of interest as food source: Biochemical composition and digestibility. *Algal Research*, *42*. Available from https://doi.org/10.1016/j.algal.2019.101617, http://www.sciencedirect.com/science/journal/aip/22119264.

Pei, Y., Cai, S., Ryu, B., Zhou, C., Hong, P., & Qian, Z.-J. (2022). An ACE inhibitory peptide from Isochrysis zhanjiangensis exhibits antihypertensive effect via anti-inflammation and anti-apoptosis in HUVEC and hypertensive rats. *Journal of Functional Foods*, *92*. Available from https://doi.org/10.1016/j.jff.2022.105061.

Pellicena, P., & Schulman, H. (2014). CaMKII inhibitors: From research tools to therapeutic agents. *Frontiers in Pharmacology*, *5*. Available from https://doi.org/10.3389/fphar.2014.00021.

Sheih, I. C., Fang, T. J., Wu, T. K., & Lin, P. H. (2010). Anticancer and antioxidant activities of the peptide fraction from algae protein waste. *Journal of Agricultural and Food Chemistry*, *58*(2), 1202–1207. Available from https://doi.org/10.1021/jf903089mTaiwan, http://pubs.acs.org/doi/pdfplus/10.1021/jf903089m.

Sheih, I. C., Wu, T. K., & Fang, T. J. (2009). Antioxidant properties of a new antioxidative peptide from algae protein waste hydrolysate in different oxidation systems. *Bioresource Technology*, *100*(13), 3419–3425. Available from https://doi.org/10.1016/j.biortech.2009.02.014.

Shih, M. F., Chen, L. C., & Cherng, J. Y. (2013). Chlorella 11-peptide inhibits the production of macrophage-induced adhesion molecules and reduces endothelin-1 expression and endothelial permeability. *Marine Drugs*, *11*(10), 3861–3874. Available from https://doi.org/10.3390/md11103861, http://www.mdpi.com/1660-3397/11/10/3861/pdf.

Stack, J., Gouic, A. V. L., Tobin, P. R., Guihéneuf, F., Stengel, D. B., & FitzGerald, R. J. (2018). Protein extraction and bioactive hydrolysate generation from two microalgae, Porphyridium purpureum and Phaeodactylum tricornutum. *Journal of Food Bioactives*, *1*. Available from https://doi.org/10.31665/jfb.2018.1134.

Suttisuwan, R., Phunpruch, S., Saisavoey, T., Sangtanoo, P., Thongchul, N., & Karnchanatat, A. (2019). Isolation and characterization of anti-inflammatory peptides derived from trypsin hydrolysis of microalgae protein (Synechococcus sp. VDW). *Food Biotechnology*, *33*(4), 303–324. Available from https://doi.org/10.1080/08905436.2019.1673171.

Tzachor, A., Rozen, O., Khatib, S., Jensen, S., & Avni, D. (2021). Photosynthetically controlled spirulina, but not solar spirulina, inhibits TNF-α secretion: Potential implications for COVID-19-related cytokine storm therapy. *Marine Biotechnology*, *23*(1), 149–155. Available from https://doi.org/10.1007/s10126-021-10020-z.

Vo, T. S., & Kim, S. K. (2013). Down-regulation of histamine-induced endothelial cell activation as potential anti-atherosclerotic activity of peptides from Spirulina maxima. *European Journal of Pharmaceutical Sciences*, *50*(2), 198–207. Available from https://doi.org/10.1016/j.ejps.2013.07.001, http://www.elsevier.com/locate/ejps.

Vo, T. S., Ryu, B. M., & Kim, S. K. (2013). Purification of novel anti-inflammatory peptides from enzymatic hydrolysate of the edible microalgal Spirulina maxima. *Journal of Functional Foods*, *5*(3), 1336–1346. Available from https://doi.org/10.1016/j.jff.2013.05.001.

Wang, X., Wang, H., Pierre, J. F., Wang, S., Huang, H., Zhang, J., Liang, S., Zeng, Q., Zhang, C., Huang, M., Ruan, C., Lin, J., & Li, H. (2018). Marine microalgae bioengineered Schizochytrium sp. meal hydrolysates inhibits acute inflammation. *Scientific Reports*, *8*(1). Available from https://doi.org/10.1038/s41598-018-28064-y, http://www.nature.com/srep/index.html.

Wild, K. J., Steingaß, H., & Rodehutscord, M. (2018). Variability in nutrient composition and in vitro crude protein digestibility of 16 microalgae products. *Journal of Animal Physiology and Animal Nutrition*, *102*(5), 1306–1319. Available from https://doi.org/10.1111/jpn.12953, http://onlinelibrary.wiley.com/journal/10.1111/(ISSN)1439-0396.

Wu, H., He, H. L., Chen, X. L., Sun, C. Y., Zhang, Y. Z., & Zhou, B. C. (2008). Purification and identification of novel angiotensin-I-converting enzyme inhibitory peptides from shark meat hydrolysate. *Process Biochemistry*, *43*(4), 457–461. Available from https://doi.org/10.1016/j.procbio.2008.01.018.

Xia, E., Zhai, L., Huang, Z., Liang, H., Yang, H., Song, G., Li, W., & Tang, H. (2019). Optimization and identification of antioxidant peptide from underutilized Dunaliella salina protein: Extraction, in vitro gastrointestinal digestion, and fractionation. *BioMed Research International*, *2019*, 1–9. Available from https://doi.org/10.1155/2019/6424651.

Xiao, F., & Guo, F. (2022). Impacts of essential amino acids on energy balance. *Molecular Metabolism*, *57*. Available from https://doi.org/10.1016/j.molmet.2021.101393.

Xie, J., Chen, X., Wu, J., Zhang, Y., Zhou, Y., Zhang, L., Tang, Y.-J., & Wei, D. (2018). Antihypertensive effects, molecular docking study, and isothermal titration calorimetry assay of angiotensin I-converting enzyme inhibitory peptides from Chlorella vulgaris. *Journal of Agricultural and Food Chemistry*, *66*(6), 1359–1368. Available from https://doi.org/10.1021/acs.jafc.7b04294.

Zhang, B., & Zhang, X. (2013). Separation and nanoencapsulation of antitumor polypeptide from Spirulina platensis. *Biotechnology Progress*, *29*(5), 1230–1238. Available from https://doi.org/10.1002/btpr.1769.

Zhou, L., Li, K., Duan, X., Hill, D., Barrow, C., Dunshea, F., Martin, G., & Suleria, H. (2022). Bioactive compounds in microalgae and their potential health benefits. *Food Bioscience*, *49*. Available from https://doi.org/10.1016/j.fbio.2022.101932.

Zhu, Q., Chen, X., Wu, J., Zhou, Y., Qian, Y., Fang, M., Xie, J., & Wei, D. (2017). Dipeptidyl peptidase IV inhibitory peptides from Chlorella vulgaris: In silico gastrointestinal hydrolysis and molecular mechanism. *European Food Research and Technology*, *243*(10), 1739–1748. Available from https://doi.org/10.1007/s00217-017-2879-1.

CHAPTER

Bioactive peptides released from seaweeds during gastrointestinal digestion 13

María Fraga-Corral[1], Aurora Silva[1,2], Javier Echave[1], Lucia Cassani[1], Jianbo Xiao[1,3], Jesus Simal-Gandara[1] and Miguel Angel Prieto[1]

[1]*Department of Analytical Chemistry and Food Science, Universidade de Vigo, Nutrition and Bromatology Group, Instituto de Agroecoloxía e Alimentación (IAA) – CITEXVI, Vigo, España*

[2]*REQUIMTE/LAQV, Instituto Superior de Engenharia do Porto, Instituto Politécnico do Porto, Porto, Portugal*

[3]*International Joint Research Laboratory of Intelligent Agriculture and Agri-products Processing, Jiangsu University, Zhenjiang, P.R. China*

List of Abbreviations

General	
AA	amino acid
EAA	essential amino acid
NEAA	non−essential amino acid
PH	protein hydrolysates
GI	gastrointestinal
BAPs	bioactive peptides
ORAC	oxygen radical absorption capacity
FRAP	ferric reducing antioxidant power
DPPH	2,2-Diphenyl-1-picrylhydrazyl
ACE	angiotensin-I-converting enzyme
ROS	reactive oxygen species
IL	interleukin
TNF-α	tumor necrosis factor-α
COX-2	cyclooxygenase-2
NO	nitric oxide
iNOS	nitric oxide synthase
NF-κB	nuclear factor κappa beta
MAPK	mitogen-activated protein kinase
mTOR	mammalian target of rapamycin
IGF-IR	insulin-like growth factor-I receptor
PI3K	phosphatidylinositol 3-kinase
DPP-IV	dipeptidyl peptidase-IV
GLP-1	glucagon-like peptide-1
Amino acid	**Three-letter code** (Single-letter code)

Protein Digestion-Derived Peptides. DOI: https://doi.org/10.1016/B978-0-443-19141-1.00013-3

Alanine	Ala (A)
Arginine	Arg (R)
Asparagine	Asn (N)
Aspartic acid	Asp (D)
Cysteine	Cys (C)
Glutamic acid	Glu (E)
Glutamine	Gln (Q)
Glycine	Gly (G)
Histidine	His (H)
Isoleucine	Ile (I)
Leucine	Leu (L)
Lysine	Lys (K)
Methionine	Met (M)
Phenylalanine	Phe (F)
Proline	Pro (P)
Serine	Ser (S)
Threonine	Thr (T)
Tryptophan	Trp (W)
Tyrosine	Tyr (Y)
Valine	Val (V)

13.1 Introduction

Seaweed is considered a relevant nutritional resource since its chemical composition includes high amounts of key macronutrients, especially proteins and lipids, but it also provides micronutrients, such as vitamins and minerals. Besides, macroalgae are a well-known source of dietary fiber and several minor but crucial phytochemicals associated with health benefits, such as polyphenols. Therefore kelps represent a natural, sustainable, and rich source of bioactive molecules which may serve as substrates for the development of added-value products by industrial sectors, such as food, cosmetics, pharmaceutical, or feeding, among others (Carpena et al., 2021). Regarding the global food crisis, macroalgae are revealed as a source of proteins which may be a potential and partial solution to the future increasing protein demand derived as a result of the estimated steady rise of the global population (García-Oliveira et al., 2022). In fact, kelps have been demonstrated to proportionally overpass the production of protein estimated between 2.5 and 7.5 tons/ha/year when compared to terrestrial plants known to represent important sources of proteins, such as soybean or legumes with estimated production rates between 1 and 2 tons/ha/year (Bleakley & Hayes, 2017). Nevertheless, algae protein proportions vary among the three classes Rhodophyta (red), Ochrophyta (brown), and Chlorophyta (green). In this sense, Rhodophyta is the class with the highest contents of protein followed by Chlorophyta and Ochrophyta (Pliego-Cortés et al., 2020). In this regard, red and green algae are considered to have contents of protein similar to terrestrial plants considered to possess high amounts of protein (Ohanenye et al., 2020). However, it is clear that these protein estimations are highly dependent on several considerations, such as species variations, geographical origin, climatic parameters, or seasonal conditions (Lorenzo et al., 2017). Regarding seasonal peak productions, the highest amounts of protein have been quantified at the end of the winter and the commencement of spring and so the lowest levels were observed by the end of summer and the beginning of autumn (Denis et al., 2010). Other parameters that affect protein production are related to stress due to biotic and abiotic factors since they are strongly involved in the regulation of the expression and biosynthesis of protein and defense metabolites.

To determine the protein quality is essential to know their amino acid (AA) profile as well as their digestibility. High-quality proteins are considered those with high relative amounts of essential AA (EAAs) since human beings are not capable of synthesizing them and so they represent limiting factors. In this sense, macroalgae proteins have been described to contain important amounts of EAA. The other parameter regarded as key to considering the quality of protein is the protein digestibility. However, this condition is highly dependent on side factors, such as the simultaneous intake of antinutrients, such as tannins or some polysaccharides. These molecules may get attached to the proteins or amino acids structure and modify it which ultimately may hamper their absorption (Bleakley & Hayes, 2017; Fleurence et al., 2017). Nevertheless, macroalgae proteins have an amino acid profile described to possess digestibility with values reaching 70% which are comparable to rates for vegetables (74%), grains (76%), and legumes and fruits (82%) (Bleakley & Hayes, 2017). Therefore including macroalgae as dietary ingredients together with rich protein foods is considered a strategy to enhance the intake of high-quality proteins (Fleurence et al., 2017).

Macroalgae has been consumed historically, especially in Asiatic countries, and in recent decades its consumption has strongly increased over the world. Indeed, as demonstrated by their chemical composition, macroalgae represent sustainable and highly nutritional ingredients for humans and animals, especially regarding their protein content. Seaweed is currently used both as raw ingredients but also to improve the nutritional or organoleptic properties of different foodstuffs. Macroalgae extracts may be used as additives to increment the protein content of foodstuff with low concentrations or poor amino acidic profile or as natural additives or preservatives to substitute synthetic molecules with associated negative side effects (Dumay et al., 2014). In this context, in recent years the search for natural potential therapeutics has gained attention both in academia and the industry, to which bioactive peptides (BAPs) derived from various protein sources are being increasingly studied. BAPs with diverse sequences have been described as potent antioxidant, antidiabetic, hypotensive, and antiinflammatory molecules (Cermeño et al., 2020). BAPs with these bioactive properties have been described from both animal (e.g., meat discards, casein), plant sources (e.g., fruit discards, seeds, legumes) and also seaweeds, being obtained by protein extraction and hydrolysis by chemical and enzymatic methods (Nongonierma &

FitzGerald, 2017). At the time of the development of this work, there is not a great number of reports available detailing the bioactive properties of seaweed BAPs obtained by gastrointestinal (GI) digestion. To compensate for this limitation, works detailing significant bioactivities following protein hydrolysis methods with various proteases have been included in order to highlight the potential of seaweeds as a source of potentially valuable BAPs.

13.2 Protein content and composition of seaweeds

The current dietetic models established usually imply the consumption of ready and ultra-processed foodstuffs with very high caloric intake and very low nutritional value. This adopted diet pattern is prompting the rise of pathologies directly related to diet. Among the most concerning cardiovascular diseases, obesity, and diabetes, both types I and II stand out for their constant increment and their repercussions on health and so on the public health systems. However, in the last years, it seems the social concern regarding nutritional and lifestyle issues is growing. This trend can be observed through the switch to commercialized products which has shown a swing through the presence of healthier, natural and sustainable options. This tendency is probably due to consumers' awareness regarding the preventive effect that may exert on health. In this sense, macroalgae presence as dietary products is gaining space in Western markets likely for their well-known nutritional properties backed up by their high content of bioactive compounds. Currently, macroalgae represent a product demanded by industrial sectors, such as food, pharmaceutical, and cosmetics, due to their biologically active compounds strongly claimed by consumers (Dawczynski et al., 2007).

Macroalgae stand out for their protein content but especially for their AA profile (Table 13.1). As pointed out before, the proportion of protein is highly dependent on the taxonomic class of the macroalgae. In this sense, scientific literature shows very variable data even though Rhodophyta seems to be the class with the highest protein content between 12.5% and 44.0%, followed by Chlorophyta with content in the range of 9.6%–23.3% and Ochrophyta with the lower values from 4.5% to 16.8% (Gómez-Ordóñez et al., 2010; Pliego-Cortés et al., 2020) (Table 13.1). These protein contents, especially those of red macroalgae, are comparable to those reported for fish, eggs, cereals, and soybean (Machado et al., 2020). Indeed, red and green seaweed protein proportion is comparable or even slightly higher than values observed for legumes and nuts are considered to possess around 25% or cereals with an estimated protein content of 13% (Ohanenye et al., 2020). But as explained above, proteins need to meet some requirements to determine their nutritional value and to be considered high-quality. Indeed, high relative amounts of EAA ensure an appropriate daily intake (Cherry et al., 2019). Therefore the first criterion to meet is regards the AA composition. In this sense, macroalgae proteins have been described to contain important amounts of glycine, alanine, glutamic acid, and aspartic acid,

Table 13.1 Protein content and amino acid (AA) composition of macroalgae.

Species	Protein content (% dw)	Free AA composition[a]	Free AA (mg/g dw)	EAA (%)	EAA: NEAA	References
Rhodophyceae						
Gracilaria chilensis	18.9	Met, Glu, His, Asp, Phe, Ile, Val, Cys, Ser, Ala, Lys, Thr, Arg, Leu, Gly, Leu, Tyr, Pro	81.7	59	0.85	Ortiz et al. (2009)
Gracilaria domingensis	10.6	Asp, Glu, Ala, Arg, Gly, Ser, Tyr, Pro, Hyp, Phe, His, Ile, Leu, Lys, Met, Thr, Trp, Val	106.6	40.1	0.67	Machado et al. (2020)
Gracilaria gracilis	18.7	Glu, Asp, Arg, Lys, Gly, Leu, Ala, Ser, Val, Thr, Ile, Phe, Pro, Tyr, Met, His, Cys	187	48	0.92	Ferreira et al. (2021)
	30	Val, Asp, Glu, Ile, Leu, Ala, Thr, Phe, Ser, Lys, Arg, Tyr, Gly, Pro, Cys, His, Met	262	43	1.27	Batista et al. (2020)
Gracilariopsis lemaneiformis	20.87	Ala, Val, Leu, Iso, Met, Pro, Phe, Gly, Thr, Ser, Tyr, Asp, Glu, Lys, Arg, His	272	40.7	0.68	Wen et al. (2006)
Palmaria palmata	12.5	Thr, Val, Iso, Leu, Tyr, Phe, His, Lys, Met, Cys, Tau, Asp, Ser, Glu, Gly, Ala, Arg, Pro	112.18	37.7	0.61	Connor et al. (2020)
Porphyra spp.	27.4	Arg, Tyr, Thr, Val, Met, Lys, Ile, Leu, Phe, Ala, Asp, Glu, His, Ser, Gly, Gln, Tau, Trp, Hyp	106.3	44.3	1.32	Vieira et al. (2018)

(Continued)

Table 13.1 Protein content and amino acid (AA) composition of macroalgae. *Continued*

Species	Protein content (% dw)	Free AA composition[a]	Free AA (mg/g dw)	EAA (%)	EAA: NEAA	References
Porphyra dioica	28.7	Asp, Glu, Ala, Arg, Gly, Ser, Tyr, Pro, Hyp, Phe, His, Ile, Leu, Lys, Met, Thr, Trp, Val	286.6	39.8	0.66	Machado et al. (2020)
Porphyra columbina	24.61	Asp, Glu, Ser, His, Gly, Thr, Arg, Ala, Pro, Tyr, Phe, Val, Met, Cys, Ile, Trp, Leu, Lys	246.1	35.0	0.54	Cian et al. (2014)
Chlorophyceae						
Caulerpa racemosa	19.9	Thr, Val, Lys, Iso, Leu, Phe, His, Met, Asp, Ser, Glu, Gly, Ala, Tyr, Pro, Hyp	n.a.	45.3	0.83	Magdugo et al. (2020); Biancarosa et al. (2017)
Cladophora rupestris	12	Asp, Glu, Lys, Leu, Gly, Arg, Pro, Val, Ala, Thr, Phe, Ser, Tyr, Ile, Met, His	95.9	38.1	0.65	
Codium fragile	11	Val, Glu, Met, Asp, Leu, Ala, Thr, Lys, Gly, Ser, Phe, Arg, Ile, Cys, His, Pro	53.5	49.6	0.98	Ortiz et al. (2009)
Ulva spp.	23.3	Arg, Tyr, Thr, Val, Met, Lys, Ile, Leu, Phe, Ala, Asp, Glu, His, Ser, Gly, Gln, Tau, Trp, Hyp	116.2	43.6	1.3	Vieira et al. (2018)
Ulva capensis	19.4	Ile, Leu, Lys, Met, Cys, Phe, Tyr, Thr, Val, His, Asp, Glu, Pro, Ser, Gly, Ala, Arg	n.a.	34.2	0.52	Shuuluka et al. (2013)

(Continued)

Table 13.1 Protein content and amino acid (AA) composition of macroalgae. *Continued*

Species	Protein content (% dw)	Free AA composition[a]	Free AA (mg/g dw)	EAA (%)	EAA: NEAA	References
Ulva fasciata	11.1	Thr, Val, Lys, Ile, Leu, Phe, His, Met, Asp, Ser, Glu, Gly, Ala, Tyr, Pro, Hyp	111	42.2	0.73	Magdugo et al. (2020)
Ulva lactuca	18.1	Ile, Leu, Lys, Met, Cys, Phe, Tyr, Thr, Val, His, Asp, Glu, Pro, Ser, Gly, Ala, Arg	181	32.9	0.49	Shuuluka et al. (2013)
Ulva rigida	21.3	Ile, Leu, Lys, Met, Cys, Phe, Tyr, Thr, Val, His, Asp, Glu, Pro, Ser, Gly, Ala, Arg	213	28.8	0.40	
	9.6	Asp, Glu, Ala, Arg, Gly, Ser, Tyr, Pro, Hyp, Phe, His, Ile, Leu, Lys, Met, Thr, Trp, Val	96.22	40.8	0.69	
Phaeophyceae						
Bifurcaria bifurcata	8.9	Thr, Val, Met, Iso, Leu, Phe, Lys, His, Arg, Tyr, Asn, Ser, Glu, Gly, Ala, Pro, Cys	73.20	40.0	0.67	Lorenzo et al. (2017)
Fucus vesiculosus	12.99	Thr, Val, Met, Iso, Leu, Phe, Lys, His, Arg, Tyr, Asn, Ser, Glu, Gly, Ala, Pro, Cys	119.04	41.0	0.69	
Fucus spiralis	11.8	Arg, Tyr, Thr, Val, Met, Lys, Ile, Leu, Phe, Ala, Asp, Glu, His, Ser, Gln, Tau, Trp, Hyp	47.4	38.7	1.08	Vieira et al. (2018)

(Continued)

Table 13.1 Protein content and amino acid (AA) composition of macroalgae. *Continued*

Species	Protein content (% dw)	Free AA composition[a]	Free AA (mg/g dw)	EAA (%)	EAA: NEAA	References
Macrocystis pyrifera	10.8	Glu, Asp, Val, Met, Arg, Ser, Thr, Gly, Ala, Phe, Ile, Tyr, Leu, Lys, Cys, His, Pro	51.3	38.9	0.21	Ortiz et al. (2009)
Sargasum polycistum	8.6	Arg, Tyr, Thr, Val, Met, Lys, Ile, Leu, Phe, Ala, Asp, Glu, Ser, Gly, Pro	74.90	50.0	1.00	Nazarudin et al. (2021)
Sargassum fusiforme	10.9	Glu, Asp, Leu, Val, Gly, Ala, Phe, Arg, Thr, Ser, Ile, Pro, Lys, Tyr, His, Met, Cys, Trp, Tau	90.9	54	0.6	Dawczynski et al. (2007)
Undaria pinnatifida	18.9	Glu, Asp, Leu, His, Lys, Val, Gly, Phe, Ala, Thr, Ile, Ser, Arg, Pro, Tyr, Met, Cys, Trp, Tau	87.3	35.5	0.7	

AA, *Amino acids;* dw, *dry weight;* EAA, *essential amino acids;* NEAA, *non–essential amino acids;* n.a., *not analyzed.*
[a]*Amino acids are listed following decreasing concentration.*

whereas tryptophan, cysteine, and lysine are less abundant (Table 13.1) (Cian et al., 2014; Shuuluka et al., 2013). The content of EAA in macroalgae may reach values between 30% and 40%, indeed, in some species just the content of glutamic acid and aspartic acid in macroalgae reach values of 30% (Kazir et al., 2019). The EAA content of macroalgae is comparable to other foodstuffs considered relevant sources of EAA, such as casein with values around 44%, ovalbumin about 52%, and legumes with ranges around 45%, of total AA (Fleurence, 1999; Paiva et al., 2014). Apart from the relevance of seaweed as a source of EAA, it may be noted that monosodium glutamate, derived from the glutamic acid naturally present in algae is the main responsible for the typical umami taste (Poojary et al., 2019) while the taste associated to algae has been described to be due to the presence of glycine and alanine (Bocanegra et al., 2021). Besides, few macroalgae were described to possess important amounts of taurine, a required EAA implied in bile digestion and mainly associated with meat intake (Harnedy & Fitzgerald, 2011). Regarding specific AA, it appears that brown seaweeds generally account for a higher content of methionine, while green algae tend to present higher threonine concentrations. On the other hand, red macroalgae display higher alanine and valine concentrations (Table 13.1).

Therefore considering all these facts, macroalgae represent a promising source of proteins not just as raw ingredient but as matrix for the protein extraction for their further utilization in the development of functional products especially for their application as food and feed ingredients (Cermeño et al., 2020).

13.3 Seaweed protein hydrolysis

Protein can be processed using extreme conditions, such as acidic or heat treatments, which prompt the breakage of the peptide bonds and so it releases AA. This process of obtaining free AA residues or partial protein hydrolysates (PHs) is known as protein hydrolysis (Fig. 13.1). In the case of seaweed several steps are required to obtain BAPs and different approaches may be followed. Nevertheless, the application of pretreatments is highly recommended since it facilitates the extraction and improves the recovery yield. The relevance of protein hydrolysis relies on the biological activities reported for both obtained products. Macroalgae are regarded as an important source of proteins that need to be first extracted, precipitated, and then isolated for their further hydrolysis into bioactive smaller peptide sequences (Fig. 13.1). Therefore seaweed is also a very appreciated resource for the extraction and isolation of PHs and BAPs (Kim & Wijesekara, 2010).

Regarding the hydrolysis processes required to obtain PHs and BAPs, there are approaches which can be classified into two categories: chemical or biological (enzyme-based). Chemical approaches imply a denaturalization of the structure through bound breakage. The denaturalization can be accomplished by the application of temperature above 40°C, by switching the pH using alkali or acid conditions, depending on the isoelectric point of the solution to treat or by applying a combination of both approaches. Even though this is an easy and cost-effective strategy results are not totally predictable. Full control of this technique would require very precise conditions not always reachable since the peptide sequence may present variations, indeed working with biological products usually implies the assumption of an inherent range of variability. Therefore the composition of the PHs may vary providing diverse peptide sequences with modifications in their amino acidic profile (Tavano, 2013). In this sense, the use of strong acids is recognized as an effective hydrolytic agent of hydrophobic bonds. For instance, a solution containing two parts of chlorohydric acid and one of trichloroacetic acid was found to reduce the time of protein hydrolysis from 24 h to 25 min when applied at 166°C and to even fully break down tryptophan. On the other hand, the

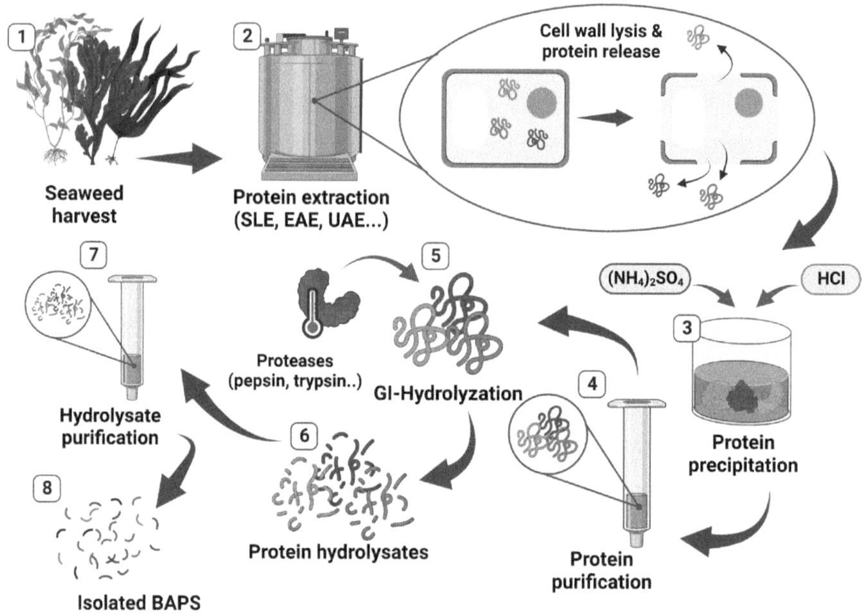

FIGURE 13.1

Diagram of seaweed protein extraction, purification, and hydrolysis process. (1) Macroalgae selection. (2) Extraction of proteins using disruptive methods. (3) Protein precipitation using salts or acids. (4) Protein purification using ultrafiltration. (5 and 6) Hydrolysis of proteins by chemical or biological methods. (7) Purification of PHs. (8) Isolation of functional BAPs.

use of strong alkalis, such as sodium hydroxide, has other effects on the AA since it may reduce arginine, cystine, isoleucine, serine or threonine; it also can create alternative residues, such as lanthionine and lysinoalanine; and it may modify the optic isomer and create the opposite enantiomer, for example, L-lysine maybe switch into D-lysine (Provansal et al., 1975). Even though the use of high temperature has been described as an effective hydrolytic agent by itself it requires long incubation times (Suetsuna et al., 2004).

Regarding biological hydrolysis, it mostly implies the use of enzymes to control the breakage. Even though biological agents are used for this type of hydrolysis it usually requires additional conditioning factors to fully optimize the process, such as the application of temperature, agitation, adjustments of pH conditions, control of enzyme-to-substrate ratio, and incubation times. Indeed, it is well recognized that the application of low temperatures boosts the enzymatic hydrolysis of proteins or peptide sequences, and it avoids the denaturalization of the enzyme involved in the process which permits its re-utilization for further hydrolysis. The use of enzymes as hydrolytic agents presents two important advantages against the chemical approaches. Firstly, outcomes are much more precise since each enzyme attack to specific target bonds. Second, enzymatic hydrolysis does not require the use of organic solvents which facilitates the purification stages and so increases the peptide yield (Castro et al., 2011; Hammed et al., 2013). Nevertheless, different factors may alter yield results, such as the ratio substrate: enzyme or the time and the steps required to fully develop the hydrolytic process, and these factors are strongly determined by the enzyme selected (Cian, Martínez-Augustin, et al., 2012b; Hammed et al., 2013). The finalization of the hydrolytic reaction requires the additional application of heat or pH modifications. This step may represent a drawback since the thermal deactivation of the enzymes may induce undesired side reactions (Cao et al., 2017). To solve this issue, enzymes can be immobilized on solid surfaces which facilitates their removal from the substrate and so the finalization of the reaction by performing a physical separation. Besides, this approach additionally prevents the aggregation of the enzymes and the peptides (Homaei et al., 2013). Alternatively, it has been suggested that the presence of coextracted molecules, such as polysaccharides or phenolic compounds, may be an inhibitory option to finalize the hydrolytic activity although further studies must be carried out to fully understand this approach (Admassu et al., 2018a).

Regarding the type of enzymes mainly used for obtaining PHs and BAPs from proteins obtained from macroalgae, it is clear that proteases are the most used enzymes to achieve these reactions. Among them, the typical ones include trypsin, pepsin, papain, and α-chymotrypsin and they can be applied individually or combined. Pepsin has been one of the most recurrent enzymes applied in recent years, since it represents a cheap and accessible reagent, besides it has the capacity to emulate gastrointestinal digestion (Indumathi & Mehta, 2016). Currently, another class of enzymes apart from the proteases has been incorporated into hydrolytic enzyme cocktails, the carbohydrases. Each class of proteolytic enzyme provides different BAPs mixtures and so different bioactivities. Some commercial enzymatic products named after flavourzyme or corolase as well as some fungal and bacterial proteases have been demonstrated to provide high yields for BAPs (Harnedy et al., 2015). Some works have been developed to compare the hydrolytic capacity of several enzymes but also the bioactivity of the obtained BAPs mixtures. For instance, proteins isolated from *Gracilariopsis lemaneiformis* were submitted to hydrolytic processes using up to five classes of proteases, namely trypsin, pepsin, papain, α-chymotrypsin, and alcalase. Among the tested enzymes, the use of α-chymotrypsin yields the most active hydrolysates regarding their antioxidant capacity (Zhang et al., 2019). Different studies demonstrated that proteins extracted from *Palmaria palmata* and hydrolyzed with papain, alcalase, flavourzyme, or corolase yield BAPs with renin inhibitor capacity and so were considered antihypertensive (Fitzgerald et al., 2012; Harnedy et al., 2015)

To obtain efficient and feasible results when using enzymatic hydrolysis for treating proteins obtained from macroalgae it is required to previously known the algal proteome. This is a requirement to perform an adequate selection of the enzymes based on their target bonds. It represents a major drawback since it may increase the final cost of the yield PHs and BAPs. Besides, the application of pretreatments that prompt algal cell disruption, such as pulse electric fields or bead milling, is key to maximizing the yield production and the purity (Akaberi et al., 2019; Alavijeh et al., 2020).

Independently of the strategy applied, the hydrolytic solution obtained includes a mixture of active and inactive peptides which required purification steps (Fig. 13.1). Besides, the purification steps are also aimed to concentrate the amounts of PHs and BAPs in the final solution. Sequential purifications are preferred to ensure the best results (Kazir et al., 2019). Among the purification approaches utilized to separate BAPs from those inactive the most common techniques are the use of filtration membranes and reverse-phase chromatography (Chabeaud et al., 2009; Harnedy & Fitzgerald, 2011). When using filtration membranes, they are selected based on their cut-off masses, since BAPs usually range between 2 and 20 AA, the most used ones are those from 5 to 30 kDa although modifications in the hydrolytic process usually require modifications in the BAP profile and so in the selection of the membrane cut-off (Harnedy & FitzGerald, 2013; Jeon & Samarakoon, 2013). PHs and BAPs may be purified by the above-mentioned techniques to ensure a higher concentration of peptides with low molecular weight or to obtain a more specific and purified batch of BAPs.

13.4 Seaweed protein digestibility

As mentioned, the protein content of seaweed generally ranges from 10% up to around 40% of their dry weight, making them an attractive alternative to conventional plant protein sources, such as soybeans or legumes. However, the digestibility of seaweed protein is an important factor to consider when evaluating its potential as a protein source for human consumption, but also to assess the potential release of bioavailable BAPs. The digestibility of protein refers to the extent to which the protein is broken down into smaller peptides and AA during the digestive process, and it is a crucial aspect of protein quality. The factors that can affect the digestibility of seaweed protein include species, harvesting time, growing conditions (light exposure, temperature, nutrient availability), processing methods, and the presence of antinutritional compounds, such as phenolic acids, tannins, and lectins.

In vitro methods, such as the pepsin-pancreatin digestion assay, are commonly used to determine the digestibility of seaweed protein. In this method, a sample of seaweed protein is mixed with pepsin and pancreatin, which are digestive enzymes that mimic the digestive process in the human gut. The amount of protein that is digested and the number of peptides and AA that are released during the digestion are then measured (Mišurcová, 2011). In vitro methods can provide further evidence by analyzing endothelial absorption, transportation, and metabolization on cell cultures derived from intestinal epithelial tissue, such as Caco-2 or HT-29 cells (Bleakley & Hayes, 2017). In vivo methods, employing animal models, can also be used to determine the digestibility of seaweed protein. In this method, rats are fed a diet containing a test protein source, and the amount of protein that is absorbed and utilized by the body is measured. The results of both in vitro and in vivo methods can be used to calculate the apparent digestibility coefficient (ADC) of the protein, which provides an estimate of its digestibility.

The k-Protein Digestibility-Corrected AA Score (k-PDCAAS) is a widely used method recommended due to its accuracy by the Food and Agriculture Organization (FAO) for evaluating the nutritional quality of proteins. It takes into account both the digestibility of the protein and its AA composition, with a higher score indicating a higher quality protein. The k-PDCAAS method thus considers both the degree of hydrolysis of a protein following in vitro GI, as well as the EAA composition of the released peptides. The corrected score is then compared to a reference protein (usually egg yolk protein) to determine the quality of the protein. A score of 1.0 indicates that the protein is of the same quality as the reference protein, while a score less than 1.0 indicates a lower-quality protein (De Bhowmick & Hayes, 2022). The k-PDCAAS method is widely used because it provides a simple and standardized way to evaluate protein quality and it has been used to evaluate the quality of many protein sources, including soybeans, legumes, and animal products. However, the use of the k-PDCAAS method to evaluate seaweed protein has been limited, with only a few studies being published to date (De Bhowmick & Hayes, 2022).

Seaweed protein digestibility and bioavailability may vary across species since it is highly dependent upon the AA composition, folding, and potential cross-link of the proteins with cell wall polysaccharides. For instance, *Ulva lactuca* protein isolates were subjected to complete GI-digestion, and their nitrogen digestibility was assessed. The authors found that the extracted protein achieved a higher score (84.7%) than the whole unprocessed seaweed (79.9%), yet it was lower than a soybean protein meal (98%) (Bikker et al., 2016). Likewise, other authors have studied the protein in vitro digestibility of various red (*Palmaria palmata*, *Chondrus crispus*, *Meristotheca papulosa*, *Sarcodiotheca gaudichaudii*) and brown (*Ascophyllum nodosum*, *Saccharina latissima*, *Fucus vesiculosus*, *Alaria esculenta*) seaweeds, finding that it was between 78% and 86%, being generally lower in brown seaweeds compared to the red ones (Tibbetts et al., 2016). As mentioned, the authors suggested that cell wall polysaccharides found in brown algae could hinder protein digestibility and did as well establish a negative correlation between total phenolic content and protein digestibility (Tibbetts et al., 2016). Another study performed a digestibility analysis following the k-PDCAAS method on *A. esculenta*, *Fucus serratus*, *F. vesiculosus*, *U. lactuca*, *P. palmata*, and *Asparagopsis taxiformis*(De Bhowmick & Hayes, 2022). The authors compared the k-PDCAAS scores with the classical in vitro digestibility methods (which only take into account the degree of protein hydrolysis), finding that every studied alga decreased its score by k-PDCAAS. Some species, such as *P. palmata* or *F. serratus*, achieved scores over 0.6, whereas *F. vesiculosus*, with a digestibility of 0.82, only achieved a k-PDCAAS of 0.08, due to this sample having only an AA score of 0.101, which as mentioned is used to calculate the k-PDCAAS. With some exceptions, the k-PDCAAS of these studied algae was similar to that of rice, barley or wheat (De Bhowmick & Hayes, 2022).

13.5 Bioactive properties of seaweed-derived peptides

Different classes of macroalgae PHs and BAPs have been tested for their potential bioactivities. Most of the current knowledge has been developed through the assessment of their in vivo and in vitro capacities, while very scarce data has been collected from clinical trials. Among the tested bioactivities the antioxidant, antiobesity, antidiabetic, hypotensive, and antiinflammatory stand out while the anticancer one has been slightly explored (Table 13.2).

13.5.1 Antioxidant

The constant presence of stressful factors is involved in the overproduction of reactive oxygen species (ROS). Excessive ROS in cellular environments lead to their interaction with key structures, such as proteins or lipids attached to the membrane, and so implied in cellular transport or DNA. The attachment of ROS to cellular components strongly alters not just the cellular functionality inducing cell death or unregulated proliferation but also the organic functions. Indeed, oxidative stress is widely recognized as a disruptive health condition and one of the most relevant underground symptoms of most chronic diseases since it implies the development of inflammatory processes, metabolic disorders, and even cancer (Reuter et al., 2010). In this context, antioxidants represent relevant agents to combat the prooxidative conditions induced by the overproduction of ROS. Antioxidants reduce the impact of ROS and prevent the damage they may induce (Yu et al., 2011). The chemical properties of antioxidant molecules are directly involved in their capacity to quench ROS, such as in the case of BAPs and PHs obtained from macroalgae whose conformation and AA profile are relevant to overturn ROS damages (Wu et al., 2016). In this sense, BAPs are considered to exert stronger antioxidant capacities than PHs with higher molecular weights or proteins likely due to their size. BAPs are usually formed by shorter AA chains which facilitate their cellular absorption and their chemical process, instead more complex molecular structures hint at their accessibility.

In the last decade, several works have demonstrated the antioxidant capacity of BAPs extracted from several macroalgae species. For instance, a protein fraction extracted from *Palmaria palmata* and treated with chymotrypsin generated BAPs of different molecular sizes. BAPs with lower molecular weight, specifically under 10 kDa were demonstrated to possess stronger antioxidant properties when determined with a 2,2-diphenyl-1-picrylhydrazyl (DPPH) assay. Similarly, in another work using *P. palmata* a BAP consisting of 9 amino acid residues showed

Table 13.2 Biological activities of seaweed BAPs and PHs.

Species	Protein hydrolysis/ digestion	Peptide sequence (s)	Bioactivity	Results	References
Rhodophyta					
Gracilariopsis lemaneiformis	α-chymotrypsin; E:S (1:4), pH 8, 2 h, 37°C	ELWKTF	In vitro Antioxidant	DPPH radical scavenging. $EC_{50} = 1.51$ mg/mL	Zhang et al. (2019)
	Trypsin; E:S (1:4), pH 8, 8 h, 37°C	QVEY	In vitro Hypotensive	ACE inhibitory activity $IC_{50} = 0.25$ mg/mL	Cao et al. (2017)
Porphyra sp.	Pepsin; E:S (1:8), pH 2, 4 h, 37°C	GGSK, ELS	In vitro Antidiabetic	α-Amylase inhibition IC_{50} (GGSK) = 0.8 mg/mL, IC_{50} (ELS) = 0.9 mg/mL	Admassu et al. (2018b)
Neopyropia yezoensis	Pepsin; E:S (1:40), pH 2, 45°C, 5 h	NMEKGSSSVVSSRM	*Ex vivo* Anticoagulant activity	Blood clotting retardation; $IC_{50} = 4.49$ μg/mL	Indumathi and Mehta (2016)
Palmaria palmata	Chymotrypsin; E:S (1:20), pH 8, 30°C, 24 h	33 BAPs <10 kDa	In vitro Hypotensive	ACE inhibition: 0.75 mg/mL = 24% 5 mg/mL = 68%	
			In vitro Antioxidant	DPPH radical scavenging: 0.75 mg/mL = 33% 5 mg/mL = 66%	
	Thermolysin; 1% enzyme solution,70°C, and 3 h; Pepsin; 1% enzyme solution, pH 2, 37°C, and 3 h; Trypsin; 1% enzyme solution, pH 8, 37°C, and 3 h; and Chymotrypsin; 1% enzyme solution, pH 8, 37°C, and 3 h	LRY	In vitro Hypotensive	ACE inhibition $IC_{50} = 0.01$ mg/mL	Furuta et al. (2016)

(*Continued*)

Table 13.2 Biological activities of seaweed BAPs and PHs. *Continued*

Species	Protein hydrolysis/ digestion	Peptide sequence (s)	Bioactivity	Results	References
Bangia fusco-purpurea	Pepsin; E:S (1:200), pH 7.5, 37°C, and 90 min	ALLAGDPSVLEDR, VVGGTGPVDEWGIAGAR	In vitro Hypotensive	ACE inhibition IC_{50} = = 57.2 μg/mL for ALLAGDPSVLEDR IC_{50} = 66.2 μg/mL for VVGGTGPVDEWGIAGAR	Wu et al. (2017)
Chlorophyta					
Ulva intestinalis	Trypsin + Pepsin + Papain; E:S (1:4), pH 8.42, 5 h, 28.5°C	FGMPLDR, MELVLR	In vitro Hypotensive	ACE inhibition IC_{50} (FGMPLDR, MELVLR) = 0.18 mg/mL IC_{50} = 237 μM, for MELVLR	Sun et al. (2019)
Ulva rigida	Pepsin; E:S (1:100), pH 2.0, 37°C, and 20 h	IP, AFL	In vitro ACE inhibitory activity	IC_{50} (IP) = 0.023 mg/mL IC_{50} (AFL) = 0.020 mg/mL	Paiva et al. (2016)
Caulerpa lentillifera	Thermolysin; E:S (1:50), 60°C, pH 8.5, 16 h	FDGIP, AIDPVRA	In vitro Hypotensive	ACE inhibition IC_{50} (FDGIP) = 0.03 mg/mL IC_{50} (AIDPVRA) = 0.04 mg/mL	Joel et al. (2018)
Ochrophyta					
Sargassum maclurei	Pepsin (80 U/g S), pH 2, 2 h, 37°C/Papain (60 U/g S), pH 7, 3 h, 50°C	RVLSAAFNTR, IMNILEK, GGVQAIR, KAALMEK, GVFDGPCGT, SGVFDGPCGT, QNIGDPR, AYSSGVSFK, RWDISQPY, LVYIVQGR, KPGGSGR, LGLSAKNYGR, KEAWLIEK, REVADDK, ENFFFAGIDK, QEMVDK, EEEEEEQQQ	In vitro Hypotensive	Systolic blood pressure reduction from 170 to 150 mm Hg at 100 mg/kg bw; 25% endothelin-1 inhibition at 1.5 mg/mL	Zheng et al. (2020)

Laminaria japonica	Alcalase + Papain + Trypsin (E:S n.s.), time n.s., pH 7.5, 55°C	KY, GKY, STKY, AKY, AKYSY, KKFY, FY, KFKY	In vitro Hypotensive	ACE inhibition IC_{50} = 0.6 mg/mL of PHs	Chen et al. (2016)
Saccharina longicruris	Trypsin; E:S (1:20), pH 7, 30°C, and 24 h	EAESSLTGGNGCAK, IGNGGELPR, ILVLQSNQIR, ISAILPSR, ISGLIYEETR, LPDAALNR, MALSSLPR, QVHPDTGISK, TITLDVEPSDTIDGVK	In vitro Antimicrobial activity against *Staphylococcus aureus*	PH (2.5 mg/mL) reduced microbial growth (from 0.2 to 0.1 μmax/h)	
Undaria pinnatifida	Pepsin (E:S 0.03), pH 2, 5 h, 45°C	AIYK, YKYY, KFYG, YNKL	In vitro Hypotensive	ACE inhibition IC_{50} (AIYK) = 213 μM IC_{50} (YKYY) = 64.2 μM IC_{50} (KFYG) = 90 μM IC_{50} (YNKL) = 21 μM	Suetsuna and Nakano (2000)
			In vivo Hypotensive	Single oral administration of 50 mg/kg bw lowered systolic blood pressure by 40–50 mmHg	

ACE, *Angiotensin-I-converting enzyme;* DPPH, *2,2-Diphenyl-1-picrylhydrazyl;* EC_{50}, *effective concentration 50%;* E:S, *enzyme:substrate ratio;* IC_{50}, *inhibitory concentration 50%;* n.s., *not specified;* PHs, *protein hydrolysates.*

antioxidant capacity when tested with two different assays: oxygen radical absorbance capacity (ORAC, 152 nmol TE/μmol) and ferric reducing antioxidant power (FRAP, 21 nmol TE/μmol) (Harnedy et al., 2017). Other genera, such as *Neoporphyra*, have been reported as effective sources to obtain BAPs, such as *Neoorphyra haitanensis*. Proteins extracted from *N. haitanensis* and treated with pepsin demonstrated their antioxidant capacity when assayed with DPPH and FRAP assays (Lin Wen et al., 2020). Another work using *Porphyra dioica* demonstrated that BAPs obtained from different enzymatic treatments also have different antioxidant activities. Indeed, proteins from *P. dioica* treated with pepsin displayed a stronger antioxidant capacity than when treated with pepsin-Corolase. Nevertheless, independently of the applied treatment, the antioxidant capacity of the BAPs obtained was demonstrated to be directly related to their size, being those smaller than 10 kDa peptides the most effective (Pimentel, Cermeño, Kleekayai, Machado, et al., 2020b). The same authors tested the PHs of the same species, *P. dioica*, but using two different proteases: prolyve and flavourzyme. They found that this hydrolysis increased five times the antioxidant activity. This digestive approach was observed to be capable of releasing a notable amount of the content of phenolic compounds which was directly correlated with the protein hydrolysis. The simultaneous release of these two classes of molecules may create a positive synergy responsible for the trigger of the antioxidant capacity (Pimentel, Cermeño, Kleekayai, Harnedy-Rothwell, et al., 2020). Finally, another study optimized the hydrolysis of proteins from *Gracilariopsis lemaneiformis* using chymotrypsin. They maintained two fixed conditions, the concentration of substrate (10 mg/mL) and the reaction time (2 h) and optimized the remaining experimental parameters, namely enzyme concentration (1–3%, enzyme:substrate, E:S, w/w), temperature (19°C–41°C), and pH (8.5–9.5). Optimal conditions were achieved with a 1–10 E:S ratio incubated at a temperature of 46.4°C under alkaline conditions (pH 9.2). These experimental conditions provided a BAP consisting of 6 AA residues with a strong antioxidant capacity mainly attributed to its AA profile containing hydrophobic and aromatic AA and to its low molecular weight (Zhang et al., 2019).

13.5.2 Antidiabetic and antiobesity

Among the current global diseases related to nutritional imbalances and hormonal disorders the main concerns are diabetes mellitus type 2 and obesity. They consist of metabolic syndromes which usually have other associated diseases and so the design of preventive tools may avoid their development and the associated ailments. The key to ameliorating and preventing any illness is to understand its molecular mechanisms. In the case of these diseases, dipeptidyl peptidase-IV (DPP-IV), α-glucosidase, and α-amylase represent some of the key target molecules. The role of DPP-IV, an exopeptidase transmembrane enzyme, is to metabolize proteinic hormones. Among the target molecules of DPP-IV, the glucagon-like peptide-1 (GLP-1) is highly relevant for the prevention of diabetes and obesity. This hormone GLP-1 is responsible for the secretion of glucose-dependent insulin and for inducing satiety (Opinto et al., 2013).

Therefore the inhibition of DPP-IV and the boost of the secretion of GLP-1 or insulin may enhance the management of systemic glucose and so on metabolism. On the other hand, digestive enzymes act in the small intestine and their role is the degradation of polysaccharides, α-glucosidase attacks glucose while the target of the α-amylase is the starch (Mennella et al., 2014).

In this context, BAPs obtained from macroalgae have shown antidiabetic and antiobesity properties. Several authors pointed to two main factors involved in the inhibitory mechanism: the size/weight of the BAPs and the lack of competition with the substrate. For example, *N. haitanensis* extracts treated with ultrasounds and later submitted to simulated gastric digestion were demonstrated to inhibit α-glucosidase. Among the BAPs, those with lower molecular weight and higher peptide content were demonstrated to have stronger antioxidant and inhibitory properties (Wen et al., 2020). BAPs obtained from other species, *Porphyra* spp. showed similar results, PHs with lower molecular weight than 3 kDa had strong inhibition capacity against α-amylase. Their capacity, with an IC_{50} of 2.6 mM, was further demonstrated with synthetically synthesized BAPs, being a tripeptide (GLS) and tetrapeptide (GGLS) those with the highest inhibitory activity (Admassu et al., 2018b). Other activities have been reported for BAPs obtained from different macroalgae, such as *Neopyropia yezoensis*. The administration of 1 μg/mL of these BAPs triggered cell differentiation and proliferation in the small intestine of mice. This cellular response was deduced to be due to the upregulation of the signaling pathway of the insulin-like growth factor 1 receptor (IGF-1R) (Lee et al., 2015a). It was pointed out this upregulation was performed through the substrates of several enzymes which boost the activation of side pathways inducing the development and angiogenesis of the epithelial tissue of the small intestine avoiding the appearance of the associated inflammation (Lee et al., 2015a).

The most relevant species assessed for its antidiabetic and antiobesity capacities have been *P. palmata* which has been repeatedly assayed. Indeed, PHs, BAPs and proteins extracted from *P. palmata* were explored as potential preventive tools for diabetes and obesity through the inhibition of DPP-IV. In general, PHs and BAPs presented better values of IC_{50}, around 1.2 mg/mL, while proteins required a concentration 4 times higher to reach the same value (Stack et al., 2017). Protein extracts from *P. palmata* treated with Corolase were demonstrated to have strong inhibitory capacity with IC_{50} values of 1.65 mg/mL, the hydrosoluble PHs were less active with IC_{50} values of 3.2 mg/mL and the mixture with of the alkaline fraction provided an intermediate activity with IC_{50} of 2.3 mg/mL (Harnedy & FitzGerald, 2013). In the same line of experiments, PHs from *P. palmata* found to inhibit DDP-IV and angiotensin-I-converting enzyme (ACE) and to possess antioxidant properties were identified as smaller portions of the penicillin-binding proteins and the ribulose-1,5-bisphosphate carboxylase/oxygenase (RuBisCo) (Cermeño et al., 2019). The effect of BAPs from *P. palmata* as antidiabetic and antiobesity agents have been further corroborated at in vivo studies. Diabetic mice treated for 18 days with PHs from *P. palmata* reduced glycemia, improved the secretion of insulin secretion and diminished the levels of glycated hemoglobin. As side effects, PHs from *P. palmata* enhanced the glucagon-like peptide 1 level and minimized the energy intake (McLaughlin et al., 2020). In another work by the same authors, the negative effects of PHs from *P. palmata*, such as their insulinotropic capacity or their toxicity, were assessed. For this purpose, they used simulated gastric digestion based on several cell cultures, including pancreatic, adipose, and enteroendocrine cell lines. Data demonstrated no negative effects and reinforced the inhibition capacity of the PHs against DPP-IV and their boosting effect in the secretion of insulin and GLP-1. These effects seem to be due to an increment in the mobilization of the calcium together with an upregulation of cyclic adenosine monophosphate which ultimately led to a stimulation of incretin. Incretin comprises a group of hormones involved in the reduction of the glucose levels in the blood due to the increment in the secretion of insulin. Indeed, the oral administration of a dose of 100 mg/kg body weight of these PHs ameliorates the glucose tolerance of mice and induced satiety after 90 min. These outcomes point to the enhancement of the GLP-1 and the upregulation of incretin (McLaughlin et al., 2021).

13.5.3 Hypotensive

Different BAPs obtained from a few macroalgae groups demonstrated to have hypotensive properties through inhibition of the ACE or renin, an enzyme that can stimulate renal activity or produce angiotensin I, the substrate of ACE (Hayes & Tiwari, 2015; Seca & Pinto, 2018). This pathway named after the renin-ACE-aldosterone axis is well-known to regulate cardiac pressure (Paz Ocaranza et al., 2020).

Protein extracts obtained from *G. lemaneiformis* were treated with different proteases of which trypsin was demonstrated to produce a major tetrapeptide consisting of QVEY. The BAP with the strongest inhibition capacity against ACE reached values around 78% (Cao et al., 2017). A similar work assessed the BAPs obtained from the treatment of proteins of *U. intestinalis* with alcalase, α-chymotrypsin, papain, pepsin, and trypsin. Also, in this case, BAPs derived from trypsin treatment induced the highest inhibition of ACE. The most active peptides consisted of sequences of 7 AA (FGMPLDR) and 6 AA (MELVLR) (Sun et al., 2019). Another study based on proteins from *Caulerpa lentillifera* assessed the use of α-chymotrypsin, pepsin, thermolysin, and trypsin. Two BAPs derived from the application of thermolysin and consisting of 5 AA (FDGIP) and 7 AA (AIDPVRA) showed the highest ACE inhibitory activity (Joel et al., 2018). In another study, *N. yezoensis* proteins treated with pepsin were hydrolyzed into a peptide sequence consisting of NMEKGSSSVVSSRM that resulted in anticoagulant activity (Indumathi & Mehta, 2016). In a similar experiment, pepsin was used to hydrolyze *U. rigida* proteins whose obtained peptides were demonstrated to inhibit ACE (Paiva et al., 2016). Similarly, pepsin-derived peptides from proteins extracted from *Bangia fusco-purpurea*, consisting of the sequences ALLAGDPSVLEDR and VVGGTGPVDEWGIAGAR, were demonstrated to possess ACE inhibitory activity (Q. Wu et al., 2017). The sequential hydrolysis of proteins from *P. palmata* using thermolysin, pepsin, trypsin, and chymotrypsin was also analyzed in the context of the inhibition of the ACE activity. The hydrolysis using thermolysin was proved to generate BAPs with inhibitory capacity. Among the obtained BAPs, the tripeptide LRY was underlined as the most active (Furuta et al., 2016). As pointed out before, the selection of enzymes and the optimization of the experimental conditions are key on the release of peptides with biological activity.

Among the in vivo studies used to assess the antihypertensive activity of BAPs three works using three different species stand out. For instance, *P. palmata* protein extracts treated with papain reached an AA sequence (IRLIIVLMPILMA) with a strong capacity to inhibit ACE showing an IC_{50} value of 0.32 mg/mL (Fitzgerald et al., 2012). Later, this same BAP was chemically synthesized and tested on spontaneously hypertensive rats where it induced ACE inhibition accompanied by a significant reduction in systolic blood pressure (Fitzgerald et al., 2014). Same animal models, spontaneously hypertensive rats, treated with BAPs of Tyr or Trp obtained from *U. pinnatifida* were demonstrated to diminish blood pressure after a single oral administration of 1 mg/kg body weight (Sato et al., 2002). Clinical trials proving the hypotensive function of BAPs or PHs of macroalgae are scarce. However, a clinical trial demonstrated that the daily administration of 1.8 g of a PHs *Porphyra* sp. significantly reduce, from 157 to 142 mmHg, the systolic blood pressure in hypertensive subjects whereas no changes were observed in normotensive ones ($\leq$120 mmHg). To our knowledge, this is the only clinical trial published and used as the basis for the development of a patented product.

13.5.4 Antiinflammatory

The inflammatory response is developed as a reaction to several conditions mainly related to damage both at cellular or tissue levels and due to pro-oxidative environments, infections or even tumors (Crusz & Balkwill, 2015). Proinflammatory agents but also inhibitors are involved in the modulation of the inflammatory response. Cytokines are one of the most common proinflammatory mediators that involve interleukins (IL) and tumor necrosis factor-α (TNF-α). Nevertheless, prostaglandins, such as cyclooxygenase-2 (COX-2) or nitric oxide synthase (iNOS), are also key agents in the development of the inflammatory process since they induce the production of ROS and nitric oxide (NO) involved in the activation of the cellular class of the macrophages. Activated macrophages create a cascade response by which innate immune cells are promoted to release cytokines, such as interleukin (IL)-6, tumor necrosis factor (TNF)-α, or the transforming growth factor-β, involved in cellular proliferation (Grivennikov et al., 2010). Other inflammatory mediators synthesized after the activation of the macrophages are the nuclear factor κB (NF-κB) and the mitogen-activated protein kinase (MAPK) (Tornatore et al., 2012). In this sense, the application of BAPs and PHs obtained from macroalgae can trigger the effect of antiinflammatory mediators and/or diminish the expression of proinflammatory mediators.

Seaweed BAPs have been reported to ameliorate inflammatory responses, but only in vitro. PHs obtained from proteins of *Porphyra columbina* were demonstrated as antiinflammatory agents when tested in vivo models using three different cell lines isolated from rat spleen: macrophages, T lymphocytes, and splenocytes. They were used to determine the lack of toxicity of the PHs and their capacity regarding cellular proliferation and the production of inflammatory mediators, mainly cytokines. The mixture of *P. columbina* PHs with concanavalin A created a synergy that promotes the proliferation of lymphocytes. Besides, the only application of PHs reduced the secretion of TNF-α and IL-6 and increased the secretion of IL-10. This cytokine, IL-10, was pointed out as the main responsible for the antiinflammatory response since it exerts its action over macrophages and lymphocytes and inhibits the pathways triggered by MAPK and NF-κB (Cian et al., 2012b; Cian, López-Posadas, et al., 2012a). Following the same experimental approach, the same research group assessed the antiinflammatory potential of PHs obtained from proteins of several species belonging to the genus *Ulva* which showed immunomodulatory capacity in addition (Cian et al., 2018). Another work reinforced the antiinflammatory activity of macroalgae BAPs using a protein extract from *N. yezoensis* applied into a cellular line of macrophages (RAW 264.7) stimulated with lipopolysaccharide. The overexpression in this cell line of inflammatory mediators, such as iNOS, COX-2, IL-1β, and NF-κB, was downregulated when treated with *N. yezoensis* BAPs probably induced by a downregulation of the expression of p38 MAPK and c-Jun N-terminal kinase (Lee et al., 2015b).

13.5.5 Other bioactive properties

Besides these main bioactive properties attributed to seaweed BAPs, other bioactive properties have also been studied. For example, GI-digested PHs from *P. columbina* have been tested for antiplatelet aggregation properties. The authors performed an *ex vivo* assay, observing the degree of human thrombocyte aggregation induced by adenosine diphosphate (Cian et al., 2015). Their results showed that PHs could display a 44% inhibition, while this was increased to 48% after the PHs were GI-digested, also maintaining their ACE inhibitory activity (Cian et al., 2015). Interestingly, although the PHs maintained or improved their bioactive properties, GI consistently diminished the antioxidant capacity of the BAPs, as confirmed by various in vitro assays.

Other bioactivities, such as the antitumor potential of BAPs from algae, have also been poorly studied. A tripeptide (PPY) obtained from *N. yezoensis* was reported to induce apoptosis in the cancerous cell line MCF-7 (Park et al., 2014). In this work, the capacity of this tripeptide was demonstrated as an inhibitor of IGF-IR whose overexpression has been associated with the proliferation of tumors. Indeed, the tripeptide downregulated the expression of IGF-IR and the extracellular signal-regulated *kinase*, also with antiinflammatory and apoptotic functions. Thus this tripeptide seems to exert its action by reducing this cell proliferation but also by reducing the cancerous population through the activation of cellular apoptosis. This activity was described to occur

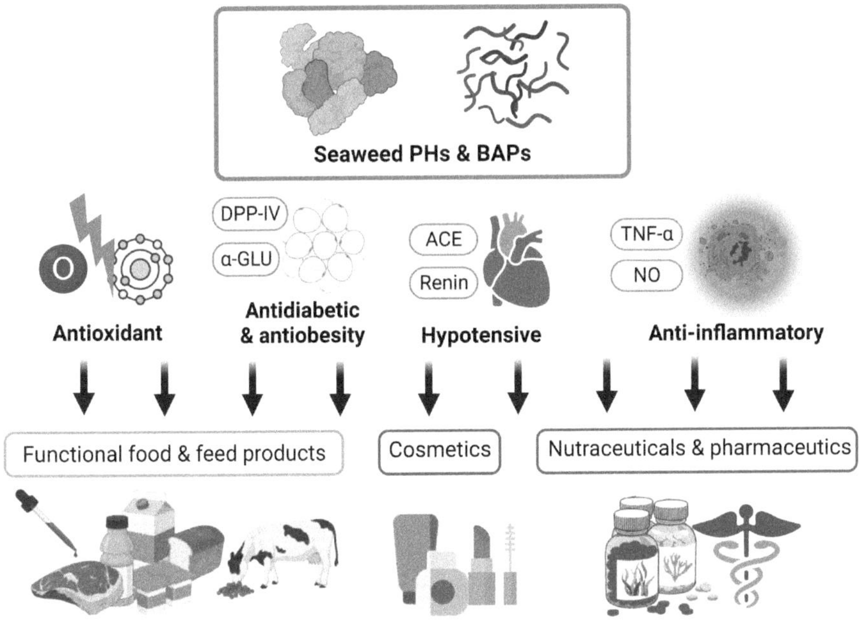

FIGURE 13.2

Main described bioactive properties of seaweed PH and isolated BAPs, pointing potential industrial applications of them.

due to the deactivation of p85, a subunit of phosphatidylinositol 3-kinases (PI3K) essential in cell proliferation and differentiation. The cell cycle was demonstrated to get slow since the expression of cyclins was reduced which was correlated with the presence of two cyclin inhibitor molecules p21 and p27 (Park et al., 2014). The same research group further analyzed the anticancer properties of this *N. yezoensis* tripeptide (PPY) demonstrating that its action was exerted through the upregulated pathway of mTOR, autophagy, and apoptotic mediators as well as the downregulation of p70S6 kinase related to metastatic processes (Park et al., 2015).

Altogether, seaweed BAPs have been reported to exert diverse positive effects on health, which could derive in various effective applications directed at the general population, although a great deal of further research is still necessary (Fig. 13.2).

13.6 Conclusion

Macroalgae represent very attractive sources of proteins, especially red and green algae with estimated contents between 10% and 40% of their weight. The most relevant feature of their protein content is their profile regarding EAA composition which may reach values in the range of 30%–40% and so comparable to other terrestrial organisms considered relevant sources of EAA, such as legumes with values around 45%. Considering these properties macroalgae have been pointed as a useful matrix to extract proteins that can be applied both as nutritional ingredients and as processed products, such as PHs and BAPs. These derived products required further treatments, including a few steps, such as sample pretreatment, chemical or biological hydrolysis, and purification. Even though longer procedures are required to obtain PHs and BAPs, they have become very demanded products for their improved digestibility. Indeed, in vitro protein digestibility for red and brown seaweeds has been established in the range of 80%, being generally lower in brown seaweeds due to the polysaccharides present in their cell wall. This high protein digestibility rate was demonstrated to be positively correlated with better amino acid and peptide bioaccessibility which ultimately may enhance their bioactivities. Among them, antioxidant, antiobesity, antidiabetic, hypotensive, and antiinflammatory stand out while anticancer has been scarcely explored. Regarding the antioxidant capacity, BAPs from four species were detailly evaluated: *Palmaria palmata*, *Neoorphyra haitanensis*, *Porphyra dioica*, and *Gracilariopsis lemaneiformis*. However, other bioactivities have been suggested to be exerted through the antioxidant capacity of the BAPs or PHs. Regarding the antidiabetic and antiobesity activities of macroalgae BAPS, they can exert these activities through different mechanisms. For instance, *N. haitanensis* inhibited the α-glucosidase, tri- and tetrapeptides from *Porphyra* inhibited α-amylase and *Neopyropia yezoensis* induced differentiation, proliferation and angiogenesis of epithelial cells of the small intestine that prevent the appearance of inflammation processes. Several works demonstrated that *P. palmata* has different mechanisms of action depending on the peptide obtained. In vitro assays displayed the capacity of PHs and BAPs of *P. palmata* to inhibit DPP-IV or ACE whereas in vivo studies showed their ability to improve the secretion of insulin, reduce glucose levels in the blood, induce satiety, boost GLP-1, and upregulate incretin. The hypotensive role of the macroalgae PHs and BAPs was always related to their ACE inhibitory activity which is known to act on the renin-ACE-aldosterone pathway involved in the cardiac pressure regulation. In vivo experiments proved their ability to reduce blood pressure after a unique oral administration, outcomes proved also in the only clinical trial found. In regards to the antiinflammatory role of macroalgae peptides from *Porphyra columbina* PHs reduced the secretion of proinflammatory cytokines, such as TNF-α and IL-6, while increasing the secretion of IL-10 involved in the inhibition of the pathways boosted by MAPK and NF-κB. Similarly, *N. yezoensis* has the ability to downregulate proinflammatory mediators, such as iNOS, COX-2, IL-1β, and NF-κB. Other bioactivities assessed were the antiplatelet aggregation of *P. columbina* or the anticancer property of *N. yezoensis* induced by mediators of apoptosis and autophagy, the downregulation of metastatic processes and its antioxidant activity.

Therefore based on the high content and high quality of the protein content of macroalgae as well as based on the biological properties of the PHs and BAPs obtained, seaweed represents a promising source for industrial sectors. Macroalgae may be used as raw ingredients but also as a matrix for protein extraction aimed to obtain PHs and BAPs which can be further used in the development of functional products. They have been proven to be useful tools to improve health status or to prevent the development of chronic and global diseases, such as obesity, diabetes, or cardiovascular affections, which strongly impact human well-being and in the public health systems.

Acknowledgments

The research leading to these results was supported by: MICINN, supporting the Ramón y Cajal grant for M.A. Prieto (RYC-2017-22891) and Jianbo Xiao (RYC-2020–030365-I); Xunta de Galicia, supporting the program EXCELENCIA-ED431F 2020/12, the program EXCELENCIA-ED431F 2022/01, and the postdoctoral grants of M. Fraga-Corral (ED481B-2019/096) and L. Cassani (ED481B-2021/152). Authors are grateful to Ibero-American Program on Science and Technology (CYTED—AQUA-CIBUS, P317RT0003), to the Bio Based Industries Joint Undertaking (JU) under grant agreement No 888003 UP4HEALTH Project (H2020-BBI-JTI-2019). The JU receives support from the European Union's Horizon 2020 research and innovation program and the Bio-Based Industries Consortium. The project SYSTEMIC Knowledge Hub on Nutrition and Food Security has received funding from national research funding parties in Belgium (FWO), France (INRA), Germany (BLE), Italy (MIPAAF), Latvia (IZM), Norway (RCN), Portugal (FCT), and Spain (AEI) in a joint action of JPI HDHL, JPI-OCEANS and FACCE-JPI launched in 2019 under the ERA-NET ERA-HDHL (n° 696295).

References

Admassu, H., Gasmalla, M. A. A., Yang, R., & Zhao, W. (2018a). Identification of bioactive peptides with α-amylase inhibitory potential from enzymatic protein hydrolysates of red seaweed (Porphyra spp). *Journal of Agricultural and Food Chemistry*, *66*(19), 4872–4882. Available from https://doi.org/10.1021/acs.jafc.8b00960, http://pubs.acs.org/journal/jafcau.

Admassu, H., Gasmalla, M. A. A., Yang, R., & Zhao, W. (2018b). Bioactive peptides derived from seaweed protein and their health benefits: Antihypertensive, antioxidant, and antidiabetic properties. *Journal of Food Science*, *83*(1), 6–16. Available from https://doi.org/10.1111/1750-3841.14011, http://www3.interscience.wiley.com/journal/118509799/issueyear?year = 2008.

Akaberi, S., Gusbeth, C., Silve, A., Senthilnathan, D. S., Navarro-López, El., Molina-Grima, E., Müller, G., & Frey, W. (2019). Effect of pulsed electric field treatment on enzymatic hydrolysis of proteins of Scenedesmus almeriensis. *Algal Research*, *43*. Available from https://doi.org/10.1016/j.algal.2019.101656.

Alavijeh, R. S., Karimi, K., Wijffels, R. H., van den Berg, C., & Eppink, M. (2020). Combined bead milling and enzymatic hydrolysis for efficient fractionation of lipids, proteins, and carbohydrates of Chlorella vulgaris microalgae. *Bioresource Technology*, *309*. Available from https://doi.org/10.1016/j.biortech.2020.123321.

Batista, S., Pereira, R., Oliveira, B., Baião, L. F., Jessen, F., Tulli, F., Messina, M., Silva, J. L., Abreu, H., & Valente, L. M. P. (2020). Exploring the potential of seaweed Gracilaria gracilis and microalga Nannochloropsis oceanica, single or blended, as natural dietary ingredients for European seabass Dicentrarchus labrax. *Journal of Applied Phycology*, *32*(3), 2041–2059. Available from https://doi.org/10.1007/s10811-020-02118-z.

De Bhowmick, G., & Hayes, M. (2022). In vitro protein digestibility of selected seaweeds. *Foods*, *11*(3). Available from https://doi.org/10.3390/foods11030289.

Biancarosa, I., Espe, M., Bruckner, C. G., Heesch, S., Liland, N., Waagbø, R., Torstensen, B., & Lock, E. J. (2017). Amino acid composition, protein content, and nitrogen-to-protein conversion factors of 21 seaweed species from Norwegian waters. *Journal of Applied Phycology*, *29*(2), 1001–1009. Available from https://doi.org/10.1007/s10811-016-0984-3.

Bikker, P., van Krimpen, M. M., van Wikselaar, P., Houweling-Tan, B., Scaccia, N., van Hal, J. W., Huijgen, W. J. J., Cone, J. W., & López-Contreras, A. M. (2016). Biorefinery of the green seaweed Ulva lactuca to produce animal feed, chemicals and biofuels. *Journal of Applied Phycology*, *28*(6), 3511–3525. Available from https://doi.org/10.1007/s10811-016-0842-3.

Bleakley, S., & Hayes, M. (2017). Algal proteins: Extraction, application, and challenges concerning production. *Foods*, *6*(5). Available from https://doi.org/10.3390/foods6050033.

Bocanegra, A., Macho-González, A., Garcimartín, A., Benedí, J., & Sánchez-Muniz, F. J. (2021). Whole alga, algal extracts, and compounds as ingredients of functional foods: Composition and action mechanism relationships in the prevention and treatment of type-2 diabetes mellitus. *International Journal of Molecular Sciences*, *22*(8). Available from https://doi.org/10.3390/ijms22083816.

Cao, D., Lv, X., Xu, X., Yu, H., Sun, X., & Xu, N. (2017). Purification and identification of a novel ACE inhibitory peptide from marine alga Gracilariopsis lemaneiformis protein hydrolysate. *European Food Research and Technology*, *243*(10), 1829–1837. Available from https://doi.org/10.1007/s00217-017-2886-2.

Carpena, M., Caleja, C., Pereira, E., Pereira, C., Ćirić, A., Soković, M., Soria-Lopez, A., Fraga-Corral, M., Simal-Gandara, J., Ferreira, I. C. F. R., Barros, L., & Prieto, M. A. (2021). Red seaweeds as a source of nutrients and bioactive compounds: Optimization of the extraction. *Chemosensors*, *9*(6). Available from https://doi.org/10.3390/chemosensors9060132, https://www.mdpi.com/2227-9040/9/6/132/pdf.

Castro, H. C., Abreu, P. A., Geraldo, R. B., Martins, R. C. A., Dos Santos, R., Loureiro, N. I. V., Cabral, L. M., & Rodrigues, C. R. (2011). Looking at the proteases from a simple perspective. *Journal of Molecular Recognition*, *24*(2), 165–181. Available from https://doi.org/10.1002/jmr.1091.

Cermeño, M., Kleekayai, T., Amigo-Benavent, M., Harnedy-Rothwell, P., & FitzGerald, R. J. (2020). Current knowledge on the extraction, purification, identification, and validation of bioactive peptides from seaweed. *Electrophoresis*, *41*(20), 1694–1717. Available from https://doi.org/10.1002/elps.202000153, http://onlinelibrary.wiley.com/journal/10.1002/(ISSN)1522-2683.

Cermeño, M., Stack, J., Tobin, P. R., O'Keeffe, M. B., Harnedy, P. A., Stengel, D. B., & Fitzgerald, R. J. (2019). Peptide identification from a: Porphyra dioica protein hydrolysate with antioxidant, angiotensin converting enzyme and dipeptidyl peptidase IV inhibitory activities. *Food and Function*, *10*(6), 3421–3429. Available from https://doi.org/10.1039/c9fo00680j, http://pubs.rsc.org/en/journals/journal/fo.

Chabeaud, A., Vandanjon, L., Bourseau, P., Jaouen, P., Chaplain-Derouiniot, M., & Guerard, F. (2009). Performances of ultrafiltration membranes for fractionating a fish protein hydrolysate: Application to the refining of bioactive peptidic fractions. *Separation and Purification Technology*, *66*(3), 463–471. Available from https://doi.org/10.1016/j.seppur.2009.02.012.

Chen, J. C., Wang, J., Zheng, B. D., Pang, J., Chen, L. J., Lin, H. T., & Guo, X. (2016). Simultaneous determination of 8 small antihypertensive peptides with tyrosine at the C-terminal in Laminaria japonica hydrolysates by RP-HPLC method. *Journal of Food Processing and Preservation*, *40*(3), 492–501. Available from https://doi.org/10.1111/jfpp.12628, http://www.blackwellpublishers.co.uk/journal_new.asp?site = 1.

Cherry, P., O'hara, C., Magee, P. J., Mcsorley, E. M., & Allsopp, P. J. (2019). Risks and benefits of consuming edible seaweeds. *Nutrition Reviews*, *77*(5), 307–329. Available from https://doi.org/10.1093/nutrit/nuy066, http://nutritionreviews.oxfordjournals.org/.

Cian, R. E., Fajardo, M. A., Alaiz, M., Vioque, J., González, R. J., & Drago, S. R. (2014). Chemical composition, nutritional and antioxidant properties of the red edible seaweed Porphyra columbina. *International Journal of Food Sciences and Nutrition*, *65*(3), 299–305. Available from https://doi.org/10.3109/09637486.2013.854746.

Cian, R. E., Garzón, A. G., Ancona, D. B., Guerrero, L. C., & Drago, S. R. (2015). Hydrolyzates from Pyropia columbina seaweed have antiplatelet aggregation, antioxidant and ACE I inhibitory peptides which maintain bioactivity after simulated gastrointestinal digestion. *LWT*, *64*(2), 881–888. Available from https://doi.org/10.1016/j.lwt.2015.06.043, https://www.journals.elsevier.com/lwt.

Cian, R. E., Hernández-Chirlaque, C., Gámez-Belmonte, R., Drago, S. R., Sánchez de Medina, F., & Martínez-Augustin, O. (2018). Green alga Ulva spp. hydrolysates and their peptide fractions regulate cytokine production in splenic macrophages and lymphocytes involving the TLR4-NFκB/MAPK pathways. *Marine Drugs*, *16*(7). Available from https://doi.org/10.3390/md16070235, http://www.mdpi.com/1660-3397/16/7/235/pdf.

Cian, R. E., López-Posadas, R., Drago, S. R., Sánchez De Medina, F., & Martínez-Augustin, O. (2012a). A Porphyra columbina hydrolysate upregulates IL-10 production in rat macrophages and lymphocytes through an NF-κB, and p38 and JNK dependent mechanism. *Food Chemistry*, *134*(4), 1982–1990. Available from https://doi.org/10.1016/j.foodchem.2012.03.134.

Cian, R. E., Martínez-Augustin, O., & Drago, S. R. (2012b). Bioactive properties of peptides obtained by enzymatic hydrolysis from protein byproducts of Porphyra columbina. *Food Research International*, *49*(1), 364–372. Available from https://doi.org/10.1016/j.foodres.2012.07.003.

Connor, J. O. ', Meaney, S., Williams, G. A., & Hayes, M. (2020). Extraction of protein from four different seaweeds using three different physical pre-treatment strategies. *Molecules (Basel, Switzerland)*, *25*(8). Available from https://doi.org/10.3390/molecules25082005, https://www.mdpi.com/1420-3049/25/8/2005/pdf.

Crusz, S. M., & Balkwill, F. R. (2015). Inflammation and cancer: Advances and new agents. *Nature Reviews Clinical Oncology*, *12*(10), 584–596. Available from https://doi.org/10.1038/nrclinonc.2015.105, http://www.nature.com/nrclinonc/archive/index.html.

Dawczynski, C., Schubert, R., & Jahreis, G. (2007). Amino acids, fatty acids, and dietary fibre in edible seaweed products. *Food Chemistry*, *103*(3), 891–899. Available from https://doi.org/10.1016/j.foodchem.2006.09.041.

Denis, C., Morançais, M., Li, M., Deniaud, E., Gaudin, P., Wielgosz-Collin, G., Barnathan, G., Jaouen, P., & Fleurence, J. (2010). Study of the chemical composition of edible red macroalgae Grateloupia turuturu from Brittany (France). *Food Chemistry, 119*(3), 913–917. Available from https://doi.org/10.1016/j.foodchem.2009.07.047.

Dumay, J., Morançais, M., Munier, M., Le Guillard, C., & Fleurence, J. (2014). Phycoerythrins: Valuable proteinic pigments in red seaweeds. *Advances in Botanical Research, 71*. Available from https://doi.org/10.1016/B978-0-12-408062-1.00011-1, http://www.elsevier.com/wps/find/bookdescription.cws_home/675801/description#description.

Ferreira, M., Teixeira, C., Abreu, H., Silva, J., Costas, B., Kiron, V., & Valente, L. M. P. (2021). Nutritional value, antimicrobial and antioxidant activities of micro- and macroalgae, single or blended, unravel their potential use for aquafeeds. *Journal of Applied Phycology, 33*(6), 3507–3518. Available from https://doi.org/10.1007/s10811-021-02549-2.

Fitzgerald, C., Aluko, R. E., Hossain, M., Rai, D. K., & Hayes, M. (2014). Potential of a renin inhibitory peptide from the red seaweed palmaria palmata as a functional food ingredient following confirmation and characterization of a hypotensive effect in spontaneously hypertensive rats. *Journal of Agricultural and Food Chemistry, 62*(33), 8352–8356. Available from https://doi.org/10.1021/jf500983n, http://pubs.acs.org/journal/jafcau.

Fitzgerald, C., Mora-Soler, L., Gallagher, E., O'Connor, P., Prieto, J., Soler-Vila, A., & Hayes, M. (2012). Isolation and characterization of bioactive pro-peptides with in vitro renin inhibitory activities from the macroalga Palmaria palmata. *Journal of Agricultural and Food Chemistry, 60*(30), 7421–7427. Available from https://doi.org/10.1021/jf301361c.

Fleurence, J. (1999). Seaweed proteins: Biochemical, nutritional aspects and potential uses. *Trends in Food Science and Technology, 10*(1), 25–28. Available from https://doi.org/10.1016/S0924-2244(99)00015-1.

Fleurence, J., Morançais, M., & Dumay, J. (2017). *Seaweed proteins* 2nd ednin *food processing* (pp. 245–262). Elsevier. Available from https://www.sciencedirect.com/book/9780081007228, 10.1016/B978-0-08-100722-8.00010-3.

Furuta, T., Miyabe, Y., Yasui, H., Kinoshita, Y., & Kishimura, H. (2016). Angiotensin I converting enzyme inhibitory peptides derived from phycobiliproteins of dulse Palmaria palmata. *Marine Drugs, 14*(2). Available from https://doi.org/10.3390/md14020032.

García-Oliveira, P., Fraga-Corral, M., Pereira, A. G., Prieto, M. A., & Simal-Gandara, J. (2022). Solutions for the sustainability of the food production and consumption system. *Critical Reviews in Food Science and Nutrition, 62*(7), 1765–1781. Available from https://doi.org/10.1080/10408398.2020.1847028.

Grivennikov, S. I., Greten, F. R., & Karin, M. (2010). Immunity, inflammation, and cancer. *Cell, 140*(6), 883–899. Available from https://doi.org/10.1016/j.cell.2010.01.025.

Gómez-Ordóñez, Eva, Jiménez-Escrig, Antonio, & Rupérez, Pilar (2010). Dietary fibre and physicochemical properties of several edible seaweeds from the northwestern Spanish coast. *Food Research International, 43*(9), 2289–2294. Available from https://doi.org/10.1016/j.foodres.2010.08.005.

Hammed, A. M., Jaswir, I., Amid, A., Alam, Z., Asiyanbi-H, T. T., & Ramli, N. (2013). Enzymatic hydrolysis of plants and algae for extraction of bioactive compounds. *Food Reviews International, 29*(4), 352–370. Available from https://doi.org/10.1080/87559129.2013.818012.

Harnedy, P. A., & FitzGerald, R. J. (2013). In vitro assessment of the cardioprotective, anti-diabetic and antioxidant potential of Palmaria palmata protein hydrolysates.

Journal of Applied Phycology, *25*(6), 1793–1803. Available from https://doi.org/10.1007/s10811-013-0017-4.

Harnedy, P. A., & Fitzgerald, R. J. (2011). Bioactive proteins, peptides, and amino acids from macroalgae. *Journal of Phycology*, *47*(2), 218–232. Available from https://doi.org/10.1111/j.1529-8817.2011.00969.x.

Harnedy, P. A., O'Keeffe, M. B., & Fitzgerald, R. J. (2015). Purification and identification of dipeptidyl peptidase (DPP) IV inhibitory peptides from the macroalga Palmaria palmata. *Food Chemistry*, *172*, 400–406. Available from https://doi.org/10.1016/j.foodchem.2014.09.083.

Harnedy, P. A., O'Keeffe, M. B., & FitzGerald, R. J. (2017). Fractionation and identification of antioxidant peptides from an enzymatically hydrolysed Palmaria palmata protein isolate. *Food Research International*, *100*, 416–422. Available from https://doi.org/10.1016/j.foodres.2017.07.037.

Hayes, M., & Tiwari, B. K. (2015). Bioactive carbohydrates and peptides in foods: An overview of sources, downstream processing steps and associated bioactivities. *International Journal of Molecular Sciences*, *16*(9), 22485–22508. Available from https://doi.org/10.3390/ijms160922485, http://www.mdpi.com/1422-0067/16/9/22485/pdf.

Homaei, A. A., Sariri, R., Vianello, F., & Stevanato, R. (2013). Enzyme immobilization: An update. *Journal of Chemical Biology*, *6*(4), 185–205. Available from http://www.springer.com/chemistry/physical/journal/12154?cm_mmc = AD-_-enews-_-PSE1674-_-0&, 10.1007/s12154-013-0102-9.

Indumathi, P., & Mehta, Alka (2016). A novel anticoagulant peptide from the Nori hydrolysate. *Journal of Functional Foods*, *20*, 606–617. Available from https://doi.org/10.1016/j.jff.2015.11.016.

Jeon, Y. J., & Samarakoon, K. (2013). *Recovery of proteins and their biofunctionalities from marine algae. Marine proteins and peptides: Biological activities and applications* (pp. 253–269). South Korea: John Wiley and Sons. Available from http://onlinelibrary.wiley.com/book/10.1002/9781118375082, 10.1002/9781118375082.ch12.

Joel, C. H., Sutopo, C. C. Y., Prajitno, A., Su, J. H., & Hsu, J. L. (2018). Screening of angiotensin-I converting enzyme inhibitory peptides derived from Caulerpa lentillifera. *Molecules (Basel, Switzerland)*, *23*(11). Available from https://doi.org/10.3390/molecules23113005, https://www.mdpi.com/1420-3049/23/11/3005/pdf.

Kazir, M., Abuhassira, Y., Robin, A., Nahor, O., Luo, J., Israel, A., Golberg, A., & Livney, Y. D. (2019). Extraction of proteins from two marine macroalgae, Ulva sp. and Gracilaria sp., for food application, and evaluating digestibility, amino acid composition and antioxidant properties of the protein concentrates. *Food Hydrocolloids*, *87*, 194–203. Available from https://doi.org/10.1016/j.foodhyd.2018.07.047.

Kim, S. K., & Wijesekara, I. (2010). Development and biological activities of marine-derived bioactive peptides: A review. *Journal of Functional Foods*, *2*(1), 1–9. Available from https://doi.org/10.1016/j.jff.2010.01.003.

Lee, M. K., Kim, I. H., Choi, Y. H., Choi, J. W., Kim, Y. M., & Nam, T. J. (2015a). The proliferative effects of Pyropia yezoensis peptide on IEC-6 cells are mediated through the epidermal growth factor receptor signaling pathway. *International Journal of Molecular Medicine*, *35*(4), 909–914. Available from https://doi.org/10.3892/ijmm.2015.2111, http://www.spandidos-publications.com/ijmm/35/4/909.

Lee, H. A., Kim, I. H., & Nam, T. J. (2015b). Bioactive peptide from Pyropia yezoensis and its anti-inflammatory activities. *International Journal of Molecular Medicine*, *36*

(6), 1701–1706. Available from https://doi.org/10.3892/ijmm.2015.2386, http://www.spandidos-publications.com/ijmm/36/6/1701/download.

Lorenzo, J. M., Agregán, R., Munekata, P. E. S., Franco, D., Carballo, J., Şahin, S., Lacomba, R., & Barba, F. J. (2017). Proximate composition and nutritional value of three macroalgae: Ascophyllum nodosum, Fucus vesiculosus and Bifurcaria bifurcata. *Marine Drugs*, *15*(11). Available from https://doi.org/10.3390/md15110360, http://www.mdpi.com/1660-3397/15/11/360/pdf.

Machado, M., Machado, S., Pimentel, F. B., Freitas, V., Alves, R. C., & Oliveira, M. B. P. P. (2020). Amino acid profile and protein quality assessment of macroalgae produced in an integrated multi-trophic aquaculture system. *Foods*, *9*(10). Available from https://doi.org/10.3390/foods9101382, https://www.mdpi.com/2304-8158/9/10/1382.

Magdugo, R. P., Terme, N., Lang, M., Pliego-Cortés, H., Marty, C., Hurtado, A. Q., Bedoux, G., & Bourgougnon, N. (2020). An analysis of the nutritional and health values of Caulerpa racemosa (Forsskål) and Ulva fasciata (Delile)—Two chlorophyta collected from the Philippines. *Molecules (Basel, Switzerland)*, *25*(12). Available from https://doi.org/10.3390/molecules25122901, https://www.mdpi.com/1420-3049/25/12/2901/pdf.

McLaughlin, C. M., Harnedy-Rothwell, P. A., Lafferty, R. A., Sharkey, S., Parthsarathy, V., Allsopp, P. J., McSorley, E. M., FitzGerald, R. J., & O'Harte, F. P. M. (2021). Macroalgal protein hydrolysates from Palmaria palmata influence the 'incretin effect' in vitro via DPP-4 inhibition and upregulation of insulin, GLP-1 and GIP secretion. *European Journal of Nutrition*, *60*(8), 4439–4452. Available from https://doi.org/10.1007/s00394-021-02583-3.

McLaughlin, C. M., Sharkey, S. J., Harnedy-Rothwell, P., Parthsarathy, V., Allsopp, P. J., McSorley, E. M., FitzGerald, R. J., & O'Harte, F. P. M. (2020). Twice daily oral administration of Palmaria palmata protein hydrolysate reduces food intake in streptozotocin induced diabetic mice, improving glycaemic control and lipid profiles. *Journal of Functional Foods*, *73*. Available from https://doi.org/10.1016/j.jff.2020.104101, http://www.elsevier.com/wps/find/journaldescription.cws_home/717426/description#description.

Mennella, I., Fogliano, V., & Vitaglione, P. (2014). Salivary lipase and α-amylase activities are higher in overweight than in normal weight subjects: Influences on dietary behavior. *Food Research International*, *66*, 463–468. Available from https://doi.org/10.1016/j.foodres.2014.10.008.

Mišurcová, L. (2011). *Seaweed digestibility and methods used for digestibility determination. Handbook of marine macroalgae: Biotechnology and applied phycology* (pp. 285–301). John Wiley and Sons Czech Republic. Available from http://onlinelibrary.wiley.com/book/10.1002/9781119977087, 10.1002/9781119977087.ch13.

Nazarudin, M. F., Alias, N. H., Balakrishnan, S., Wan Hasnan, W. N. I., Noor Mazli, N. A. I., Ahmad, M. I., Md Yasin, I. S., Isha, A., & Aliyu-Paiko, M. (2021). Chemical, nutrient and physicochemical properties of brown seaweed, Sargassum polycystum c. Agardh (phaeophyceae) collected from Port Dickson, Peninsular Malaysia. *Molecules (Basel, Switzerland)*, *26*(17). Available from https://doi.org/10.3390/molecules26175216, https://www.mdpi.com/1420-3049/26/17/5216/pdf.

Nongonierma, A. B., & FitzGerald, R. J. (2017). Strategies for the discovery and identification of food protein-derived biologically active peptides. *Trends in Food Science and*

Technology, *69*, 289–305. Available from https://doi.org/10.1016/j.tifs.2017.03.003, http://www.elsevier.com/wps/find/journaldescription.cws_home/601278/description#description.

Ohanenye, I. C., Tsopmo, A., Ejike, C. E. C. C., & Udenigwe, C. C. (2020). Germination as a bioprocess for enhancing the quality and nutritional prospects of legume proteins. *Trends in Food Science and Technology*, *101*, 213–222. Available from https://doi.org/10.1016/j.tifs.2020.05.003, http://www.elsevier.com/wps/find/journaldescription.cws_home/601278/description#description.

Opinto, G., Natalicchio, A., & Marchetti, P. (2013). Physiology of incretins and loss of incretin effect in type 2 diabetes and obesity. *Archives of Physiology and Biochemistry*, *119*(4), 170–178. Available from https://doi.org/10.3109/13813455.2013.812664.

Ortiz, J., Uquiche, E., Robert, P., Romero, N., Quitral, V., & Llantén, C. (2009). Functional and nutritional value of the Chilean seaweeds Codium fragile, Gracilaria chilensis and Macrocystis pyrifera. *European Journal of Lipid Science and Technology*, *111*(4), 320–327. Available from https://doi.org/10.1002/ejlt.200800140.

Paiva, L., Lima, E., Neto, A. I., & Baptista, J. (2016). Isolation and characterization of angiotensin I-converting enzyme (ACE) inhibitory peptides from Ulva rigida C. Agardh protein hydrolysate. *Journal of Functional Foods*, *26*, 65–76. Available from https://doi.org/10.1016/j.jff.2016.07.006, http://www.elsevier.com/wps/find/journaldescription.cws_home/717426/description#description.

Paiva, L., Lima, E., Patarra, R. F., Neto, A. I., & Baptista, J. (2014). Edible Azorean macroalgae as source of rich nutrients with impact on human health. *Food Chemistry*, *164*, 128–135. Available from https://doi.org/10.1016/j.foodchem.2014.04.119.

Park, S. J., Ryu, J., Kim, I. H., Choi, Y. H., & Nam, T. J. (2015). Activation of the mTOR signaling pathway in breast cancer MCF-7 cells by a peptide derived from Porphyra yezoensis. *Oncology Reports*, *33*(1), 19–24. Available from https://doi.org/10.3892/or.2014.3557, http://www.spandidos-publications.com/or/33/1/19.

Park, S. J., Ryu, J., Kim, I. H., Choi, Y. H., & Nam, T. J. (2014). Induction of apoptosis by a peptide from Porphyra yezoensis: Regulation of the insulin-like growth factor I receptor signaling pathway in MCF-7 cells. *International Journal of Oncology*, *45*(3), 1011–1016. Available from https://doi.org/10.3892/ijo.2014.2509, http://www.spandidos-publications.com/ijo/45/3/1011/download.

Paz Ocaranza, M., Riquelme, J. A., García, L., Jalil, J. E., Chiong, M., Santos, R. A. S., & Lavandero, S. (2020). Counter-regulatory renin–angiotensin system in cardiovascular disease. *Nature Reviews Cardiology*, *17*(2), 116–129. Available from https://doi.org/10.1038/s41569-019-0244-8, http://www.nature.com/nrcardio/archive/index.html.

Pimentel, F. B., Cermeño, M., Kleekayai, T., Harnedy-Rothwell, P. A., Fernandes, E., Alves, R. C., Beatriz, P. P. O. M., & FitzGerald, R. J. (2020). Enzymatic modification of Porphyra dioica-derived proteins to improve their antioxidant potential. *Molecules (Basel, Switzerland)*, *25*(12). Available from https://doi.org/10.3390/molecules25122838, https://www.mdpi.com/1420-3049/25/12/2838/pdf.

Pimentel, F. B., Cermeño, M., Kleekayai, T., Machado, S., Rego, A., Fernandes, E., Alves, R. C., Oliveira, M. B. P. P., & FitzGerald, R. J. (2020). Contribution of in vitro simulated gastrointestinal digestion to the antioxidant activity of Porphyra dioica conchocelis. *Algal Research*, *51*. Available from https://doi.org/10.1016/j.algal.2020.102085, http://www.sciencedirect.com/science/journal/aip/22119264.

Pliego-Cortés, H., Wijesekara, I., Lang, M., Bourgougnon, N., & Bedoux, G. (2020). Current knowledge and challenges in extraction, characterization and bioactivity of seaweed protein and seaweed-derived proteins. *Advances in Botanical Research, 95*, 289–326. Available from https://doi.org/10.1016/bs.abr.2019.11.008, http://www.elsevier.com/wps/find/bookdescription.cws_home/675801/description#description.

Poojary, M. M., Orlien, V., & Olsen, K. (2019). Conventional and enzyme-assisted green extraction of umami free amino acids from Nordic seaweeds. *Journal of Applied Phycology, 31*(6), 3925–3939. Available from https://doi.org/10.1007/s10811-019-01857-y.

Provansal, M. M. P., Cuq, J. L. A., & Cheftel, J. C. (1975). Chemical and nutritional modifications of sunflower proteins due to alkaline processing. Formation of amino acid cross-links and isomerization of lysine residues. *Journal of Agricultural and Food Chemistry, 23*(5), 938–943. Available from https://doi.org/10.1021/jf60201a030.

Reuter, S., Gupta, S. C., Chaturvedi, M. M., & Aggarwal, B. B. (2010). Oxidative stress, inflammation, and cancer: How are they linked? *Free Radical Biology and Medicine, 49* (11), 1603–1616. Available from https://doi.org/10.1016/j.freeradbiomed.2010.09.006.

Sato, M., Hosokawa, T., Yamaguchi, T., Nakano, T., Muramoto, K., Kahara, T., Funayama, K., Kobayashi, A., & Nakano, T. (2002). Angiotensin I-converting enzyme inhibitory peptides derived from Wakame (Undaria pinnatifida) and their antihypertensive effect in spontaneously hypertensive rats. *Journal of Agricultural and Food Chemistry, 50*(21), 6245–6252. Available from https://doi.org/10.1021/jf020482t.

Seca, A. M. L., & Pinto, D. C. G. A. (2018). Overview on the antihypertensive and anti-obesity effects of secondary metabolites from seaweeds. *Marine Drugs, 16*(7). Available from https://doi.org/10.3390/md16070237, http://www.mdpi.com/1660-3397/16/7/237/pdf.

Shuuluka, D., Bolton, J. J., & Anderson, R. J. (2013). Protein content, amino acid composition and nitrogen-to-protein conversion factors of Ulva rigida and Ulva capensis from natural populations and Ulva lactuca from an aquaculture system, in South Africa. *Journal of Applied Phycology, 25*(2), 677–685. Available from https://doi.org/10.1007/s10811-012-9902-5.

Stack, J., Tobin, P. R., Gietl, A., Harnedy, P. A., Stengel, D. B., & FitzGerald, R. J. (2017). Seasonal variation in nitrogenous components and bioactivity of protein hydrolysates from Porphyra dioica. *Journal of Applied Phycology, 29*(5). Available from http://www.wkap.nl/journalhome.htm/0921-8971.

Suetsuna, K., Maekawa, K., & Chen, J. R. (2004). Antihypertensive effects of Undaria pinnatifida (wakame) peptide on blood pressure in spontaneously hypertensive rats. *Journal of Nutritional Biochemistry, 15*(5), 267–272. Available from https://doi.org/10.1016/j.jnutbio.2003.11.004.

Suetsuna, K., & Nakano, T. (2000). Identification of an antihypertensive peptide from peptic digest of wakame (Undaria pinnatifida). *Journal of Nutritional Biochemistry, 11*(9), 450–454. Available from https://doi.org/10.1016/S0955-2863(00)00110-8.

Sun, S., Xu, X., Sun, X., Zhang, X., Chen, X., & Xu, N. (2019). Preparation and identification of ACE inhibitory peptides from the marine macroalga Ulva intestinalis. *Marine Drugs, 17* (3). Available from https://doi.org/10.3390/md17030179.

Tavano, O. L. (2013). Protein hydrolysis using proteases: An important tool for food biotechnology. *Journal of Molecular Catalysis B: Enzymatic, 90*, 1–11. Available from https://doi.org/10.1016/j.molcatb.2013.01.011.

Tibbetts, S. M., Milley, J. E., & Lall, S. P. (2016). Nutritional quality of some wild and cultivated seaweeds: Nutrient composition, total phenolic content and in vitro digestibility. *Journal of Applied Phycology, 28*(6), 3575–3585. Available from https://doi.org/10.1007/s10811-016-0863-y.

Tornatore, L., Thotakura, A. K., Bennett, J., Moretti, M., & Franzoso, G. (2012). The nuclear factor kappa B signaling pathway: Integrating metabolism with inflammation. *Trends in Cell Biology*, *22*(11), 557–566. Available from https://doi.org/10.1016/j.tcb.2012.08.001.

Vieira, E. F., Soares, C., Machado, S., Correia, M., Ramalhosa, M. J., Oliva-teles, M. T., Paula Carvalho, A., Domingues, V. F., Antunes, F., Oliveira, T. A. C., Morais, S., & Delerue-Matos, C. (2018). Seaweeds from the Portuguese coast as a source of proteinaceous material: Total and free amino acid composition profile. *Food Chemistry*, *269*, 264–275. Available from https://doi.org/10.1016/j.foodchem.2018.06.145.

Wen, X., Peng, C., Zhou, H., Lin, Z., Lin, G., Chen, S., & Li, P. (2006). Nutritional composition and assessment of Gracilaria lemaneiformis Bory. *Journal of Integrative Plant Biology*, *48*(9), 1047–1053. Available from https://doi.org/10.1111/j.1744-7909.2006.00333.x.

Wen, L., Tan, S., Zeng, L., Wang, Z., Ke, X., Zhang, Z., Tang, H., Guo, H., & Xia, E. (2020). Ultrasound-assisted extraction and in vitro simulated digestion of Porphyra haitanensis proteins exhibiting antioxidative and α-glucosidase inhibitory activity. *Journal of Food Measurement and Characterization*, *14*(6), 3291–3298. Available from https://doi.org/10.1007/s11694-020-00564-4.

Wu, Q., Cai, Q. F., Yoshida, A., Sun, L. C., Liu, Y. X., Liu, G. M., Su, W. J., & Cao, M. J. (2017). Purification and characterization of two novel angiotensin I-converting enzyme inhibitory peptides derived from R-phycoerythrin of red algae (Bangia fuscopurpurea). *European Food Research and Technology*, *243*(5), 779–789. Available from https://doi.org/10.1007/s00217-016-2792-z, http://link.springer.de/link/service/journals/00217/index.htm.

Wu, M., Tong, C., Wu, Y., Liu, S., & Li, W. (2016). A novel thyroglobulin-binding lectin from the brown alga Hizikia fusiformis and its antioxidant activities. *Food Chemistry*, *201*, 7–13. Available from https://doi.org/10.1016/j.foodchem.2016.01.061.

Yu, Z., Yin, Y., Zhao, W., Wang, F., Yu, Y., Liu, B., Liu, J., & Chen, F. (2011). Characterization of ACE-inhibitory peptide associated with antioxidant and anticoagulation properties. *Journal of Food Science*, *76*(8), C1149–C1155. Available from https://doi.org/10.1111/j.1750-3841.2011.02367.x.

Zhang, X., Cao, D., Sun, X., Sun, S., & Xu, N. (2019). Preparation and identification of antioxidant peptides from protein hydrolysate of marine alga Gracilariopsis lemaneiformis. *Journal of Applied Phycology*, *31*(4), 2585–2596. Available from https://doi.org/10.1007/s10811-019-1746-9.

Zheng, Y., Zhang, Y., & San, S. (2020). Efficacy of a novel ACE-inhibitory peptide from Sargassum maclurei in hypertension and reduction of intracellular endothelin-1. *Nutrients*, *12*(3). Available from https://doi.org/10.3390/nu12030653.

CHAPTER

Bioactive peptides released from edible insects during gastrointestinal digestion

14

Joaquín Navarro del Hierro[1], Blanca Hernández-Ledesma[2] and Diana Martin[3,4]

[1]*Sección Departamental de Farmacia Galénica y Tecnología Alimentaria, Facultad de Veterinaria, Universidad Complutense de Madrid, Madrid, Spain*

[2]*Department of Bioactivity and Food Analysis, Institute of Food Science Research (CIAL, CSIC–UAM, CEI UAM1 + CSIC), Madrid, Spain*

[3]*Sección Departamental de Ciencias de la Alimentación, Facultad de Ciencias, Universidad Autónoma de Madrid, Madrid, Spain*

[4]*Departamento de Producción y Caracterización de Nuevos Alimentos, Instituto de Investigación en Ciencias de la Alimentación (CIAL) (CSIC–UAM), Madrid, Spain*

14.1 Introduction

In 2013 the Food and Agriculture Organization (FAO) and other world authorities recommended the inclusion of insects in our diets (van Huis et al., 2013). This advice was motivated by the expected insufficient animal protein in the following decades to feed an increasing world population, a situation that required the search for alternative protein sources. In this respect, edible insects have been identified as one of the best candidates due to their nutritional composition being able of covering the quantitative and qualitative diet requirements for proteins, together with lipids and other micronutrients. Additionally, the low environmental impact of insect rearing is currently considered one of the most sustainable animal protein production alternatives (van Huis & Oonincx, 2017; van Huis, 2020). Therefore the recommendations made by the FAO, as well as the nutritional and environmental advantages boosted the safety evaluation and regulation of edible insects for food and feed in Europe since 2015. The production of whole insects and their parts was regulated for the first time by including edible insects under the term of novel foods, under the Regulation (EU) 2015/2283 (European Parliament & Council of the European Union, 2015). Following this regulation, the European Food Safety Authority (EFSA) ruled for the first time in January 2021 on the first application for a novel food made from *Tenebrio molitor* (mealworm). The EFSA considered that the intake of this edible insect was safe under the proposed uses and levels (Turck et al., 2021), being approved later by the European Commission, and considered one of the most relevant milestones in the recently emerged field of edible insects for food (European Commission,

Protein Digestion-Derived Peptides. DOI: https://doi.org/10.1016/B978-0-443-19141-1.00014-5

Commission Implementing Regulation, 2021). Since then, three more insect species have been approved by the European Commission: *Locusta migratoria* (migratory locust), *Acheta domesticus* (house cricket), and *Alphitobius diaperinus* (lesser mealworm). It is expected that throughout 2023, EFSA will issue more positive opinions regarding requests made for other insect species, such as *Gryllodes sigillatus* (cricket), *Hermetia illucens* (black soldier fly), or *Apis mellifera* (western honeybee). In addition, the authorization of edible insects as a protein source for feed was already approved for the specific case of aquaculture feeding in 2017, having been extended for pigs and poultry feeding during 2021.

Due to the scientific and public interest in insect protein, the current available research on this macronutrient is considerable, including the composition, amino acid profile, techno-functional properties, technological processes to obtain such proteins, as well as the impact of processing conditions on their functionality. In this sense, most of the technological efforts are focused on the production of protein concentrates from different edible species. The obtention of protein hydrolysates is another strong research field to offer more advantageous products, such as improved solubility and technological behavior, as well as better desirable values of digestibility or palatability (Martínez-Alvarez et al., 2015). At the same time, the hydrolysis of proteins may lead to the production of peptides with potential bioactivity. Those described bioactive peptides from edible insects show similar biological activities as those confirmed for peptides from other animal proteins, such as milk, eggs, or meat. Thus antioxidant, antihypertensive, or antidiabetic activities, among others, have been reported for hydrolyzed proteins from different edible insects (Rivero Pino et al., 2020). The main mechanisms involved in such biological activities from insect peptides would be related to the scavenging of radicals and/or chelation of metal ions, inhibition of angiotensin-converting enzyme-I (ACE), or inhibition of dipeptidyl peptidase-IV (DPP-IV), respectively.

Considering the evidence on the role of hydrolyzed insect proteins on the production of bioactive peptides, it should be remarked that, similar to other food proteins, proteins from edible insects might also have the ability to produce bioactive peptides during the digestive hydrolysis process. However, the current available information at this respect is scarce, as most of the current efforts are related to the study of the digestion and digestibility process of insect proteins, which needs to be elucidated in order to take advantage of the nutritional potential of these "novel" proteins. Therefore the study of the release of bioactive peptides after the gastrointestinal digestion of edible insect protein has just emerged but is already showing promising results.

This chapter will show the current state of the art of the potential of edible insects to produce bioactive peptides during the gastrointestinal digestion process. Considering that more than 1900 different species of edible insects have been recognized (Weru et al., 2021), this chapter was initially focused on those suggested by the EFSA Scientific Committee to be used as food and feed in Europe (Hardy et al., 2015), or that are also currently authorized as food. These species are *T. molitor*, *L. migratoria*, *A. domesticus*, *A. diaperinus*, *H. illucens*, *G. sigillatus*,

Gryllus assimilis, *Musca domestica*, and *Bombyx mori*. However, a lack of information regarding this topic was found for the species *A. domesticus*, *G. assimilis*, and *H. illucens* during the writing of this chapter, whereas for *L. migratoria* the available information was scarce. A larger variety of studies was found for the species *T. molitor*, *A. diaperinus*, *G. sigillatus*, *M. domestica*, and *B. mori* nonetheless.

14.2 Edible insects

14.2.1 *Tenebrio molitor*

The species *T. molitor* or mealworm, is one of the most popular edible insects for food and feed, mainly consumed under the larvae form. Concerning its composition, protein content has been described in the range of 40%–70% of dry weight, and 15%–40% content for lipids (dry weight), depending on the insect feeding and processing conditions (Rumbos et al., 2020).

As previously described, bioactive peptides from insects, including *T. molitor*, have shown abilities as ACE inhibitors. The ACE is a metabolic enzyme that converts angiotensin I into the vasoconstrictor angiotensin II. Therefore ACE inhibition has been targeted for blood pressure reduction (Dai et al., 2013). The ACE inhibitory peptides are usually small-chain peptides (2–20 amino acids) containing hydrophobic residues, such as proline (P), phenylalanine (F), or tyrosine (Y), in the tripeptide sequence at the C-terminal end. This structure facilitates the interaction with the active site of ACE (Li et al., 2004).

Rivero Pino et al. (2020) simulated the hydrolysis of *T. molitor* with different enzymes, including those similar to digestive hydrolysis, such as trypsin. The raw protein before hydrolysis exhibited low ACE inhibition ($IC_{50} > 100$ mg/mL), whereas the enzymatic treatment led to a significant improvement in this activity, with an average IC_{50} of 0.27 mg/mL. The high levels of ACE inhibition attained were tentatively attributed to the specific proteolytic mechanisms related to small-chain peptides containing hydrophobic residues in the tripeptide sequence at the C-terminal end. Specifically, such released peptides usually show strong ACE inhibition when arginine (A) and lysine (K) residues are placed at the C-terminus.

After *in vitro* gastrointestinal digestion of *T. molitor*, Zielińska et al. (2018) and Zielińska, Baraniak, et al. (2017) showed that the released peptides had high antiradical activity and ability to chelate iron ions. These authors measured the antioxidant activity by different methods (2,2′-azino-bis(3-ethylbenzothiazoline-6-sulfonic acid), ABTS; 2,2-diphenyl-1-picrylhydrazyl, DPPH; ferrous iron chelating activity and ferric-reducing antioxidant power, FRAP) and from samples from different technological pretreatments, such as boiling or baking. Specifically, selected fractions separated by gel filtration derived from the digestion were evaluated. Interestingly, both processing methods of boiling and baking improved the antioxidant activities compared to raw samples, suggesting that changes in the sequence of insect peptide and denaturation might take place during heat treatment, causing modifications in sites available to react with radicals. Concerning characterization, the active peptide derived from insect protein were identified as NYVADGLG (cuticle protein), AAAPVAVAK (cuticle protein), YDDGSYKPH (ADFb/cofilin), and AGDDAPR (actin) (Zielińska et al., 2018). In addition, in a previous study, the same authors (Zielińska et al., 2017), also showed that the highest concentration of peptides after digestion was found for boiled *T. molitor* and that the heat treatment caused a slight reduction in the content of peptides. After digestion, the hydrolysates were also submitted to a simulated absorption process by dialysis with a membrane tube. The results demonstrated that the absorbed peptide fractions had a higher antiradical activity than the hydrolysates released during digestion (Zielińska et al., 2017).

When Rivero Pino et al. (2020) simulated the hydrolysis of *T. molitor* with trypsin, a significant improvement in the antioxidant activity was observed. In addition, a low hydrolysis degree showed a weak radical-scavenging activity, which improved with increasing hydrolysis until an optimum was reached. Thus further hydrolysis was detrimental to both DPPH scavenging and FRAP. Specifically, the trypsin hydrolysis showed maximal DPPH scavenging activity at a 10% degree of hydrolysis (IC_{50} of 0.93 mg/mL). Contrarily, the ferrous chelating activity improved continuously with the degree of hydrolysis. These authors showed that the peptide fraction of 0.5–2 kDa was the most abundant in the hydrolysates displaying better DPPH and Fe^{3+} reducing power, which corresponded to peptide chains from 2–10 amino acid residues.

Rivero Pino et al. (2020) showed for the first time that only when *T. molitor* was trypsin-hydrolyzed exhibited DPP-IV inhibitory activity unlike the nonhydrolyzed sample and that such antidiabetic activity increased with the degree of hydrolysis. Additionally, Dávalos Terán et al. (2020) showed the ability of pepsin to release DPP-IV inhibitory peptides from the proteins of *T. molitor* by *in silico* tools. Most of the high-scoring structural proteins were cuticular proteins, which mainly contain hydrophobic amino acids, known as potent DPP-IV inhibiting peptides (Dávalos Terán et al., 2020).

Chen et al. (2019) produced hydrolysates from *T. molitor* with pepsin and trypsin. Then, the amino acid sequences of potential antithrombotic peptides were identified and their potential binding interactions within the active site of thrombin were determined by *in silico* tools through molecular docking simulation. These authors identified the peptides SLVDAIGMGP and AGFAGDDAPR as potential antithrombotic peptides. Taking into account the physicochemical properties of these peptides, the authors suggested that they might inhibit thrombin activity by interacting with thrombin at exosite 1.

The bioactive effects of peptides released from *T. molitor* after digestion have been summarized in Table 14.1.

14.2.2 *Alphitobius diaperinus*

A. diaperinus, or lesser mealworm is also typically produced as larvae, like mealworms. In addition, *A. diaperinus* stands out for being more easily produced for

Table 14.1 Bioactive effects of peptides released from *T. molitor* after digestion.

Bioactivity	Digestive enzyme(s)	Peptides involved	Outcome	References
Antihypertensive	Trypsin	Small-chain peptides with hydrophobic residues in the tripeptide sequence at the C-terminal end	ACE $IC_{50} = 0.27 \pm 0.01$ mg/mL	Rivero Pino et al. (2020)
Antioxidant	Simulated oral-gastric-intestinal digestion (α-amylase, pepsin, pancreatin, and bile extract)	Peptide fractions of <3.5 kDa previously submitted to heat treatments	DPPH: $IC_{50} = 18.4-85.6$ μg/mL ABTS: $IC_{50} = 13.1-24.3$ μg/mL Fe^{2+} chelating activity: $IC_{50} = 9.9-19.5$ μg/mL Reducing power: Abs700 = 0.202–0.280	ZielÂinska et al. (2017)
	Trypsin	Peptide chains from 2–10 amino acid residues (0.5–2 kDa)	DPPH $IC_{50} = \sim 0.9-\sim 1.5$ mg/mL Fe^{2+} chelating activity $IC_{50} = \sim 0.9-\sim 1.75$ mg/mL	Rivero Pino et al. (2020)
Antidiabetic	Trypsin	Peptides (0.5–2 kDa) submitted to different degrees of hydrolysis	DPP-IV $IC_{50} = \sim 8-\sim 16$ mg/mL	
	Pepsin	Cuticular protein hydrolysate	DPP-IV ~20% inhibition at 0.375 mg/mL	Dávalos Terán et al. (2020)
		Noncuticular protein hydrolysate	DPP-IV ~6% inhibition at 0.375 mg/mL	
Antithrombotic	Pepsin followed by trypsin	SLVDAIGMGP and AGFAGDDAPR	Thrombin inhibitory activity of 28% at 0.2 mg/mL	Chen et al. (2019)

ABTS, *2,20-azino-bis(3-ethylbenzothiazoline-6-sulfonic acid);* ACE, *angiotensin I converting enzyme;* DPP-IV, *dipeptidyl peptidase IV;* DPPH, *2,2-diphenyl-1-picrylhydrazyl;* IC_{50}, *half maximal inhibitory concentration.*

protein-rich flours when compared to *T. molitor* (Ricciardi & Baviera, 2016). These larvae can reach a protein content in the range of 50%–60% and lipids around 20%–30% (Bjørge et al., 2018).

It is known that dietary compounds can modulate enterohormone secretion by the enteroendrocrine cells, modulating different physiological processes, such as appetite regulation, gastric motility, or glucose homeostasis. As an example, different dietary proteins and protein hydrolysates can induce the secretion of satiety hormones, such as glucagon-like peptide-1 (GLP-1), cholecystokinin (CCK), and peptide tyrosine tyrosine (PYY) (Miguéns-Gómez et al., 2020). Miguéns-Gómez et al. (2020) compared the modulation of different proteins (insects, beef, and almonds) on the enteroendocrine system and food intake regulation by *ex vivo* models of human and pig intestine treated with *in vitro* digests of such protein sources, and assayed the satiety capacity of raw insect, beef, and almond in rats. The used insect was *A. diaperinus*, both raw powder and protein concentrate. In the case of pig duodenal enteroendocrine secretions, these authors showed that digested *A. diaperinus* and beef stimulated the secretion of CCK more than almond, and that the insect and beef showed a similar effect on PYY secretion. All the digested samples significantly reduced ghrelin secretion, but *A. diaperinus* was the least effective. In the case of human colon samples, digested *A. diaperinus* and beef were the most effective in inducing the secretion of GLP-1. The result for the insect protein concentrate was similar to the raw insect. Additionally, both raw insect and protein concentrate from *A. diaperinus* limited ghrelin secretion in human colon samples compared to the other protein sources. Contrarily to expected, the performed *in vivo* study showed that rats fed *A. diaperinus* ate more compared to the control and beef groups, suggesting that insect proteins might stimulate food intake, regardless of the contrary anorexigenic effect observed *ex vivo* and in cells.

The DPP-IV inhibitory activity of *in vitro* digested protein isolates and concentrates from *A. diaperinus* was evaluated by Lacroix et al. (2019) and the effect of digestion time, type of enzyme, and peptide fractions were assessed. During the first 60 minutes of digestion with pepsin, the ability of the samples to inhibit the activity of DPP-IV increased rapidly. However, during the pancreatic phase of the digestion (120 minutes), the inhibitory activity increased more slowly and plateaued during the last hour of digestion.

14.2.3 *Gryllodes sigillatus*

G. sigillatus or banded cricket is typically consumed in its adult stage. Contrary to those insects under larvae form, the protein content of insects consumed as adults can reach higher values. Thus in case of *G. sigillatus*, protein contents in the range of 60%–70% of dry matter have been described (Hall et al., 2017).

Hall et al. (2020) performed the proteolysis of *G. sigillatus* proteins with alcalase and this protein hydrolysate was followed by *in vitro* simulated gastrointestinal digestion. This process generated a complex mixture of peptides that mainly consisted of <1 kDa peptides. Then, peptides were isolated according to their size in a gel filtration column and four peak fractions were collected (S1–S4), which were tested against DPP-IV, ACE, and α-glucosidase. The fraction S4 showed superior bioactivity against all the assayed enzymes. Thus this fraction was separated by reverse phase chromatography in six fractions (F1–F6). Not only the fraction with the most hydrophobic peptides was the most bioactive one, but also the strongest peptides were present in the cationic fraction. Most of the identified peptides within the cationic fraction were effectively predicted to possess antihypertensive capabilities and were used in molecular docking studies against human ACE. These authors evidenced that the pentapeptides YKPRP, PHGAP, and VGPPQ had binding properties similar to those of captopril, the reference drug used to block ACE activity. Finally, the concentrated cationic fraction of digested peptides was tested for antiinflammatory activity on nuclear factor-kappa-light-chain-enhancer of activated B cells (NF-κB) Luciferase reporter RAW264.7 macrophage cells. These authors showed a nonconcentration-dependent inhibition of inflammation.

After *in vitro* gastrointestinal digestion of *G. sigillatus*, Zielińska et al. (2018) observed a high antioxidant activity (as determined by ABTS, DPPH, ferrous iron chelating activity and FRAP assays) of the released peptides and from samples from different technological pretreatments, such as boiling or baking. These authors compared the antioxidant activity of digests from different species (*G. sigillatus*, *T. molitor*, and *Schistocerca gregaria*), and the highest antioxidant activity against ABTS and DPPH radicals formation was found for the peptide fraction obtained from the *G. sigillatus* protein preparation. Concerning characterization, the antioxidant active peptides derived from insect proteins were identified as IIAPPER (actin fragment), LAPSTIK (actin fragment), VAPEEHPV (actin fragment), and KVEGDLK (myosin fragment). Additionally, the fraction from the *G. sigillatus* protein preparation, which showed the highest antiradical activity, was also the most effective inhibitor of lipoxygenase (LOX) and cyclooxygenase 2 (COX 2) by *in vitro* assays, two enzymes commonly used for the evaluation of the antiinflammatory activity. These methods consist of assessing the activity of LOX or the peroxidase activity of COXs (monitoring the appearance of oxidized *N,N,N′,N′*-tetramethyl-*p*-phenylenediamine) in the presence of inhibitors (Zielińska et al., 2018). Furthermore, the heat treatment process had a significant general effect on the improvement of the assayed bioactivities.

Hall et al. (2018) evaluated the antioxidant capacity of *G. sigillatus* hydrolyzed proteins before and after *in vitro* gastrointestinal digestion. The scavenging activity of both ABTS and DPPH radicals, as well as the reducing effect by the FRAP assay and chelating activity of ferrous iron increased after digestion. Similarly, the measurement of DPP-IV inhibition showed that before digestion, proteins had low bioactivity, but it increased after gastrointestinal hydrolysis. This result agreed with the same evidence found by Nongonierma et al. (2018) related to the improvement of DPP-IV inhibitory activity after *in vitro* gastrointestinal digestion of hydrolyzed proteins from *G. sigillatus*. Nevertheless, the evaluation of the ACE inhibitory activity by Hall et al. (2018) showed that extensive hydrolysis of the proteins could decrease this bioactivity after digestion.

The bioactive effects of peptides released from *G. gryllodes* after digestion have been summarized in Table 14.2.

14.2.4 *Musca domestica*

M. domestica, or housefly, is a common insect found worldwide. This insect is traditionally consumed under its larval form, also known as maggots. Housefly larvae contain around 60% protein and 25% fat (Hwangbo et al., 2009; Pieterse & Pretorius, 2014).

Sun et al. (2019) identified and characterized the peptides DFTPVCTTELGR (Dr12, 1338.48 Da) and ARFEELCSDLFR (AR12, 1485.66 Da) from the alcalase-hydrolysate of *M. domestica* and evaluated the effect of pH and simulated gastrointestinal digestion on their antioxidant activity by the ABTS cation radical scavenging ability. Regarding the influence of pH on the peptides, the half maximal effective concentration (EC_{50}) values of Dr12 progressively increased (worsened antioxidant activity) at acidic conditions (pH 3 and 1) compared to pH 7, which was related to the changes of charge in the peptide, specifically at N- or C-terminal. Contrarily, the ABTS cation radical scavenging ability of AR12 increased in acidic conditions (decreased EC_{50}). During the simulated gastrointestinal digestion of the peptides, which was assayed by a pepsin-pancreatin hydrolysis treatment, Dr12 displayed a slight decrease in the antioxidant activity compared to the untreated peptide, possibly due to the degradation of the structure. AR12 showed great antioxidant stability during the simulated gastrointestinal digestion compared to the untreated peptide (EC_{50} values between 0.35 and 0.39 mM).

In animal models of diabetes, a protein-rich extract of *M. domestica* larvae was administered by gastric gavage to rats and the beneficial effects on the metabolic disorder were investigated (Mei et al., 2018). The peptides presumably released from the protein-rich extract after digestion were able of lowering blood glucose levels in streptozotocin-induced type 2 diabetic rats, the mechanism possibly being related to the regulation of hyperlipidemia. Specifically, the protein extract was able of lowering serum levels of total cholesterol, low-density lipoproteins (LDL) and triglycerides, as well as increasing the levels of high-density lipoproteins (HDL). It also alleviated oxidative stress, partly restoring β-cell protective functions. In addition, rats treated with the extract had higher levels of superoxide dismutase (SOD) activity, whilst it reversed the decreased expression of glucose transporter type 4 (GLUT4), phospho-antimicrobial peptide (AMP)-activated protein kinase, SOD1, and catalase (CAT) in the liver, skeletal muscle and pancreatic tissue in diabetic rats. The expression of the peroxisome proliferator-activated receptor γ (PPARγ) was also increased in the liver and adipose tissue of the diabetic rats.

Table 14.2 Bioactive effects of peptides released from *G. gryllodes* after digestion.

Bioactivity	Digestive enzyme(s)	Peptides involved	Outcome	References
Antihypertensive	Pepsin and pancreatin (simulated gastrointestinal digestion)	Purified cationic peptide fractions (obtained after size exclusion, reverse-phase and ion-exchange)	ACE $IC_{50} = 1.92$ μg/mL	Hall et al. (2020)
		Purified anionic peptide fractions (obtained after size exclusion, reverse-phase and ion-exchange)	ACE $IC_{50} = 0.5$ mg/mL	
Antidiabetic	Pepsin and pancreatin (simulated gastrointestinal digestion)	Purified cationic peptide fractions (obtained after size exclusion, reverse-phase and ion-exchange)	α-amylase $IC_{50} = 18.5$ μg/mL	
			α-glucosidase $IC_{50} = 13.9$ μg/mL	
		Purified anionic peptide fractions (obtained after size exclusion, reverse-phase and ion-exchange)	α-amylase $IC_{50} = 5$ mg/mL	
			α-glucosidase $IC_{50} = 76.6$ μg/mL	
	Pepsin and porcine pancreatic enzyme preparation (simulated gastrointestinal digestion)	Protein isolate	DPP-IV $IC_{50} = 0.78$ mg/mL	Nongonierma et al. (2018)
		Hydrolysate fraction obtained with Protamex (H5)	DPP-IV $IC_{50} = 0.71$ mg/mL	
	Pepsin and pancreatin (simulated gastrointestinal digestion)	Alcalase-hydrolysate fractions based on degree of hydrolysis	DPP-IV: ∼35–∼65% inhibition at 1.25 mg/mL	Hall et al. (2018)

(*Continued*)

Table 14.2 Bioactive effects of peptides released from *G. gryllodes* after digestion. *Continued*

Bioactivity	Digestive enzyme(s)	Peptides involved	Outcome	References
Antioxidant	Pepsin and pancreatin (simulated gastrointestinal digestion)	Alcalase-hydrolysate fractions based on degree of hydrolysis	ABTS: 689–799 μmol TE/mg sample DPPH: 1575–1926 μmol TE/mg sample FRAP: 1074–1278 μmol TE/mg sample MIC: 72.7%–82.9% MIC inhibition	
Antioxidant	α-Amylase, pepsin, pancreatin, and bile extract (Simulated oral-gastric-intestinal digestion)	Peptide fractions of <3.5 kDa previously submitted to heat treatments	DPPH: EC_{50} = 6.9–31.0 μg/mL ABTS: EC_{50} = 2.75–30.1 μg/mL Fe^{2+} chelating activity: EC_{50} = 2.8–166 μg/mL Reducing power: Abs700 = 0.138–0.180	ZielÂnska et al. (2018)
Antiinflammatory	α-Amylase, pepsin, pancreatin and bile extract (Simulated oral-gastric-intestinal digestion)	Peptide fractions of <3.5 kDa previously submitted to heat treatments	LOX IC_{50} = 0.13–20.7 μg/mL COX-2 IC_{50} = 0.26–16.5 μg/mL	

ABTS, *2,20-azino-bis(3-ethylbenzothiazoline-6-sulfonic acid)*; ACE, *angiotensin I converting enzyme*; COX, *cyclooxygenase*; DPP-IV, *dipeptidyl peptidase IV*; DPPH, *2,2-diphenyl-1-picrylhydrazyl*; EC_{50}, *half maximal effective concentration*; FRAP, *ferric-reducing ability of plasma*; IC_{50}, *half maximal inhibitory concentration*; LOX, *lipoxygenase*; MIC, *metal ion chelating ability*.

Cecropin, a widely studied AMP from *M. domestica*, has also shown promising results for ulcerative colitis (UC) therapy in mice. Mesoporous carbon nanoparticles loaded with cecropin were intragastrically administered to mice. An improved colonic injury in the UC mice was observed by effectively inhibiting inflammation and oxidative stress, as well as maintaining colonic tight junctions, and regulating intestinal flora (Zhang, Gui, et al., 2021).

Low molecular weight peptides (LMWP) from *M. domestica* larvae have shown antidiarrhea effects in mice after oral administration and subsequent digestion, the regulation of the intestinal microbiota being possibly one of the potential antidiarrhea mechanisms of the LMWP (Chu et al., 2019).

14.2.5 *Bombyx mori*

B. mori or silkworm is a popular insect, due to their traditional use for obtaining the silk. In addition, as edible insect, pupae are the most used forms for food or feed, as a typical by-product of the silk industry. Protein content of silkworm can be found in the range of 40%–70% and lipids can reach value higher than 30% (Tassoni et al., 2022)

One of the most studied insect species in terms of the biological activities of peptides released after digestion is *B. mori*. Specifically, the ACE inhibitory activity of this edible insect has been very heavily assessed. Vercruysse et al. (2005) gave the first report of increased ACE inhibitory activity of *B. mori* protein hydrolysates produced after complete gastrointestinal digestion (pepsin, trypsin, and α-chymotrypsin). Gastrointestinal digestion led to peptides with the strongest ACE IC_{50} (0.6–0.7 mg/mL, depending on the amount of enzyme used) compared to the IC_{50} of the nonhydrolyzed fraction (17–72 mg/mL). Later, Li et al. (2014) hydrolyzed *B. mori* pupae with alcalase and after different purification methods, they obtained a fraction (α-P_3-6-b) with an ACE IC_{50} of 28 μg/mL, a value that was higher than the reported values for the synthetic ACE inhibitory drug captopril. The analysis revealed that the peptide contained in the active fraction was VEIS. This fraction was further subjected to gastrointestinal digestion and, whilst the ACE inhibitory rates of pupa α-P_3-6-b samples were significantly reduced by 9.21% after hydrolysis using pepsin, the ACE inhibitory rate increased by 4.94% after addition of trypsin. Similarly, Jia et al. (2015) purified the ACE inhibitory peptide KHV from alcalase hydrolysate of ultrasonic-pretreated *B. mori* pupa protein, with an IC_{50} of 12.82 μM and an extraordinary stability against gastrointestinal pepsin, trypsin, and α-chymotrypsin. In this case, docking simulation of the peptide at the ACE active site revealed Zn(II) interaction and hydrogen bond interactions with ACE residues N^{277}, Q^{281}, T^{282}, H^{383}, D^{415}, K^{454}, S^{526}, F^{527}, and Q^{530}. Another ACE inhibitory peptide (ASL) from *B. mori* pupa was identified by Wu et al. (2015) after hydrolysis with gastrointestinal endopeptidases pepsin, trypsin, and α-chymotrypsin. Lineweaver–Burk plot suggested competitive inhibition of the peptide against ACE, with an IC_{50} of 102.15 μM. Molecular docking studies showed very strong hydrogen bonds with ACE residues A^{354}, Q^{281}, and H^{353}. More recently Tao et al. (2017) showed that the peptide GNPWM obtained after the alcalase digestion of *B. mori* pupa protein exhibited strong ACE inhibitory activity (IC_{50}, 21.70 μM) that was not affected by the action of pepsin and pancreatin due to its resistance to the digestion process (relative content of intact GNPWM after the digestion process was around 92%) and unaffected ACE inhibitory activity after gastrointestinal digestion with pepsin and pancreatin.

The antioxidant activity of peptides from *B. mori* after gastrointestinal digestion has also been addressed in several studies. Wu et al. (2011) produced protein hydrolysates from the fifth larval instar of *B. mori* after gastrointestinal digestion with pepsin, trypsin, and α-chymotrypsin. The IC_{50} value of the hydrolysate for scavenging the DPPH radical was 57.9 μg/mL. The hydrolysate also had a noticeable effect on reducing power as well as an effective capacity for chelating ferrous ions ($IC_{50} = 2$ mg/mL). Zhang, Wang, et al. (2021) obtained more recently antioxidant peptides from *B. mori* pupae protein hydrolysates produced by acidic and neutral proteases and further digested with pepsin, pancreatin, and chymotrypsin. The two identified peptides were FKGPACA and SVLGTGC, with the main active sites possibly located in the ACA and TGC fragments, respectively. The ABTS radical scavenging activity of FKGPACA was stable (around 55% ABTS radical scavenging activity) after digestion with pepsin, while that of SVLGTGC experienced a 10.5% reduction (from around 70.5% scavenging activity of the untreated peptide to around 63%). The digestion with pancreatin reduced scavenging activities by 10.59% for FKGPACA and 43.56% for SVLGTGC, while after digestion with chymotrypsin, the antioxidant activity of both peptides remained stable compared to the nondigested peptides. Very recently Khammuang et al. (2022) showed that the protein hydrolysate obtained from *B. mori* pupa with trypsin exhibited a strong ability for scavenging ABTS ($IC_{50} = 15.35 \pm 0.17$ μg/mL) and DPPH radicals ($IC_{50} = 25.59 \pm 0.85$ μg/mL), even though the exact peptide sequence from this hydrolysate was not elucidated.

Immunomodulatory peptides, which have a positive effect on body functions thanks to the enhancement of immune cell functions, antibody synthesis, natural killer (NK) cell activity, and cytokine regulation, have also been identified in *B. mori*. Li et al. (2019) isolated the hexapeptide PNPNTN from the alcalase hydrolysate of ultramicro-pretreated *B. mori* pupa and assessed the splenocyte proliferation rate (proliferation of T- and/or B-lymphocytes) of the peptide after treatment with gastrointestinal proteases. The peptide-induced splenocyte proliferation response after pepsin and trypsin digestion was kept constant compared to the nondigested peptide (84%–85% proliferation rate). Authors linked the stability of PNPNTN to the presence of short-chain peptides that are less susceptible to degradation by gastrointestinal proteases. Later, the same authors identified another bioactive peptide (DHAV) with immunomodulatory activity Li et al. (2020). Gastrointestinal digestion with pepsin and trypsin had no effect on the activity of DHAV, as there was no change in splenocyte proliferation rates (around 75%). The peptide also exhibited the ability to stimulate the expression of immune-related factors including interleukin-6 (IL-6) and IL-12, NF-κB, cyclin D1 (CD1), and cyclin-dependent kinase 4 (CDK-4). In addition, Pyo et al. (2013) confirmed an increase in the activation markers CD1 lb and CD40 in RAW264.7 cells induced by trypsin-digested peptides (> 30 kDa) from *B. mori*. Further analysis revealed the presence of fibroin-derived peptides.

Very recently, the DPP-IV inhibitory activity from *B. mori* pupae peptides has been also confirmed. Luo et al. (2022) identified the peptide LPPEHDWR from the <1 kDa pepsin-hydrolysate fraction as the main peptide responsible for the strong DPP-IV inhibitory activity, as confirmed by the IC_{50} value (267 μM) of the chemically synthesized peptide. The peptide was subjected to simulated gastrointestinal digestion and no variations were observed in the DPP-IV inhibitory activity before and after digestion. In addition, Lineweaver–Burk curves revealed a competitive/noncompetitive mixed-type DPP-IV inhibition mode, whilst molecular docking assays showed the formation of four hydrogen bonds between the peptide and DPP–IV residues (R^{523}, Y^{715}, Y^{548}, and I^{368}).

The effects of cooking and baking on *B. mori* flour in the presence and absence of fructose have been recently assessed by David-Birman et al. (2019). Authors found that Maillard glycation with fructose improves the techno-functionality of the flour and modulates the possible generation of bioactive peptides in the gut lumen. The flour was also demonstrated to be a source for different antimicrobial and acyl-CoA binding proteins, such as DVTWDKQVGGGKVF and FKQATVGDADPANRPGLLDL, respectively. Furthermore, authors also demonstrated that after the *in vitro* simulation of a "senior" gastric digestion model with pepsin, 14 novel peptides were identified by the PeptideRanker with a probability of over 80% to carry some biological activity, such as YFPYNFR, DPVFWMIMK, GSGGGGGGGWCAVAETSVAHGAAPA CGPR, and KYFPYNFR.

The bioactive effects of peptides released from *B. mori* after digestion have been summarized in Table 14.3.

14.2.6 *Locusta migratoria*

L. migratoria, known as migratory locust, is known to frequently cause severe pest problems worldwide, mostly in warm regions. Despite the complexity of its rearing production under controlled conditions, this insect was already approved in Europe for food and feed, mainly due to the protein content closer to 60% and lipids around 20% (Van Peer et al., 2021). Nevertheless, only one study was found during the writing of this chapter on the potential bioactivity of peptides during gastrointestinal digestion. Thus Zielińska, Karaś, et al. (2017) evaluated different bioactive properties of *L. migratoria* hydrolysates obtained after simulated gastrointestinal digestion. The products of digestion effectively showed effective antioxidant activities (DPPH and ABTS), chelating activities and reducing power. Interestingly, these authors suggested that, after digestion, the different tested edible insects had higher antioxidant activity than some protein hydrolysates obtained from plants or other animal products.

Table 14.3 Bioactive effects of peptides released from *B. mori* after digestion.

Bioactivity	Digestive enzyme(s)	Peptide(s) involved	Outcome	References
Antihypertensive	Pepsin, trypsin, and α-chymotrypsin (simulated gastrointestinal digestion)	Protein hydrolysates (peptides not identified)	ACE $IC_{50} = 0.6–0.7$ mg/mL	Vercruysse et al. (2005)
	Pepsin	Fraction α-P3–6-b (peptide VEIS)	ACE: 79% inhibition at 0.1 mg/mL	Li et al. (2014)
	Pepsin and trypsin	Fraction α-P3–6-b (peptide VEIS)	ACE: 83.5% inhibition at 0.1 mg/mL	Li et al. (2014)
	Pepsin, α-chymotrypsin, and trypsin	KHV	ACE: 65.5% inhibition at 7.5 μg/mL	Jia et al. (2015)
	Pepsin, trypsin, and α-chymotrypsin	ASL	ACE $IC_{50} = 102.15$ μM	Wu et al. (2015)
	Pepsin and pancreatin	GNPWM	ACE $IC_{50} = 21.70$ μM	Tao et al. (2017)
Antioxidant	Pepsin, trypsin, and α-chymotrypsin	Protein hydrolysate (peptides not identified)	DPPH $IC_{50} = = 57.9$ μg/mL	Wu et al. (2011)
			Reducing power: Abs700 = 0.830	
			Fe^{2+} chelating activity $IC_{50} = 2.03$ mg/mL	
	Pepsin	FKGPACA	ABTS: ~55% inhibition (unknown concentration)	Zhang, Gui, et al. (2021)
	Pepsin and pancreatin	FKGPACA	ABTS: ~50% inhibition (unknown concentration)	
	Pepsin, pancreatin, and chymotrypsin	FKGPACA	ABTS: ~55% inhibition (unknown concentration)	
	Pepsin	SVLGTGC	ABTS: ~64% inhibition (unknown concentration)	
	Pepsin and pancreatin	SVLGTGC		

			ABTS: ~40% inhibition (unknown concentration)	
	Pepsin, pancreatin, and chymotrypsin	SVLGTGC	ABTS: ~74% inhibition (unknown concentration)	
	Trypsin	Protein hydrolysates (peptides not identified)	ABTS $IC_{50} = 15.35\ \mu g/mL$	Khammuang et al. (2022)
			DPPH $IC_{50} = 25.6\ \mu g/mL$	
Immunomodulatory	Pepsin and trypsin	PNPNTN	84%–85% splenocyte proliferation rate	Li et al. (2019)
	Pepsin and trypsin	DHAV	75% splenocyte proliferation rates	Li et al. (2020)
			↑Expression of IL-6, IL-12, NF-κB, CD1 and CDK-4	
	Trypsin	Peptides >30 kDa	↑Activation of CD1 and CD40 in RAW264.7 cells	Pyo et al. (2013)
Antidiabetic	Pepsin	LPPEHDWR	DPP-IV $IC_{50} = 267\ \mu M$	Luo et al. (2022)

ABTS, *2,20-azino-bis(3-ethylbenzothiazoline-6-sulfonic acid);* ACE, *angiotensin I converting enzyme;* CD1, *cyclin D1;* CD4, *cluster of differentiation 40;* CDK-4, *cyclin-dependent kinase 4;* DPP-IV, *dipeptidyl peptidase IV;* DPPH, *2,2-diphenyl-1-picrylhydrazyl;* IC_{50}, *half maximal inhibitory concentration;* IL, *interleukin;* NF-κB, *nuclear factor kappa-light-chain-enhancer of activated B cells.*

14.3 Conclusions

In conclusion, the study of bioactive peptides from edible insects has shown great potential in terms of their health benefits, and the gastrointestinal digestion has been identified as a key factor in the release of these peptides, which have been shown to have antihypertensive, antidiabetic, immunomodulatory, antioxidant, antiinflammatory and appetite-regulating properties, among others. This chapter has provided a comprehensive overview of the current state of knowledge on the release of bioactive peptides after gastrointestinal digestion. While multiple edible insect species have been widely explored in the literature, it is clear that many other edible insect species have not yet been fully studied, or not studied at all, in terms of how the digestion of insect protein affects the release of peptides with potential bioactivity. This presents a significant opportunity for future research, as the discovery of new bioactive peptides from edible insects could have far-reaching implications for the food and health industries. In light of these findings, it is recommended that future research should focus on the exploration of new or underexplored insect species and the application of gastrointestinal digestion models to isolate and elucidate peptides with potential effects on human health and well-being. The many benefits that these peptides have been shown to offer, combined with the growing demand for sustainable protein sources, make this an area of research that is poised to grow significantly in the coming years. The continued exploration of this topic will provide valuable insights into the potential health benefits of edible insects and will inform the development of new functional foods and nutraceuticals that can improve human health.

References

Bjørge, J. D., Overgaard, J., Malte, H., Gianotten, N., & Heckmann, L.-H. (2018). Role of temperature on growth and metabolic rate in the tenebrionid beetles *Alphitobius diaperinus* and *Tenebrio molitor*. *Journal of Insect Physiology*, *107*, 89–96. Available from https://doi.org/10.1016/j.jinsphys.2018.02.010, https://linkinghub.elsevier.com/retrieve/pii/S0022191017303797.

Chen, F., Jiang, H., Lu, Y., Chen, W., & Huang, G. (2019). Identification and *in silico* analysis of antithrombotic peptides from the enzymatic hydrolysates of *Tenebrio molitor* larvae. *European Food Research and Technology*, *245*(12), 2687–2695. Available from https://doi.org/10.1007/s00217-019-03381-2, https://link.springer.com/article/10.1007/s00217-019-03381-2.

Chu, F., Jin, X., & Ma, H. (2019). Anti-diarrhea effects and identification of *Musca domestica* larvae low molecular weight peptides (LMWP). *Journal of Pharmaceutical and Biomedical Analysis*, *173*, 162–168. Available from https://doi.org/10.1016/j.jpba.2019.05.032, https://linkinghub.elsevier.com/retrieve/pii/S0731708518322209.

Dai, C., Ma, H., Luo, L., & Yin, X. (2013). Angiotensin I-converting enzyme (ACE) inhibitory peptide derived from *Tenebrio molitor* (L.) larva protein hydrolysate. *European Food Research and Technology*, *236*(4), 681–689. Available from https://doi.org/10.1007/s00217-013-1923-z, http://link.springer.com/10.1007/s00217-013-1923-z.

David-Birman, T., Moshe, H., & Lesmes, U. (2019). Impact of thermal processing on physicochemical properties of silk moth pupae *(Bombyx mori)* flour and *in-vitro* gastrointestinal proteolysis in adults and seniors. *Food Research International*, *123*, 11–19. Available from https://doi.org/10.1016/J.FOODRES.2019.04.042, https://doi.org/10.1016/j.foodres.2019.04.042.

Dávalos Terán, I., Imai, K., Lacroix, I. M. E., Fogliano, V., & Udenigwe, C. C. (2020). Bioinformatics of edible yellow mealworm *(Tenebrio molitor)* proteome reveal the cuticular proteins as promising precursors of dipeptidyl peptidase-IV inhibitors. *Journal of Food Biochemistry*, *44*(2). Available from https://doi.org/10.1111/jfbc.13121, https://onlinelibrary.wiley.com/doi/abs/10.1111/jfbc.13121.

European Commission, Commission Implementing Regulation (EU). (2021). 2021/882 of 1 June 2021 authorising the placing on the market of dried *Tenebrio molitor larva as a novel food under Regulation (EU)* 2015/2283 of the European Parliament and of the Council, and amending Commission Implementing Regulation (EU) 2017/2470. Official Journal of the European Union. L 194, 16–20, https://eur-lex.europa.eu/legal-content/EN/TXT/?uri = CELEX%3A32021R0882.

European Parliament and Council of the European Union. (2015). Regulation (EU) 2015/2283 of the European Parliament and of the Council of 25 November 2015 on novel foods, amending Regulation (EU) No 1169/2011 of the European Parliament and of the Council and repealing Regulation (EC) No 258/97 of the European Parliament and of the Council and Commission Regulation (EC) No 1852/2001. *Official Journal of the European Union. L, 327*, 1–22, https://eur-lex.europa.eu/legal-content/en/TXT/?uri = CELEX%3A32015R2283.

Hall, F., Johnson, P. E., & Liceaga, A. (2018). Effect of enzymatic hydrolysis on bioactive properties and allergenicity of cricket *(Gryllodes sigillatus)* protein. *Food Chemistry*, *262*, 39–47. Available from https://doi.org/10.1016/J.FOODCHEM.2018.04.058.

Hall, F. G., Jones, O. G., O'Haire, M. E., & Liceaga, A. M. (2017). Functional properties of tropical banded cricket *(Gryllodes sigillatus)* protein hydrolysates. *Food Chemistry*, *224*, 414–422. Available from https://doi.org/10.1016/j.foodchem.2016.11.138, https://linkinghub.elsevier.com/retrieve/pii/S0308814616319872.

Hall, F., Reddivari, L., & Liceaga, A. M. (2020). Identification and characterization of edible cricket peptides on hypertensive and glycemic *in vitro* inhibition and their anti-inflammatory activity on RAW 264.7 macrophage cells. *Nutrients*, *12*(11), 3588. Available from https://doi.org/10.3390/nu12113588, –3588, https://www.mdpi.com/2072-6643/12/11/3588.

Hardy, A., Benford, D., Noteborn, H. P. J. M., Halldorsson, T. I., Schlatter, J., Solecki, R. A., Jeger, M., Knutsen, H. K., More, S., Mortensen, A., Naegeli, H., Ockleford, C., Ricci, A., Rychen, G., Silano, V., Turck, D., Baggesen, D. L., Bonsall, M., Charlton, A., … Vlak, J. M. (2015). Risk profile related to production and consumption of insects as food and feed. *EFSA Journal*, *13*(10), 4257. Available from https://doi.org/10.2903/J.EFSA.2015.4257, https://efsa.onlinelibrary.wiley.com/doi/10.2903/j.efsa.2015.4257.

Hwangbo, J., Hong, E., Jang, A., Kang, H. K., Oh, J. S., Kim, B. W., & Park, B. S. (2009). Utilization of house fly-maggots, a feed supplement in the production of broiler chickens. *Journal of Environmental Biology*, *30*(4), 609–614.

Jia, J., Wu, Q., Yan, H., & Gui, Z. (2015). Purification and molecular docking study of a novel angiotensin-I converting enzyme (ACE) inhibitory peptide from alcalase hydrolysate of ultrasonic-pretreated silkworm pupa *(Bombyx mori)* protein. *Process Biochemistry*, *50*(5), 876–883. Available from https://doi.org/10.1016/J.PROCBIO.2014.12.030.

Khammuang, S., Sarnthima, R., & Sanachai, K. (2022). Purification and identification of novel antioxidant peptides from silkworm pupae *(Bombyx mori)* protein hydrolysate and molecular docking study. *Biocatalysis and Agricultural Biotechnology, 42*102367. Available from https://doi.org/10.1016/j.bcab.2022.102367, https://linkinghub.elsevier.com/retrieve/pii/S1878818122000949.

Lacroix, I. M. E., Dávalos Terán, I., Fogliano, V., & Wichers, H. J. (2019). Investigation into the potential of commercially available lesser mealworm *(A. diaperinus)* protein to serve as sources of peptides with DPP-IV inhibitory activity. *International Journal of Food Science & Technology, 54*(3), 696–704. Available from https://doi.org/10.1111/IJFS.13982, https://ifst.onlinelibrary.wiley.com/doi/10.1111/ijfs.13982.

Li, G. H., Le, G. W., Hui Shi, Y., & Shrestha, S. (2004). Angiotensin I–converting enzyme inhibitory peptides derived from food proteins and their physiological and pharmacological effects. *Nutrition Research, 24*(7), 469–486. Available from https://doi.org/10.1016/J.NUTRES.2003.10.014.

Li, X., Li, Y., Huang, X., Zheng, J., Zhang, F., & Kan, J. (2014). Identification and characterization of a novel angiotensin I-converting enzyme inhibitory peptide (ACEIP) from silkworm pupa. *Food Science and Biotechnology, 23*(4), 1017–1023. Available from https://doi.org/10.1007/s10068-014-0138-9, https://link.springer.com/article/10.1007/s10068-014-0138-9.

Li, Z., Zhao, S., Xin, X., Zhang, B., Thomas, A., Charles, A., Lee, K. S., Jin, B. R., & Gui, Z. (2019). Purification and characterization of a novel immunomodulatory hexapeptide from alcalase hydrolysate of ultramicro-pretreated silkworm *(Bombyx mori)* pupa protein. *Journal of Asia-Pacific Entomology, 22*(3), 633–637. Available from https://doi.org/10.1016/j.aspen.2019.04.005, https://linkinghub.elsevier.com/retrieve/pii/S1226861518306861.

Li, Z., Zhao, S., Xin, X., Zhang, B., Thomas, A., Charles, A., Lee, K. S., Jin, B. R., & Gui, Z. (2020). Purification, identification and functional analysis of a novel immunomodulatory peptide from silkworm pupa protein. *International Journal of Peptide Research and Therapeutics, 26*(1), 243–249. Available from https://doi.org/10.1007/s10989-019-09832-4, http://link.springer.com/10.1007/s10989-019-09832-4.

Luo, F., Fu, Y., Ma, L., Dai, H., Wang, H., Chen, H., Zhu, H., Yu, Y., Hou, Y., & Zhang, Y. (2022). Exploration of dipeptidyl peptidase-IV (DPP-IV) inhibitory peptides from silkworm pupae *(Bombyx mori)* proteins based on in silico and in vitro. *Journal of Agricultural and Food Chemistry, 70*(12), 3862–3871. Available from https://doi.org/10.1021/acs.jafc.1c08225, https://pubs.acs.org/doi/10.1021/acs.jafc.1c08225.

Martínez-Alvarez, O., Chamorro, S., & Brenes, A. (2015). Protein hydrolysates from animal processing by-products as a source of bioactive molecules with interest in animal feeding: A review. *Food Research International, 73*, 204–212. Available from https://doi.org/10.1016/J.FOODRES.2015.04.005.

Mei, H., Xu, J., He, Y., Yang, X., Liu, W., Tian, W., Zeng, Y., & Zhu, J. (2018). Protein-rich extract of *Musca domestica* larvae alleviated metabolic disorder in STZ-induced type 2 diabetic rat model via hepatoprotective and pancreatic β-cell protective activities. *Journal of Biosciences, 43*(5), 969–983. Available from https://doi.org/10.1007/s12038-018-9804-z, https://link.springer.com/article/10.1007/s12038-018-9804-z.

Miguéns-Gómez, A., Grau-Bové, C., Sierra-Cruz, M., Jorba-Martín, R., Caro, A., Rodríguez-Gallego, E., Beltrán-Debón, R., Blay, M. T., Terra, X., Ardévol, A., & Pinent, M. (2020). Gastrointestinally digested protein from the insect *Alphitobius diaperinus* stimulates a different intestinal secretome than beef or almond, producing a differential response in food intake in rats. *Nutrients*, *12*(8), 2366. Available from https://doi.org/10.3390/NU12082366, https://www.mdpi.com/2072-6643/12/8/2366.

Nongonierma, A. B., Lamoureux, C., & FitzGerald, R. J. (2018). Generation of dipeptidyl peptidase IV (DPP-IV) inhibitory peptides during the enzymatic hydrolysis of tropical banded cricket (Gryllodes sigillatus) proteins. *Food & Function*, *9*(1), 407–416. Available from https://doi.org/10.1039/C7FO01568B, http://xlink.rsc.org/?DOI = C7FO01568B.

Pieterse, E., & Pretorius, Q. (2014). Nutritional evaluation of dried larvae and pupae meal of the housefly *(Musca domestica)* using chemical- and broiler-based biological assays. *Animal Production Science*, *54*(3), 347. Available from https://doi.org/10.1071/AN12370, http://www.publish.csiro.au/?paper = AN12370.

Pyo, K.H., Kim, M.K., Shin, K.S., Chun, H.S., & Shin, E.H. (2013). Involvement of trypsin-digested silk peptides in the induction of RAW264.7 macrophage activation. 8 (12), 1755–1758 https://doi.org/10.1177/1934578X1300801226. https://journals.sagepub.com/doi/abs/10.1177/1934578X1300801226.

Wu, Q. Y., Guang-Xiu Tan, J.-Q. J., & Zhong-Zheng G., J.-L. X. (2011). Physicochemical properties of silkworm larvae protein isolate and gastrointestinal hydrolysate bioactivities. *African Journal of Biotechnology*, *10*(32), 6145–6153. Available from https://academicjournals.org/journal/AJB/article-abstract/425DA1431433.

Ricciardi, C., & Baviera, C. (2016). Role of carbohydrates and proteins in maximizing productivity in *Alphitobius diaperinus (Coleoptera: Tenebrionidae)*. *Redia-Giornale Di Zoologia*, *99*, 97–105. Available from https://doi.org/10.19263/REDIA-99.16.13.

Rivero Pino, F., Pérez Gálvez, R., Espejo Carpio, F. J., & Guadix, E. M. (2020). Evaluation of *Tenebrio molitor* protein as a source of peptides for modulating physiological processes. *Food & Function*, *11*(5), 4376–4386. Available from https://doi.org/10.1039/D0FO00734J, http://xlink.rsc.org/?DOI = D0FO00734J.

Rumbos, C. I., Karapanagiotidis, I. T., Mente, E., Psofakis, P., & Athanassiou, C. G. (2020). Evaluation of various commodities for the development of the yellow mealworm, *Tenebrio molitor*. *Scientific Reports*, *10*(1)11224. Available from https://doi.org/10.1038/s41598-020-67363-1, http://www.nature.com/articles/s41598-020-67363-1.

Sun, T., Zhang, S., Yang, W., Zhao, Z., & Yang, D. (2019). Housefly pupae-derived antioxidant peptides exerting neuroprotective effects on hydrogen peroxide-induced oxidative damage in PC12 cells. *Molecules (Basel, Switzerland)*, *24*(24), 4486. Available from https://doi.org/10.3390/MOLECULES24244486, https://www.mdpi.com/1420-3049/24/24/4486.

Tao, M., Wang, C., Liao, D., Liu, H., Zhao, Z., & Zhao, Z. (2017). Purification, modification and inhibition mechanism of angiotensin I-converting enzyme inhibitory peptide from silkworm pupa *(Bombyx mori)* protein hydrolysate. *Process Biochemistry*, *54*, 172–179. Available from https://doi.org/10.1016/J.PROCBIO.2016.12.022.

Tassoni, L., Cappellozza, S., Dalle Zotte, A., Belluco, S., Antonelli, P., Marzoli, F., & Saviane, A. (2022). Nutritional composition of *bombyx mori* pupae: A systematic review. *Insects*, *13*(7), 644. Available from https://doi.org/10.3390/insects13070644, https://www.mdpi.com/2075-4450/13/7/644.

Turck, D., Castenmiller, J., De Henauw, S., Hirsch-Ernst, K. I., Kearney, J., Maciuk, A., Mangelsdorf, I., McArdle, H. J., Naska, A., Pelaez, C., Pentieva, K., Siani, A., Thies, F., Tsabouri, S., Vinceti, M., Cubadda, F., Frenzel, T., Heinonen, M., Marchelli, R., ...

Knutsen, H. K. (2021). Safety of dried yellow mealworm *(Tenebrio molitor larva)* as a novel food pursuant to Regulation (EU) 2015/2283. *EFSA Journal*, *19*(1). Available from https://doi.org/10.2903/j.efsa.2021.6343, http://doi.wiley.com/10.2903/j.efsa.2021.6343.

van Huis, A. (2020). Insects as food and feed, a new emerging agricultural sector: A review. *Journal of Insects as Food and Feed*, *6*(1), 27–44. Available from https://doi.org/10.3920/JIFF2019.0017, https://www.wageningenacademic.com/doi/10.3920/JIFF2019.0017.

van Huis, A., & Oonincx, D. G. A. B. (2017). The environmental sustainability of insects as food and feed. A review. *Agronomy for Sustainable Development*, *37*(5), 1–14. Available from https://doi.org/10.1007/s13593-017-0452-8, http://www.bugsinthepicture.com.

van Huis, A., Van Itterbeeck, J., Klunder, H., Mertens, E., Halloran, A., Muir, G., & Vantomme, P. (2013). *Edible insects: Future prospects for food and feed security* (p. 187) Food and Agriculture Organization of the United Nations (FAO).

Van Peer, M., Frooninckx, L., Coudron, C., Berrens, S., Álvarez, C., Deruytter, D., Verheyen, G., & Van Miert, S. (2021). Valorisation potential of using organic side streams as feed for *Tenebrio molitor, Acheta domesticus* and *Locusta migratoria*. *Insects*, *12*(9), 796. Available from https://doi.org/10.3390/insects12090796, https://www.mdpi.com/2075-4450/12/9/796.

Vercruysse, L., Smagghe, G., Herregods, G., & Van Camp, J. (2005). ACE inhibitory activity in enzymatic hydrolysates of insect protein. *Journal of Agricultural and Food Chemistry*, *53*(13), 5207–5211. Available from https://doi.org/10.1021/jf050337q, https://pubs.acs.org/doi/full/10.1021/jf050337q.

Weru, J., Chege, P., & Kinyuru, J. (2021). Nutritional potential of edible insects: A systematic review of published data. *International Journal of Tropical Insect Science*. Available from https://doi.org/10.1007/s42690-021-00464-0, http://link.springer.com/10.1007/s42690-021-00464-0.

Wu, Q., Jia, J., Yan, H., Du, J., & Gui, Z. (2015). A novel angiotensin-I converting enzyme (ACE) inhibitory peptide from gastrointestinal protease hydrolysate of silkworm pupa *(Bombyx mori)* protein: Biochemical characterization and molecular docking study. *Peptides*, *68*, 17–24. Available from https://doi.org/10.1016/j.peptides.2014.07.026.

Zhang, L., Gui, S., Xu, Y., Zeng, J., Wang, J., Chen, Q., Su, L., Wang, Z., Deng, R., Chu, F., Liu, W., Jin, X., & Lu, X. (2021). Colon tissue-accumulating mesoporous carbon nanoparticles loaded with *Musca domestica* cecropin for ulcerative colitis therapy. *Theranostics*, *11*(7), 3417–3438. Available from https://doi.org/10.7150/thno.53105, https://www.thno.org/v11p3417.htm.

Zhang, Y., Wang, J., Zhu, Z., Li, X., Sun, S., Wang, W., & Sadiq, F. A. (2021). Identification and characterization of two novel antioxidant peptides from silkworm pupae protein hydrolysates. *European Food Research and Technology*, *247*(2), 343–352. Available from https://doi.org/10.1007/s00217-020-03626-5, https://link.springer.com/10.1007/s00217-020-03626-5.

Zielińska, E., Baraniak, B., & Karaś, M. (2017). Antioxidant and anti-inflammatory activities of hydrolysates and peptide fractions obtained by enzymatic hydrolysis of selected heat-treated edible insects. *Nutrients*, *9*(9), 970. Available from https://doi.org/10.3390/nu9090970, http://www.mdpi.com/2072-6643/9/9/970.

Zielińska, E., Baraniak, B., & Karaś, M. (2018). Identification of antioxidant and anti-inflammatory peptides obtained by simulated gastrointestinal digestion of three edible insects species *(Gryllodes sigillatus, Tenebrio molitor, Schistocerca gragaria)*.

International Journal of Food Science & Technology, *53*(11), 2542–2551. Available from https://doi.org/10.1111/ijfs.13848, http://doi.wiley.com/10.1111/ijfs.13848.

Zielińska, E., Karaś, M., & Jakubczyk, A. (2017). Antioxidant activity of predigested protein obtained from a range of farmed edible insects. *International Journal of Food Science & Technology*, *52*(2), 306–312. Available from https://doi.org/10.1111/ijfs.13282, http://doi.wiley.com/10.1111/ijfs.13282.

CHAPTER

Bioactive peptides from fruit food waste after simulated gastrointestinal digestion 15

Ester Hernández-Corroto[1], María Luisa Marina[1,2] and María Concepción García[1,2]

[1]*Departamento de Química Analítica, Química Física e Ingeniería Química, Universidad de Alcalá, Madrid, Spain*

[2]*Instituto de Investigación Química "Andrés M. del Río", Universidad de Alcalá, Madrid, Spain*

List of abbreviations

AA	Acetic acid
ABTS	2,2′-Azino-bis(3-ethylbenzothiazoline-6-sulfonic acid) diammonium
ACE	Angiotensin-I converting enzyme
AU	Activity unit
BB	Borate buffer
Cease	Cholesterol esterase
ChCl	Choline chloride
DPPH	2,2-Diphenyl-1-(2,4,6-trinitrophenyl) hydrazyl
DTT	Dithiothreitol
E/S ratio	Enzyme substrate
EC_{50}	Effective concentration at 50% of inhibition
GID	Gastrointestinal digestion
HDL-C	High-density lipoproteins
HIFU	High-intensity focused ultrasounds
HILIC	Hydrophilic interaction chromatography
HMG CoA	3-Hydroxy-3-methylglutaryl coenzyme A
IC_{50}	The half maximum inhibitory concentration
LDL-C	Low-density lipoproteins
NADES	Natural deep eutectic solvent
PB	Phosphate buffer
PL	Pancreatic lipase
PLE	Pressurized liquid extraction
ROS	Reactive oxygen species
RP-HPLC	Reversed-phase high-performance liquid chromatography
SBP	Systolic blood pressure
SCX	Strong cation exchange
SDS	Sodium dodecyl sulfate
SHR	Spontaneously hypertensive rats
S–L extraction	Solid–liquid extraction

Protein Digestion-Derived Peptides. DOI: https://doi.org/10.1016/B978-0-443-19141-1.00015-7

SPE	Solid phase extraction
TC	Total cholesterol
TCA	Trichloroacetic acid
TEAC	Trolox equivalent antioxidant capacity
WHO	World health organization

15.1 Introduction

Fruits are very significant to human nutrition and are the base of a healthy diet. The World Health Organization (WHO) recommends consuming, at least, 400 g of fruits and vegetables each day (Fruit & Vegetables – Your Dietary Essentials, 2020). The regular consumption of fruits, as part of a balanced diet, has been associated with health benefits due to their content in bioactive compounds. Among them are vitamins, carbohydrates, minerals, proteins, fibers, phenolic compounds, etc. (Yahia et al., 2018). The consumption of these molecules has been associated to the prevention of different types of cancer, cardiovascular and neurodegenerative diseases, diabetes, and respiratory conditions, among others (Yahia et al., 2018).

Fruit processing generates huge quantities of waste with a negative environmental impact. In fact, fruit industry produces and discards at least 25%–30% of every fruit (Patra et al., 2022). The reutilization of this waste would contribute to a circular economy transforming fruit waste into valuable bioproducts (Ng et al., 2020). Many researchers have demonstrated that, like the edible part of fruits, fruit waste contains important phytochemicals. Moreover, in some cases, these compounds are present in a higher amount than in the edible tissues (Wadhwa et al., 2015). Therefore fruit waste can be a potential low-cost source of bioactive compounds (De Ancos et al., 2015; Medina-Meza & Ganjyal Medina-Meza & Ganjyal2018, 2022;).

Proteins can be a source of bioactive peptides. In fact, these molecules are usually encrypted within a protein sequence becoming active when they are released by gastrointestinal digestion (GID) or food processing (enzymatic treatment, cooking, fermentation, ripening, etc.) (Daliri et al., 2017). Fig. 15.1 shows a flow chart for the release of bioactive peptides after GID from fruit waste proteins and from fruit waste protein hydrolysates.

15.2 Protein content in fruit waste

Fig. 15.2 shows fruit processing products (flesh and waste) and the protein content of the obtained waste. Peels, pomace, and seeds are the main waste from fruit and represent 16% of the total food waste (Cassani & Gomez-Zavaglia, 2022). Pomace mainly contains seeds, peels, and small amount of fruit pulp. *Prunus* fruits, pomegranate, pumpkin, watermelon, tomato, and pineapple are some examples of fruits whose residues range from 3 to 50% of the fruit weight and contain up to 44% of proteins.

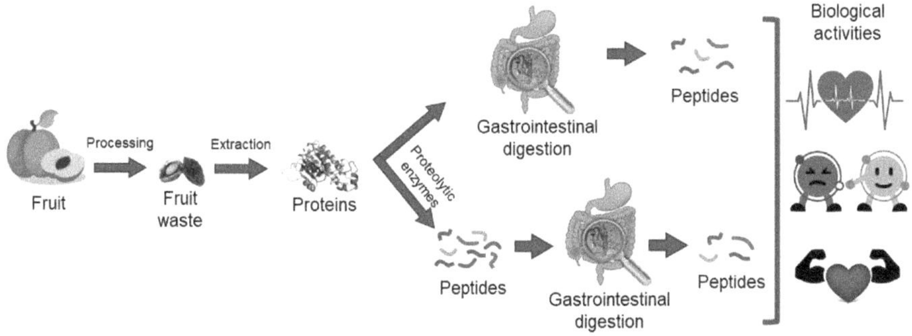

FIGURE 15.1 Flow chart for the release of bioactive peptides after GID.

Bioactive peptides obtained from fruit waste proteins and from fruit waste protein hydrolysates.

Prunus fruits belong to the Rosaceae family. They are composed of more than 400 species where peach, plum, nectarine, apricot, and cherry are the most highlighted (Munekata et al., 2022). These fruits are worldwide consumed resulting in huge amounts of waste (25%–40%). Peel, pomace, and seeds contain between 5% and 27% of proteins (Fig. 15.2) (Rudke et al., 2022). The valorization of these fruit residues represents an opportunity to develop food products with health-promoting properties.

Pomegranate belongs to the Punicaceae family. Pomegranates are widely known for their antioxidant properties, which has promoted the consumption of fresh pomegranates and processed products like the juice. After extraction of the pomegranate juice, large amounts of waste, mainly constituted by seeds and peels, are generated (Cano-Lamadrid et al., 2022; Venkitasamy et al., 2019). These residues can be employed for the fortification of foods to improve their properties, quality, and shelf-life (Das et al., 2021). Macronutrients like proteins have been found in all pomegranate residues highlighting seeds with up to 17% (Fig. 15.2). Main reported pomegranate proteins are albumins, globulins, prolamins, and glutelins (Elfalleh et al., 2012).

Pumpkin belongs to the Cucurbitaceae family, and it comprises several species. Pumpkin has a high nutritional value and it is considered a healthy food (Batool et al., 2022; Dhiman et al., 2009; Rico et al., 2020). It is used for the preparation of food products, such as pies, soups, stews, jam, pickle, beverages, candy, bakery, and confectionery products (Dhiman et al., 2009; Rico et al., 2020). The processing of pumpkin results in a waste made of seeds and peels. These residues are rich in proteins, especially seeds that could contain up to a 44% (Fig. 15.2).

Watermelon also belongs to the *Cucurbitaceae* family, and it is mainly processed to produce juices, nectars, and fruit cocktails. The amount of fruit waste ranged from 32% to 48% of fruit weight (Rico et al., 2020). The number of seeds per fruit varies from 500 to 900, depending on the fruit size, and they can contain up to

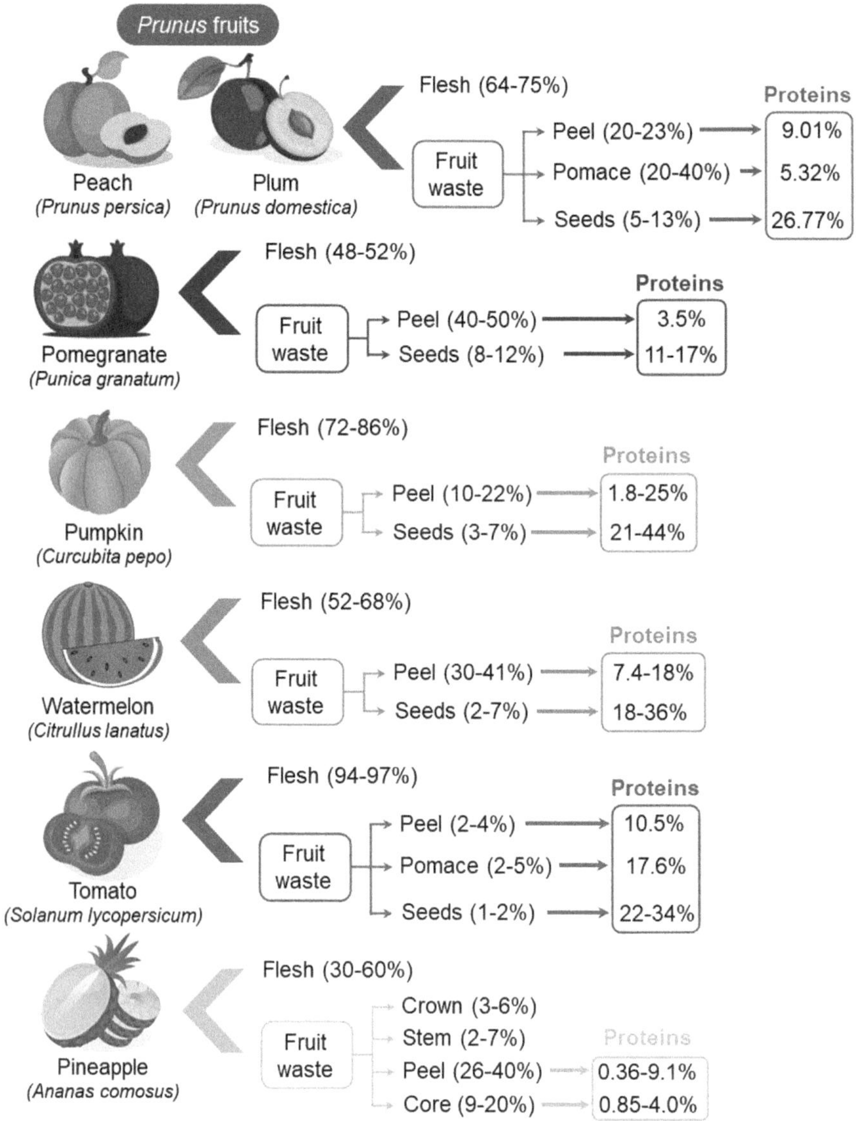

FIGURE 15.2 Fruit processing by-products and protein content from fruit wastes.

Source: *Data obtained from Das et al. (2021), Rajan et al. (2022), Rico et al. (2020), Rudke et al. (2022), and Venkitasamy et al. (2019).*

36% of proteins, depending on the specie (Fig. 15.2). Watermelon seeds are considered a rich source of proteins. Among them are globulins (63.7%) followed by albumins (18.6%), glutelins (14%), and prolamins (3.7%) (Nkoana et al., 2022).

Tomato belongs to the Solanaceae family. Tomato is widely consumed and there is a huge variety of products containing this fruit (soup, sauces, ketchup, tomato preserves (juice, purée, and paste), and dried tomatoes). Seeds, pomace, and peels are generated as residues in the processing of tomatoes (Rajan et al., 2022). Like in other fruits, tomato seeds present the highest protein content among tomato waste (Fig. 15.2).

Pineapple belongs to the Bromeliaceae family (Salve & Ray, 2020). Pineapple is very appreciated for its high content in bromelain, a very expensive mixture of proteolytic enzymes with several biotechnological applications in the pharmaceutical, food, and cosmetic industries (Rico et al., 2020). The food industry also uses pineapple to produce juices and canned fruit. Main waste from the processing of this fruit includes the crown (3%–6%), the stems (2%–7%), the peels (26%–40%), and the core (9%–20%) (Fig. 15.2). The amount of proteins in these residues is lower than in other fruits.

Among all fruit waste, the seeds are the residues with the highest amount of proteins. The sustainable extraction of these proteins would allow to revalorize this waste contributing to the circular economy. Moreover, these proteins could contain bioactive peptides that would be released during GID, processing, or both, exerting interesting properties.

15.3 Fruit waste bioactive peptides released in gastrointestinal digestion

Protein digestion begins in the stomach under acidic conditions. Gastric juice contains pepsins that degrade proteins to polypeptides. Polypeptides are subsequently hydrolyzed by carboxypeptidase A and B, α-chymotrypsin, trypsin, elastase, and pancreatic proteases in the lumen of the small intestine leading to smaller oligopeptides and single amino acids. After that, released oligopeptides undergo a second hydrolysis in the intestinal villi by peptidases producing dipeptides, tripeptides, and free amino acids (Ahmed et al., 2022; Udenigwe et al., 2017). These peptides can cross the epithelium of the small intestine, enter the bloodstream, and be transported throughout the body. Because of these processes, proteins, and polypeptides can release peptides that could exert biological functions. This fact has made very attractive the evaluation of the potential of GID to obtain bioactive peptides and different approaches have been carried out to simulate it. GID mainly involves two steps: a gastric and intestinal digestion. Simulated gastric digestion is performed by using pepsin (a moderately specific protease) which preferentially cleaves proteins at the carboxyl side of hydrophobic residues, such as phenylalanine (F) and leucine (L) (Ahmed et al., 2022). As different types of proteases act in the intestinal digestion, its simulation has been carried out in different ways. In some cases, this step has been performed using trypsin, which cleaves on the carboxy side of arginine (R) and lysine (K) residues (Toldrá et al., 2018). In others, chymotrypsin, which cleaves on the carboxy side of aromatic or hydrophobic amino acids, has been employed (Toldrá et al., 2018). A third approach involves the use of a mixture of enzymes, called pancreatin, which include the previous enzymes, among others.

The works evaluating the release of bioactive peptides from fruit waste by GID can be divided into three types. In some of them, bioactive peptides were obtained from fruit waste proteins using simulated GID. In other cases, proteins were first processed with food enzymes and then, submitted to a simulated GID. A third group of works reported in vivo studies to evaluate the release of bioactive peptides from fruit waste. Despite the interest of in vitro studies, in vivo studies are very important to confirm exerted ability since the immune function of the organism, blood pressure, interactions with other food compounds in the bloodstream, and the insulin and glucose level can modulate some peptide properties (Karaś, 2019).

15.3.1 Bioactive peptides obtained from fruit waste proteins by in vitro gastrointestinal digestion

The release of bioactive peptides after GID depends on various factors, such as the amount of proteins, the extraction technique and extracting solvent, and the amino acid composition of the proteins. Table 15.1 summarizes the extraction conditions and purification of proteins obtained from fruit waste, the GID conditions employed for the release of peptides, the identification of peptides in these hydrolysates, and their bioactivities.

Bioactive peptides have been obtained from proteins present in fruit seeds or in cakes remaining after extraction of other compounds (mainly oil). A previous extraction of proteins was carried out from waste employing a solid–liquid extraction, high-intensity focused ultrasounds (HIFU), or pressurized liquid extraction (PLE). After extraction, a protein purification step was performed by isoelectric precipitation or using cold acetone or ammonium sulfate. Afterwards, bioactive peptides were released by in vitro GID.

In the case of *prunus* fruit seeds, proteins were extracted using HIFU and, then, purified by precipitation with cold acetone (García et al., 2016). Released peptides by a simulated GID showed high antioxidant and angiotensin-converting enzyme (ACE) inhibitory capacities although there were differences among *prunus* seeds. Plum seeds showed higher antioxidant activity than peach seeds followed by apricot seeds, while peach seeds exhibited higher antihypertensive activity than apricot and plum seeds. Plum and apricot cake proteins were obtained by alkaline extraction and isoelectric precipitation (Čakarević, Vidović, Vladić, Gavarić, et al., 2019; Čakarević, Vidović, Vladić, Jokić, et al., 2019). Released peptides from plum cake proteins showed lower ABTS inhibition (37%)

Table 15.1 Release of bioactive peptides from fruit waste proteins by in vitro GID.

Fruit waste	Protein extraction conditions	Protein purification	GID conditions	Peptide sequence	Biological activity	References
Peach seeds	HIFU (100 mM Tris–HCl (pH 7.5) containing SDS (0.5%, w/v) and DTT (0.5%, w/v), 1 minute, 30% amplitude)	Cold acetone	Pepsin (pH 2, 37°C, 1 hour, 1:35 E/S ratio), pancreatin (pH 8, 37°C, 2 hours, 1:25 E/S ratio)	–	Antioxidant: 58% ABTS inhibition ACE-inhibitory: 130 μg/mL (IC_{50})	García et al. (2016)
Apricot seeds	HIFU (100 mM Tris–HCl (pH 7.5) containing SDS (0.5%, w/v) and DTT (0.5%, w/v), 1 minute, 30% amplitude)	Cold acetone	Pepsin (pH 2, 37°C, 1 hour, 1:35 E/S ratio), pancreatin (pH 8, 37°C, 2 hours, 1:25 E/S ratio)	–	Antioxidant: 45% ABTS inhibition ACE-inhibitory: 140 μg/mL (IC_{50})	García et al. (2016)
Plum seeds	HIFU (100 mM Tris–HCl (pH 7.5) containing SDS (0.5%, w/v) and DTT (0.5%, w/v), 1 minute, 30% amplitude)	Cold acetone	Pepsin (pH 2, 37°C, 1 hour, 1:35 E/S ratio), pancreatin (pH 8, 37°C, 2 hours, 1:25 E/S ratio)	–	Antioxidant: 63% ABTS inhibition ACE-inhibitory: 150 μg/mL (IC_{50})	García et al., 2016
Plum oil cake	S–L extraction under alkaline conditions	Isoelectric point	Pepsin (pH 2.5, 37°C, 2 hours, 1:25 E/S ratio), pancreatin (pH 7, 37°C, 2 hours, 1:50 E/S ratio)	–	Antioxidant: 30% DPPH inhibition, 37% ABTS inhibition ACE-inhibitory: 106 μg/mL (IC_{50})	Čakarević, Vidović, Vladić, Jokić, et al. (2019)

(Continued)

Table 15.1 Release of bioactive peptides from fruit waste proteins by in vitro GID. *Continued*

Fruit waste	Protein extraction conditions	Protein purification	GID conditions	Peptide sequence	Biological activity	References
Apricot kernel cake	S–L extraction under alkaline conditions	Isoelectric point	Pepsin (pH 2.5, 37°C, 2 hours, 1:25 E/S ratio), pancreatin (pH 7, 37°C, 2 hours, 1:50 E/S ratio)	–	Antioxidant: 65% DPPH inhibition, 94% ABTS inhibition ACE-inhibitory: 92 μg/mL (IC_{50})	Čakarević, Vidović, Vladić, Gavarić, et al. (2019)
Olive seeds	HIFU (100 mM Tris–HCl (pH 7.5) containing SDS (0.5%, w/v) and DTT (0.5%, w/v), 1 minute, 30% amplitude)	Cold acetone	Pepsin (pH 2, 37°C, 1 hour, 1:35 E/S ratio), pancreatin (pH 8, 37°C, 2 hours, 1:25 E/S ratio)	–	Antioxidant: 38% ABTS inhibition ACE-inhibitory: 150 μg/mL (IC_{50})	García et al. (2016)
Pomegranate seeds	HIFU (50 mM bicarbonate buffer (pH 11) containing SDS (1.5%, w/v), 5 minutes, 20% amplitude)	Purified and nonpurified extracts with cold acetone	Pepsin (pH 2, 37°C, 1 hour, 1:35 E/S ratio), pancreatin (pH 8, 37°C, 2 hours, 1:25 E/S ratio)	PSGRP, YNPR, TPPK, REGDL, NDPY, WFLP, LSLH, EERVL, QLLA, and REGDLL[a]	Antioxidant: 10% ABTS inhibition, 30% H_2O_2 scavenging Hypocholesterolemic: 47% CEase inhibition, 48% micellar solubility inhibition ACE-inhibitory: 75%	Guzmán-Lorite et al. (2022a)
	PLE (20 mM bicarbonate buffer (pH 11), 170°C, 5 cycles, 12 min/cycle)			PSGRP, ENPK, YNPR, TPPK, RADF, FRFE, WEER, QLPQ, ERVLQ, RQLV, QYSW, RALP, AGRLSV, QLRV, EERVL, RDFV, RLVP, RALPLQ, RALPL,	Antioxidant: 55% ABTS inhibition, 12% H_2O_2 scavenging Hypocholesterolemic: 85% CEase inhibition, 55% micellar solubility inhibition ACE-inhibitory: 45%	

				QELR, RGRPPPR, and RDKDR[a]		
	HIFU (ChCl:AA, 15 minutes, 80% amplitude, 60°C)	–		MVKGH, SSKKK, ERERPSRP, YNPR, QLVR, RLKY, RLHVLR, RDFVR, RALP, LLSSPRR, WSPR, REGDLL, QLRVPY, KKTD, RPSPR, QELR, AYLAPH, RGRPPPR, RDKDR	Antioxidant: 20% ABTS inhibition, 10% DPPH inhibition, 65% H_2O_2 scavenging	Guzmán-Lorite et al. (2022b)
	PLE (20 mM bicarbonate buffer (pH 11), 170°C, 3 cycles, 12 minutes/cycle)			TGPT, PSGRP, SGRAPP, QSDL, NDPY, WDAN, WFLP, EERVL, SEDW, HVVLP, DPLV, QYPL, WQLT, KKTD, AVDV, EVTE	Antioxidant: 7% ABTS inhibition, 50% DPPH inhibition	
Longan seeds	S–L extraction (PBS (pH 7.2), overnight, 40°C)	Cold ammonium sulfate (80%)	Pepsin (pH 2.5, 37°C, 3 hours, 1:20 E/S ratio), pancreatin (pH 7.5, 37°C, 3 hours, 1:20 E/S ratio)	ETSGMKPTEL, ISSMGILVCL	ACE-inhibitory: 250 ± 4 μg/mL (IC_{50} fraction <3 kDa), 18.22 ± 0.03 μg/mL (IC_{50} RP-HPLC fraction)	Nuchprapha et al. (2020)
Gac seeds	HIFU (SDS (1%, w/v), 3 minutes, 30% amplitude, 0°C)	Cold TCA (10%) in acetone	Pepsin (pH 2, 37°C, 24 hours, 250 AU/mg), trypsin (250 AU/mg) and chymotrypsin (40 AU/mg) (pH 8, 37°C, 24 hours)	LPGY, LLPGY, **LLVY**, **ALVY**, VPAVL, **LLAPHY**, LGVF, ALLY, LLNY, ALEAY, VLPPLE, AEVF, LPPPPGL, **LSTSTDVR**	ACE-inhibitory: 70 μg/mL (IC_{50} fraction <3 kDa), 26% (RP-HPLC fraction)	Ngamsuk et al. (2020)

(Continued)

Table 15.1 Release of bioactive peptides from fruit waste proteins by in vitro GID. *Continued*

Fruit waste	Protein extraction conditions	Protein purification	GID conditions	Peptide sequence	Biological activity	References
Pumpkin oil cake	S–L extraction (water and NaCl solution (10%), 3 hours)	Water	Pepsin (pH 2.0, 37°C, 120 minutes, 1:25 E/S ratio)	–	ACE-inhibitory: 650 ± 20 μg/mL (IC_{50}) Antioxidant: 6.8% DPPH inhibition, 5.7% ABTS inhibition	Popović et al. (2017)
			Pepsin (pH 2.0, 37°C, 120 minutes, 1:25 E/S ratio), α-chymotrypsin (1:50 E/S ratio), and trypsin (1:12.5 E/S ratio) (pH 7.5, 37°C, 120 minutes)	–	ACE-inhibitory: 300 ± 40 μg/mL (IC_{50}) Antioxidant: 6.2% DPPH inhibition, 7.9% ABTS inhibition	
Watermelon seeds	S–L extraction under alkaline conditions (1 hour)	Isoelectric point	Pepsin (pH 2.2, 37°C, 5 hours, 1:100 E/S ratio)	–	ACE-inhibitory: 1.4 mg/mL (IC_{50}) Antioxidant: 1.4 mg/mL DPPH inhibition (EC_{50}), 0.95 mg/mL H_2O_2-scavenging (EC_{50})	Arise et al. (2016)
			Trypsin (pH 8.0, 37°C, 5 hours, 1:100 E/S ratio)	–	ACE-inhibitory: 1.4 mg/mL (IC_{50}) Antioxidant: 2.9 mg/mL DPPH inhibition (EC_{50}), 0.37 mg/mL H_2O_2-scavenging (EC_{50})	

Tomato paste	S–L extraction (PBS (pH 7.2), 40 minutes)	Ammonium sulfate (85%)	Pepsin (pH 1.2, 37°C, 60 minutes, 1:25 E/S ratio) and pancreatin (pH 7.6, 37°C, 60 minutes, 1:21 E/S ratio)	–	Antiangiogenic	Treggiari et al. (2017)

GID, gastrointestinal digestion; HIFU, high-intensity focused ultrasounds; PLE, pressurized liquid extraction; S–L extraction, solid–liquid extraction; SDS, sodium dodecyl sulfate; DTT, dithiothreitol; ChCl, choline chloride; AA, acetic acid; PBS, phosphate buffer saline; TCA, trichloroacetic acid; E/S ratio, enzyme/substrate ratio; ABTS, 2,2′-azino-bis(3-ethylbenzothiazoline-6-sulfonic acid) diammonium; ACE, angiotensin-I converting enzyme; CEase, cholesterol esterase; RP-HPLC, reversed-phase high-performance liquid chromatography; AU, activity unit; DPPH, 2,2-diphenyl-1-(2,4,6-trinitrophenyl) hydrazyl, IC_{50}, the half maximal inhibitory concentration; EC_{50}, effective concentration at 50% of inhibition.

Bold peptide sequences (LLVY, ALVY, LLAPHY, and LSTSTDVR) in Gac seeds were considered the most biologically active.

[a]Peptide sequences common to purified and nonpurified protein hydrolysates.

than *prunus* seeds (45%–63%) (García et al., 2016), but a higher ACE-inhibitory capacity, reaching up to 106 μg/mL (Čakarević, Vidović, Vladić, Jokić, et al., 2019). Higher activities (antioxidant capacity of 94% and $IC_{50} = 92$ μg/mL for antihypertensive capacity) were observed by the same research group for peptides released from apricot kernel cake (Čakarević, Vidović, Vladić, Gavarić, et al., 2019). Antioxidant and ACE-inhibitory peptides have been also obtained from olive seeds. In this case, activities (38% for ABTS inhibition and $IC_{50} = 150$ μg/mL for ACE-inhibitory activity) were lower than the obtained from some *prunus* fruits (García et al., 2016).

Pomegranate seeds have also been employed to release bioactive peptides after simulated GID. For that purpose, authors proposed different sustainable strategies for the extraction of proteins: PLE (Guzmán-Lorite et al., 2022a), HIFU under alkaline conditions (Guzmán-Lorite et al., 2022a), HIFU with a natural deep eutectic solvent (NADES) composed of choline chloride and acetic acid (Guzmán-Lorite et al., 2022b), and a successive extraction employing HIFU and PLE (Guzmán-Lorite et al., 2022b). In some cases, extracted proteins were purified with cold acetone before GID and results were compared with the obtained when proteins were nonpurified (Guzmán-Lorite et al., 2022a). Peptides released from HIFU-purified proteins and from PLE-purified proteins showed higher reducing ability than those obtained from nonpurified proteins. However, the ABTS inhibition was lower than the obtained in nonpurified protein hydrolysates. Authors justified these results due to the coextraction of phenolic compounds. The same group also studied the effect of the extracting solvent and the combination of extraction techniques on the antioxidant activity of released peptides. The extracting solvent affected to the antioxidant properties observing a higher inhibition (up to 65% of inhibition) in hydrolysates from proteins extracted with NADES than in those from proteins extracted with an alkaline solution (in both cases HIFU was employed). Moreover, the use of a different technique (PLE) to extract the proteins remaining in the pellet from the HIFU extraction enabled to have additional peptides with antioxidant properties (Guzmán-Lorite et al., 2022b). In addition to antioxidant properties, hydrolysates obtained from PLE and HIFU proteins also showed hypocholesterolemic and antihypertensive activities (Guzmán-Lorite et al., 2022a). Despite hydrolysates presented a high hypocholesterolemic capacity (85% inhibition), it was not clear the contribution of peptides since no significant differences were found, in most cases, when comparing hydrolysates obtained from purified and nonpurified proteins. Even some hypocholesterolemic inhibitions of hydrolysates were lower than the observed for intact samples (before GID). Regarding antihypertensive activity, peptides released from HIFU-purified proteins showed a high ACE inhibition (75%) although there were not significant differences with the protein extracts. Authors also employed mass spectrometry for the identification of peptides in hydrolysates. A total of 41 peptides were identified in the HIFU purified protein hydrolysate and 49 peptides in the PLE purified protein hydrolysate, where 11 peptides were common to both extraction techniques (YNPR, TPPK, REGDL, REEL,

QLPQ, WFLP, RQLV, EERVL, RLVP, REGDLL, and PSGRP). A lower amount of peptides were identified in the nonpurified extracts (see common peptides to purified and nonpurified protein hydrolysates included in Table 15.1). Only 2 (YNPR and REGDLL) out of the 41 peptides identified in the purified protein hydrolysate obtained by HIFU, under alkaline conditions, were common to the identified when using NADES as extracting solvent (Guzmán-Lorite et al., 2022a; Guzmán-Lorite et al., 2022b). As expected, PLE extraction after HIFU extraction using NADES, allowed to identify 15 additional peptides (Guzmán-Lorite et al., 2022b).

Another seeds that have been used to obtain antihypertensive peptides by simulated GID are those from longan fruit (Nuchprapha et al., 2020). Proteins were previously obtained using solid–liquid extraction followed by precipitation with ammonium sulfate. Peptides were fractionated by ultrafiltration observing the highest ACE inhibition in the fraction below 3 kDa ($IC_{50} = 250 \pm 4$ μg/mL). A subsequent fractionation by reversed-phase high-performance liquid chromatography (RP-HPLC) enabled to concentrate antihypertensive peptides in a fraction having a IC_{50} value of 18.22 ± 0.03 μg/mL. Two peptides (ETSGMKPTEL and ISSMGILVCL) were identified as responsible for this activity.

Proteins from gac seed and pumpkin oil cake were extracted by HIFU and conventional extraction, respectively. GID using pepsin, trypsin, and α-chymotrypsin was carried out to obtain peptides from these proteins. Gac seed protein hydrolysates were also ultrafiltered observing the highest ACE inhibition in the fraction below 3 kDa (Ngamsuk et al., 2020). This fraction showed higher antihypertensive activity ($IC_{50} = 70 \pm 4$ μg/mL) than the fraction obtained from longan seeds ($IC_{50} = 250 \pm 4$ μg/mL) (Nuchprapha et al., 2020). A total of 14 peptides were identified using both hydrophilic interaction chromatography (HILIC) and RP-HPLC coupled to mass spectrometry. Four of these peptides (ALVY, LLVY, LSTSTDVR, and LLAPHY) were considered ACE inhibitory peptide candidates after this verification using synthetic peptides (Ngamsuk et al., 2020). A lower antihypertensive activity was observed for the pumpkin oil cake protein hydrolysates. Moreover, authors demonstrated that peptides released in GID showed higher ACE inhibition ($IC_{50} = 300 \pm 40$ μg/mL) than peptides obtained in the first step (gastric digestion with pepsin) ($IC_{50} = 650 \pm 20$ μg/mL) (Popović et al., 2017). Furthermore, peptides in pumpkin oil cake hydrolysates also presented antioxidant properties associated to the radical scavenging activity.

Pepsin and trypsin were used for the digestion of proteins in watermelon seeds (Arise et al., 2016). They firstly extracted the proteins under alkaline conditions and purified them by isoelectric precipitation. They compared the capability of both digestion steps to release bioactive peptides. The hydrolysates obtained by gastric digestion showed a slightly higher ACE inhibition ($IC_{50} = 80 \pm 1\%$) than the obtained by intestinal digestion ($IC_{50} = 73.6 \pm 0.9\%$), but no significant differences were found in the 50% ACE-inhibitory concentrations (1.4 mg/mL for both hydrolysates). Furthermore, both hydrolysates presented antioxidant properties highlighting that from the pepsin digestion that showed the highest antiradical ability (EC_{50} = 1.4 mg/mL) and that from trypsin digestion with the strongest reducing ability (EC_{50} = 0.37 mg/mL).

In the case of the tomato waste, proteins were extracted by a conventional methodology and purified by precipitation with ammonium sulfate. Extracted proteins were as small as peptides (<3.5 kDa) and did not suffer any fragmentation when submitting to pepsin and pancreatin digestion (Treggiari et al., 2017). In addition to their resistance, their small size makes them to be able to pass through the intestine walls and exert their bioactivity. A further study allowed to observe the capability of this protein to inhibit vascular endothelial growth factor A (VEGFA)-induced cell migration, which is related with the prevention of pathological angiogenesis.

An additional waste of fruits is those units that do not reach the standard for their commercialization. Despite they have not been included in Table 15.1, it is important to remark the work carried out by Piscopo et al. (2020) using *Feijoa sellowiana* Berg fruit (Piscopo et al., 2020). After submitting feijoa proteins to an in vitro GID, resulting hydrolysates showed an antioxidant activity 19 times higher than the obtained for undigested proteins. Moreover, the hydrolysate also exhibited a higher antimicrobial activity than the intact proteins.

15.3.2 Bioactive peptides obtained from fruit waste hydrolysates by in vitro gastrointestinal digestion

Bioactive peptides can also be released by in vitro GID from hydrolysates previously obtained by the processing of fruit waste proteins with food-grade proteases, such as alcalase, thermolysin, pepsin, trypsin, or papain. In most cases, both processing-derived peptides and GID-derived peptides showed greater bioactivity than their parent proteins. Additionally, the previous enzymatic processing of proteins also resulted in high hydrolysis degrees and bioactivities. Moreover, results have demonstrated that the choice of the food protease is a critical factor since proteases differ in their active site specificity, catalytic mechanism, temperature, and optimal pH (Udenigwe et al., 2017). Table 15.2 summarizes the extraction conditions and purification of proteins obtained from fruit waste, the food-grade protease and conditions employed for the processing of proteins, the fractionation and purification of released peptides, the applied GID conditions, the sequences of peptides identified in the final hydrolysate, and their bioactivities.

The recovery of fruit waste proteins was always carried out by a conventional solid–liquid extraction. In some cases, the extraction was accelerated using HIFU. After extraction, proteins were purified by isoelectric precipitation or using cold acetone. Regarding food-grade proteases, alcalase followed by thermolysin were the most usually employed for the processing of fruit waste proteins. Alcalase was used for the enzymatic treatment of seacbuckthorn seed meal and pumpkin oil cake (Vaštag et al., 2013; Xiang et al., 2020). Processed residues were then submitted to the simulated GID. Twenty-two peptides were observed in

Table 15.2 Release of bioactive peptides by in vitro GID from hydrolysates obtained by the enzymatic processing of fruit waste proteins.

Fruit waste	Protein extraction conditions	Protein purification	Enzymatic processing	Fractionation/ purification	GID conditions	Peptide sequence	Biological activity	References
Seabuckthorn seed meal	S–L extraction under alkaline conditions	Isoelectric point	Alcalase (deionized water pH 10, 50°C, 8 hours, 1:400 E/S ratio)	–	Pepsin (pH 2.0, 37°C, 2 hours, 3% w/v), pancreatin (pH 6.8, 37°C, 4 hours, 4% w/v)	EEAASLR, FR, NLLHR, APYR, RDR, and VR	Antiobesity: 75% PL inhibition	Xiang et al. (2020)
Pumpkin oil cake	S–L extraction under alkaline conditions	Isoelectric point	Alcalase (pH 8, 50°C, 1 hour, 1:125 E/S ratio)	–	Pepsin (pH 2.1, 37°C, 30 minutes, 1:25 E/S ratio), α-chymotrypsin (1:50 E/S ratio) and trypsin (1:12.5 E/S ratio) (pH 7.5, 37°C, 89 minutes)	–	Antioxidant: 0.18 ± 0.03 mg TEAC (EC_{50}) ACE-inhibitory: 404 ± 14 μg/mL (IC_{50})	Vaštag et al. (2013)
Peach seeds	HIFU (100 mM Tris–HCl (pH 7.5) containing SDS (0.5%, w/v) and DTT (0.5%, w/v), 1 minute, 30% amplitude)	Cold acetone	Thermolysin (5 mM PB pH 8, 50°C, 4 hours, 0.1 g enzyme/g protein)	Ultrafiltration (cut-off filter <3 kDa)	Pepsin (pH 2, 37°C, 1 hour, 1:35 E/S ratio), pancreatin (pH 8, 37°C, 3 hours, 1:25 E/S ratio)	LYSPH, LYTPH, and HLLP	ACE-inhibitory: 18 ± 2 μg/mL (IC_{50})	Vásquez-Villanueva et al. (2015)

(Continued)

Table 15.2 Release of bioactive peptides by in vitro GID from hydrolysates obtained by the enzymatic processing of fruit waste proteins. *Continued*

Fruit waste	Protein extraction conditions	Protein purification	Enzymatic processing	Fractionation/ purification	GID conditions	Peptide sequence	Biological activity	References
Plum seeds	HIFU (100 mM Tris–HCl (pH 7.5) containing SDS (1%, w/v) and DTT (0.25%, w/v), 1 minute, 30% amplitude)	Cold acetone	Thermolysin (5 mM PB pH 8, 50°C, 4 hours, 0.1 g enzyme/g protein)	Ultrafiltration (cut-off filter <3 kDa)	Pepsin (pH 2, 37°C, 1 hour, 1:35 E/S ratio), pancreatin (pH 8, 37°C, 2 hours, 1:25 E/S ratio)	LYSPH, LYTPH, and LAGNPENE	ACE-inhibitory: 30 ± 3 μg/mL (IC_{50})	González-García et al. (2015)
			Alcalase (5 mM BB pH 8.5, 50°C, 3 hours, 0.3 AU/g protein)	–	Pepsin (pH 2, 37°C, 1 hour, 1:35 E/S ratio), pancreatin (pH 8, 37°C, 2 hours, 1:25 E/S ratio)	MLPSLPK, HLPLL, NLPLL, KGVL, HGVLQ, DQVPR, and NGGK	Antioxidant: 53 ± 4% ABTS inhibition	
Olive seeds	S–L extraction (125 mM Tris–HCl (pH 7.5) containing SDS (1%, w/v) and DTT (0.1%, w/v), 1 minute)	Cold acetone	Alcalase (5 mM PB pH 8, 50°C, 2 hours, 0.15 AU/g protein)	Ultrafiltration (cut-off filter <3 kDa)	Pepsin (pH 2, 37°C, 1 hour, 1:35 E/S ratio), pancreatin (pH 8, 37°C, 2 hours, 1:25 E/S ratio)	–	Antioxidant: TEAC value 382 ± 24	Esteve et al. (2015)
			Thermolysin (5 mM PB pH 8, 50°C, 2 hours, 0.05 g enzyme/g protein)	Ultrafiltration (cut-off filter <3 kDa)	Pepsin (pH 2, 37°C, 1 hour, 1:35 E/S ratio), pancreatin (pH 8, 37°C, 2 hours, 1:25 E/S ratio)	–	ACE-inhibitory: 36 μg/mL (IC_{50})	
	HIFU (100 mM Tris–HCl (pH	Cold acetone	Alcalase (5 mM BB pH 8.5, 50°C, 4 hours,	Ultrafiltration (cut-off filter >5 kDa) and	Pepsin (pH 2, 37°C, 1 hour, 1:35 E/S ratio),	–	Hypolipidemic: 53% CEase	Prados et al. (2018)

	7.5) containing SDS (0.5%, w/v) and DTT (0.5%, w/v), 5 minutes, 30% amplitude)		0.15 AU/g protein)	semipreparative RP-HPLC	pancreatin (pH 8, 37°C, 3 hours, 1:25 E/S ratio)		inhibition, 15% PL inhibition	
Bitter melon seeds	HIFU (PBS, 30 s pulse, 30% amplitude, 4 times)	TCA (10%) in acetone	Thermolysin (pH 8, 60°C, 12 hours, 1:100 E/S ratio)	Ultrafiltration (cut-off filter <3 kDa) and RP-HPLC	Pepsin (pH 1.5, 37°C, 120 minutes, 1:50 E/S ratio), trypsin (1:20 E/S ratio) and α-chymotrypsin (1:20 E/S ratio) (pH 8, 37°C, 5 hours)	VSGAGRY	ACE-inhibitory: 8 μM (IC_{50})	Priyanto et al. (2015)
Tomato seeds	S–L extraction under alkaline conditions (1 hour)	Isoelectric point	Alcalase (PB 1:5 (w/v) pH 8, 50°C, 138.62 minutes, 3% E/S ratio)	Ultrafiltration (cut-off filter <3 kDa) and RP-HPLC	Pepsin (pH 3, 37°C, 2 hours, 2000 U/mL) and pancreatin (pH 7, 37°C, 2 hours, 100 U/mL)	–	Antioxidant	Meshginfar et al. (2018)
Coconut cake	S–L extraction (0.4 M NaCl, 1 hour, 4°C) High pressure (400 MPa, 15 minutes, 35°C)	–	Alcalase (pH 8.0, 45°C, 2 hours, 1:200 E/S ratio), flavourzyme (pH 7.0, 50°C, 2 hours, 1:200 E/S ratio),	Ultrafiltration (cut-off filter <3 kDa), SEC, semipreparative RP-HPLC, RP-HPLC	Pepsin (pH 2, 37°C, 1 hour, 1:35 E/S ratio) and pancreatin (pH 7, 37°C, 2 hours, 1:25 E/S ratio)	PQFYW and VVLYR	ACE-inhibitory: 116–257 μg/mL (IC_{50})	Li et al. (2016)
	S–L extraction (water, 1 hour, 4°C)		pepsin (pH 2.0, 37°C, 1 hour, 1:333 E/S ratio), and trypsin (pH 7.0, 37°C,	Ultrafiltration (cut-off filter <3 kDa), SEC,	Pepsin (pH 2, 37°C, 1 hour, 1:35 E/S ratio) and pancreatin	KAQYPYV	Antioxidant: 72 μM H_2O_2-scavenging	Zheng et al. (2019)

(*Continued*)

Table 15.2 Release of bioactive peptides by in vitro GID from hydrolysates obtained by the enzymatic processing of fruit waste proteins. *Continued*

Fruit waste	Protein extraction conditions	Protein purification	Enzymatic processing	Fractionation/ purification	GID conditions	Peptide sequence	Biological activity	References
			1 hour, 1:333 E/S ratio)	semipreparative RP-HPLC	(pH 7, 37°C, 2 hours, 1:25 E/S ratio)		ACE-inhibitory: 38 μM (IC_{50})	
Palm kernel	S–L extraction (0.4 M NaCl, 1 hour, 4°C and 0.1 M NaOH, 1 hour, 4°C). High pressure (300 MPa, 20 minutes, 35°C)	–	Alcalase (pH 8.5, 45°C, 2 hours, 1:200 E/S ratio), flavourzyme (pH 7.0, 50°C, 2 hours, 1:200 E/S ratio), pepsin (pH 2.0, 37°C, 1 hour, 1:333 E/S ratio) and trypsin (pH 7.0, 37°C, 1 hour, 1:333 E/S ratio)	Ultrafiltration (cut-off filter <3 kDa), SEC, semipreparative RP-HPLC, RP-HPLC	Pepsin (pH 2, 37°C, 1 hour, 1:35 E/S ratio) and pancreatin (pH 7, 37°C, 2 hours, 1:25 E/S ratio)	VIEPDR, LPILR, ADVFNPR, VVLYK	ACE-inhibitory: 502–818 μM (IC_{50})	Zheng et al. (2017)
Jackfruit seeds	S–L extraction (water, 30 minutes at 4°C, and 30 minutes at 90°C)	Ammonium sulfate (80%)	Trypsin (50 mM PB, 37°C, 5 hours, 1:10 E/S ratio)	Ultrafiltration (cut-off filter <3 kDa), C18-SPE and SCX-SPE	Pepsin (pH 2, 37°C, 1 hour, 1:35 E/S ratio), pancreatin (pH 7, 37°C, 2 hours, 1:25 E/S ratio)	VGPWQK	Antioxidant: 0.9 mg/mL ABTS inhibition (EC_{50}).	Chai et al. (2021)

GID, gastrointestinal digestion; S–L extraction, solid–liquid extraction; HIFU, high-intensity focused ultrasounds; SDS, sodium dodecyl sulfate; DTT, dithiothreitol; PBS, phosphate buffer saline; TCA, trichloroacetic acid; E/S ratio, enzyme/substrate ratio; PL, pancreatic lipase; TEAC, trolox equivalent antioxidant; ACE, angiotensin-I converting enzyme; PB, phosphate buffer; BB, borate buffer; ABTS, 2,2′-azino-bis(3-ethylbenzothiazoline-6-sulfonic acid) diammonium; AU, activity unit; SEC, size-exclusion chromatography; RP-HPLC, reversed-phase high-performance liquid chromatography; CEase, cholesterol esterase; SPE, solid phase extraction; SCX, strong cation exchange; IC_{50}, the half maximal inhibitory concentration; EC_{50}, effective concentration at 50% of inhibition.

the hydrolysate from the seacbuckthorn seed meal. Six of them (EEAASLR, FR, NLLHR, APYR, RDR, and VR) were highlighted due to their potential antiobesity properties (Xiang et al., 2020). No peptide identification was carried out in the case of the pumpkin oil cake although they showed antioxidant and ACE-inhibitory properties, especially after intestinal digestion (Vaštag et al., 2013).

In some cases, peptides released during enzymatic processing were fractionated in order to isolate most active ones. Ultrafiltration through 3 kDa cut-off filters, semipreparative size exclusion chromatography (SEC) and reversed-phase high-performance liquid chromatography (RP-HPLC), and solid-phase extraction with reversed-phase or cation-exchange cartridges have been used with this purpose. *Prunus* fruit waste, olive seeds, bitter melon seeds, tomato seeds, coconut cake, palm kernel, and jackfruit seeds were submitted to different fractionations and purifications before GID.

The potential of *prunus* fruit waste to obtain antioxidant and antihypertensive peptides after GID had been observed (see Table 15.1) (Čakarević, Vidović, Vladić, Gavarić, et al., 2019; Čakarević, Vidović, Vladić, Jokić, et al., 2019; García et al., 2016). Nevertheless, a previous enzymatic treatment with thermolysin followed by ultrafiltration allowed to increase the ACE inhibition of peptides obtained from peach and plum seed proteins after GID ($IC_{50} = 18 \pm 2$ μg/mL for peach seeds (Vásquez-Villanueva et al., 2015) and $IC_{50} = 30 \pm 3$ μg/mL for plum seeds (González-García et al., 2015)). Only three peptides, in each case, resisted the simulated GID (Fig. 15.3A). Moreover, two of these peptides (LYSPH and LYTPH) were common to both residues. Plum seed proteins were also hydrolyzed with alcalase before GID showing a $53 \pm 4\%$ of ABTS inhibition after GID (González-García et al., 2015). Seven peptides resisted the GID (MLPSLPK, HLPLL, NLPLL, KGVL, HGVLQ, DQVPR, and NGGK), while eight additional peptides were obtained after GID (Fig. 15.3B).

Olive seeds also constitute a potential source of antioxidant and antihypertensive peptides (García et al., 2016). Like *prunus* seeds, most active peptides were obtained by the previous treatment of proteins with alcalase and thermolysin and subsequent ultrafiltration (Esteve et al., 2015). It is worth highlighting the ACE-inhibitory activity of peptides which varied from $IC_{50} = 150$ μg/mL (GID from olive seed proteins) to $IC_{50} = 36$ μg/mL (GID after thermolysin hydrolysis) (Esteve et al., 2015). Antioxidant activity was determined in the hydrolysates obtained from GID after alcalase hydrolysis. Moreover, olive seeds also presented lipid-lowering capacity (Prados et al., 2018). The fraction with peptides with molecular weights higher than 5 kDa showed a 46% increase in the capacity to inhibit cholesterol esterase after GID, while there were not significant differences in the pancreatic lipase inhibition before and after GID (15% of inhibition).

Bitter melon seeds have also been employed for obtaining peptides with high ACE inhibition ($80 \pm 4\%$) by processing with thermolysin. Despite it was possible to identify two main peptides in the hydrolysate (VSGAGRY and VDSDVVKG), only peptide VSGAGRY was considered as a potential antihypertensive candidate due to its high ACE-inhibitory capacity ($IC_{50} = 8$ μM) and its resistance to GID (Priyanto

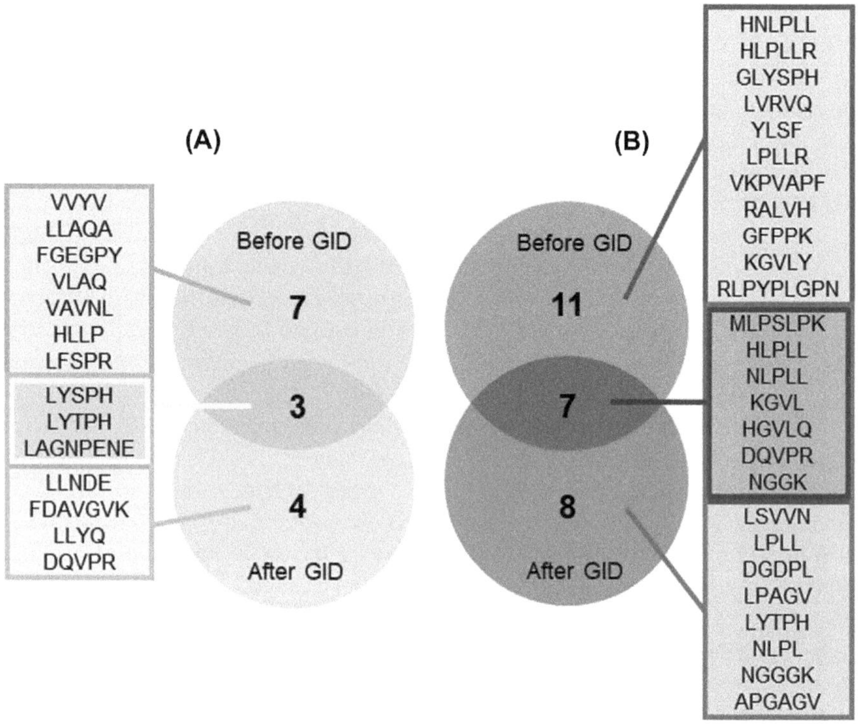

FIGURE 15.3 Venn diagram comparing the plum seed-derived peptides obtained before and after GID.

(A) Thermolysin hydrolysates and (B) alcalase hydrolysates (González-García et al., 2015).

et al., 2015). Tomato seed proteins processed with alcalase also showed a high resistance to GID keeping its high antioxidant properties (Meshginfar et al., 2018).

Coconut and palm cakes are residues from the coconut and palm oil extraction, respectively. They have been processed sequentially with different enzymes (alcalase, thermolysin, flavourzyme, pepsin, and trypsin) to evaluate the release of antioxidant and ACE-inhibitory peptides. Both residues were submitted to different pretreatments (heating, ultrasonication, and high pressure) before hydrolysis with food-grade enzymes. The highest ACE inhibition was observed when using high pressure (pretreatment shown in Table 15.2) being 52% and 80% the ACE inhibition for coconut cake and palm kernel, respectively (Li et al., 2016; Zheng et al., 2017). Moreover, the bioactivities obtained by the coconut cake hydrolysate were compared to those obtained for each hydrolysate using a single enzyme (Zheng et al., 2019). Unspecific enzymes, such as alcalase and flavourzyme, increased the hydrolysis degree (between 6% and 18%) and showed improved bioactivities (up to 25%) compared with pepsin and trypsin. Furthermore, the sequential hydrolysis with all enzymes enabled to reach a 48% of ACE inhibition and a 55% of hydroxyl radical-scavenging. Hydrolysates were also fractionated by ultrafiltration and semi-preparative SEC and RP-HPLC. After fractionation, IC_{50} values of ACE-inhibitory peptides changed up to 104 μg/mL or 37 μM for coconut cake (Li et al., 2016; Zheng et al., 2019) and 486 μM for palm kernel (Zheng et al., 2017). These results did not show a significant decrease after GID meaning that antihypertensive peptides had good stability to GID conditions (Li et al., 2016; Zheng et al. 2017, 2019). However, ACE-inhibition was lower than the obtained with *prunus* and olive seeds peptides. Some significant peptides were observed in both the coconut cake globulin (Li et al., 2016) and albumin fractions (Zheng et al., 2019) and in the palm kernel (Zheng et al., 2017). Especially interesting was peptide KAQYPYV, from the albumin fraction, since it showed simultaneously antihypertensive properties and reducing ability (Zheng et al., 2019).

Jackfruit seeds contain many small seeds that have been processed with trypsin and fractionated to obtain peptides with antioxidant ability. Nevertheless, not all peptides were resistant to GID, and, in some cases, peptides could be degraded. For example, VGPWQK peptide showed a 5.43-fold lower antioxidant capacity than the obtained before GID (Chai et al., 2021). This peptide may be more suitable as a food preservative agent than as an orally ingested promoting diet supplement.

15.3.3 Bioactive peptides obtained from fruit wastes by in vivo gastrointestinal digestion

Once demonstrated that peptides released from fruit waste proteins have bioactive properties by in vitro assays, many authors have also confirmed this bioactivity by an in vivo GID. The in vivo activity of a bioactive peptide depends on its capacity to reach target organs in an intact conformation. Therefore peptides have to resist the GID to show their bioactivity (Ahmed et al., 2022). Table 15.3 groups the in vivo experiments carried out to demonstrate the antihypertensive, hypolipidemic, and antiplatelet activities of peptides released from fruit waste.

Antihypertensive activity of peptides released from coconut cake (Li et al., 2016), palm kernel (Zheng et al., 2017), peach seeds (Vásquez-Villanueva et al., 2019), and bitter melon (Priyanto et al., 2015) were studied using spontaneously hypertensive rats (SHRs), while normotensive rats were employed in the study of antihypertensive peptides from palm kernel cake (Zarei et al., 2015). In this work, angiotensin I was injected into the rats to produce an increase in blood pressure. The number of animals (male) in every study ranged from 16 to 30. They were divided into groups of four to six animals. Animals aged from 8 to 40 weeks with a body weight up to 382 g were kept at 23°C–25°C under 12 hours light/dark cycles. Antihypertensive treatment was carried out during a time ranging from 90 minutes to 6 weeks. Administered doses were varied from 2 to 200 mg/kg and, in some case, authors tried different doses. Systolic blood pressure (SBP) reductions up to 40 mmHg observing the highest reduction with the hydrolysate from the bitter melon seeds when using the lowest dose

Table 15.3 In vivo GID experiments to evaluate the bioactivity of peptides released from fruit waste.

Fruit waste	Protein extraction conditions	Protein purification	Enzymatic processing	Fractionation and purification processes	Animal characteristics	Doses (mg/kg)	Treatment time	Biological activity	SBP or TC reduction	Reference
Coconut cake	S–L extraction (0.4 M NaCl, 1 hour, 4°C) High pressure (400 MPa, 15 minutes, 35°C)	–	Alcalase (pH 8, 45°C, 2 hours, 1:200 E/S ratio), flavourzyme (pH 7, 50 °C, 2 hours, 1:200 E/S ratio), pepsin (pH 2, 37 °C, 1 hour, 1:333 E/S ratio), and trypsin (pH 7, 37°C, 1 hour, 1:333 E/S ratio)	Ultrafiltration (cut-off filter <3 kDa) and SEC	5 male SHRs aged 8 weeks	100–200	6 weeks	ACE inhibitory	SBP: 25 mmHg	Li et al. (2016)
Palm kernel	S–L extraction (0.4 M NaCl, 1 hour, 4°C and 0.1 M NaOH, 1 hour, 4°C) High pressure (300 MPa,	–	Alcalase (pH 8, 45°C, 2 hours, 1:200 E/S ratio), flavourzyme (pH 7, 50°C, 2 hours, 1:200 E/S ratio), pepsin (pH	Ultrafiltration (cut-off filter <3 kDa) and SEC	5 male SHRs aged 8 weeks	100–200	6 weeks	ACE-inhibitory	SBP: 29 mmHg	Zheng et al. (2017)

	20 minutes, 35°C)		2, 37°C, 1 hour, 1:333 E/S ratio), and trypsin (pH 7, 37°C, 1 hour, 1:333 E/S ratio)							
Palm kernel cake	S–L extraction under alkaline conditions (2 hours)	Isoelectric point	Papain (50 mM PB, pH 6.5, 65°C, 30 hours, 1:50 E/S ratio)	–	5 male normotensive rats aged 12 weeks, and 280 ± 20 g body weight.	75	90 minutes	ACE inhibitory	SBP: no significant reduction	Zarei et al. (2015)
Peach seeds	HIFU (100 mM Tris–HCl (pH 7.5) containing SDS (0.5%, w/v) and DTT (0.5%, w/v), 1 minute, 30% amplitude)	Cold acetone	Thermolysin (5 mM PB pH 8, 50°C, 4 hours, 0.1 g enzyme/g protein)	Ultrafiltration (cut-off filter <3 kDa)	6 male SHRs aged 25 – 40 weeks and 382 ± 43 g	10	3–6 hours	ACE inhibitory	SBP: 30 mmHg	Vásquez-Villanueva et al. (2019)
Bitter melon seeds	HIFU (PBS, 30 s pulse, 30% amplitude, 4 times)	TCA (10%) in acetone	Thermolysin (pH 8, 60°C, 12 hours, 1:100 E/S ratio)	Ultrafiltration (cut-off filter <3 kDa) and semipreparative RP-HPLC	4 male SHRs, aged 10 weeks, and 180–240 g body weight	2	8 hours	ACE inhibitory	SBP: 40 mmHg	Priyanto et al. (2015)

(*Continued*)

Table 15.3 In vivo GID experiments to evaluate the bioactivity of peptides released from fruit waste. *Continued*

Fruit waste	Protein extraction conditions	Protein purification	Enzymatic processing	Fractionation and purification processes	Animal characteristics	Doses (mg/kg)	Treatment time	Biological activity	SBP or TC reduction	Reference
Olive seeds	HIFU (100 mM Tris–HCl (pH 7.5) containing SDS (0.5%, w/v) and DTT (0.5%, w/v), 5 minutes, 30% amplitude)	Cold acetone	Alcalase (5 mM BB pH 8.5, 50°C, 4 hours, 0.15 AU/g protein)	–	9 – 11 male and female mice	200	11 weeks	Hypolipidemic activity	TC: 20%	Prados et al. (2020)
					5 – 6 male and female mice	400	8 weeks	Hypolipidemic activity	TC: 25%	
Pineapple core	–	–	–	–	Male mice	0.21	3 weeks	Antiplatelet activity	–	Musfiroh et al. (2018)

S–L extraction, solid–liquid extraction; HIFU, high-intensity focused ultrasounds; SDS, sodium dodecyl sulfate; DTT, dithiothreitol; PBS, phosphate buffer saline; NaOH, sodium hydroxide; TCA, trichloroacetic acid; SBP: systolic blood pressure; TC, total cholesterol; SHRs, spontaneously hypertensive rats; E/S ratio, enzyme/substrate ratio; ACE, angiotensin-I converting enzyme; PB, phosphate buffer; BB, borate buffer; AU, activity unit; RP-HPLC, reversed-phase high-performance liquid chromatography.

(2 mg/kg) (Priyanto et al., 2015). These results demonstrated the significant potential of fruit waste for the production of bioactive peptides with high hypotensive properties.

In addition to antihypertensive properties, other abilities demonstrated by in vivo experiments were the hypolipidemic and antiplatelet capacities. The capacity of olive seed hydrolysates to reduce serum cholesterol in mice was evaluated at two different concentration levels (200 and 400 mg/kg/day) for 4–11 weeks (Prados et al., 2020). Total cholesterol (TC) level of male mice increased with the intake of a high cholesterol diet after 4 weeks; however, the administration of a 200 mg/kg dose of the olive seed hydrolysates reduced serum TC in a 17% and a 20%, after 5 and 11 weeks of administration, respectively. A higher olive seed hydrolysate dose (400 mg/kg) decreased serum TC in a 25%, after 8 weeks of treatment. This effect was attributed to the increase in high-density lipoproteins (HDL-C) and the decrease in low-density lipoproteins (LDL-C) in serum. Moreover, this work related this hypolipidemic effect to the capability showed by the olive seed hydrolysate in in vitro assays to reduce the micellar cholesterol solubility and to inhibit important enzymes (pancreatic lipase, cholesterol esterase, and 3-hydroxy-3-methylglutaryl coenzyme A (HMG-CoA) reductase) involved in the biosynthesis and absorption of cholesterol. ADLY, FLPH, KLPLL, and TLVY peptides contained in olive seed hydrolysate could be responsible for the observed hypolipidemic effect.

Antiplatelet activity exerted by bromelain enzyme extracted from pineapple core was measured according to the bleeding time of mice (Musfiroh et al., 2018). The administration of a 210 μg/kg dose of pineapple core extracts to male mice for 3 weeks increased the bleeding time in 515 ± 182%. This result was more significant than the observed by the administration of aspirin (231 ± 141%) and confirmed the potential of this waste to prevent thrombotic cerebrovascular and cardiovascular diseases.

Considering that some fruits are rejected for their commercialization and that they can also be considered waste, in vivo studies using the fruit flesh have been included in this section, although they are not in Table 15.3. The oral administration of 10 mg bitter gourd protein extracts per kg of diabetic rats allowed to significantly reduce the blood glucose up to 24% after 120 minutes. This glucose-lowering effect was attributed to the α-amylase and α-glucosidase inhibition observed in in vitro assays (Poovitha & Parani, 2016). Other example is the obtained with tomato miniproteins resisting simulated GID. The administration of these proteins (500 nM) to zebrafish embryos inhibited the formation of the subintestinal vessels associated to angiogenesis. Taking into account that angiogenesis is required for cancer development and progression, the use of these tomato miniproteins can be an important strategy to avoid it (Treggiari et al., 2015).

15.4 Bioactivity of peptides

Bioactive peptides are well-known to possess numerous functional properties that promote health. BIOPEP database reports about 4611 bioactive peptides that have

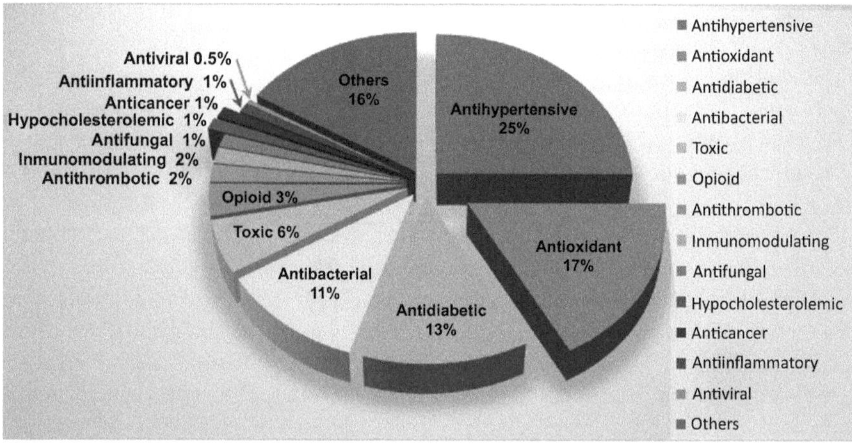

FIGURE 15.4 Distribution of bioactive peptides according to their activity.

BIOPEP database collects about 4611 bioactive peptides identified to date (Minkiewicz et al., 2008).

been identified to date (Fig. 15.4) (Minkiewicz et al., 2008). Peptides with antihypertensive, antioxidant, and antidiabetic activities are the most highlighted. In fact, ACE-inhibition and antioxidant capacity are the main bioactivities of peptides released from fruit waste proteins after GID.

15.4.1 Antihypertensive activity

Hypertension (high blood pressure) is characterized by a persistent SBP value higher than 140 mmHg and a diastolic blood pressure (DBP) value higher than 90 mmHg (Daliri et al., 2017). ACE plays a significant role in the Renin-Angiotensin System which regulates blood pressure in humans by converting angiotensin I into angiotensin II, a potent vasoconstrictor, and deactivating the vasodilator bradykinin (Jogi et al., 2022; Shahidi & Zhong, 2008). Peptides with a short size (2–12 amino acids) and a high content in hydrophobic and aromatic amino acids residues are considered as potentially antihypertensive candidates since they can inhibit the ACE. Indeed, a high presence of leucine (L), proline (P), and tyrosine (Y) amino acids was found in peptide sequences obtained from *prunus* seeds (LYSPH, LYTPH, HLLP, and LAGNPENE) (González-García et al., 2015; Vásquez-Villanueva et al., 2015). An additional hydrophobic amino acid, like alanine (A), appeared in numerous peptides from Gac seeds (LPGY, LLPGY, LLVY, ALVY, VPAVL, LLAPHY, LGVF, ALLY, LLNY, ALEAY, VLPPLE, AEVF, LPPPPGL, and LSTSTDVR) (Ngamsuk et al., 2020). Moreover, valine (V) amino acid residue was highlighted in peptides from bitter melon seeds (VSGAGRY and VDSDVVKG) (Priyanto et al., 2015), coconut cake (PQFYW and VVLYR) (Li et al., 2016), and palm kernel (VIEPDR, LPILR, ADVFNPR, and VVLYK) (Zheng et al., 2017). Other hydrophobic amino acids, such as methionine (M) and isoleucine (I), were contained in peptides from longan seeds (ETSGMKPTEL and ISSMGILVCL) (Nuchprapha et al., 2020). All these peptides released from fruit waste proteins can be considered potential ACE inhibitory peptides.

15.4.2 Antioxidant activity

Living organisms are constantly exposed to reactive oxygen species (ROS) because of the respiration and the cellular metabolism. Nevertheless, the production of these free radicals could increase by exposition to environmental pollutants, smoking, radiation, drugs, ozone, pesticides, and industrial solvents. An imbalance between ROS and endogenous antioxidant defenses can lead to oxidative stress, which has been related to aging, cancer, heart disease, stroke, arteriosclerosis, and diabetes, among other diseases. Protection against free radicals can be promoted by the intake of dietary antioxidants, such as antioxidant peptides (Gulcin, 2020). Peptides containing histidine (H), tyrosine (Y), tryptophan (W), methionine (M), cysteine (C), and proline (P) have been significantly correlated with radical quenching activity (Erdmann et al., 2008; Xiong, 2010). As examples, peptides released from plum seeds (MLPSLPK, HLPLL, NLPLL, KGVL, HGVLQ, DQVPR, and NGGK) (González-García et al., 2015), pomegranate seeds (MVKGH, ERERPSRP, RLHVLR, RALP, LLSSPRR, WSPR, QLRVPY, RPSPR, AYLAPH, and RGRPPPR, among others) (Guzmán-Lorite et al., 2022b), and coconut cake (KAQYPYV) (Zheng et al., 2019) presented antioxidant properties by multiple mechanisms, including radical scavenging and/or reducing abilities.

15.4.3 Antidiabetic activity

Diabetes is a metabolic disorder characterized by an increase of blood sugar level due to insufficiencies in insulin. This disease is classified into type I and type II. Type I diabetes mellitus is an autoimmune disease characterized by a selective and progressive loss of insulin secretion by pancreatic β-cells. Type II diabetes mellitus can be attributed to an imbalance between insulin secretion and glucose absorption (Daliri et al., 2017; Kumar et al., 2022). Antidiabetic peptides can avoid high blood glucose levels by the inhibition of specific enzymes, such as α-amylase, α-glucosidase, and dipeptidyl peptidase IV (Acquah et al., 2022). Peptides released from bitter gourd after in vivo GID demonstrated a glucose-lowering effect in diabetic rats probably due to multiple mechanisms of enzyme inhibition (Poovitha & Parani, 2016).

15.4.4 Hypocholesterolemic activity

Cholesterol is necessary for the production of vitamin D, steroid hormones, and bile acids; however, an excessive blood cholesterol can lead to cardiovascular diseases and arteriosclerosis (Daliri et al., 2017). Peptides with cholesterol-lowering ability usually present high amounts of glutamic acid (E) and aspartic acid (D) and low ratios of methionine–glycine and lysine–arginine (Erdmann et al., 2008). Reduction of blood cholesterol levels can be affordable by different mechanisms: reduction of the micellar cholesterol solubility, inhibition of pancreatic cholesterol esterase enzyme, inhibition of pancreatic lipase enzyme, binding of bile acids, and inhibition of HMG-CoA reductase. For instance, peptides obtained from seabuckthorn seed meal (EEAASLR, FR, NLLHR, APYR, RDR, and VR) showed pancreatic lipase inhibitory ability (Xiang et al., 2020). Additionally, hydrolysates obtained from pomegranate and olive seeds showed capacity to inhibit cholesterol esterase and pancreatic lipase, and/or to reduce micellar cholesterol solubility (Guzmán-Lorite et al., 2022a; Prados et al., 2018). Hypolipidemic properties of peptides from olive seeds were confirmed by in vivo experiments (Prados et al., 2020).

15.5 Concluding remarks

Fruits are sources of many residues including peel, pomace, and seeds. Due to the high protein content of fruit seeds and pomace, these residues have been preferred to obtain bioactive peptides. The release of bioactive peptides depends on the fruit waste protein content, the technique, and solvent employed for the extraction of proteins, the amino acid composition of the proteins, and the conditions of digestion. Peptides released from fruit waste proteins after simulated GID showed mainly antihypertensive and antioxidant properties. These biological activities can be improved when proteins are previously processed with food-grade proteases before GID. Alcalase and thermolysin have been the most employed proteases for this purpose. Fractionation and purification of hydrolysates have also enabled to isolate most active fractions. In many cases, the exerted biological activities could be related to specific peptides and some of them could resist GID. The oral administration of protein extracts and hydrolysates obtained from fruit waste reduced the systolic blood pressure in hypertensive rats and the serum total cholesterol level in mice, confirming the bioactivity and bioavailability of peptides released after in vivo GID. Therefore, some peptides from fruit waste can be considered bioactive with ACE-inhibitory capacity, reducing abilities, radical scavenging capacity, and/or hypocholesterolemic capacity. These peptides could be framed as functional ingredients and their incorporation into food, pharmaceutical, and cosmetic products could be very useful to promote health.

Acknowledgments

The authors thank the Spanish Ministry of Science and Innovation (ref. PID2020–114891RB-I00) and the Comunidad of Madrid and European funding from FSE and FEDER programs (S2018/BAA-4393, AVANSECAL-II-CM). E.H.C. thanks the University of Alcalá for her postdoctoral contract.

Conflict of interest

Authors declare no conflict of interest.

References

Acquah, C., Dzuvor, C. K. O., Tosh, S., & Agyei, D. (2022). Anti-diabetic effects of bioactive peptides: Recent advances and clinical implications. *Critical Reviews in Food Science and Nutrition*, *62*(8), 2158–2171. Available from https://doi.org/10.1080/10408398.2020.1851168, https://www.tandfonline.com/loi/bfsn20.

Ahmed, T., Sun, X., & Udenigwe, C. C. (2022). Role of structural properties of bioactive peptides in their stability during simulated gastrointestinal digestion: A systematic review. *Trends in Food Science and Technology*, *120*, 265–273. Available from https://doi.org/10.1016/j.tifs.2022.01.008, http://www.elsevier.com/wps/find/journaldescription.cws_home/601278/description#description.

Arise, R. O., Yekeen, A. A., Ekun, O. E., & Olatomiwa, O. J. (2016). Protein hydrolysates from Citrullus lanatus seed: Antiradical and hydrogen peroxide-scavenging properties and kinetics of angiotensin-I converting enzyme inhibition. *Ceylon Journal of Science*, *45*(2), 39. Available from https://doi.org/10.4038/cjs.v45i2.7387.

Batool, M., Ranjha, M. M. A. N., Roobab, U., Manzoor, M. F., Farooq, U., Nadeem, H. R., Nadeem, M., Kanwal, R., Abdelgawad, H., Al Jaouni, S. K., Selim, S., & Ibrahim, S. A. (2022). Nutritional value, phytochemical potential, and therapeutic benefits of pumpkin (Cucurbita sp.). *Plants*, *11*(11). Available from https://doi.org/10.3390/plants11111394, https://www.mdpi.com/2223-7747/11/11/1394/pdf?version = 1653452349.

Cano-Lamadrid, M., Martínez-Zamora, L., Castillejo, N., & Artés-Hernández, F. (2022). From pomegranate byproducts waste to worth: A review of extraction techniques and potential applications for their revalorization. *Foods*, *11*(17), 2596. Available from https://doi.org/10.3390/foods11172596.

Cassani, L., & Gomez-Zavaglia, A. (2022). Sustainable food systems in fruits and vegetables food supply chains. *Frontiers in Nutrition*, *9*. Available from https://doi.org/10.3389/fnut.2022.829061, http://journal.frontiersin.org/journal/nutrition.

Čakarević, J., Vidović, S., Vladić, J., Gavarić, A., Jokić, S., Pavlovíc, N., Blažíc, M., & Popovíc, L. (2019). Production of bio-functional protein through revalorization of apricot kernel cake. *Foods*, *8*(8). Available from https://doi.org/10.3390/foods8080318, https://www.mdpi.com/2304-8158/8/8/318/pdf.

Čakarević, J., Vidović, S., Vladić, J., Jokić, S., Pavlović, N., & Popović, L. (2019). Plum oil cake protein isolate: A potential source of bioactive peptides. *Food and Feed Research*, *46*(2), 171–178. Available from https://doi.org/10.5937/ffr1902171c.

Chai, T. T., Xiao, J., Mohana Dass, S., Teoh, J. Y., Ee, K. Y., Ng, W. J., & Wong, F. C. (2021). Identification of antioxidant peptides derived from tropical jackfruit seed and investigation of the stability profiles. *Food Chemistry*, *340*. Available from https://doi.org/10.1016/j.foodchem.2020.127876, http://www.elsevier.com/locate/foodchem.

De Ancos, B., Colina-Coca, C., González-Peña, D., & Sánchez-Moreno, C. (2015). Bioactive compounds from vegetable and fruit by-products. *Biotechnology of Bioactive Compounds: Sources and Applications*. Available from https://doi.org/10.1002/9781118733103.ch1, http://www.wiley.com/remtitle.cgi?isbn = 1118733495.

Daliri, E. B. M., Oh, D. H., & Lee, B. H. (2017). Bioactive peptides. *Foods*, *6*(5), 1–21. Available from https://doi.org/10.3390/foods6050032, https://www.mdpi.com/2304-8158/6/5/32.

Das, A. K., Nanda, P. K., Chowdhury, N. R., Dandapat, P., Gagaoua, M., Chauhan, P., Pateiro, M., & Lorenzo, J. M. (2021). Application of pomegranate by-products in muscle foods: Oxidative indices, colour stability, shelf life and health benefits. *Molecules (Basel, Switzerland)*, *26*(2). Available from https://doi.org/10.3390/MOLECULES26020467, https://www.mdpi.com/1420-3049/26/2/467/pdf.

Dhiman, A. K., Sharma, K. D., & Attri, S. (2009). Functional constituents and processing of pumpkin: A review. *Journal of Food Science and Technology*, *46*(5), 411–417.

Elfalleh, W., Hannachi, H., Guetat, A., Tlili, N., Guasmi, F., Ferchichi, A., & Ying, M. (2012). Storage protein and amino acid contents of Tunisian and Chinese pomegranate (Punica granatum L.) cultivars. *Genetic Resources and Crop Evolution*, *59*(6), 999–1014. Available from https://doi.org/10.1007/s10722-011-9739-9, http://www.wkap.nl/journalhome.htm/0925-9864.

Erdmann, K., Cheung, B. W. Y., & Schröder, H. (2008). The possible roles of food-derived bioactive peptides in reducing the risk of cardiovascular disease. *Journal of Nutritional Biochemistry*, *19*(10), 643–654. Available from https://doi.org/10.1016/j.jnutbio.2007.11.010.

Esteve, C., Marina, M. L., & García, M. C. (2015). Novel strategy for the revalorization of olive (Olea europaea) residues based on the extraction of bioactive peptides. *Food Chemistry*, *167*, 272–280. Available from https://doi.org/10.1016/j.foodchem.2014.06.090, http://www.elsevier.com/locate/foodchem.

Fruit and Vegetables – Your Dietary Essentials. (2020). *The international year of fruits and vegetables*. Food and Agriculture Organization of the United Nations. Available from http://doi.org/10.4060/cb2395en.

García, M. C., González-García, E., Vásquez-Villanueva, R., & Marina, M. L. (2016). Apricot and other seed stones: Amygdalin content and the potential to obtain antioxidant, angiotensin i converting enzyme inhibitor and hypocholesterolemic peptides. *Food and Function*, *7*(11), 4693–4701. Available from https://doi.org/10.1039/c6fo01132b, http://www.rsc.org/Publishing/Journals/FO/about.asp.

González-García, E., Puchalska, P., Marina, M. L., & García, M. C. (2015). Fractionation and identification of antioxidant and angiotensin-converting enzyme-inhibitory peptides obtained from plum (Prunus domestica L.) stones. *Journal of Functional Foods*, *19*, 376–384. Available from https://doi.org/10.1016/j.jff.2015.08.033, http://www.elsevier.com/wps/find/journaldescription.cws_home/717426/description#description.

Gulcin, İ. (2020). Antioxidants and antioxidant methods: An updated overview. *Archives of Toxicology*, *94*(3), 651–715. Available from https://doi.org/10.1007/s00204-020-02689-3, http://link.springer.de/link/service/journals/00204/index.htm.

Guzmán-Lorite, M., Marina, M. L., & García, M. C. (2022a). Successive extraction using natural deep eutectic solvents and pressurized liquids for a greener and holistic recovery of proteins from pomegranate seeds. *Food Research International*, *161*111862. Available from https://doi.org/10.1016/j.foodres.2022.111862.

Guzmán-Lorite, M., Marina, M. L., & García, M. C. (2022b). Pressurized liquids vs. high intensity focused ultrasounds for the extraction of proteins from a pomegranate seed waste. *Innovative Food Science & Emerging Technologies*, *77*102958. Available from https://doi.org/10.1016/j.ifset.2022.102958.

Jogi, N., Yathisha, U. G., Bhat, I., & Mamatha, B. S. (2022). Antihypertensive activity of orally consumed ACE-I inhibitory peptides. *Critical Reviews in Food Science and*

Nutrition, *62*(32), 8986–8999. Available from https://doi.org/10.1080/10408398.2021.1938508, https://www.tandfonline.com/loi/bfsn20.

Karaś, M. (2019). Influence of physiological and chemical factors on the absorption of bioactive peptides. *International Journal of Food Science and Technology*, *54*(5), 1486–1496. Available from https://doi.org/10.1111/ijfs.14054, http://onlinelibrary.wiley.com/journal/10.1111/(ISSN)1365-2621.

Kumar, R., Kaur, N., Bansal, N., & Mehta, K. (2022). Potential anti-inflammatory role of anti-diabetic agents. *JK Science (New York, N.Y.)*, *24*(4), 216–220. Available from http://journal.jkscience.org/index.php/JK-Science/article/download/150/141.

Leong, Y. K., & Chang, J. S. (2022). Valorization of fruit wastes for circular bioeconomy: Current advances, challenges, and opportunities. *Bioresource Technology*, *359*. Available from https://doi.org/10.1016/j.biortech.2022.127459, http://www.elsevier.com/locate/biortech.

Li, Y., Zheng, Y., Zhang, Y., Liu, L., & Zhao, S. (2016). Purification, characterization, synthesis,: In vivo and in vitro antihypertensive activity of bioactive peptides derived from coconut (Cocos nucifera L.) cake globulin hydrolysates. *RSC Advances*, *6*(95), 92688–92698. Available from https://doi.org/10.1039/c6ra19971b, http://pubs.rsc.org/en/journals/journalissues.

Medina-Meza, I. G., & Ganjyal, G. (2018). Food processing by-products and their utilizationfruit processing by-products: A rich source for bioactive compounds and value added products. In A. K. Anal (Ed.), *Food processing by-products and their utilization*. John Wiley & Sons Ltd, 9781118432921 doi: 10.1002/9781118432921.ch2.

Meshginfar, N., Sadeghi Mahoonak, A., Hosseinian, F., Ghorbani, M., & Tsopmo, A. (2018). Production of antioxidant peptide fractions from a by-product of tomato processing: Mass spectrometry identification of peptides and stability to gastrointestinal digestion. *Journal of Food Science and Technology*, *55*(9), 3498–3507. Available from https://doi.org/10.1007/s13197-018-3274-z, http://www.springerlink.com/content/121580/.

Minkiewicz, P., Dziuba, J., Iwaniak, A., Dziuba, M., & Darewicz, M. (2008). BIOPEP database and other programs for processing bioactive peptide sequences. *Journal of AOAC International*, *91*(4), 965–980. Available from https://doi.org/10.1093/jaoac/91.4.96510603271, https://academic.oup.com/jaoac/issue.

Munekata, P. E. S., Yilmaz, B., Pateiro, M., Kumar, M., Domínguez, R., Shariati, M. A., Hano, C., & Lorenzo, J. M. (2022). Valorization of by-products from Prunus genus fruit processing: Opportunities and applications. *Critical Reviews in Food Science and Nutrition*. Available from https://doi.org/10.1080/10408398.2022.2050350, https://www.tandfonline.com/loi/bfsn20.

Musfiroh, F.F., Setiasih, S., Handayani, S., Hudiyono, S., & Ilyas, N.M. (2018). In vivo antiplatelet activity aggregation assay of bromelain fractionate by ethanol from extract pineapple core (Ananas comosus [l.] merr.). In *IOP conference series: Materials science and engineering* (Vol. 299). Institute of Physics Publishing Indonesia. http://www.iop.org/EJ/journal/mse.

Ngamsuk, S., Huang, T. C., & Hsu, J. L. (2020). ACE inhibitory activity and molecular docking of gac seed protein hydrolysate purified by HILIC and RP-HPLC. *Molecules (Basel, Switzerland)*, *25*(20). Available from https://doi.org/10.3390/molecules25204635, https://www.mdpi.com/1420-3049/25/20/4635.

Ng, H. S., Kee, P. E., Yim, H. S., Chen, P. T., Wei, Y. H., & Chi-Wei Lan, J. (2020). Recent advances on the sustainable approaches for conversion and reutilization of food wastes to valuable bioproducts. *Bioresource Technology*, *302*. Available from https://doi.org/10.1016/j.biortech.2020.122889, http://www.elsevier.com/locate/biortech.

Nkoana, D. K., Mashilo, J., Shimelis, H., & Ngwepe, R. M. (2022). Nutritional, phytochemical compositions and natural therapeutic values of citron watermelon (Citrullus lanatus var. citroides): A review. *South African Journal of Botany*, *145*, 65–77. Available from https://doi.org/10.1016/j.sajb.2020.12.008, http://www.elsevier.com.

Nuchprapha, A., Paisansak, S., Sangtanoo, P., Srimongkol, P., Saisavoey, T., Reamtong, O., Choowongkomon, K., & Karnchanatat, A. (2020). Two novel ACE inhibitory peptides isolated from longan seeds: Purification, inhibitory kinetics and mechanisms. *RSC Advances*, *10*(22), 12711–12720. Available from https://doi.org/10.1039/d0ra00093k, http://pubs.rsc.org/en/journals/journal/ra.

Patra, A., Abdullah, S., & Pradhan, R. C. (2022). Review on the extraction of bioactive compounds and characterization of fruit industry by-products. *Bioresources and Bioprocessing*, *9*(1). Available from https://doi.org/10.1186/s40643-022-00498-3, https://bioresourcesbioprocessing.springeropen.com.

Piscopo, M., Tenore, G. C., Notariale, R., Maresca, V., Maisto, M., de Ruberto, F., Heydari, M., Sorbo, S., & Basile, A. (2020). Antimicrobial and antioxidant activity of proteins from Feijoa sellowiana Berg. fruit before and after in vitro gastrointestinal digestion. *Natural Product Research*, *34*(18), 2607–2611. Available from https://doi.org/10.1080/14786419.2018.1543686, http://www.tandf.co.uk/journals/titles/14786419.asp.

Poovitha, S., & Parani, M. (2016). In vitro and in vivo α-amylase and α-glucosidase inhibiting activities of the protein extracts from two varieties of bitter gourd (Momordica charantia L.). *BMC Complementary and Alternative Medicine*, *16*. Available from https://doi.org/10.1186/s12906-016-1085-1, http://www.biomedcentral.com/bmccomplementalternmed/.

Popović, L., Stolić, Ž., Čakarević, J., Torbica, A., Tomić, J., & Šijački, M. (2017). Biologically active digests from pumpkin oil cake protein: Effect of cross-linking by transglutaminase. *Journal of the American Oil Chemists' Society*, *94*(10), 1245–1251. Available from https://doi.org/10.1007/s11746-017-3041-8.

Prados, I. M., Marina, M. L., & García, M. C. (2018). Isolation and identification by high resolution liquid chromatography tandem mass spectrometry of novel peptides with multifunctional lipid-lowering capacity. *Food Research International*, *111*, 77–86. Available from https://doi.org/10.1016/j.foodres.2018.05.009, http://www.elsevier.com/inca/publications/store/4/2/2/9/7/0.

Prados, I. M., Orellana, J. M., Marina, M. L., & García, M. C. (2020). Identification of peptides potentially responsible for in vivo hypolipidemic activity of a hydrolysate from olive seeds. *Journal of Agricultural and Food Chemistry*, *68*(14), 4237–4244. Available from https://doi.org/10.1021/acs.jafc.0c01280, http://pubs.acs.org/journal/jafcau.

Priyanto, A. D., Doerksen, R. J., Chang, C. I., Sung, W. C., Widjanarko, S. B., Kusnadi, J., Lin, Y. C., Wang, T. C., & Hsu, J. L. (2015). Screening, discovery, and characterization of angiotensin-I converting enzyme inhibitory peptides derived from proteolytic hydrolysate of bitter melon seed proteins. *Journal of Proteomics*, *128*, 424–435. Available from https://doi.org/10.1016/j.jprot.2015.08.018, http://www.elsevier.com.

Rajan, A., Kumar, S., Sunil, C. K., Radhakrishnan, M., & Rawson, A. (2022). Recent advances in the utilization of industrial byproducts and wastes generated at different stages of tomato processing: Status report. *Journal of Food Processing and Preservation*, *46*(11). Available from https://doi.org/10.1111/jfpp.17063, http://onlinelibrary.wiley.com/journal/10.1111/(ISSN)1745-4549.

Rico, X., Gullón, B., Alonso, J. L., & Yáñez, R. (2020). Recovery of high value-added compounds from pineapple, melon, watermelon and pumpkin processing by-products: An overview. *Food Research International*, *132*. Available from https://doi.org/

10.1016/j.foodres.2020.109086, http://www.elsevier.com/inca/publications/store/4/2/2/9/7/0.

Rudke, C. R. M., Zielinski, A. A. F., & Ferreira, S. R. S. (2022). From biorefinery to food product design: Peach (Prunus persica) by-products deserve attention. *Food and Bioprocess Technology*. Available from https://doi.org/10.1007/s11947-022-02951-9, https://www.springer.com/journal/11947.

Salve, R. R., & Ray, S. (2020). Comprehensive study of different extraction methods of extracting bioactive compounds from pineapple waste—A review. *The Pharma Innovation Journal*, *9*, 327–340.

Shahidi, F., & Zhong, Y. (2008). Bioactive peptides. *Journal of AOAC International, 91*, 914–931.

Toldrá, F., Reig, M., Aristoy, M. C., & Mora, L. (2018). Generation of bioactive peptides during food processing. *Food Chemistry*, *267*, 395–404. Available from https://doi.org/10.1016/j.foodchem.2017.06.119, http://www.elsevier.com/locate/foodchem.

Treggiari, D., Zoccatelli, G., Chignola, R., Molesini, B., Minuz, P., & Pandolfini, T. (2017). Tomato cystine-knot miniproteins possessing anti-angiogenic activity exhibit in vitro gastrointestinal stability, intestinal absorption and resistance to food industrial processing. *Food Chemistry*, *221*, 1346–1353. Available from https://doi.org/10.1016/j.foodchem.2016.11.020, http://www.elsevier.com/locate/foodchem.

Treggiari, D., Zoccatelli, G., Molesini, B., Degan, M., Rotino, G. L., Sala, T., Cavallini, C., Macrae, C. A., Minuz, P., & Pandolfini, T. (2015). A cystine-knot miniprotein from tomato fruit inhibits endothelial cell migration and angiogenesis by affecting vascular endothelial growth factor receptor (VEGFR) activation and nitric oxide production. *Molecular Nutrition and Food Research*, *59*(11), 2255–2266. Available from https://doi.org/10.1002/mnfr.201500267, http://www.mnf-journal.de.

Udenigwe, C. C., Nwachukwu, I. D., & Yada, R. Y. (2017). *Advances on the production and application of peptides for promoting human health and food security. Global Food Security and Wellness* (pp. 195–219). Springer New York. Available from http://doi.org/10.1007/978-1-4939-6496-3.

Vaštag, Z., Popović, L., Popović, S., Peričin-Starčević, I., & Krimer-Malešević, V. (2013). In vitro study on digestion of pumpkin oil cake protein hydrolysate: Evaluation of impact on bioactive properties. *International Journal of Food Sciences and Nutrition*, *64*(4), 452–460. Available from https://doi.org/10.3109/09637486.2012.749837.

Venkitasamy, C., Zhao, L., Zhang, R., & Pan, Z. (2019). *Pomegranate. Integrated processing technologies for food and agricultural by-products* (pp. 181–216). Elsevier. Available from https://www.sciencedirect.com/book/9780128141380.

Vásquez-Villanueva, R., Marina, M. L., & García, M. C. (2015). Revalorization of a peach (Prunus persica (L.) Batsch) byproduct: Extraction and characterization of ACE-inhibitory peptides from peach stones. *Journal of Functional Foods*, *18*, 137–146. Available from https://doi.org/10.1016/j.jff.2015.06.056, http://www.elsevier.com/wps/find/journaldescription.cws_home/717426/description#description.

Vásquez-Villanueva, R., Orellana, J. M., Marina, M. L., & García, M. C. (2019). Isolation and characterization of angiotensin converting enzyme inhibitory peptides from peach seed hydrolysates: In vivo assessment of antihypertensive activity. *Journal of Agricultural and Food Chemistry*, *67*(37), 10313–10320. Available from https://doi.org/10.1021/acs.jafc.9b02213, http://pubs.acs.org/journal/jafcau.

Wadhwa, M., Bakshi, M. P. S., & Makkar, H. P. S. (2015). Wastes to worth: Value added products from fruit and vegetable wastes. *CAB Reviews: Perspectives in Agriculture, Veterinary*

Science, Nutrition and Natural Resources., *10*. Available from https://doi.org/10.1079/PAVSNNR201510043, http://www.cabi.org/cabreviews/FullTextPDF/2016/20163022692.pdf.

Xiang, H., Waterhouse, D. S., Liu, P., Waterhouse, G. I. N., Li, J., & Cui, C. (2020). Pancreatic lipase-inhibiting protein hydrolysate and peptides from seabuckthorn seed meal: Preparation optimization and inhibitory mechanism. *LWT*, *134*. Available from https://doi.org/10.1016/j.lwt.2020.109870, https://www.journals.elsevier.com/lwt.

Xiong, Y. L. (2010). *Antioxidant peptides bioactive proteins and peptides as functional foods and nutraceuticals* (pp. 29−42). Wiley-Blackwell. Available from http://onlinelibrary.wiley.com/book/10.1002/9780813811048, http://doi.org/10.1002/9780813811048.ch3.

Yahia, E. M., García-Solís, P., & Maldonado Celis, M. E. (2018). *Contribution of fruits and vegetables to human nutrition and health. Postharvest physiology and biochemistry of fruits and vegetables* (pp. 19−45). Elsevier. Available from http://www.sciencedirect.com/science/book/9780128132784, http://doi.org/10.1016/B978-0-12-813278-4.00002-6.

Zarei, M., Forghani, B., Ebrahimpour, A., Abdul-Hamid, A., Anwar, F., & Saari, N. (2015). In vitro and in vivo antihypertensive activity of palm kernel cake protein hydrolysates: Sequencing and characterization of potent bioactive. *Industrial Crops and Products*, *76*, 112−120. Available from https://doi.org/10.1016/j.indcrop.2015.06.040, http://www.elsevier.com/inca/publications/store/5/2/2/8/2/5.

Zheng, Y., Li, Y., & Li, G. (2019). ACE-inhibitory and antioxidant peptides from coconut cake albumin hydrolysates: Purification, identification and synthesis. *RSC Advances*, *9* (11), 5925−5936. Available from https://doi.org/10.1039/c8ra10269d, http://pubs.rsc.org/en/journals/journal/ra.

Zheng, Y., Li, Y., Zhang, Y., Ruan, X., & Zhang, R. (2017). Purification, characterization, synthesis, in vitro ACE inhibition and in vivo antihypertensive activity of bioactive peptides derived from oil palm kernel glutelin-2 hydrolysates. *Journal of Functional Foods*, *28*, 48−58. Available from https://doi.org/10.1016/j.jff.2016.11.021, http://www.elsevier.com/wps/find/journaldescription.cws_home/717426/description#description.

Index

Note: Page numbers followed by "*f*" and "*t*" refer to figures and tables, respectively.

D

J

K

L

T

U